Springer Texts in Business and Economics

More information about this series at http://www.springer.com/series/10099

Susheng Wang

Microeconomic Theory

Fourth Edition

Susheng Wang
Department of Economics
Hong Kong University of Science
  and Technology
Clear Water Bay, Hong Kong
China

ISSN 2192-4333   ISSN 2192-4341   (electronic)
Springer Texts in Business and Economics
ISBN 978-981-13-0040-0   ISBN 978-981-13-0041-7   (eBook)
https://doi.org/10.1007/978-981-13-0041-7

Library of Congress Control Number: 2018940428

1st edition: © Wang Susheng, China Renmin University Press 2006
2nd edition: © Wang Susheng, McGill-Hill 2012
3rd edition: © Wang Susheng, free online 2016
4th edition: © Springer Nature Singapore Pte Ltd. 2018
This work is subject to copyright. All rights are reserved by the Publisher, whether the whole or part of the material is concerned, specifically the rights of translation, reprinting, reuse of illustrations, recitation, broadcasting, reproduction on microfilms or in any other physical way, and transmission or information storage and retrieval, electronic adaptation, computer software, or by similar or dissimilar methodology now known or hereafter developed.
The use of general descriptive names, registered names, trademarks, service marks, etc. in this publication does not imply, even in the absence of a specific statement, that such names are exempt from the relevant protective laws and regulations and therefore free for general use.
The publisher, the authors and the editors are safe to assume that the advice and information in this book are believed to be true and accurate at the date of publication. Neither the publisher nor the authors or the editors give a warranty, express or implied, with respect to the material contained herein or for any errors or omissions that may have been made. The publisher remains neutral with regard to jurisdictional claims in published maps and institutional affiliations.

Printed on acid-free paper

This Springer imprint is published by the registered company Springer Nature Singapore Pte Ltd.
The registered company address is: 152 Beach Road, #21-01/04 Gateway East, Singapore 189721, Singapore

Preface

This book covers microeconomic theory at the master's and Ph.D. levels for students in business schools and economics departments. It concisely covers major mainstream microeconomic theories today, including neoclassical microeconomics, game theory, information economics, and contract theory.

The coverage. Microeconomics is a required subject for all students in a business school, especially for students in economics, accounting, finance, management, and marketing. It is a standard practice nowadays that a Ph.D. program in a business school requires a one-year course in microeconomic theory. This one-year course is generally divided into two one-semester courses. Chapters 1–4 are the core subjects for the first semester. These chapters cover neoclassical economics, which includes demand and supply theories, risk theory, and general equilibrium. It is the foundation of modern microeconomics. Chapters 7–11 are the core subjects for the second semester. Chapters 7–9 cover game theory. Chapters 7 and 8 cover noncooperative games, dealing respectively with imperfect and incomplete information games. Chapter 9 covers cooperative games, which is useful for advanced topics in information economics and organization theory. Chapters 10–12 cover information economics, including asymmetric information and incentive contracts. Chapters 5 and 6 are optional; these chapters are the foundations of some field courses, which may or may not be covered depending on the emphasis of the instructor. Chapter 5 covers some basic theories in financial economics, especially asset pricing. Chapter 6 contains the basic theory of industrial organizations, which can be covered in either semester.

Special features of this book. Mas-Colell et al. (1995) is the standard textbook for Ph.D. microeconomics today. In terms of its contents, our book covers all the important materials in Mas-Colell et al. (1995). However, we do not include materials that we consider to be out of the mainstream. Those materials that we think should belong to field courses (such as social choice and public finance) are also left out. On the other hand, we strengthen many topics that we think are important, particularly in Chaps. 3, 5, 6, 10, 11, and 12. Roughly half of the materials in this book can be found in Mas-Colell et al. (1995) and the rest are from many different sources. Some advanced materials exist only in papers. Our book is intended to be a more focused and complete coverage on mainstream topics in microeconomic theory. In particular, this book is intended for postgraduate students in business schools, rather than just for students in economics departments.

This book is intended as a handbook for graduate students and professors. It is concise with many of detailed explanations deliberately left out. Such a book is good for students who will learn most of the details and explanations from lecture. It is also good for instructors who would want to fill in the details by themselves in their own words. Teachers generally do not like a textbook to have too many details. With a detailed textbook, some students may skip class and the instructor may even feel like reading the book aloud in class. This book can also serve as a reference book for those who have already learned all the materials. When such a person wants to refresh some materials, he or she can quickly find the key points of the material in the book without going through all the details.

The main enhancement of this edition over the second edition published by McGraw-Hill in 2012 is the addition of one chapter, Chap. 8, on incomplete information games. Such games are essential for information economics, but are not included in Mas-Colell et al. (1995). There are also many minor revisions in all chapters.

Supporting materials. Exercises and their solutions are available in a PDF file at www.bm.ust.hk/~sswang/; click the link in Part VI to download. Errors are inevitable in such a book with so many diverse materials. We will offer corrections at the site. We may also provide more materials such as more exercises, new sections, and chapters on that site.

Acknowledgement. I would like to thank Dr. Virginia A. Unkefer for professional English editing.

Hong Kong, China
December 2017

Susheng Wang

Contents

Part I Neoclassical Economics

1 Producer Theory 3
  1 Technology 3
  2 The Firm's Problem. 14
  3 Short-Run and Long-Run Cost Functions 24
     3.1 Definitions. 24
     3.2 Relationship Between AC and MC Curves 26
     3.3 Relationship Between SR and LR Cost Functions. 27
  4 Properties. 29
  5 Aggregation. 35

2 Consumer Theory 37
  1 Existence of the Utility Function 37
  2 The Consumer's Problem 43
  3 Properties. 47
  4 Aggregation. 56
  5 Integrability 57
  6 Revealed Preferences. 58
  7 Intertemporal Analysis. 61

3 Risk Theory 69
  1 Introduction 69
  2 Expected Utility Theory. 70
  3 Mean-Variance Utility 77
  4 Measurement of Risk Aversion 79
  5 Stochastic Dominance 81
  6 Demand for Insurance 86
  7 Demand for Risky Assets 88
  8 Portfolio Analysis 91

4 General Equilibrium Theory 95
  1 The General Equilibrium Concept 95
  2 GE in a Pure Exchange Economy 96

3	Pareto Optimality	103
4	Welfare Theorems	109
5	General Equilibrium with Production	114
	5.1 General Equilibrium with Production	114
	5.2 Efficiency of General Equilibrium	118
6	General Equilibrium with Uncertainty	123

Part II Micro-Foundation of Markets

5 Micro-foundation of Finance ... 129

1	Security Markets	129
	1.1 Contingent Markets	129
	1.2 Security Markets	130
2	Static Asset Pricing	134
3	Representative Agent Pricing	136
4	The Capital Asset Pricing Model	138
5	Dynamic Asset Pricing	140
	5.1 The Euler Equation	140
	5.2 Dynamic CAPM	143
6	Continuous-Time Stochastic Programming	144
	6.1 Continuous-Time Random Variables	144
	6.2 Continuous-Time Stochastic Programming	146
7	The Black-Scholes Pricing Formula	149

6 Micro-foundation of Industry ... 153

1	A Competitive Output Market	153
2	A Monopoly	161
	2.1 A Single-Price Monopoly	161
	2.2 A Price-Discriminating Monopoly	163
	2.3 Monopoly Pricing Under Asymmetric Information	164
3	Allocative Efficiency	166
4	Monopolistic Competition	170
5	Oligopoly	172
	5.1 Bertrand Equilibrium	173
	5.2 Cournot Equilibrium	175
	5.3 Stackelberg Equilibrium	178
	5.4 Cooperative Equilibrium	178
	5.5 Competition Versus Cooperation	180
	5.6 Cooperation in a Repeated Game	181
6	Production Differentiation	182
7	Location Equilibrium	183
8	Entry Barriers	186
9	Strategic Deterrence Against Potential Entrants	189

	10	Competitive Input Markets	191
		10.1 Demand and Supply	192
		10.2 Equilibrium and Welfare.	193
	11	A Monopsony	197
	12	Vertical Relationships	200
		12.1 Independent Firms.	200
		12.2 An Integrated Firm	201
		12.3 Explanation.	202

Part III Game Theory

7 Imperfect Information Games 209
   1 Two Game Forms 209
      1.1 The Extensive Form 209
      1.2 The Normal Form. 212
      1.3 Mixed Strategy 214
   2 Equilibria in Normal-Form Games. 217
      2.1 Nash Equilibrium 217
      2.2 Dominant-Strategy Equilibrium. 220
      2.3 Trembling-Hand Perfect Nash Equilibrium 223
      2.4 Reactive Equilibrium. 225
   3 Equilibria in Extensive-Form Games 227
      3.1 Nash Equilibrium 227
      3.2 Subgame Perfect Nash Equilibrium. 230
      3.3 Bayesian Equilibrium 235
   4 Refinements of Bayesian Equilibrium 248
      4.1 Perfect Bayesian Equilibrium 248
      4.2 Sequential Equilibrium 254
      4.3 BE Under Complete Dominance: CDBE. 261
      4.4 BE Under Equilibrium Dominance: EDBE 264
      4.5 Alternative Versions of PBE. 268

8 Incomplete Information Games 271
   1 Bayesian Nash Equilibrium 272
   2 Signalling Games. 279
      2.1 Pure Strategies in Signalling 279
      2.2 Mixed Strategies in Signalling 280
      2.3 Cheap Talk 286

9 Cooperative Games 299
   1 The Nash Bargaining Solution. 299
      1.1 The Nash Solution 300
      1.2 Implementation of the Nash Solution 302

		2	The Alternating-Offer Bargaining Solution	303
		2.1	The Alternating-Offer Solution	303
		2.2	Extensions.	305
	3		The Core.	306
	4		The Shapley Value	311
		4.1	The Balanced Contributions Property	311
		4.2	The Shapley Value	312

Part IV Information Economics

10 Market Information 319

	1		The Akerlof Model	319
		1.1	The Used Car Market with Incomplete Information	319
		1.2	The Used Car Market with Complete Information	323
		1.3	The Used Car Market with Symmetric Information	323
		1.4	Discussion.	323
	2		The Rothschild-Stiglitz Model.	324
		2.1	Insurance with Symmetric Information: A Contingent Market.	325
		2.2	Insurance with Symmetric Information: Insurance Market.	326
		2.3	Insurance with Asymmetric Information	329
		2.4	Extensions.	334
	3		Job Market Without Signals	335
		3.1	The Model	336
		3.2	Bayesian Equilibrium	336
		3.3	Subgame Perfect Nash Equilibrium.	338
		3.4	Constrained Pareto Optimum	338
	4		The Spence Model: Job Market Signalling	340
		4.1	The Complete Information Solution	344
		4.2	The No-Signalling Solution.	344
		4.3	Separating Equilibria.	344
		4.4	Pooling Equilibria.	349
		4.5	Partial Separating Equilibrium.	351
		4.6	Government Intervention.	353
		4.7	Equilibrium Refinement	354
		4.8	Questions	356
	5		The Spence Model: Job Market Screening.	357
		5.1	Pooling Equilibrium	357
		5.2	Separating Equilibrium	358

11 Mechanism Design ... 363
1. A Story of Mechanism Design ... 363
   - 1.1 Market Mechanism Versus Direct Mechanism ... 363
   - 1.2 The Optimal Allocation: A Graphic Illustration ... 365
   - 1.3 The Optimal Allocation: Mathematical Presentation ... 366
   - 1.4 The Optimal Allocation: A General Case ... 368
   - 1.5 Market Mechanisms Versus the Direct Mechanism ... 370
2. The Revelation Principle ... 370
3. Examples of Allocation Schemes ... 376
4. IC Conditions in Linear Environments ... 383
5. IR Conditions and Efficiency Criteria ... 386
   - 5.1 Individual Rationality Conditions ... 386
   - 5.2 Efficiency Criteria ... 386
6. Optimal Allocation Schemes ... 388
   - 6.1 Monopoly Pricing ... 388
   - 6.2 A Buyer-Seller Model with Linear Utility Functions ... 390
   - 6.3 Labor Market ... 394
   - 6.4 Optimal Auction ... 397
   - 6.5 A Buyer-Seller Model with Quasi-Linear Utility ... 400

12 Incentive Contracts ... 405
1. The Standard Agency Model ... 405
   - 1.1 Verifiable Effort: The First Best ... 406
   - 1.2 Nonverifiable Effort: The Second Best ... 408
2. Two-State Agency Models ... 414
   - 2.1 Verifiable Effort ... 415
   - 2.2 Unverifiable Effort ... 416
   - 2.3 Example: Insurance ... 417
3. Linear Contracts Under Risk Neutrality ... 419
   - 3.1 Single Moral Hazard ... 420
   - 3.2 Double Moral Hazard ... 422
4. A Conditional Fixed Contract ... 429
   - 4.1 The Model ... 429
   - 4.2 The Optimal Contract ... 430
5. A Suboptimal Linear Contract ... 433

Appendix: Optimization Methods ... 435

References ... 455

Index ... 459

Notation and Terminology

The word 'iff' and symbol '⇔' mean 'if and only if.'

'w.r.t.' means 'with respect to.'

$U[a,b]$ denotes the uniform distribution over an interval $[a,b]$.

$\mathbb{R}$ is the set of all real numbers. Denote $\mathbb{R}_+ \equiv \{x \in \mathbb{R} | x \geq 0\}$, $\mathbb{R}_{++} \equiv \{x \in \mathbb{R} | x \geq 0\}$, $\mathbb{R}_- \equiv \{x \in \mathbb{R} | x \leq 0\}$.

$x \cdot y = \sum_i x_i y_i$ denotes the inner product of two vectors $x, y \in \mathbb{R}^n$.

FOC stands for 'first-order condition.' SOC stands for 'second-order condition.' FOA stands for 'first-order approach.'

A function $f : \mathbb{R}_+^n \to \mathbb{R}$ is homogeneous of degree α if $f(\lambda x) = \lambda^\alpha f(x)$, $\forall x \in \mathbb{R}_+^n$, $\lambda > 0$. In particular, when $\alpha = 1$, we also say that f is linearly homogeneous, and when $\alpha = 0$, we also say that f is zero homogeneous. See the Appendix.

For $f : \mathbb{R}^n \to \mathbb{R}$, $f'(x)$ is the gradient, i.e., $f'(x) = (f_{x_1}(x), \ldots, f_{x_n}(x))$, $f''(x)$ is the Hessian matrix. For $f : \mathbb{R}^n \to \mathbb{R}^m$, $f'(x)$ is the Jacobian matrix. See the Appendix.

If an $n \times n$ square symmetric matrix A is negative semi-definite, we denote $A \leq 0$; if A is negative definite, we denote $A < 0$. We denote positive semi-definiteness and positive definiteness by the same way.

For any two vectors $x, y \in \mathbb{R}^n$, denote

$$x \geq y \quad \text{if} \quad x_i \geq y_i, \; \forall i,$$
$$x > y \quad \text{if} \quad x \geq y, \; x \neq y,$$
$$x \gg y \quad \text{if} \quad x_i > y_i, \; \forall i.$$

Part I
Neoclassical Economics

Producer Theory

1 Technology

Efficient Production

Suppose that a firm has n goods to serve as inputs and/or outputs. If the firm uses y_i^- units of good i as an input and produces y_i^+ units of the good, where $y_i^-, y_i^+ \in \mathbb{R}_+$, then $y_i \equiv y_i^+ - y_i^-$ is the net output of good i. The firm's <u>production plan</u> $y = (y_1, y_2, \ldots, y_n) \in \mathbb{R}^n$ is a list of net outputs of all the goods that it produces as outputs and/or uses as inputs. For convenience, we can simply treat positive numbers in a net output vector as outputs and negative numbers as inputs.

A firm is defined by a set $\mathbb{Y} \subset \mathbb{R}^n$ of production plans. These production plans are technologically feasible production plans for the firm. Hence, we call $\mathbb{Y}$ the firm's <u>production possibility set</u>. The production possibility set is assumed to have the free disposal property: if $y \in \mathbb{Y}$, then $y - z \in \mathbb{Y}$ for any $z \in \mathbb{R}_+^n$. That is, output can be freely thrown away at no cost.

A production plan $y \in \mathbb{Y}$ is <u>technologically efficient</u> if there is no $\hat{y} \in \mathbb{Y}$ such that $\hat{y} > y$.[1] A production plan is <u>economically efficient</u> if it maximizes the firm's profits $\pi = p \cdot y$ for a given price vector p over the production possibility set $\mathbb{Y}$. Technological efficiency is dependent only on the firm's technological characteristics; economic efficiency is further dependent on the market.

We can easily see that economic efficiency implies technological efficiency. Technological efficiency has nothing to do with the market. No matter what the prices of inputs and outputs are, the firm needs to achieve technological efficiency in order to achieve economic efficiency; see Fig. 1. We can think of the firm's choice problem in two steps: the firm has to achieve technological efficiency first

---

[1] A weak version of technological efficiency is as follows: a production plan $y \in \mathbb{Y}$ is technologically efficient if there is no $\hat{y} \in \mathbb{Y}$ such that $\hat{y} \gg y$. See the definition of inequalities $>$ and $\gg$ in Notation and Terminology.

Fig. 1 Technological efficiency versus economic efficiency

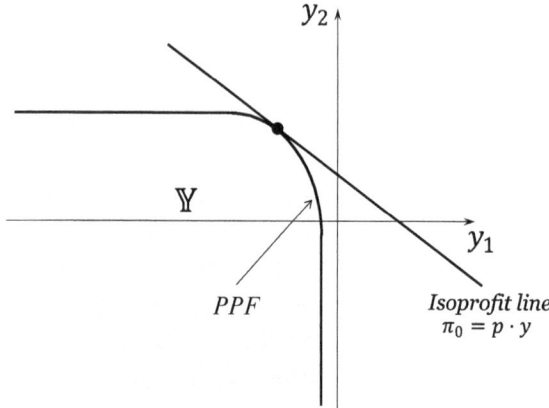

and then achieve economic efficiency on the production frontier, where the production frontier (PPF) is defined to be the set of all technologically efficient production plans.

Proposition 1.1 *Economic efficiency implies technological efficiency.* ∎

To investigate more details of a firm's technology, we now use a function to define a firm's technology. There is a twice differentiable function $G : \mathbb{R}^m \to \mathbb{R}$ that defines the production possibility set: $\mathbb{Y} \equiv \{y \in \mathbb{R}^m | G(y) \leq 0\}$. In order for this definition to make sense, we need to impose two assumptions on G.

Assumption 1.1 (*Strict Monotonicity*) $G_{y_i}(y) > 0, \forall i, y$.

By Assumption 1.1, since $G_{y_i}(y) > 0$, any y such that $G(y) < 0$ is technologically inefficient: one can increase the production (or decrease the use) of one product y_i while keeping others constant (some of the y_j's are inputs). Therefore, $\{y \in \mathbb{R}^m | G(y) = 0\}$ is the technologically efficient production set or the production frontier.

Proposition 1.2 *Set* $\left\{y \in \mathbb{R}^m | G(y) = 0\right\}$ *is the production frontier.* ∎

We now define the marginal rate of transformation (MRT) as

$$MRT_{ij} \equiv -\frac{\partial y_j}{\partial y_i}\bigg|_{G(y)=0} = \frac{G_{y_i}}{G_{y_j}} > 0. \tag{1}$$

MRT_{12} is the slope of the production frontier when there are only two goods y_1 and y_2. Assumption 1.1 implies a positive MRT.

Assumption 1.2 G is quasi-convex. ∎

Proposition 1.3 *Under Assumption 1.1, for $n = 2$, G is quasi-convex iff*[2]

$$\begin{vmatrix} 0 & G_{y_1}(y) & G_{y_2}(y) \\ G_{y_1}(y) & G_{y_1 y_1}(y) & G_{y_1 y_2}(y) \\ G_{y_2}(y) & G_{y_2 y_1}(y) & G_{y_2 y_2}(y) \end{vmatrix} \leq 0, \quad \text{for all } y.$$

Proof By Theorem A.7, this inequality is necessary. Let us show that it is also sufficient. Let the unique solution from $G(y_1, y_2) = 0$ be $y_2 = y_2(y_1)$, which is guaranteed by the Implicit Function Theorem. Then,[3]

$$\begin{aligned} \frac{d^2 y_2}{dy_1^2} &= \frac{d}{dy_1}\left(\frac{dy_2}{dy_1}\right) = \frac{d}{dy_1}\left(-\frac{G_{y_1}[y_1, y_2(y_1)]}{G_{y_2}[y_1, y_2(y_1)]}\right) = \frac{\partial}{\partial y_1}\left(-\frac{G_{y_1}}{G_{y_2}}\right) + \frac{\partial}{\partial y_2}\left(-\frac{G_{y_1}}{G_{y_2}}\right)\frac{dy_2}{dy_1} \\ &= -\frac{G_{y_1 y_1} G_{y_2} - G_{y_1} G_{y_2 y_1}}{G_{y_2}^2} - \frac{G_{y_1 y_2} G_{y_2} - G_{y_1} G_{y_2 y_2}}{G_{y_2}^2}\left(-\frac{G_{y_1}}{G_{y_2}}\right) \\ &= -\frac{G_{y_1 y_1} G_{y_2}^2 - 2 G_{y_2} G_{y_1} G_{y_2 y_1} + G_{y_1}^2 G_{y_2 y_2}}{G_{y_2}^3} \\ &= \frac{1}{G_{y_2}^3}\begin{vmatrix} 0 & G_{y_1} & G_{y_2} \\ G_{y_1} & G_{y_1 y_1} & G_{y_1 y_2} \\ G_{y_2} & G_{y_2 y_1} & G_{y_2 y_2} \end{vmatrix} \leq 0. \end{aligned}$$

This means that the production frontier is concave, implying a convex set $\{y \in \mathbb{R}^2 | G(y) \leq 0\}$. Since we can go through the same derivation with $H(y) \equiv G(y) - a, \forall a \in \mathbb{R}$, by definition, G is quasi-convex. ∎

Proposition 1.3 offers us a convenient tool to verify Assumption 1.2. It also relates quasi-convexity of G to the second-order condition (SOC) of the firm's problem. To see this, consider the profit-maximizing problem:

$$\pi(p) \equiv \max_{y \in \mathbb{R}^m} p \cdot y \qquad (2)$$
$$\text{s.t.} \quad G(y) \leq 0,$$

where $p \in \mathbb{R}_+^m$ is given, $p \gg 0$. By Assumption 1.1, the production frontier is the set of efficient production. Since technological efficiency is necessary for economic efficiency (profit maximization), the problem is equivalent to

---

[2] This is from Arrow and Enthoven (1961, p. 796). For sufficient conditions of quasi-convexity when $n > 2$, see Wang (2008, Chap. 3).

[3] Denote $F(y_1, y_2) \equiv -\frac{G_{y_1}(y_1, y_2)}{G_{y_1}(y_1, y_2)}$. Then, $\frac{d}{dy_1} F[y_1, y_2(y_1)] = F_{y_1} + F_{y_2}\frac{dy_2}{dy_1}$.

$$\pi(p) \equiv \max_{y \in \mathbb{R}^m} p \cdot y$$
$$\text{s.t.} \quad G(y) = 0.$$

Let $L(y, \lambda) \equiv p \cdot y + \lambda G(y)$. Then, by the Lagrange Theorem, the first-order condition (FOC) is

$$p + \lambda G'(y^*) = 0 \tag{3}$$

and the SOC is

$$h^T L''_y(y^*, \lambda) h \leq 0, \quad \text{for all } h \text{ satisfying } G'(y^*) \cdot h = 0.$$

We have $L''_y(y^*, \lambda) = \lambda G''(y^*)$. By the FOC, since $G'(y^*) \geq 0$ and $p > 0$, we have $\lambda < 0$. Therefore, the SOC becomes

$$h^T G''(y^*) h \leq 0, \quad \text{for all } h \text{ satisfying } G'(y^*) \cdot h = 0. \tag{4}$$

If G is strictly quasi-convex, the inequalities in Proposition 1.3 will be strict. Then, by Theorem A.8, (4) will be satisfied. Therefore, under Assumptions 1.1 and 1.2, a solution y^* from the FOC (3) must be the solution of the firm's problem (2). This is illustrated in Fig. 2.

We now consider the case in which the firm produces only one output; we may think of this output as an index of the firm's outputs. Denote $y \in \mathbb{R}_+$ as the firm's output and $x \in \mathbb{R}^n_+$ as the firm's inputs. Then, a typical production plan is $(y, -x) \in \mathbb{Y}$, where $\mathbb{Y}$ is a set in $\mathbb{R}^{n+1}$. Given a vector $x \in \mathbb{R}^n_+$ of inputs, denote the maximum technologically feasible output as $f(x)$:

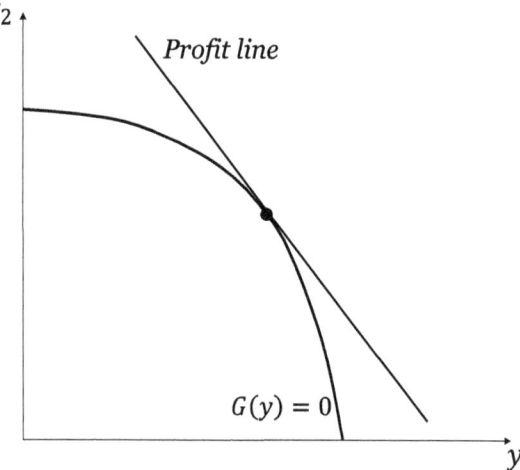

Fig. 2 Profit maximization with implicit production technology

1 Technology

$$f(x) \equiv \max_{(y,-x)\in \mathbb{Y}} y.$$

We call this function $f : \mathbb{R}^n_+ \to \mathbb{R}_+$ the <u>production function</u>. We can easily verify that $y = f(x)$ if $(y, -x)$ is technologically efficient. That is, the production function represents technologically efficient plans.

Proposition 1.4 *For the production function f, if $(y, -x)$ is technologically efficient, then $y = f(x)$.* ∎

Given a production function f, an <u>isoquant</u> is a curve defined by

$$Q(y) \equiv \{x \in \mathbb{R}^n_+ \mid y = f(x)\}.$$

The isoquant for output y is the set of all inputs that produces y as the maximum technologically feasible output.

Example 1.1 Cobb-Douglas Technology. For $0 \leq \alpha \leq 1$, let

$$\mathbb{Y} \equiv \{(y, -x_1, -x_2) \in \mathbb{R}_+ \times \mathbb{R}^2_- \mid y \leq x_1^\alpha x_2^{1-\alpha}\}.$$

By definition, as shown in Fig. 3, we have

$$f(x_1, x_2) = x_1^\alpha x_2^{1-\alpha}, \quad Q(y) = \{(x_1, x_2) \in \mathbb{R}^2_+ \mid y = x_1^\alpha x_2^{1-\alpha}\}. \quad ∎$$

Technological Characteristics

We are interested in some key characteristics of the firm's technology, from which economic insight can be obtained. First, for production function

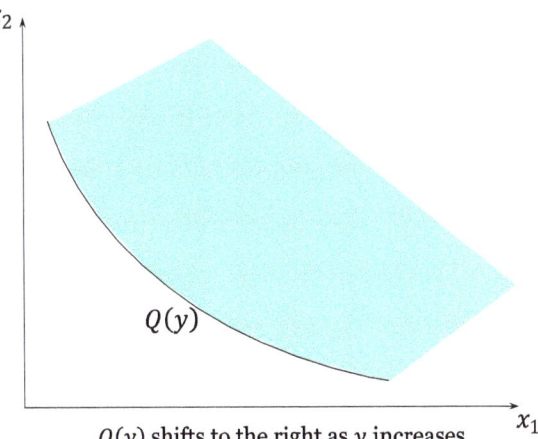

Fig. 3 Cobb-Douglas technology

$Q(y)$ shifts to the right as y increases

$y = f(x_1, x_2, \ldots, x_n)$, given output y_0, with one unit increase in x_1, by how much do we need to decrease x_2 to maintain the same level of output y_0? This amount is $\frac{\Delta x_2}{\Delta x_1}\big|_{f(x)=y_0}$. That is, $\frac{\Delta x_2}{\Delta x_1}\big|_{f(x)=y_0}$ measures the certain substitutability of the two inputs. By differentiating both sides the following equation

$$f(x_1, x_2, \ldots, x_n) = y_0,$$

and using the fact that $dy_0 = 0$ and $dx_i = 0$ for $i \geq 3$, we have

$$\lim_{\Delta x_1 \to 0} \frac{\Delta x_2}{\Delta x_1}\bigg|_{f(x)=y_0} = \frac{\partial x_2}{\partial x_1} = -\frac{f_{x_1}(x)}{f_{x_2}(x)}.$$

Hence, the <u>marginal rate of transformation (MRT)</u> between x_i and x_j at x is defined as

$$MRT_{ij}(x) \equiv \frac{f_{x_i}(x)}{f_{x_j}(x)}. \qquad (5)$$

In Fig. 4, we can see that the MRT is the slope of the isoquant.

The definition of MRT in (5) is actually a special case of that in (1). For production function $y = f(x_1, x_2, \ldots, x_n)$, let

$$G(y_1, \ldots, y_{n+1}) \equiv y - f(x_1, \ldots, x_n) = y_{n+1} - f(-y_1, \ldots, -y_n),$$

where $y_{n+1} \equiv y$ and $y_i \equiv -x_i$ for $i = 1, \ldots, n$. Then, according to (1), the MRT between two inputs is

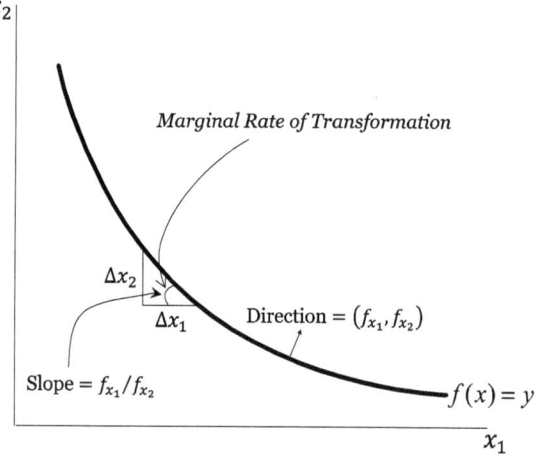

Fig. 4 The marginal rate of transformation

1 Technology

$$MRT_{ij} = \frac{G_{y_i}}{G_{y_j}} = \frac{f_{x_i}(x)}{f_{x_j}(x)}.$$

Example 1.2 For $f(x_1, x_2) = x_1^a x_2^{1-a}$, we have

$$\frac{\partial f(x)}{\partial x_1} = a x_1^{a-1} x_2^{1-a}, \quad \frac{\partial f(x)}{\partial x_2} = (1-a) x_1^a x_2^{-a},$$

implying $MRT_{12} = \frac{a}{1-a} \frac{x_2}{x_1}$. ∎

Second, we say that a production function $f : \mathbb{R}_+^n \to \mathbb{R}_+$ has global <u>constant returns to scale (CRS)</u> if $f(tx) = tf(x)$ for all $x \in \mathbb{R}_+^n$ and $t > 1$. With this production technology, the firm can double its output by doubling its inputs. Similarly, we say that the production function has global <u>increasing returns to scale (IRS)</u> or <u>decreasing returns to scale (DRS)</u> if, respectively, $f(tx) > tf(x)$ or $f(tx) < tf(x)$ for all $x \in \mathbb{R}_+^n$ and $t > 1$. Note also that, if f is globally IRS, then $f(tx) < tf(x)$ for all $t \in (0,1)$. This also applies to global CRS and DRS.

Example 1.3 For the Cobb-Douglas production function $f(x_1, x_2) = A x_1^a x_2^b$, we have

$$f(tx_1, tx_2) = A(tx_1)^a (tx_2)^b = t^{a+b} f(x_1, x_2).$$

Hence, this production function has global IRS if $a+b > 1$, global CRS if $a+b = 1$, and global DRS if $a+b < 1$. ∎

Third, the <u>elasticity of scale</u> at x measures the percentage increase in output due to one percentage increase in scale:

$$e(x) \equiv \left. \frac{d\log[f(tx)]}{d\log(t)} \right|_{t=1}.$$

In other words, $e(x)$ is the percentage increase in output when all inputs have expanded proportionally by 1%. We say that the technology exhibits <u>local increasing, constant, or decreasing returns to scale</u> if $e(x)$ is greater, equal, or less than 1, respectively.

Proposition 1.5 (Returns to Scale)

1. For $x \in \mathbb{R}_+^n$, we have

$$\begin{array}{ll} \text{global IRS} \Rightarrow \text{local IRS or CRS}, & \forall x; \\ \text{global CRS} \Rightarrow \text{local CRS}, & \forall x; \\ \text{global DRS} \Rightarrow \text{local DRS or CRS}, & \forall x. \end{array}$$

2. *For $x \in \mathbb{R}_+^n$, we have*

$$e(x) = \frac{x \cdot f'(x)}{f(x)},$$

where $f'(x) = (f_{x_1}(x), \ldots, f_{x_n}(x))^T$. Hence, for $x \in \mathbb{R}_+$, we have

$$\begin{aligned}
\text{local IRS} &\Leftrightarrow f'(x) > \tfrac{f(x)}{x}; \\
\text{local CRS} &\Leftrightarrow f'(x) = \tfrac{f(x)}{x}; \\
\text{local DRS} &\Leftrightarrow f'(x) < \tfrac{f(x)}{x}.
\end{aligned} \qquad (6)$$

3. *For $x \in \mathbb{R}_+^n$ and $y = f(x)$, we have*

$$e(x) = \frac{AC(y)}{MC(y)},$$

implying

$$\begin{aligned}
\text{local IRS} &\Leftrightarrow AC > MC; \\
\text{local CRS} &\Leftrightarrow AC = MC; \\
\text{local DRS} &\Leftrightarrow AC < MC.
\end{aligned} \qquad (7)$$

Proof 1 If f is globally IRS, then, for $t > 1$, $\Delta t > 0$ and any x, we have

$$\frac{f[(t+\Delta t)x] - f(tx)}{\Delta t} \frac{t}{f(tx)} > \frac{(1 + \frac{\Delta t}{t})f(tx) - f(tx)}{\Delta t} \frac{t}{f(tx)} = 1,$$

implying

$$\frac{df(tx)}{dt} \frac{t}{f(tx)} \geq 1.$$

Conversely, if, for all $t > 1$ and x,

$$\frac{df(tx)}{dt} \frac{t}{f(tx)} > 1,$$

then we have

$$d\log f(tx) > d\log t,$$

implying

$$\log f(tx) - \log f(x) > \log t,$$

implying

$$f(tx) > tf(x),$$

implying that f is globally IRS. We have similar results for global DRS. In summary, we have

$$\begin{aligned} \text{global IRS} &\Rightarrow \frac{df(tx)}{dt}\frac{t}{f(tx)} \geq 1, \quad \forall t > 1, x \in \mathbb{R}^n_+; \\ \text{global DRS} &\Rightarrow \frac{df(tx)}{dt}\frac{t}{f(tx)} \leq 1, \quad \forall t > 1, x \in \mathbb{R}^n_+. \end{aligned} \quad (8)$$

And conversely,

$$\begin{aligned} \frac{df(tx)}{dt}\frac{t}{f(tx)} > 1, \quad \forall t > 1, x \in \mathbb{R}^n_+ &\Rightarrow \text{global IRS}; \\ \frac{df(tx)}{dt}\frac{t}{f(tx)} < 1, \quad \forall t > 1, x \in \mathbb{R}^n_+ &\Rightarrow \text{global DRS}. \end{aligned}$$

By (8), global IRS immediately implies $e(x) \geq 1$ at any point x. Similarly, global DRS immediately implies $e(x) \leq 1$ at any point x. Hence, part 1 is proven.

2. For $x \in \mathbb{R}^n_+$, by definition, we immediately have $e(x) = \frac{1}{f(x)} x \cdot f'(x)$. This immediately implies (6).
3. The cost function is defined by

$$c(y) \equiv \min\{w \cdot x | y = f(x)\}.$$

The Lagrange function is $\mathcal{L} \equiv w \cdot x + \lambda[y - f(x)]$. The FOC is

$$w = \lambda f'(x^*),$$

and the Envelope Theorem is

$$c'(y) = \lambda.$$

We then have

$$c(y) = w \cdot x^* = \lambda f'(x^*) \cdot x^* = c'(y)[f'(x^*) \cdot x^*];$$

Fig. 5 Local returns to scale

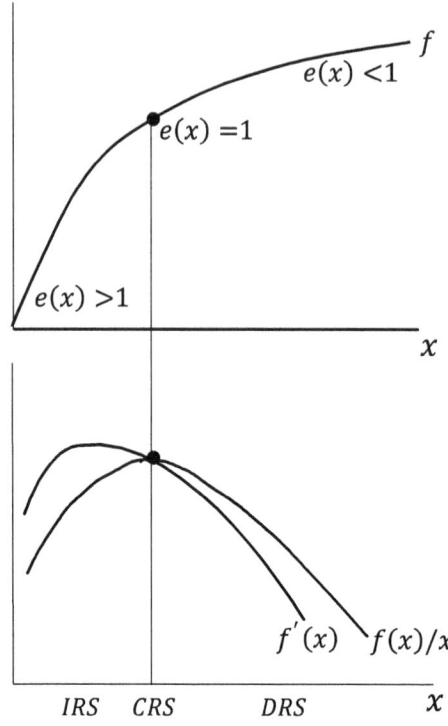

hence,

$$e(x^*) = \frac{1}{f(x)}f'(x^*) \cdot x^* = \frac{1}{y}f'(x^*) \cdot x^* = \frac{c(y)}{y}\frac{1}{c'(y)} = \frac{AC}{MC}.$$

Hence, (7) holds and it can be illustrated in Fig. 6. ∎

The property in (6) is illustrated in the following figure.
The property in (7) is illustrated in the following figure.
By the argument for the local returns to scale and Fig. 5, we immediately have the diagrams for the global returns to scale in Fig. 7.[4]
By the argument for the local returns to scale and Fig. 6, we immediately have the diagrams for the global returns to scale in Fig. 8.
Finally, let $x(w, y)$ be the cost-minimizing vector of inputs for a given price vector w of inputs and output y. The <u>elasticity of substitution</u> between the two inputs $x_1(w, y)$ and $x_2(w, y)$ is defined to be the ratio of percentage change in $\frac{x_1(w,y)}{x_2(w,y)}$ to the percentage change in $\frac{w_1}{w_2}$:

---

[4]The equation $f'(x) = f(x)/x$ gives the general solution $f(x) = Ax$, where A is an arbitrary constant, implying $f'(x) = A$.

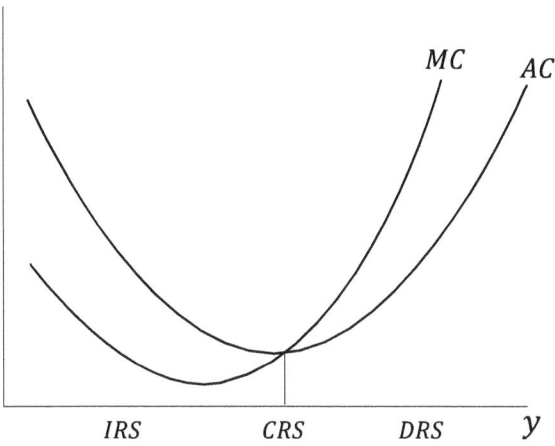

Fig. 6 Local returns to scale

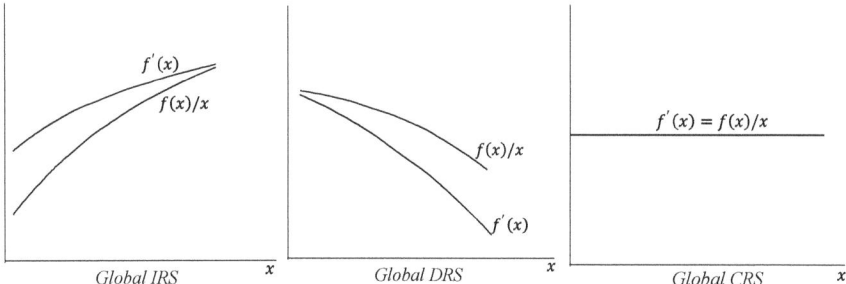

Fig. 7 Global returns to scale

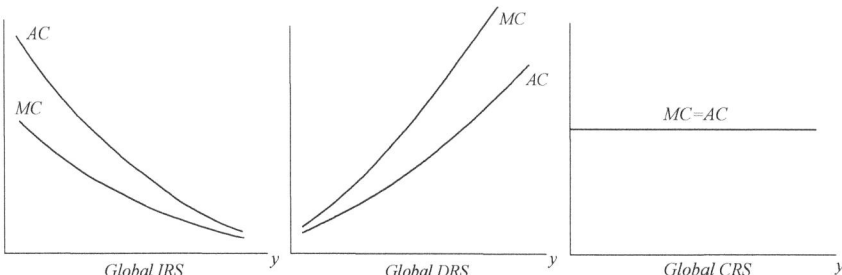

Fig. 8 Global returns to scale

$$\sigma \equiv -\partial \log \frac{x_1(w,y)}{x_2(w,y)} / \partial \log \left(\frac{w_1}{w_2}\right).$$

Note that this variable is well defined only if $\frac{x_1(w,y)}{x_2(w,y)}$ is a function of $\frac{w_1}{w_2}$. As shown later, since $x_i(w,y)$ is zero homogenous,[5] $x_i(w,y)$ is a function of $\frac{w_1}{w_2}$. Therefore, $\frac{x_1(w,y)}{x_2(w,y)}$ is a function of $\frac{w_1}{w_2}$. By the optimality of $x(w,y)$, using (11), we can alternatively write this elasticity as

$$\sigma = -\partial \log \frac{x_1}{x_2} / \partial \log \frac{f_{x_1}}{f_{x_2}},$$

which is evaluated at the optimal points. Note that while MRT is based on technological efficiency, elasticity of substitution is based on economic efficiency.

2 The Firm's Problem

The firm is assumed to maximize its profits. Profits consist of two distinct parts:

$$\text{profits} = \text{revenue} - \text{costs}.$$

<u>Revenue</u> is the money received from the sales of the firm's products. Costs are the <u>economic cost</u> or more popularly called the <u>opportunity cost</u>, which typically includes three components: the cost of labor and raw materials, the cost of capital (including depreciation), and the cost of land and natural resources.

In general, suppose that a firm takes n actions accomplished by choosing a vector $a \in \mathbb{R}^n$. The actions may include output levels, labor inputs, capital inputs, and even prices. Suppose that the firm gains revenue $R(a)$ and pays costs $C(a)$ from actions $a = (a_1, \ldots, a_n)$. The profit-maximizing firm will choose its optimal actions from

$$\pi \equiv \max_a \ R(a) - C(a).$$

By the FOC, the optimal action a^* is the solution of

$$\frac{\partial R(a^*)}{\partial a_i} = \frac{\partial C(a^*)}{\partial a_i}, \quad i = 1, \ldots, n. \tag{9}$$

This is the well-known condition of "marginal revenue = marginal cost" ($MR = MC$); see Fig. 9. The intuition is clear. By increasing a small amount Δa_i of a_i at a^*, we have

---

[5]See the Appendix for the definition of homogenous functions.

Fig. 9 MR = MC

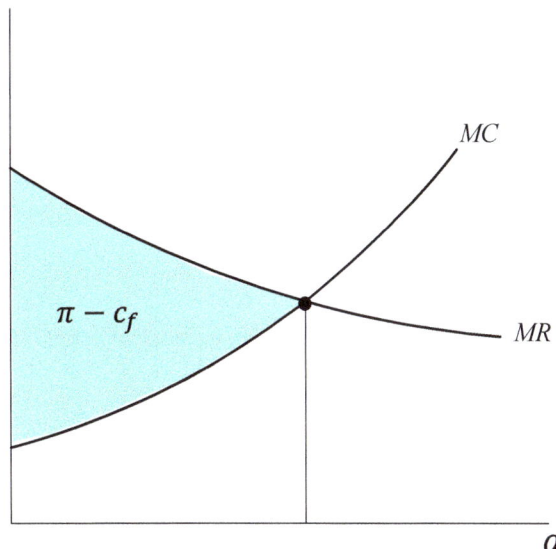

$$R(a^* + \Delta a) - R(a^*) \approx \frac{\partial R(a^*)}{\partial a_i} \cdot \Delta a_i, \quad C(a^* + \Delta a) - C(a^*) \approx \frac{\partial C(a^*)}{\partial a_i} \cdot \Delta a_i.$$

That is, the firm approximately gains $\frac{\partial R(a^*)}{\partial a_i} \cdot \Delta a_i$ and loses $\frac{\partial C(a^*)}{\partial a_i} \cdot \Delta a_i$. Hence, if $\frac{\partial R(a^*)}{\partial a_i} > \frac{\partial C(a^*)}{\partial a_i}$, the firm should increase the level of a_i, and if $\frac{\partial R(a^*)}{\partial a_i} < \frac{\partial C(a^*)}{\partial a_i}$, the firm should decrease the level of a_i. The firm will be satisfied when (9) holds.

We will only deal with competitive firms until Chap. 6. A <u>competitive firm</u> is not able to manipulate prices. It takes market prices as given (including the output and input prices). Under this assumption, we can now state the firm's problem more specifically. The firm's <u>profit function</u> $\pi : \mathbb{R}_+ \times \mathbb{R}_+^n \to \mathbb{R}$ is defined by

$$\pi(p, w) \equiv \max_{x \in \mathbb{R}_+^n} pf(x) - w \cdot x, \tag{10}$$

where p is the output price and w the vector of input prices. Here, the action variables are the quantity variables. In Chap. 6, when the firm is not competitive, both prices and quantities can be action variables. Denote the optimal choice of factors x^* from (10) as $x^* = x(p, w)$ and call it the <u>demand function</u> of the firm. Then, $y(p, w) \equiv f[x(p, w)]$ is called the <u>supply function</u>.

Example 1.4 Given a production function $y = f(x)$, the production possibility set is then defined by

$$\mathbb{P} = \{(-x, y) \in \mathbb{R}^{n+1} | f(x) \geq y\}.$$

Let $y_1 \equiv -x$, $y_2 \equiv y$ and $G(y_1, y_2) \equiv y_2 - f(-y_1)$. Then, the production possibility set is

$$\mathbb{P} = \left\{ (y_1, y_2) \in \mathbb{R}^{n+1} | G(y_1, y_2) \leq 0, y_1 \leq 0, y_2 \geq 0 \right\},$$

and

$$G_{y_1}(y_1, y_2) = f'(-y_1) = f'(x), \quad G_{y_2}(y_1, y_2) = 1,$$

and

$$\begin{vmatrix} 0 & G_{y_1} & G_{y_2} \\ G_{y_1} & G_{y_1 y_1} & G_{y_1 y_2} \\ G_{y_2} & G_{y_2 y_1} & G_{y_2 y_2} \end{vmatrix} = \begin{vmatrix} 0 & f'(x) & 1 \\ f'(x) & -f''(x) & 0 \\ 1 & 0 & 0 \end{vmatrix} = f''(x).$$

Therefore, if $f' > 0$ and $f'' < 0$, by Theorem A.7, Assumptions 1.1 and 1.2 are satisfied. Since conditions $f' > 0$ and $f'' < 0$ are usually imposed for single-output production functions, this example shows that the two assumptions are not restrictive as they do not impose unnecessary restrictions on production functions so that single-output production functions are just a special case of the more general production functions. ∎

Let us now analyze the profit maximization problem (10). The FOC for the profit maximization problem (10) is

$$pf'(x^*) = w, \tag{11}$$

where

$$w = \begin{pmatrix} w_1 \\ w_2 \\ \vdots \\ w_n \end{pmatrix}, \quad f'(x) \equiv \begin{pmatrix} f_{x_1}(x) \\ f_{x_2}(x) \\ \vdots \\ f_{x_n}(x) \end{pmatrix}.$$

The FOC implies a tangency point in Fig. 10. We can rewrite problem (10) into the following form:

$$\pi(p, w) \equiv \max_{x \in \mathbb{R}^n_+, y \geq 0} py - w \cdot x$$
$$\text{s.t.} \quad y = f(x). \tag{12}$$

This problem can be illustrated graphically in Fig. 10.

The second-order condition (SOC) for (10) is

$$f''(x^*) \equiv \left(\frac{\partial^2 f(x^*)}{\partial x_i \partial x_j} \right)_{n \times n} \leq 0.$$

Fig. 10 Profit maximization

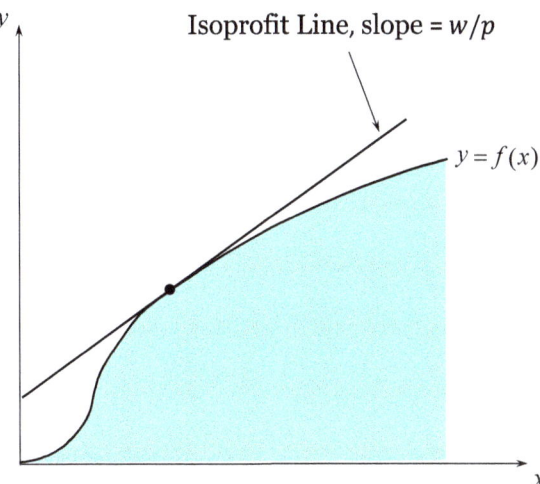

As can be seen from Fig. 10, this condition can be used to rule out the other tangency point, which is the other solution of the FOC equation.

The <u>cost function</u> is defined by

$$c(w, y) \equiv \min_{x}\{w \cdot x | y \leq f(x)\}, \tag{13}$$

which is the minimum expenditure on producing y, given price vector w. If $f(x)$ is strictly increasing, we can rewrite 13 as

$$c(w, y) \equiv \min_{x \in \mathbb{R}^n_+} w \cdot x \tag{14}$$
$$\text{s.t. } y = f(x).$$

Denote the optimal choice of factors x^* from (13) as $x^* = x(w, y)$, and call it the <u>conditional demand function</u>. Notice that this demand function is different from the previous demand function. It is conditional because the output is given instead of being optimal. For problem (13), consider the Lagrange function

$$\mathcal{L}(x, \lambda) = w \cdot x + \lambda[y - f(x)],$$

where $\lambda \geq 0$ is some constant. By the Lagrange Theorem in the Appendix, the FOC is

$$w = \lambda f'(x^*)$$

or

Fig. 11 Cost minimization

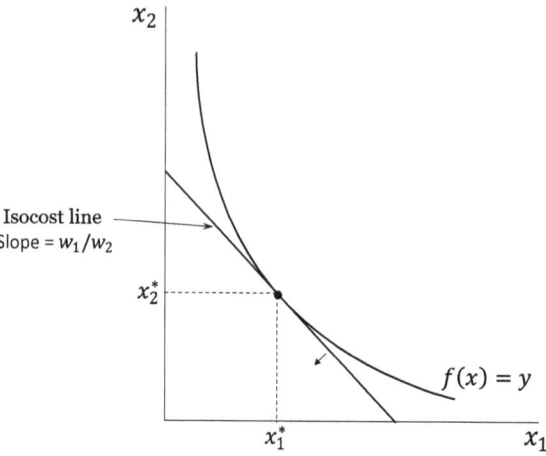

$$\frac{w_i}{w_j} = \frac{f_{x_i}(x^*)}{f_{x_j}(x^*)}, \quad i,j = 1,2,\ldots,n. \tag{15}$$

Condition (15) can also be derived graphically using Fig. 11.

As shown, the slope w_1/w_2 of the isocost line is equal to the slope of the isoquant $f_{x_i}(x)/f_{x_j}(x)$ at the optimal solution. Therefore, (15) must be true.

Condition (15) can also be derived by economic intuition. Both sides of (15) represent economic rates of substitution. The right-hand side is the internal rate of substitution, at which factor j can be exchanged for factor i while maintaining technological efficiency. The left-hand side is the external/market rate of substitution, at which factor j can be exchanged for factor i while maintaining constant expenditure. When

$$\frac{w_i}{w_j} > \frac{f_{x_i}(x^*)}{f_{x_j}(x^*)},$$

factor i is too costly relative to factor j at x^* compared to the relative benefit f_{x_i}/f_{x_j}. In this case, the firm should reduce the input of x_i relative to the input of x_j. The same argument holds for the opposite inequality. The best choice is therefore where (15) holds.

By the Lagrange Theorem, the SOC for (13) is

$$h^T L''_x(x^*, \lambda) h \geq 0, \quad \text{for all } h \text{ satisfying } f'(x^*) \cdot h = 0.$$

That is, (since $\lambda > 0$),

$$h^T f''(x^*) h \leq 0, \quad \text{for all } h \text{ satisfying } f'(x^*) \cdot h = 0, \tag{16}$$

2 The Firm's Problem

which can be dealt with using Theorem A.8 in the Appendix. This SOC can be guaranteed by quasi-concavity of f; see Theorem A.13. Notice that the conditions used to verify quasi-concavity and this SOC are very similar; see Theorems A.7 and A.8.

What is the intuition for the SOC? Concavity of $f(x)$ at x^* means that $x^T f''(x^*) x \leq 0$ for any $x \in \mathbb{R}^n$. Hence, condition (16) means that $f(x)$ is concave at x^* relative to the vector space orthogonal to $f'(x^*)$. In Fig. 11, we can see that, when $\{x | f(x) \geq y\}$ is convex, $f(x)$ is concave relative to the vector space orthogonal to $f'(x^*)$ (which is actually the isocost line). Hence, convexity of $\{x | f(x) \geq y\}$ implies (16). At the same time, we also know that quasi-concavity of f implies convexity of $\{x | f(x) \geq y\}$. Therefore, quasi-concavity of f implies the SOC (16).

After we find the cost function $c(w, y)$, we can proceed to solve the following problem of profit maximization:

$$\max_{y \geq 0} py - c(w, y). \tag{17}$$

The FOC for this problem is the well-known condition $MR = MC$ for a competitive firm, i.e.,

$$p = \frac{\partial c(w, y^*)}{\partial y}.$$

The SOC is the convexity of the cost function in y at y^*:

$$\frac{\partial^2 c(w, y^*)}{\partial y^2} \geq 0.$$

We will establish later that $c(w, y)$ is always concave in w. Hence, (w, y^*) is a so-called saddle point.

In the short run, some inputs $z \in \mathbb{R}^m_+$ (such as capital stock) are fixed. Let w_f be the price vector for the fixed inputs. We then have the short-run total profit function:

$$\pi(p, w, z) \equiv \max_x pf(x, z) - w_v \cdot x - w_f \cdot z,$$

the short-run (variable) profit function:

$$\pi_v(p, w, z) \equiv \max_x pf(x, z) - w_v \cdot x,$$

and the short-run total cost function:

$$c(w, y, z) \equiv \min_x \{w_v \cdot x | y = f(x, z)\} + w_f \cdot z,$$

and the short-run (variable) cost function:

$$c_v(w, y, z) \equiv \min_x \{w_v \cdot x | y = f(x, z)\},$$

Example 1.5 The Cost Function of the Cobb-Douglas Technology. Consider

$$c(w, y) = \min_{x_1, x_2} w_1 x_1 + w_2 x_2$$

$$\text{s.t. } A x_1^a x_2^b = y.$$

By solving x_2 from the production function and substituting it into the objective function, the problem becomes

$$\min_{x_1} w_1 x_1 + w_2 A^{-\frac{1}{b}} x_1^{-\frac{a}{b}} y^{\frac{1}{b}}.$$

By the FOC, we can solve for the conditional demand function:

$$x_1(w_1, w_2, y) = A^{-\frac{1}{a+b}} \left(\frac{a w_2}{b w_1}\right)^{\frac{b}{a+b}} y^{\frac{1}{a+b}}.$$

Given $x_1(w_1, w_2, y)$, the other conditional demand function can be solved from the production function or by symmetry[6]:

$$x_2(w_1, w_2, y) = A^{-\frac{1}{a+b}} \left(\frac{b w_1}{a w_2}\right)^{\frac{a}{a+b}} y^{\frac{1}{a+b}}.$$

The cost function is then

$$c(w_1, w_2, y) = w_1 x_1(w_1, w_2, y) + w_2 x_2(w_1, w_2, y)$$

$$= A^{-\frac{1}{a+b}} \left[\left(\frac{a}{b}\right)^{\frac{b}{a+b}} + \left(\frac{a}{b}\right)^{-\frac{a}{a+b}}\right] w_1^{\frac{a}{a+b}} w_2^{\frac{b}{a+b}} y^{\frac{1}{a+b}}. \blacksquare$$

Example 1.6 The Short-run Profit Function for the Cobb-Douglas Technology. Suppose that in the short-run $x_2 = k$. Consider

$$\pi(p, w_1, w_2, k) = \max_{x_1, y} p y - w_1 x_1$$

$$\text{s.t. } A x_1^a k^b = y.$$

---

[6]This works since, after changes $w_1 \leftrightarrow w_2$, $x_1 \leftrightarrow x_2$, and $a \leftrightarrow b$, the original problem is the same.

2 The Firm's Problem

This is equivalent to:

$$\pi(p, w_1, w_2, k) = \max_{x_1} pAx_1^a k^b - w_1 x_1.$$

By the FOC, we can find the demand function:

$$x_1(p, w_1, w_2, k) = \left(\frac{aAp}{w_1}\right)^{\frac{1}{1-a}} k^{\frac{b}{1-a}},$$

the short-run supply function:

$$y(p, w_1, w_2, k) = A^{\frac{1}{1-a}} \left(\frac{ap}{w_1}\right)^{\frac{a}{1-a}} k^{\frac{b}{1-a}},$$

and the short-run profit function:

$$\pi(p, w_1, w_2, k) = \left(\frac{1}{a} - 1\right)(aAp)^{\frac{1}{1-a}} w_1^{-\frac{a}{1-a}} k^{\frac{b}{1-a}}. \quad \blacksquare$$

Example 1.7 The Profit Function for the Cobb-Douglas Technology. Let us write the cost function from Example 1.5 as

$$c(w_1, w_2, y) \equiv c(w_1, w_2) y^{\frac{1}{a+b}}.$$

The profit maximization problem is:

$$\max_y py - c(w_1, w_2) y^{\frac{1}{a+b}}.$$

If $a + b \neq 1$, the supply function is:

$$y(p, w_1, w_2) = \left[p \frac{a+b}{c(w_1, w_2)}\right]^{\frac{a+b}{1-a-b}},$$

and the profit function is:

$$\pi(p, w_1, w_2) = \left(\frac{1}{a+b} - 1\right)(a+b)^{\frac{1}{1-a-b}} p^{\frac{1}{1-a-b}} c(w_1, w_2)^{-\frac{a+b}{1-a-b}}.$$

When $a + b = 1$ (CRS), the profit maximization problem is

$$\max_{y \geq 0} [p - c(w_1, w_2)] y,$$

which implies the following supply function:

$$y^s = \begin{cases} \infty & \text{if } p > c(w_1, w_2), \\ [0, \infty) & \text{if } p = c(w_1, w_2), \\ 0 & \text{if } p < c(w_1, w_2), \end{cases}$$

where, when $p = c(w_1, w_2)$, y^s can take any value in $[0, \infty)$. In this case, since the supply curve is horizontal, the only possible equilibrium price is $c(w_1, w_2)$. Hence, if the demand is downward sloping, there is a market equilibrium, in which the price is $p^* = c(w_1, w_2)$ and the quantity is solely determined by demand. In this sense, we say that this market is demand-driven. ∎

When $f(x)$ has global CRS, the FOC (11) does not lead to a solution. For example, for the case with two goods, since $f'(x)$ is zero homogenous, the FOCs in (11) can be written as

$$f_{x_1}\left(\frac{x_1}{x_2}, 1\right) = \frac{w_1}{p}, \quad f_{x_2}\left(\frac{x_1}{x_2}, 1\right) = \frac{w_2}{p}.$$

That is, both equations determine x_1/x_2 and these two equations may contradict each other. For example, if $f(x) = x_1^\alpha x_2^{1-\alpha}$, then the two equations imply

$$\frac{x_1}{x_2} = \left(\frac{w_1}{\alpha p}\right)^{\frac{1}{\alpha-1}}, \quad \frac{x_1}{x_2} = \left(\frac{w_2}{(1-\alpha)p}\right)^{\frac{1}{\alpha}},$$

which implies

$$\left(\frac{w_1}{\alpha p}\right)^{\frac{1}{\alpha-1}} = \left(\frac{w_2}{(1-\alpha)p}\right)^{\frac{1}{\alpha}}.$$

That is, unless this equation happens to hold, the FOCs lead to a contradiction. The cause is that the FOCs require the solution to be an interior solution. When $f(x)$ has global CRS, the solution may not be an interior solution. To see this, consider a special case. With one input x, a production function with CRS is $f(x) = ax$, where a is a constant. Then, the situation in Fig. 10 becomes the following situation, which clearly indicates the possibility of a corner solution. Thus, with a CRS technology, we need to derive the cost function first, from which we can then determine the supply function, as we did in Example 1.7 (Fig. 12).

Example 1.8 The CES Production Function. Consider production function:

$$f(x_1, x_2) = \left(a_1 x_1^\rho + a_2 x_2^\rho\right)^{1/\rho},$$

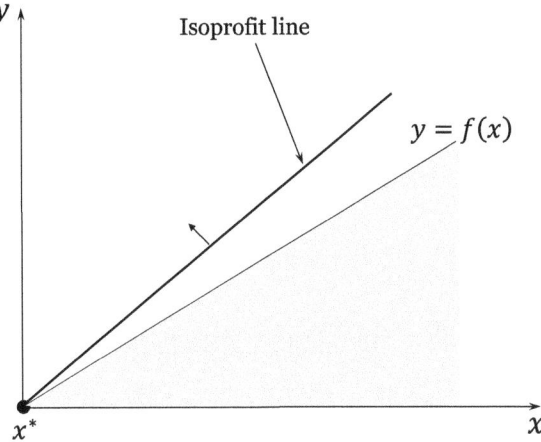

Fig. 12 A homogenous production function

where $a_1 > 0$, $a_2 > 0$ and $\rho \in [-\infty, 1]$ are constants. By (15),

$$\frac{w_1}{w_2} = \frac{(a_1 x_1^\rho + a_2 x_2^\rho)^{\frac{1}{\rho}-1} a_1 x_1^{\rho-1}}{(a_1 x_1^\rho + a_2 x_2^\rho)^{\frac{1}{\rho}-1} a_2 x_2^{\rho-1}} = \frac{a_1}{a_2} \left(\frac{x_1}{x_2}\right)^{\rho-1}.$$

Thus,

$$\frac{x_1}{x_2} = \left(\frac{a_1}{a_2}\right)^{\frac{1}{1-\rho}} \left(\frac{w_1}{w_2}\right)^{-\frac{1}{1-\rho}}.$$

Using the fact that

$$y = Ax^a \quad \Rightarrow \quad a = \frac{\partial \log y}{\partial \log x},$$

we then have

$$\sigma \equiv -\frac{\partial(x_1/x_2)}{\partial(w_1/w_2)} \frac{(w_1/w_2)}{(x_1/x_2)} = \frac{1}{1-\rho}.$$

Therefore, this function has constant elasticity of substitution (CES): $\sigma = \frac{1}{1-\rho}$. ∎

The CES function contains several well-known production functions as special cases. When $\rho = 1$ or $\sigma = \infty$, we have a linear production function:

$$y = a_1 x_1 + a_2 x_2.$$

When $\rho = 0$ or $\sigma = 1$, we further assume $a_1 + a_2 = 1$. Since, by L'Hopital's rule,

$$\lim_{\rho \to 0} \frac{\ln(a_1 x_1^\rho + a_2 x_2^\rho)}{\rho} = \lim_{\rho \to 0} \frac{a_1 x_1^\rho \ln x_1 + a_2 x_2^\rho \ln x_2}{a_1 x_1^\rho + a_2 x_2^\rho} = a_1 \ln x_1 + a_2 \ln x_2,$$

we have

$$y = \lim_{\rho \to 0} (a_1 x_1^\rho + a_2 x_2^\rho)^{\frac{1}{\rho}} = \lim_{\rho \to 0} e^{\frac{\ln(a_1 x_1^\rho + a_2 x_2^\rho)}{\rho}} = x_1^{a_1} x_2^{a_2},$$

which is the <u>Cobb-Douglas function</u>. When $\rho = -\infty$ or $\sigma = 0$, using the fact that

$$1 \leq \left[\frac{a_1^{1/\rho} x_1}{\min\left(a_1^{1/\rho} x_1, a_2^{1/\rho} x_2\right)} \right]^\rho + \left[\frac{a_2^{1/\rho} x_2}{\min\left(a_1^{1/\rho} x_1, a_2^{1/\rho} x_2\right)} \right]^\rho \leq 2,$$

for $\rho < 0$, we have

$$y = \lim_{\rho \to -\infty} \min\left(a_1^{1/\rho} x_1, a_2^{1/\rho} x_2\right) \left\{ \left[\frac{a_1^{1/\rho} x_1}{\min\left(a_1^{1/\rho} x_1, a_2^{1/\rho} x_2\right)} \right]^\rho + \left[\frac{a_2^{1/\rho} x_2}{\min\left(a_1^{1/\rho} x_1, a_2^{1/\rho} x_2\right)} \right]^\rho \right\}^{\frac{1}{\rho}}$$
$$= \min(x_1, x_2),$$

which is the <u>Leontief Production Function</u>. Three isoquants of these three special production functions are shown in Fig. 13.

For changing market conditions, or more precisely, changing price ratios, the two special technologies with $\sigma = 0$ and $\sigma = \infty$ have special reactions. As shown in Fig. 14, if the technology is Leontief, the firm typically does not react to changing market conditions at all; if the technology is linear, the firm often reacts in a dramatic fashion—from demanding only x_1 to demanding only x_2, or vice versa.

3 Short-Run and Long-Run Cost Functions

3.1 Definitions

In the short run, suppose that a vector z of factors is fixed. Let $w = (w_v, w_f)$ be the vector of prices for the variable and fixed factors, respectively. Then, the short-run total cost is:

$$c(w, y, z) = w_v \cdot x_v(w, y, z) + w_f \cdot z.$$

3 Short-Run and Long-Run Cost Functions

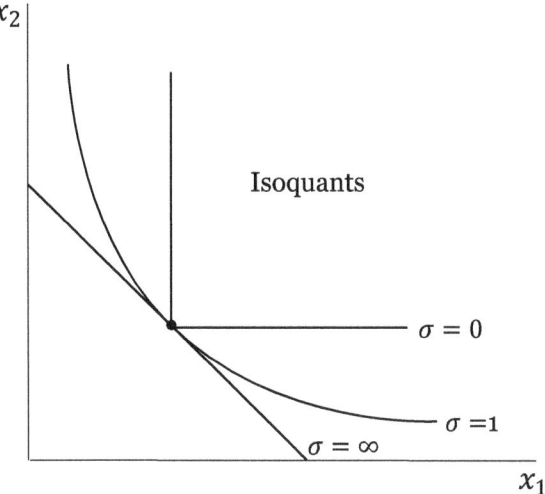

Fig. 13 Isoquants for the CES technology

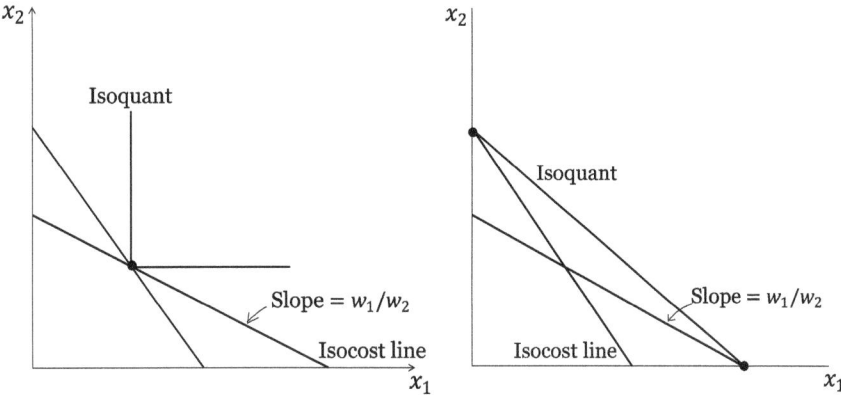

Fig. 14 The two special technologies

Define

$$\text{the short-run average variable cost (SAVC)} \equiv \frac{w_v \cdot x_v(w,y,z)}{y};$$
$$\text{the short-run average cost (SAC)} \equiv \frac{c(w,y,z)}{y};$$
$$\text{the short-run marginal cost (SMC)} \equiv \frac{\partial c(w,y,z)}{\partial y}.$$

In the long run, when all factors are variable, the firm will choose an optimal z: $z^* = z(w,y)$. Then, we will have the long-run cost (LC):

$$c(w, y) = c[w, y, z(w, y)].$$

Similarly, define

the long-run average cost (LAC) $\equiv \frac{c(w,y)}{y}$;

the long-run marginal cost (LMC) $\equiv \frac{\partial c(w,y)}{\partial y}$.

3.2 Relationship Between AC and MC Curves

Let us first find the relation between the AC and MC curves. Suppose that $c(y, z)$ is a short-run cost function with fixed input vector z, and that $c_v(y, z)$ is the short-run variable cost function. We have

$$\frac{d}{dy}\text{SAVC}(y, z) = \left[\frac{c_v(y, z)}{y}\right]' = \frac{1}{y}\left[\frac{d}{dy}c_v(y, z) - \frac{c_v(y, z)}{y}\right]$$
$$= \frac{1}{y}[\text{SMC}(y, z) - \text{SAVC}(y, z)].$$

This means that

$\text{SMC}(y, z) \leq \text{SAVC}(y, z)$ if $\frac{d}{dy}\text{SAVC}(y, z) \leq 0$;
$\text{SMC}(y, z) \geq \text{SAVC}(y, z)$ if $\frac{d}{dy}\text{SAVC}(y, z) \geq 0$;
$\text{SMC}(y, z) = \text{SAVC}(y, z)$ if $\text{SAVC}(y, z)$ reaches minimum.

Furthermore, since $c_v(0, z) = 0$, by L'Hopital's rule, we have

$$\text{SAVC}(0, z) = \lim_{y \to 0} \frac{c_v(y, z)}{y} = \lim_{y \to 0} \frac{\partial c_v(y, z)}{\partial y} = \text{SMC}(0, z).$$

Similarly,

$$\frac{d}{dy}\text{SAC}(y, z) = \frac{1}{y}[\text{SMC}(y, z) - \text{SAC}(y, z)].$$

These are also true for long-run cost functions except in that case when the variable cost is the same as the total cost. We thus have the general graphic situations for SMC, SAC and SAVC and for LMC and LAC in Fig. 15.

3 Short-Run and Long-Run Cost Functions

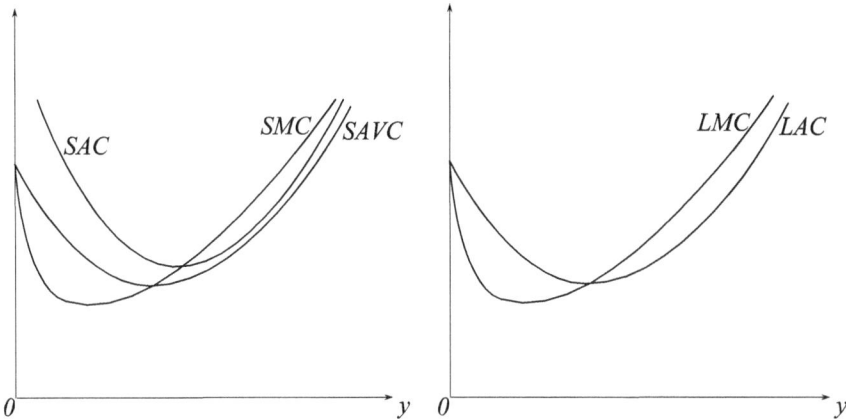

Fig. 15 AC, AVC, MC curves

3.3 Relationship Between SR and LR Cost Functions

Let us now find the relations between short-run and long-run cost functions. Suppose that, in the short run, a vector of factors z is fixed and, in the long run, the minimum point of $c(y, z)$ is $z(y) : c[y, z(y)] = c(y)$. Then,

$$c(y) = \min_z c(y, z) \tag{18}$$

implying

$$\frac{c(y)}{y} = \min_z \frac{c(y, z)}{y}.$$

That is,

$$\text{LAC}(y) = \min_z \text{SAC}(y, z) \equiv \text{SAC}[y, z(y)]. \tag{19}$$

By the Envelope Theorem, we have, for any $\hat{y} \geq 0$,

$$\frac{d}{dy} \text{LAC}(\hat{y}) = \frac{\partial}{\partial y} \text{SAC}(\hat{y}, z)|_{z=z(\hat{y})}. \tag{20}$$

The above two Eqs. (19) and (20) mean that, for any point $\hat{y}$,

(a) the short-run average cost $\text{SAC}[y, z(\hat{y})]$ must be at least as great as the long-run average cost $\text{LAC}(y)$ for all output levels $y : \text{SAC}[y, z(\hat{y})] \geq \text{LAC}(y), \forall y \geq 0$;
(b) the short-run average cost is equal to the long-run average cost at output $\hat{y}$: $\text{LAC}(\hat{y}) = \text{SAC}[\hat{y}, z(\hat{y})]$; and

(c) the long- and short-run cost curves are tangent at $\hat{y}$.

These mean that the long-run cost curve is simply the lower envelope of the short-run curves, as shown in Fig. 16.

Also, by the Envelope Theorem, from (18), for any output level $\hat{y}$, we have

$$\text{LMC}(\hat{y}) = \left.\frac{\partial c(\hat{y}, z)}{\partial y}\right|_{z=z(\hat{y})} = \text{SMC}[\hat{y}, z(\hat{y})].$$

This is illustrated in Fig. 17.

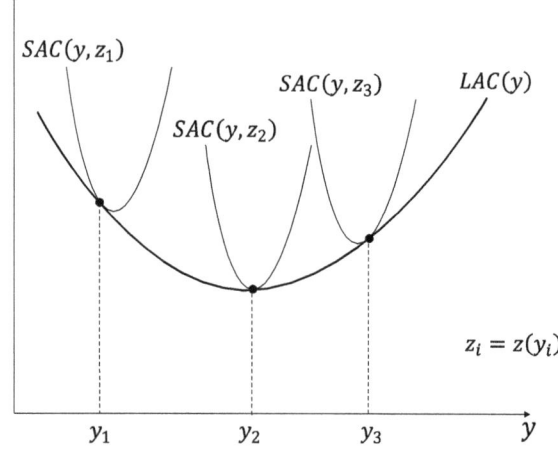

Fig. 16 Long-run and short-run average cost curves

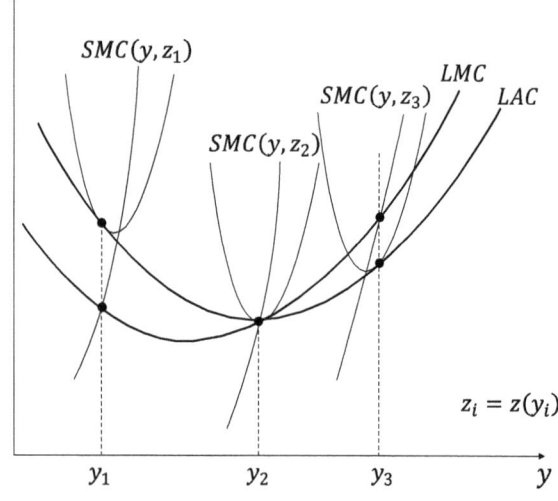

Fig. 17 Long-run and short-run marginal cost curves

4 Properties

This section discusses the properties of the various functions involved in production.

Proposition 1.6 *If the production function is homogeneous of degree α, then*

$$c(w, y) = y^{\frac{1}{\alpha}} c(w, 1),$$

i.e., the cost function is homogenous of degree $1/\alpha$ in output.

Proof We have

$$c(w,y) \equiv \min_x \left\{ w \cdot x \,\middle|\, y = f(x) \right\} = \min_x \left\{ w \cdot x \,\middle|\, 1 = f\left(\frac{x}{y^{1/\alpha}}\right) \right\}$$

$$= y^{1/\alpha} \min_x \left\{ w \cdot \frac{x}{y^{1/\alpha}} \,\middle|\, 1 = f\left(\frac{x}{y^{1/\alpha}}\right) \right\} = y^{1/\alpha} \min_z \left\{ w \cdot z \,\middle|\, 1 = f(z) \right\}$$

$$= y^{1/\alpha} c(w, 1). \qquad \blacksquare$$

See Example 1.5 for a confirmation of Proposition 1.6.

Proposition 1.7 (Properties of Cost Functions). *The cost function $c(w, y)$ has the following properties*:

(1) *increasing in w,*
(2) *linearly homogeneous in w,*
(3) *concave in w.*

Furthermore, if $c(w, y)$ is continuous, then the three conditions are sufficient for $c(w, y)$ to be a cost function.[7]

Proof We show the necessity of the three conditions only. First, given $w' \geq w$, for any $x \geq 0$ satisfying $f(x) \geq y$, we must have $w' \cdot x \geq w \cdot x$. Hence,

$$\min_x \left\{ w' \cdot x | f(x) \geq y \right\} \geq \min_x \{ w \cdot x | f(x) \geq y \},$$

i.e., $c(w', y) \geq c(w, y)$.
The second property is obvious.
For the third property, by the definition of $x(w, y)$, for any $\lambda \in [0, 1]$, we have

---

[7] A function $c(w, y)$ is a cost function if there exists an increasing and concave function f such that $c(w, y) \equiv \min_x \left\{ w \cdot x | y = f(x) \right\}$, $\forall (w, y) \in \mathbb{R}_+^{n+1}$.

$$c(w,y) = w \cdot x(w,y) \leq w \cdot x[\lambda w + (1-\lambda)w', y],$$
$$c(w',y) = w' \cdot x(w',y) \leq w' \cdot x[\lambda w + (1-\lambda)w', y].$$

Then,

$$\lambda c(w,y) + (1-\lambda)c(w',y)$$
$$\leq \lambda w \cdot x[\lambda w + (1-\lambda)w', y] + (1-\lambda)w' \cdot x[\lambda w + (1-\lambda)w', y]$$
$$= [\lambda w + (1-\lambda)w'] \cdot x[\lambda w + (1-\lambda)w', y] = c[\lambda w + (1-\lambda)w', y]. \blacksquare$$

What are the assumptions that we have used in the proof of the concavity? Only the cost-minimizing behavior! Hence, the concavity of the cost function comes solely from cost-minimizing behavior. The concavity of the cost function means that the cost rises less than proportionally when the prices increase. The intuition is clear: when the price of a factor rises, as one factor becomes relatively more expensive than other factors, the cost-minimizing firm will shift away from that factor to other factors, which results in a less than proportional increase in cost. In other words, the cost increases at a decreasing rate. This is what concavity means.

We can explain the concavity of cost graphically. In Fig. 18, at an input point x^* for price vector w^*, when w_1 changes, the firm can always spend $w_1 x_1^* + \sum_{i=2}^{n} w_i^* x_i^*$ to maintain input x_1^*. However, this may not be optimal. The optimal cost will generally be less than $w_1 x_1^* + \sum_{i=2}^{n} w_i^* x_i^*$. Since the optimal cost curve is always below the straight line, it must be concave.

Similarly, we have the following result for profit functions:

Proposition 1.8 (Properties of Profit Functions). *The profit function $\pi(p, w)$ has the following properties*:

(1) *increasing in p, decreasing in w;*
(2) *linearly homogeneous in (p, w);*
(3) *convex in (p, w).*

Fig. 18 Concavity of the cost function

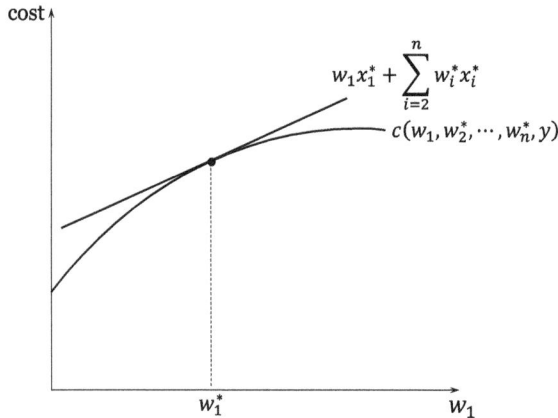

4 Properties

Proof By comparing (12) with (14), we can see that $\pi(p,w)$ is defined just like $c(w,y)$, with linear objective functions and the same constraint, except that one is minimization and the other is maximization. Hence, the properties of $\pi(p,w)$ can be proven just like those of $c(w,y)$. Alternatively, we can borrow the concavity of $c(w,y)$ to prove the convexity of $\pi(p,w)$. By the definition of $\pi(p,w)$, for any $\lambda \in [0,1]$, let

$$\hat{y} \equiv y[\lambda p + (1-\lambda)p', \lambda w + (1-\lambda)w'].$$

We have

$$\pi(p,w) \geq p\hat{y} - c(w,\hat{y}), \quad \pi(p',w') \geq p'\hat{y} - c(w',\hat{y}),$$

implying

$$\begin{aligned}
\lambda \pi(p,w) + (1-\lambda)\pi(p',w') &\geq [\lambda p + (1-\lambda)p']\hat{y} - [\lambda c(w,\hat{y}) + (1-\lambda)c(w',\hat{y})] \\
&\geq [\lambda p + (1-\lambda)p']\hat{y} - c[\lambda w + (1-\lambda)w', \hat{y}] \\
&= \pi[\lambda p + 1(1-\lambda)p', \lambda w + (1-\lambda)w'].
\end{aligned}$$
∎

The following Hotelling's lemma establishes a relationship between demand and profit functions.

Proposition 1.9 (Hotelling's Lemma). *Let $x_i(p,w)$ be the firm's demand function for factor i. Assume that it is an interior solution of the profit maximization problem. Let $y(p,w)$ and $\pi(p,w)$ be respectively the supply and profit functions. Then,*

$$y(p,w) = \frac{\partial \pi(p,w)}{\partial p}, \quad x_i(p,w) = -\frac{\partial \pi(p,w)}{\partial w_i}, \quad \forall i.$$

Proof Let $f(x)$ be the production function. Then, $x(p,w)$ is the solution of

$$\pi(p,w) \equiv \max_x pf(x) - w \cdot x.$$

By the Envelope Theorem,

$$\frac{\partial \pi(p,w)}{\partial p} = f[x(p,w)] \equiv y(p,w).$$

Similarly,

$$\frac{\partial \pi(p,w)}{\partial w_i} = -x_i(p,w). \blacksquare$$

The following Shephard's lemma establishes a relationship between cost and conditional demand functions.

Proposition 1.10 (Shephard's Lemma). *Let $x_i(w, y)$ be the firm's conditional demand function for factor i. Assume that it is an interior solution of the cost minimization problem. Let $c(w, y)$ be the cost function. Then,*

$$x_i(w, y) = \frac{\partial c(w, y)}{\partial w_i}, \quad i = 1, 2, \ldots, n.$$

Proof $x_i(w, y)$ is the solution of the following problem:

$$c(w, y) \equiv \min_x w \cdot x + \lambda[y - f(x)].$$

By the Envelope Theorem, we immediately have

$$\frac{\partial c(w, y)}{\partial w_i} = x_i(w, y). \quad \blacksquare$$

We already know the properties of cost and profit functions. We can now find the properties of demand and supply functions through Hotelling's and Shephard's lemmas.

Proposition 1.11 (Properties of Conditional Demand). *Suppose that $x(w, y)$ is twice continuously differentiable. Then,*

(1) *$x(w, y)$ is zero homogeneous in w.*
(2) *Cross-price effects are symmetric: $\frac{\partial x_i(w,y)}{\partial w_j} = \frac{\partial x_j(w,y)}{\partial w_i}$.*
(3) *<u>Substitution matrix</u> $x'_w(w, y)$ is negative semidefinite.*
(4) *$x_i(w, y)$ is decreasing in w_i.*

Proof

(1) By Theorem A.16, since $c(w, y)$ is linearly homogeneous in w $x(w, y)$ is zero homogeneous in w. See Fig. 11 for a graphic explanation.
(2) By Shephard's lemma, we have

$$c'_w(w, y) = x(w, y), \quad c''_w(w, y) = x'_w(w, y) \equiv \begin{pmatrix} \frac{\partial x_1}{\partial w_1} & \frac{\partial x_1}{\partial w_2} & \cdots & \frac{\partial x_1}{\partial w_n} \\ \frac{\partial x_2}{\partial w_1} & \frac{\partial x_2}{\partial w_2} & \cdots & \frac{\partial x_2}{\partial w_n} \\ \vdots & \vdots & & \vdots \\ \frac{\partial x_n}{\partial w_1} & \frac{\partial x_n}{\partial w_2} & \cdots & \frac{\partial x_n}{\partial w_n} \end{pmatrix}.$$

4 Properties

By continuous differentiability of $c(w, y)$,

$$\frac{\partial^2 c(w, y)}{\partial w_i \partial w_j} = \frac{\partial^2 c(w, y)}{\partial w_j \partial w_i},$$

implying

$$\frac{\partial x_j(w, y)}{\partial w_i} = \frac{\partial x_i(w, y)}{\partial w_j}.$$

That is, $x'_w(w, y)$ is symmetric.

(3) By the concavity of $c(w, y)$, $c''_w(w, y) \leq 0$. Therefore, the substitution matrix $x'_w(w, y)$ is symmetric and negative semidefinite.
(4) Since $c''_w(w, y)$ is symmetric and negative semi-definite, its diagonal entries must be negative, i.e.,

$$\frac{\partial^2 c(w, y)}{\partial w_i^2} \leq 0, \quad \forall i.$$

This immediately implies property (4). ∎

Using a similar proof, we have

Proposition 1.12 (Properties of Demand and Supply)
Let $x(p, w)$ and $y(p, w)$ be twice continuously differentiable demand and supply functions. Then,

(1) $x(p, w)$ and $y(p, w)$ are zero homogeneous in (p, w)[8];
(2) $x_i(p, w)$ is decreasing in w_i; $y(p, w)$ is increasing in p;
(3) cross-price effects are symmetric: $\frac{\partial x_i(p,w)}{\partial w_j} = \frac{\partial x_j(p,w)}{\partial w_i}$. ∎

Notice that property (2) in the above proposition says that the firm's demand function is always downward sloping. This is different from the Marshallian demand for consumers.

Proposition 1.13 (The LeChatelier Principle)
Firms will respond more to price change in the long run than in the short run. More precisely, suppose that z is a fixed vector of inputs in the short run, and $z(p, w)$ is the profit maximization choice of z in the long run for given interior prices (p, w). Then,

---
[8]See Fig. 10 for a graphic explanation.

$$\frac{\partial y(p,w)}{\partial p} \geq \frac{\partial y(p,w,z)}{\partial p}\bigg|_{z=z(p,w)}, \quad \left|\frac{\partial x_i(p,w)}{\partial w_i}\right| \geq \left|\frac{\partial x_i(p,w,z)}{\partial w_i}\right|_{z=z(p,w)}, \quad \forall i.$$

Proof Let $\pi(p,w)$ be the long-run profit function, $\pi(p,w,z)$ the short-run profit function, $y(p,w)$ and $x(p,w)$ the long-run supply and demand functions, respectively, $y(p,w,z)$ and $x(p,w,z)$ the short-run supply and demand functions, respectively, and $z^* \equiv z(p,w)$ the solution of $\max_z \pi(p,w,z)$. Given any price vector (p_0, w_0), define

$$\varphi(p,w) \equiv \pi(p,w) - \pi[p,w,z(p_0,w_0)].$$

By definition, we have

$$\varphi(p,w) \geq 0, \quad \varphi(p_0,w_0) = 0.$$

Thus, (p_0, w_0) is the minimum point of $\varphi(p,w)$. This implies $\varphi'(p_0,w_0) = 0$ and $\varphi''(p_0,w_0) \geq 0$. By $\varphi''(p_0,w_0) \geq 0$, the diagonal elements of $\varphi''(p_0,w_0)$ must be positive, that is,

$$\frac{\partial^2 \varphi(p_0,w_0)}{\partial p^2} \geq 0, \quad \frac{\partial^2 \varphi(p_0,w_0)}{\partial w_i^2} \geq 0, \quad \forall i.$$

Then, by Hotelling's lemma, we have

$$\frac{\partial y(p_0,w_0)}{\partial p} \geq \frac{\partial y[p_0,w_0,z(p_0,w_0)]}{\partial p}, \quad \frac{\partial x_i(p_0,w_0)}{\partial w_i} \leq \frac{\partial x_i[p_0,w_0,z(p_0,w_0)]}{\partial w_i}, \quad \forall i. \blacksquare$$

Example 1.9 Given a function $c(w,y) \equiv A w_1^a w_2^b y^c$, where $A, c > 0$, under what conditions, is this function a cost function? Using Proposition 1.7, increasingness in w implies

$$a, b \geq 0. \tag{21}$$

Linear homogeneity implies that

$$a + b = 1. \tag{22}$$

We have

$$c''(w,y) = \begin{pmatrix} a(a-1)A w_1^{a-2} w_2^b y^c & ab A w_1^{a-1} w_2^{b-1} y^c \\ ab A w_1^{a-1} w_2^{b-1} y^c & b(b-1) A w_1^a w_2^{b-2} y^c \end{pmatrix}.$$

Since $a+b=1$, this implies that

$$|c''(w,y)| = A^2 w_1^{2(a-1)} w_2^{2(b-1)} y^{2c} ab(1-a-b) = 0.$$

Then, concavity requires that

$$a \leq 1, \quad b \leq 1.$$

These conditions are implied by (21) and (22). Hence, (21) and (22) are necessary and sufficient conditions for $c(w,y)$ to be a cost function. The corresponding regression model is

$$\ln(c) = \beta_0 + a\ln(w_1) + b\ln(w_2) + c\ln(y).$$

Using condition (22), the regression model becomes

$$\ln\left(\frac{c}{w_2}\right) = \beta_0 + \beta_1 \ln\left(\frac{w_1}{w_2}\right) + \beta_2 \ln(y) + \varepsilon.$$

By (21), we require $\beta_1 \in [0,1]$. ∎

5 Aggregation

In applied economics, data are often from observations of aggregate variables, such as the total output of an industry, the industry demand for a certain factor, and the total profit in a sector. We have so far only discussed a firm's production, input and profits. There is a need to extend these discussions to aggregate variables. A classical approach to this problem is to show that aggregate variables can be viewed as being generated by a single firm—a fictional firm. If so, all the results that I have obtained so far become readily applicable to aggregate variables.

Specifically, given n independent production units, can the aggregate supply and profit functions from these units be viewed respectively as supply and profit functions of a firm? The answer is yes under certain conditions.

Suppose that there are m production units (firms or plants) in an economy, having production possibility sets $\mathbb{Y}_1, \ldots, \mathbb{Y}_m$ in $\mathbb{R}^n$, respectively. Assume that each $\mathbb{Y}_i$ is nonempty, closed and with the free disposal property. Let $\pi_i(p)$ be the profit function for the unit with $\mathbb{Y}_i$ and let $y_i(p)$ be the supply function. Here, $y_i(p)$ can be a set-valued function, if, for each p, $y_i(p)$ is a subset of $\mathbb{Y}_i$ instead of a single value. The aggregate supply function is

$$y(p) = \sum_{i=1}^{m} y_i(p),$$

where the sum is generally a sum of m sets. The aggregate production set is

$$\mathbb{Y} = \sum_{i=1}^{m} \mathbb{Y}_i.$$

Let $\pi^A(p)$ and $y^A(p)$ be respectively the profit and supply functions of $\mathbb{Y}$.

Proposition 1.14 *For all $p \gg 0$, we have*

$$\pi^A(p) = \sum_{i=1}^{m} \pi_i(p), \quad y^A(p) = \sum_{i=1}^{m} y_i(p).$$

In other words, the aggregate variables $\pi^A(p)$ and $y^A(p)$ can be viewed respectively as the profit and supply functions of the fictional firm defined by production possibility set $\mathbb{Y}$. ∎

Notes

Materials in this chapter are fairly standard. Many books cover them. Good references for this chapter are Varian (1992), Jehle (2001), Dixit (1990), and Mas-Colell et al. (1995).

Consumer Theory

There are two distinct approaches to modeling individual behaviors. The first approach starts with preferences and proceeds to rationalizing the preferences and then showing the existence of a utility representation. The second approach is based on observations of individual choices and proceeds to reveal individual preferences with the aid of some consistency assumptions on individual preferences. We will focus on the first approach. Although the second approach has many attractive features, the existence of a utility representation gives us a convenient tool for the development of demand theory. We briefly cover the second approach in Sect. 6.

1 Existence of the Utility Function

Consider a consumer faced with possible consumption bundles in a closed and convex set $\mathbb{X} \subset \mathbb{R}^k$, called the <u>consumption set</u>. When a consumption set is not mentioned, we implicitly assume that $\mathbb{R}^k_+$ is the consumption set.

The consumer is assumed to have <u>preferences</u> for the consumption bundles of $\mathbb{X}$. When the consumer thinks that bundle x is at least as good as bundle y, we denote $x \succsim y$. We call $\succsim$ a <u>preference relation</u> (a preorder in mathematics) on $\mathbb{X}$ if it satisfies the following two axioms:

Axiom 1 (*Reflexivity*). $\forall x \in \mathbb{X}, x \succsim x$.

Axiom 2 (*Transitivity*). $\forall x, y, z \in \mathbb{X}$, if $x \succsim y$, $y \succsim z$, then $x \succsim z$.

These two axioms are necessary for any consumer who is considered as rational.[1]

---

[1]The collective preferences of a group of rational consumers may not have such preferences at all.

© Springer Nature Singapore Pte Ltd. 2018
S. Wang, *Microeconomic Theory*, Springer Texts in Business and Economics,
https://doi.org/10.1007/978-981-13-0041-7_2

Example 2.1 The relation $\geq$ on $\mathbb{R}^n$ is a preference relation, but the relation $>$ is not a preference relation. ∎

We introduce a bit more notation:

$$x \precsim y \quad \text{if} \quad y \succsim x.$$
$$x \sim y \quad \text{if} \quad x \precsim y \quad \text{and} \quad x \succsim y.$$
$$x \succ y \quad \text{if} \quad x \succsim y \quad \text{and} \quad x \nsim y.$$

It will be convenient to have a function representing individual preferences. Given the function, we can solve the maximization problem using the Lagrange theorem and discuss the solution using existing mathematical tools. In order to represent preferences by a function, we need the following two axioms:

Axiom 3 (*Completeness*). $\forall x, y \in \mathbb{X}$, either $x \succsim y$ or $x \precsim y$ or $x \sim y$. That is, any two consumption bundles can be compared.

We say that a preference relation is <u>rational</u> if it is transitive and complete.

Axiom 4 (*Continuity*). $\forall x_0 \in \mathbb{X}$, $\{x \in \mathbb{X} | x \succsim x_0\}$ and $\{x \in \mathbb{X} | x \precsim x_0\}$ are closed sets in $\mathbb{X}$. That is, if $x_n \succsim x_0$ and $x_n \to x^*$ as $n \to \infty$, then $x^* \succsim x_0$.

It is obvious that if $\succsim$ is complete, then $\{x \in \mathbb{X} | x \succsim x_0\} = \{x \in \mathbb{X} | x \prec x_0\}^c$, where the superscript c indicates the complement set. Also, by Arrow & Intriligator (1981, Theorem 4.1), we have $\{x \in \mathbb{X} | x \succsim x_0\} = \{x \in \mathbb{X} | x \prec x_0\}^c$ iff $\succsim$ is complete.

Further, the following concepts are often used in our discussion of consumer preferences.

Axiom 5 (*Monotonicity*). $x \geq y \Rightarrow x \succsim y$.

Axiom 6 (*Strict Monotonicity*). $x > y \Rightarrow x \succ y$.

Axiom 7 (*Local Nonsatiation*). $\forall x \in \mathbb{X}$, $\varepsilon > 0$, $\exists y \in \mathbb{X}$, $\|x - y\| < \varepsilon$, such that $y \succ x$.

Axiom 8 (*Convexity*). $x \succsim z, y \succsim z \Rightarrow \lambda x + (1-\lambda) y \succsim z$, $\forall \lambda \in [0,1]$.

Axiom 9 (*Strict Convexity*). $x \succsim z, y \succsim z, x \neq y \Rightarrow \lambda x + (1-\lambda) y \succ z$, $\forall \lambda \in (0,1)$.

Here $\|x - y\|$ is the distance between x and y. Local nonsatiation says that one can always find a better choice y in any neighborhood of a consumption point x. Thus, strong monotonicity implies local nonsatiation, but not vice versa.

Denote $(\succsim, \mathbb{X})$ as the consumption space that has the preference relation $\succsim$ over the set $\mathbb{X}$. A function $u : \mathbb{X} \to \mathbb{R}$ is said to represent the preferences if

- $u(x) \geq u(y)$ if and only if $x \succsim y$, and
- $u(x) = u(y)$ if and only if $x \sim y$.

Such a function is called a <u>utility function</u> on $(\succsim, \mathbb{X})$ or a utility representation of preferences.

Convexity implies that the consumer prefers averages to extremes, which implies quasi-concavity of utility functions. The convexity is the neoclassical assumption of "diminishing marginal rates of substitution." By definition, we obviously have the following result.

Proposition 2.1

- *Continuity of preferences $\Leftrightarrow$ continuity of the utility function.*
- *(Strict) monotonicity of preferences $\Leftrightarrow$ (strict) monotonicity of the utility function.*
- *(Strict) convexity of preferences $\Leftrightarrow$ (strict) quasi-concavity of the utility function.* ∎

We now prepare to prove the existence theorem of the utility function. A set $\mathbb{X} \subset \mathbb{R}$ is said to be <u>connected</u> if it cannot be expressed as a union of two disjointed nonempty open sets in $\mathbb{X}$. That means that if $\mathbb{X}$ is connected and there are two open sets A and B in $\mathbb{X}$ such that $A \cap B = \emptyset$ and $\mathbb{X} = A \cup B$, then either $A = \emptyset$ or $B = \emptyset$.

Lemma 2.1 *A subset $A \subset \mathbb{R}$ is connected $\Leftrightarrow$ A is an interval.* ∎

Lemma 2.2 *Continuous functions map a connected set to a connected set.* ∎

Lemma 2.3 $f : \mathbb{R}_+^k \to \mathbb{R}$ *is continuous $\Leftrightarrow f^{-1}(a,b)$ is open in $\mathbb{R}_+^k$, $\forall a, b$.* ∎

Lemmas 2.1 and 2.2 are well known and can be found in Armstrong (1983, Theorems 3.19 and 3.21). Lemma 2.3 can be easily shown from the definition of continuity.

Theorem 2.1 (Existence of the Utility Function). *If the preferences $\succsim$ are reflexive, transitive, complete, and continuous, then there exists a continuous utility function u on $(\succsim, \mathbb{X})$.*

Proof For the simplicity of the proof, we assume that the preferences are strictly monotonic. Let $e \equiv (1, 1, \ldots, 1)$. For any $x \in \mathbb{R}_+^k$, we want to find a unique number $u(x)$ such that $x \sim u(x)e$. As shown in Fig. 1, the idea is to use one point $u(x)e$ on the 45°-line to represent an indifference curve $I(x)$. Define

$$A \equiv \{t \in \mathbb{R}_+ \mid te \succsim x\}, \quad B \equiv \{t \in \mathbb{R}_+ \mid te \precsim x\}.$$

Fig. 1 Functional representation of preferences

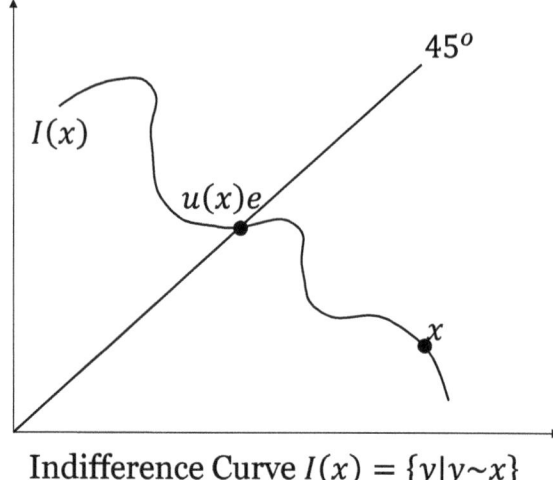

Indifference Curve $I(x) = \{y|y\sim x\}$

By the continuity axiom, A and B are closed sets in $\mathbb{R}_+$. By the monotonicity axiom, $A \neq \emptyset$. Since $0 \in B$, we know $B \neq \emptyset$. By the completeness axiom, $A \cup B = \mathbb{R}_+$. If $A \cap B = \emptyset$, then $A^c \equiv \mathbb{R}_+ \setminus A = B$ and $B^c \equiv \mathbb{R}_+ \setminus B = A$. This means that A and B are open sets in $\mathbb{R}_+$. By Lemma 2.1, $\mathbb{R}_+$ is connected. Thus, A or B must be empty. This is a contradiction. Therefore, $A \cap B \neq \emptyset$.

Suppose $t_1, t_2 \in A \cap B$. Then, $t_1 e \succsim x \succsim t_2 e \succsim x \succsim t_1 e$, implying $t_1 e \sim t_2 e$. By the strong monotonicity axiom, $t_1 = t_2$. Therefore, $A \cap B$ can contain only a single point. Denote this point as $u(x)$. Then, $x \sim u(x)e$.

We now show that this function $u(x)$ represents the preferences, i.e., $u(x) \geq u(y)$ if and only if $x \succsim y$; and $u(x) = u(y)$ if and only if $x \sim y$. By definition and strong monotonicity,

$$u(x) > u(y) \Leftrightarrow u(x)e \gg u(y)e \Leftrightarrow u(x)e \succ u(y)e \Leftrightarrow x \succ y,$$

and

$$u(x) = u(y) \Leftrightarrow x \sim u(x)e = u(y)e \sim y \Leftrightarrow x \sim y.$$

These relationships imply that u represents the preferences.

We now only need to show the continuity. For any $a, b \in \mathbb{R}$, we have[2]

$$u^{-1}(a,b) = u^{-1}[\{(-\infty, a] \cup [b, \infty)\}^c] = \{u^{-1}[(-\infty, a] \cup [b, \infty)]\}^c$$
$$= \{u^{-1}(-\infty, a] \cup u^{-1}[b, \infty)\}^c = \left\{u^{-1}(-\infty, a]\right\}^c \cap \{u^{-1}[b, \infty)\}^c$$
$$= \{x \in \mathbb{R}_+^k | u(x) \leq a\}^c \cap \{x \in \mathbb{R}_+^k | u(x) \geq b\}^c.$$

---

[2]For any two sets A and B, and any function f, we have
$f^{-1}(A^c) = [f^{-1}(A)]^c$, $f^{-1}(A \cup B) = f^{-1}(A) \cup f^{-1}(B)$.

1 Existence of the Utility Function

Since $u(x) \leq a \Leftrightarrow u(x)e \precsim ae \Leftrightarrow x \precsim ae$, we have

$$u^{-1}(a,b) = \{x \in \mathbb{R}_+^k \,|\, x \precsim ae\}^c \cap \{x \in \mathbb{R}_+^k \,|\, x \succsim be\}^c.$$

Thus, by the continuity axiom, $u^{-1}(a,b)$ is open. Therefore, by Lemma 2.3, $u(x)$ is continuous. ∎

What about the necessity of the conditions in Theorem 2.1? The first two are rationality axioms, which are necessary for the rationality of preferences. Also, since any two real values can be compared, completeness is obviously necessary. Let us now show that continuity is also necessary in an example.

Example 2.2 Nonexistence of Utility Functions for the Dictionary Order. The dictionary order on $\mathbb{R}_+^2$ is defined by

$$x \succ y \quad \text{if either} \quad x_1 > y_1 \quad \text{or} \quad x_1 = y_1 \text{ but } x_2 > y_2.$$

Define $x \succsim y$ if $x \succ y$ or $x = y$. This order is obviously complete. By definition,

$$\{x | x \succsim y\} = \{x | x_1 > y_1\} \cup \{x | x_1 = y_1 \text{ and } x_2 \geq y_2\},$$

which is not closed. This means that the dictionary order violates the axiom of continuity (Fig. 2).

We now show that the dictionary order cannot be represented by a continuous utility function. Suppose that there exists a continuous utility function u. Then, by

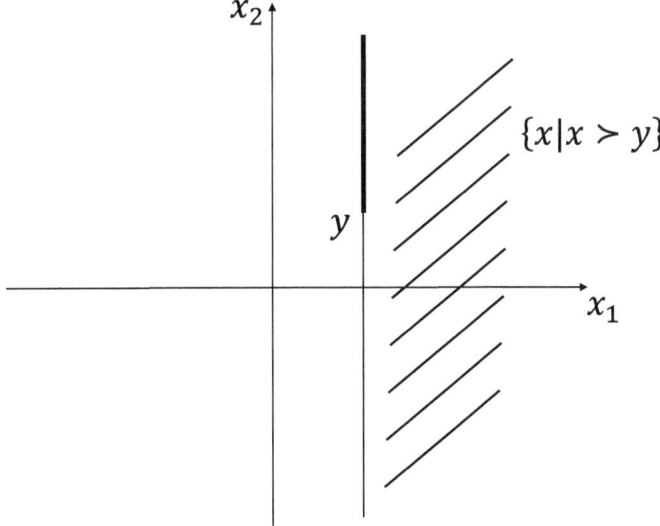

Fig. 2 Discontinuity of the dictionary order

Lemmas 2.1 and 2.2, $S_a \equiv \{x|x_1 = a\}$ and thus $u(S_a) = \{u(x)|x_1 = a\}$ are connected sets. By Lemma 2.1 again, $\{u(x)|x_1 = a\}$ must be an interval in $\mathbb{R}_+$. For $u_0 \equiv u(a, x_2^0)$ for some $x_2^0 \in \mathbb{R}$, by the definition of this dictionary order, for $x_2^1 > x_2^0$, $u_1 \equiv u(a, x_2^1) > u_0$. Since $u(S_a)$ is an interval and $u_1, u_0 \in u(S_a)$, we must have interval $(u_0, u_1) \subset u(S_a)$. Therefore, for any $a \in \mathbb{R}_+$, $u(S_a)$ is an interval with a nonempty interior.

We also know that $u(S_a) \cap u(S_b) = \emptyset$ for $b \ne a$. Hence, each real number $a \in \mathbb{R}_+$ can associate a distinct interval with a nonempty interior. Thus, there are an uncountable number of such intervals. Since each interval on $\mathbb{R}_+$ with an nonempty interior contains at least one rational number, each interval S_a will contain at least one rational number. Since these intervals are disjointed, an uncountable number of such intervals will then imply an uncountable number of rational numbers. We know that there can only be a countable number of rational numbers in $\mathbb{R}_+$. This is a contradiction. Therefore, such a function u must not exist for the dictionary order. ∎

Utility representation is not unique. In fact, if u is a utility function, then, for any strictly increasing function $\varphi : \mathbb{R} \to \mathbb{R}$, $v \equiv \varphi \circ u$ is also a utility function, where $\varphi \circ u(x) = \varphi[u(x)]$ for any x. On the other hand, we find that the utility function is unique up to a strictly increasing transformation, meaning that for any two strictly increasing utility functions $v(x)$ and $u(x)$ of the same preferences there is a strictly increasing function $\varphi : \mathbb{R} \to \mathbb{R}$ such that $v = \varphi \circ u$.

Proposition 2.2 *If u is a utility function, for any strictly increasing $\varphi : \mathbb{R} \to \mathbb{R}$, $v \equiv \varphi \circ u$ is also a utility function for the same preference relation. The utility function is unique up to a strictly increasing transformation.*

Proof We need to prove the uniqueness only. Given two strictly increasing utility functions $v(x)$ and $u(x)$, if they represent the same preferences, we look for a strictly increasing function $\varphi : \mathbb{R} \to \mathbb{R}$ such that

$$v(x) = \varphi[u(x)], \quad \text{for any} \quad x \in \mathbb{X}.$$

We identify φ as

$$\varphi(t) = v\left[u^{-1}(t)\right].$$

More rigorously, for each $t > 0$ within the range $u(\mathbb{X})$ of u, there is $x_t \in \mathbb{X}$ such that $t = u(x_t)$. Define $\varphi(t) = v(x_t)$. We now show that this φ is well defined. If there is another $y \in \mathbb{X}$ such that $u(y) = u(x_t)$, since u and v represent the same preferences, then $v(y) = v(x_t)$, implying $\varphi(t) = v(y)$. That is, although x_t may not be unique for each t, the so-defined value $\varphi(t)$ is unique. Then, for any x and y, we have

$$\varphi[u(x)] = v\left[u^{-1}(u(x))\right] = v(x), \quad \varphi[u(y)] = v\left[u^{-1}(u(y))\right] = v(y). \tag{1}$$

1 Existence of the Utility Function

If φ is not strictly increasing, then we can find two values t_1 and t_2, $t_1 < t_2$, such that $\varphi(t_1) \geq \varphi(t_2)$. We can then find two arbitrary consumption bundles x_1 and x_2 such that $u(x_1) = t_1$ and $u(x_2) = t_2$. We have $x_1 \prec_u x_2$. But, by (1), we have

$$v(x_1) = \varphi[u(x_1)] = \varphi(t_1) \geq \varphi(t_2) = \varphi[u(x_2)] = v(x_2),$$

i.e., $x_1 \succsim_v x_2$, which contradicts the fact that u and v represent the same preferences. Hence, φ must be strictly increasing. ∎

2 The Consumer's Problem

The consumer's problem is

$$v(p, I) \equiv \max_{x \in \mathbb{R}^n_+} u(x)$$
$$\text{s.t.} \quad p \cdot x \leq I.$$

That is, given price vector p and income I, the consumer chooses his consumption bundle $x \geq 0$ to maximize his utility subject to his budget constraint $p \cdot x \leq I$. The solution from this problem is denoted as $x^* = x^*(p, I)$ and called the <u>ordinary demand function</u> or the <u>Marshallian demand function</u>. $v(p, I)$ is called the <u>indirect utility function</u>.

Remark 2.1 If $p \gg 0$ and $I > 0$, the consumption set $\mathbb{X} \equiv \{x \in \mathbb{R}^n_+ \mid p \cdot x \leq I\}$ is a compact set. By the Weierstrass Theorem, if u is continuous, x^* exists.

Remark 2.2 Strict convexity of preferences will imply a unique optimal consumption choice x^*. When the existence and uniqueness of the optimal choice x^* are guaranteed, the consumer demand function is well defined.

Remark 2.3 The maximum choice x^* is independent of the particular choice of utility function for given preferences. No matter what utility function is chosen to represent the preferences, the optimal choice x^* has to satisfy $x^* \succsim x$ for any x in the consumption set. Hence, for any two utility representations u and v,

$$u(x^*) \geq u(x), \quad \text{for} \quad p \cdot x \leq I \quad \Rightarrow \quad x^* \succsim x, \quad \text{for} \quad p \cdot x \leq I$$
$$\Rightarrow \quad v(x^*) \geq v(x), \quad \text{for} \quad p \cdot x \leq I.$$

That is, any utility function for the preferences must have this x^* as a solution.

By the local nonsatiation or strict monotonicity, the budget constraint must be binding (i.e., it must be an equality). Hence, the consumer's problem becomes

$$v(p, I) \equiv \max_{x \in \mathbb{R}^n_+} u(x) \quad \text{s.t.} \quad p \cdot x = I. \tag{2}$$

Things become easier to handle with an equality constraint. The Lagrange function for the consumer's problem is $\mathcal{L}(x, \lambda) \equiv u(x) + \lambda(I - p \cdot x)$, where λ is the Lagrange multiplier. The FOC is

$$Du(x^*) = \lambda p.$$

This condition can be rearranged to give:

$$\frac{u_{x_i}(x^*)}{u_{x_j}(x^*)} = \frac{p_i}{p_j}.$$

As the explanation for (1.15), the left-hand side is the relative gain in utility at x^* and the right-hand side is the relative cost for the consumption choice x^*. By this interpretation, we understand that this condition must hold in optimality. Denote the marginal rate of substitution (MRS) at a point x as

$$MRS_{ij}(x) \equiv \frac{u_{x_i}(x)}{u_{x_j}(x)},$$

where $MRS_{ij}(x)$ is the slope of an indifference curve at point x. Figure 3 shows that the optimal solution is at the point where the slope p_1/p_2 of the budget line is equal to the slope $MRS_{12}(x^*)$ of the indifference curve at x^*.

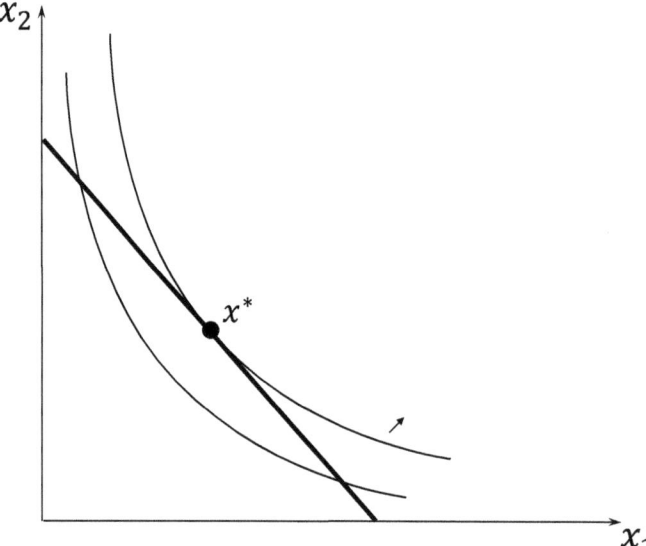

Fig. 3 Utility maximization

By Theorem A.11 in the Appendix, the SOC is

$$h^T \cdot u''(x^*) \cdot h \leq 0 \quad \text{for all } h \text{ satisfying } p \cdot h = 0.$$

This condition can be checked using Theorem A.8. This SOC can be understood by the following argument. If x^* is the maximum point, given any $h \in \mathbb{R}^n$, $f(t) \equiv u(x^* + th)$ takes the maximum value at $t = 0$. The SOC for $f(t)$ at $t = 0$ is exactly the SOC for u in the above. However, to satisfy the budget constraint $p \cdot (x^* + th) = I$, we must require h to satisfy $p \cdot h = 0$. In other words, as long as h satisfies $p \cdot h = 0$, the SOC is the concavity of $f(t)$ at $t = 0$.

The dual problem of the utility maximization is the following problem of expenditure minimization:

$$e(p,u) \equiv \min_{x \in \mathbb{R}^n_+} p \cdot x$$
$$\text{s.t.} \quad u(x) \geq u.$$

$e(p,u)$ is called the <u>expenditure function</u>, which is the minimum expenditure of achieving the utility level u. The solution from this problem is denoted as $\bar{x} = \bar{x}(p,u)$ and is called the <u>compensated demand function</u> or <u>Hicksian demand function</u>. The expenditure function and the compensated demand function are analogous respectively to the cost function and the conditional demand function in Chap. 1. Here, $\bar{x}(p,u)$ is "compensated" simply because when there is a change in prices, the income has to be changed or compensated accordingly to maintain the same level of utility. The Marshallian demand, on the other hand, keeps income constant but allows the level of utility to change. Compensated demand functions are not estimable since they depend on the unobservable utility value. Ordinary demand functions are, on the other hand, estimable because they depend on observable prices and income.[3]

Similar to the consumer's problem (2), by the local nonsatiation or strict monotonicity, the expenditure problem becomes

$$e(p,u) \equiv \min_{x \in \mathbb{R}^n_+} p \cdot x \qquad (3)$$
$$\text{s.t.} \quad u(x) = u.$$

The graphic presentation of this problem is also in Fig. 3, just as for problem (2). The FOC and SOC are therefore the same as the ones for (2).

The duality of utility maximization and expenditure minimization can be easily seen from Fig. 4.

Given a budget line, we can shift the indifference curve to get the optimal solution x^*; conversely, given an indifference curve going through x^*, we can shift

---

[3]For example, if we believe that ordinary demand has the form $x^* = A p^{\alpha_1} I^{\alpha_2}$ with positive constants A, α_1 and α_2, we can try to collect data on (x^*, p, I) and use regression model $\ln x^* = \alpha_0 + \alpha_1 \ln p + \alpha_2 \ln I + \varepsilon$ to estimate the parameters α_0, α_1 and α_2.

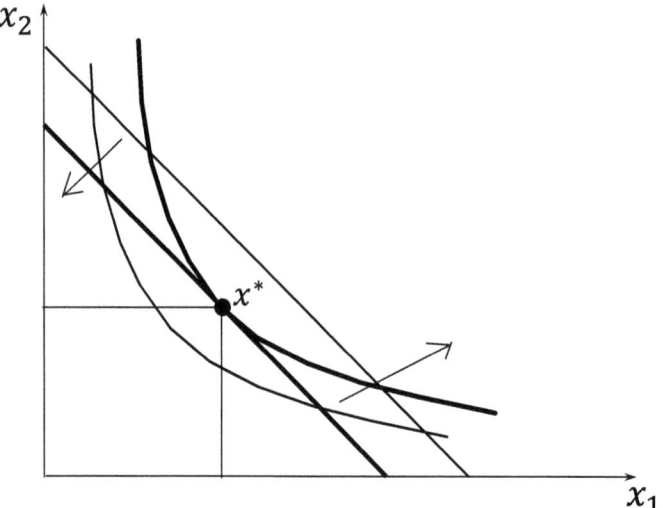

Fig. 4 Duality of utility maximization and expenditure minimization

the expenditure line parallelly to get x^*. The two processes both yield the same solution x^*. This is the so-called <u>duality</u>. Duality theory talks about the duality of the following pairs[4]:

one is min; the other is max;
one requires convexity of objective $p \cdot x$; the other requires concavity of objective $u(x)$;
one has quasi − concavity of $e(p, u)$; the other has quasi − convexity of $v(p, I)$.

Given the two types of demand functions $x^*(p, I)$ and $\bar{x}(p, u)$, we have some well-known terminologies for describing demand properties:

normal good: $\frac{\partial x_i^*}{\partial I} \geq 0$,   inferior good: $\frac{\partial x_i^*}{\partial I} < 0$;

luxury good: $\frac{I}{x_i^*} \frac{\partial x_i^*}{\partial I} > 1$,   necessary good: $0 \leq \frac{I}{x_i^*} \frac{\partial x_i^*}{\partial I} < 1$;

gross substitutes: $\frac{\partial x_i^*}{\partial p_j} > 0$,   gross complenents: $\frac{\partial x_i^*}{\partial p_j} < 0$;

net substitutes: $\frac{\partial \bar{x}_i}{\partial p_j} > 0$,   net complenents: $\frac{\partial \bar{x}_i}{\partial p_j} < 0$;

Giffen good: $\frac{\partial x_i^*}{\partial p_i} > 0$,   usual good: $\frac{\partial x_i^*}{\partial p_i} \leq 0$;

elastic: $-\frac{p_i}{x_i^*} \frac{\partial x_i^*}{\partial p_i} > 1$,   inelastic: $0 \leq -\frac{p_i}{x_i^*} \frac{\partial x_i^*}{\partial p_i} < 1$.

---

[4]The general duality is the equivalence between
$$U = \max u(x) \quad \text{and} \quad V = \min v(x)$$
$$s.t. \, v(x) \leq 0 \qquad\qquad s.t. \, u(x) \geq 0,$$
for some arbitrary functions u and v. Since for our case we have a linear function $v(x) = p \cdot x$, we have stronger properties.

Typically, if p_i increases, we expect x_i^d to decrease; if goods i and j are substitutes, then generally x_j^d increases. Hence, an increase in p_i results in an increase in x_j^d if the two goods are substitutes, i.e.,

$$\begin{array}{cc} \downarrow x_i & x_j \uparrow \\ \uparrow p_i & p_j \end{array}$$

If goods i and j are complements, then we expect

$$\begin{array}{cc} \downarrow x_i & x_j \downarrow \\ \uparrow p_i & p_j \end{array}$$

These relationships explain some of the above definitions.

3 Properties

This section discusses the properties of the various functions involved in consumer theory.

Proposition 2.3 (Continuity of Ordinary Demand Functions). *Suppose that preferences are strictly convex, strictly monotonic, and continuous. Then, the ordinary demand function $x^*(p, I)$ is continuous.*

Proof We first show the continuity of $x^*(p, I)$ in p. Let $\{p_m\}$ be a sequence of price vectors converging to $p_0 \gg 0$ as $m \to \infty$. We know that when m is large enough, $p_m \geq p_0/2$ and thus $x^*(p_m, I) \in \{x \in \mathbb{R}_+^n \,|\, 0.5p_0 \cdot x \leq I\}$. Since $\{x \in \mathbb{R}_+^n \,|\, 0.5p_0 \cdot x \leq I\}$ is a compact set, there is a subsequence $\{p_{m_k}\}$ of $\{p_m\}$ and a point x_0 such that $x^*(p_{m_k}, I) \to x_0$ as $k \to \infty$. Since $p_{m_k} \cdot x^*(p_{m_k}, I) \leq I$, taking limit $k \to \infty$, we have $p_0 \cdot x_0 \leq I$.

We want to show that $x_0 = x^*(p_0, I)$. If $x_0 \succsim x^*(p_0, I)$ and $x_0 \neq x^*(p_0, I)$, then, by the strict convexity, $0.5x_0 + 0.5x^*(p_0, I) \succ x^*(p_0, I)$. Since $p_0 \cdot 0.5x_0 + 0.5x^*(p_0, I)] \leq I$, this contradicts the fact that $x^*(p_0, I)$ is by definition the best choice among those x's satisfying $p_0 \cdot x \leq I$. On the other hand, if $x_0 \prec x^*(p_0, I)$, by the continuity of preferences, we have $t \in (0, 1)$ such that $x_0 \prec tx^*(p_0, I)$. Also by the continuity of preferences, since $x^*(p_{m_k}, I) \to x_0$, when k is large enough $x^*(p_{m_k}, I) \prec tx^*(p_0, I)$ and thus $I = p_{m_k} \cdot x^*(p_{m_k}, I) < tp_{m_k} \cdot x^*(p_0, I)$. Here, strict monotonicity of preferences is used to guarantee an equality budget constraint; it also used to imply that a preferred consumption bundle costs more. Letting $k \to \infty$, we then have $I \leq tp_0 \cdot x^*(p_0, I)$ and thus $p_0 \cdot x^*(p_0, I) > I$. This is again a contradiction. Therefore, we must have $x_0 = x^*(p_0, I)$.

Since any subsequence of $\{x^*(p_m, I)\}$ converges to the same point $x^*(p_0, I)$, sequence $\{x^*(p_m, I)\}$ must itself be convergent and the limit is $x^*(p_0, I)$. Since for

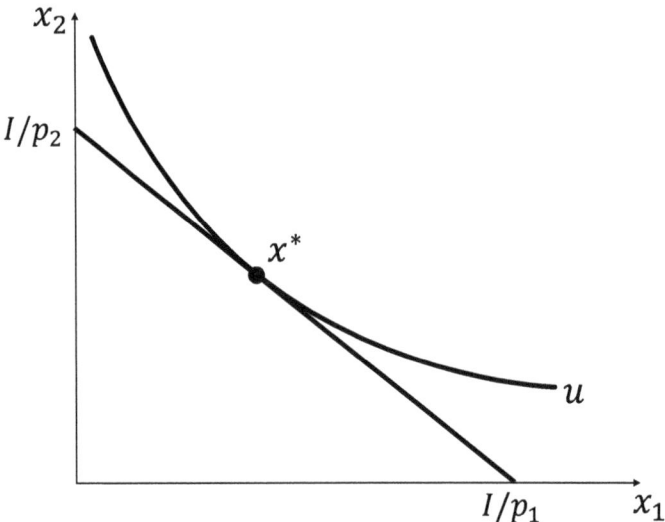

Fig. 5 Utility maximization

any convergent sequence $\{p_m\}$, $p_m \to p_0$, we have shown $x^*(p_m, I) \to x^*(p_0, I)$, $x^*(p, I)$ is continuous in p.

We now show the continuity of $x^*(p, I)$ in I. Let $\{I_m\}$ be a sequence of incomes converging to $I_0 \gg 0$ as $m \to \infty$. By zero homogeneity of $x^*(p, I)$ in (p, I), we have $x^*(p, I_m) = x^*(p/I_m, 1)$. By the continuity in p, $x^*(p/I_m, 1) \to x^*(p/I_0, 1) = x^*(p, I_0)$, which shows the continuity of $x^*(p, I)$ in I. ∎

The result in Proposition 2.3 can be easily observed in Fig. 5. With strictly quasi-concave indifference curves, when the prices and income change slightly, x^* changes slightly too.

Proposition 2.4 (Properties of Indirect Utility Functions). *The indirect utility function $v(p, I)$ has the following properties*:

(1) *decreasing in p, increasing in I;*
(2) *zero homogeneous in (p, I);*
(3) *quasi-convex in p.*

Proof Since $\{x \in \mathbb{R}_+^n \mid p' \cdot x \leq I\} \subset \{x \in \mathbb{R}_+^n \mid p \cdot x \leq I\}$ for $p' \geq p$, by the definition of v, $v(p', I) \leq v(p, I)$. Similarly, $\{x \in \mathbb{R}_+^n \mid p \cdot x \leq I'\} \supset \{x \in \mathbb{R}_+^n \mid p \cdot x \leq I\}$ for $I' \geq I$ implies that $v(p, I') \geq v(p, I)$. These results can also be easily observed in Fig. 5.

When the prices and income are both multiplied by a positive number, the budget constraint does not change at all. Therefore, $v(tp, tI) = v(p, I)$ for $t > 0$. This can also be seen in Fig. 5.

Suppose $v(p,I) \le k$ and $v(p',I) \le k$. For any $0 \le \lambda \le 1$, consider $x^* = x[\lambda p + (1-\lambda)p', I]$. With $[\lambda p + (1-\lambda)p'] \cdot x^* \le I$, we must have either $p \cdot x^* \le I$ or $p' \cdot x^* \le I$. Hence, we have either $u(x^*) \le v(p,I)$ or $u(x^*) \le v(p',I)$. Thus, $u(x^*) \le k$, i.e., $v[\lambda p + (1-\lambda)p', I] \le k$. Therefore, $v(p,I)$ is quasi-convex in p. ∎

Expenditure function $e(p,u)$ is the minimum expenditure needed to obtain a given level of utility u. The expenditure function is completely analogous to the cost function $c(w,y)$ in Chap. 1. They therefore have the same properties.

Proposition 2.5 (Properties of Expenditure Functions). *The expenditure function $e(p,u)$ has the following properties:*

(1) *increasing in p;*
(2) *linearly homogeneous in p;*
(3) *concave in p.* ∎

The following proposition provides some important identities that tie up the expenditure function $e(p,u)$, the indirect utility function $v(p,I)$, the ordinary demand function $x^*(p,I)$, and the compensated demand function $\bar{x}(p,u)$.

Proposition 2.6 (Duality Equalities).

(1) $e[p, v(p,I)] = I$;
(2) $v[p, e(p,u)] = u$;
(3) $x_i^*(p,I) = \bar{x}_i[p, v(p,I)]$;
(4) $\bar{x}_i(p,u) = x_i^*[p, e(p,u)]$. ∎

The proof of these identities is simple, which can be easily observed in Fig. 6.

The duality equalities are very useful. For example, given an indirect utility function $v(p,I)$, if we let $v(p,I)$ be u and I be e, we can get $e(p,u)$. If we have the compensated demand function $\bar{x}_i(p,u)$, by letting u be $v(p,I)$, we get the ordinary demand function $x_i^*(p,I)$.

The following is an analog of Shephard's lemma for the cost function.

Proposition 2.7 (Shephard's Lemma). *For expenditure function $e(p,u)$ and compensated demand function $\bar{x}(p,u)$, we have*

$$\bar{x}_i(p,u) = \frac{\partial e(p,u)}{\partial p_i}, \quad \forall i.$$

Proposition 2.8 (Roy's Identity). *For ordinary demand function $x_i^*(p,I)$ and indirect utility function $v(p,I)$, we have*

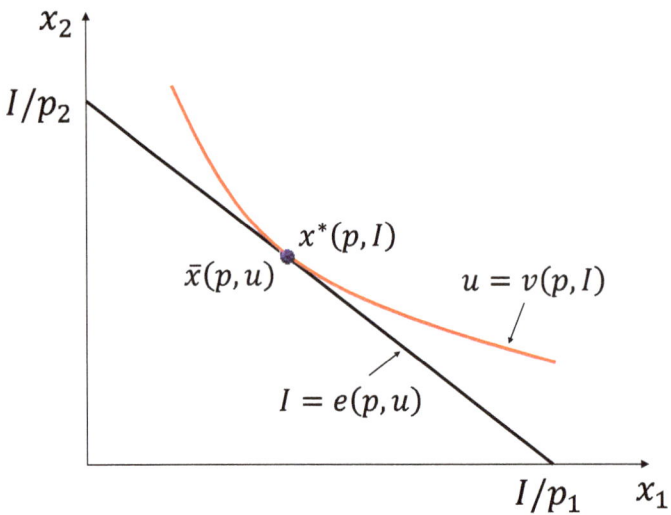

Fig. 6 The identities in Proposition 2.6

$$x_i^*(p,I) = -\frac{v_{p_i}(p,I)}{v_I(p,I)}, \quad \forall i,$$

provided that the right-hand side is well defined and that $p \gg 0$ and $I > 0$.

Proof We have

$$v(p,I) = \max_x \ u(x) + \lambda(I - p \cdot x).$$

Then, by the Envelope Theorem,

$$v_{p_i}(p,I) = -\lambda x_i, \quad v_I(p,I) = \lambda, \qquad (4)$$

which then implies Roy's identity. ∎

By $v_I(p,I) = \lambda$ in (4), λ is the utility value per unity of money. When price p_i goes up by one unity, since the consumer keeps x_i unit of consumption, the consumer loses x_i unit of money, which means the loss of λx_i units of utility value. This explains why $v_{p_i}(p,I) = -\lambda x_i$.

Example 2.3 A utility function of Constant Elasticity of Substitution (CES) is given by $u(x_1, x_2) = (x_1^\rho + x_2^\rho)^{1/\rho}$. Consider the expenditure minimization problem:

$$e(p_1, p_2, u) \equiv \min \ p_1 x_1 + p_2 x_2$$
$$\text{s.t.} \ x_1^\rho + x_2^\rho = u^\rho,$$

which gives compensated demand functions:

$$\bar{x}_1(p_1,p_2,u) = \frac{p_1^{r-1}}{(p_1^r+p_2^r)^{1-\frac{1}{r}}}u, \qquad \bar{x}_2(p_1,p_2,u) = \frac{p_2^{r-1}}{(p_1^r+p_2^r)^{1-\frac{1}{r}}}u, \qquad (5)$$

where $r \equiv \frac{\rho}{\rho-1}$. We then have the expenditure function:

$$e(p,u) = (p_1^r+p_2^r)^{1/r}u. \qquad (6)$$

By letting $I = e(p,u)$, Eq. (6) implies the indirect utility function:

$$v(p,I) = (p_1^r+p_2^r)^{-1/r}I. \qquad (7)$$

Using Roy's identity, we then have

$$x_1^*(p_1,p_2,I) = -\frac{v_{p_1}(p,I)}{v_I(p,I)} = \frac{p_1^{r-1}}{p_1^r+p_2^r}I.$$

We can also apply $u = v(p,I)$ in (5) to get this consumer demand function. ∎

Since the compensated demand function $\bar{x}(p,u)$ is mathematically the same as the conditional demand function $\bar{x}(w,y)$ in Chap. 1, they have the same properties.

Proposition 2.9 (Properties of Compensated Demand Functions). *The compensated demand function $\bar{x}(p,u)$ has the following properties:*

(1) *$\bar{x}(p,u)$ is zero homogeneous in p;*
(2) *substitution matrix $\bar{x}_p'(p,u)$ is negative semi-definite;*
(3) *the cross-price effects are symmetric: $\frac{\partial \bar{x}_j(p,u)}{\partial p_i} = \frac{\partial \bar{x}_i(p,u)}{\partial p_j}$;*
(4) *a downward-sloping demand curve: $\frac{\partial \bar{x}_i(p,u)}{\partial p_i} \leq 0$.* ∎

Proposition 2.10 (Properties of Ordinary Demand Functions). *The ordinary demand function $x^*(p,I)$ has the following properties:*

(1) *Homogeneity: $x^*(p,I)$ is zero homogeneous in (p,I).*
(2) *Compensated Symmetry: $\frac{\partial x_j^*}{\partial p_i} + x_i^* \frac{\partial x_j^*}{\partial I} = \frac{\partial x_i^*}{\partial p_j} + x_j^* \frac{\partial x_i^*}{\partial I}$.*
(3) *Adding-up condition: $\sum p_i x_i^*(p,I) = I$.*

Proof

(1) Since $x_i^*(p,I) = \bar{x}_i[p,v(p,I)]$ and $\bar{x}_i(p,u)$ is zero homogeneous in p and $v(p,I)$ is zero homogeneous in (p,I), we find that $x^*(p,I)$ is zero homogeneous in (p,I).

(2) The second result comes directly from the Slutsky equation in the next proposition and the fact that $\bar{x}(p, u)$ has symmetric cross-price effects.
(3) The third result is just the budget constraint. ∎

Proposition 2.11 (The Slutsky Equation). *For ordinary demand function $x^*(p, I)$, compensated demand function $\bar{x}(p, u)$, and indirect utility function $v(p, I)$, we have*

$$\frac{\partial x_j^*(p, I)}{\partial p_i} = \frac{\partial \bar{x}_j[p, v(p, I)]}{\partial p_i} - \frac{\partial x_j^*(p, I)}{\partial I} \cdot x_i^*, \quad \forall i, j.$$

Proof By Proposition 2.6 (4), we have

$$\bar{x}_j(p, u) = x_j^*[p, e(p, u)].$$

By differentiating this equation on both sides w.r.t. p_i and using Proposition 2.7, we have

$$\begin{aligned}\frac{\partial \bar{x}_j(p, u)}{\partial p_i} &= \frac{\partial x_j^*[p, e(p, u)]}{\partial p_i} + \frac{\partial x_j^*[p, e(p, u)]}{\partial I} \frac{\partial e(p, u)}{\partial p_i} \\ &= \frac{\partial x_j^*[p, e(p, u)]}{\partial p_i} + \frac{\partial x_j^*[p, e(p, u)]}{\partial I} \cdot \bar{x}_i(p, u).\end{aligned}$$

Let $u = v(p, I)$, which implies $I = e(p, u)$. By substituting them into above equation, we then have

$$\frac{\partial \bar{x}_j[p, v(p, I)]}{\partial p_i} = \frac{\partial x_j^*(p, I)}{\partial p_i} + \frac{\partial x_j^*(p, I)}{\partial I} \cdot x_i^*(p, I).$$

This immediately implies the Slutsky equation. ∎

The Slutsky equation says that when price p_i increases by one unit, the change $\frac{\partial x_i^*(p,I)}{\partial p_i}$ in x_i^* can be divided into two parts: the substitution effect $\frac{\partial \bar{x}_i(p,u)}{\partial p_i}$ and the income effect $-\frac{\partial x_i^*(p,I)}{\partial I} \cdot x_i^*$. The substitution effect is the response of demand to a price change along the indifference curve. Since the expenditure on x_i^* is $p_i x_i^*$, if the price is increased by one unit, the expenditure is increased by approximately x_i^* units. Hence, if the price increases one unit, then the purchasing power is reduced by approximately x_i^* units. Hence, the income effect is multiplied by x_i^*.

These effects are shown in Fig. 7, where the price goes up from p_1 to p_1'. The income level for the artificial AB line maintains the same utility level for the price change. To an increase in price, by the concavity of $e(p, u)$ in prices, the substitution effect is always negative; but the income effect can be negative or positive. For a normal good, the income effect is negative for the price increase and thus the total change in the ordinary demand x_i^* is negative; for an inferior good, however,

3 Properties

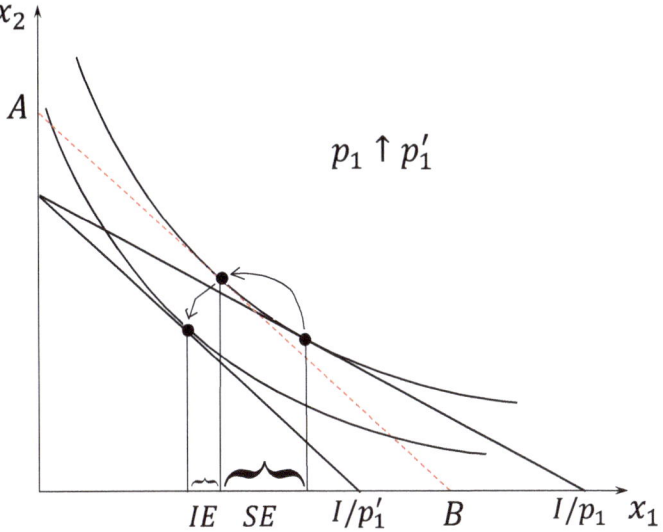

Fig. 7 Income and substitution effects

the income effect is positive and thus the total change in the ordinary demand can be either negative or positive depending on which effect is dominant. From this analysis, we can see that a Giffen good must be an inferior good.

Example 2.4 Given utility function $u(x_1, x_2) = A x_1^a x_2^{1-a}$, where $A \equiv a^a (1-a)^{1-a}$, the ordinary demand functions are:

$$x_1^*(p_1, p_2, I) = \frac{aI}{p_1}, \qquad x_2^*(p_1, p_2, I) = \frac{(1-a)I}{p_2}.$$

Thus,

$$v(p_1, p_2, I) = p_1^{-a} p_2^{a-1} I,$$

and

$$\frac{\partial x_1^*(p_1, p_2, I)}{\partial p_1} = -\frac{aI}{p_1^2}, \qquad \frac{\partial x_1^*(p_1, p_2, I)}{\partial I} = \frac{a}{p_1}.$$

Using the indirect utility function, we can solve $u = p_1^{-a} p_2^{a-1} e$ for the expenditure function:

$$e(p_1, p_2, u) = p_1^a p_2^{1-a} u.$$

By Shephard's Lemma, we get the compensated demand function:

$$\bar{x}_1(p_1,p_2,u) = a\left(\frac{p_2}{p_1}\right)^{1-a} u.$$

Then,

$$\frac{\partial \bar{x}_1(p_1,p_2,u)}{\partial p_1} = a(a-1)p_1^{a-2}p_2^{1-a}u.$$

We then have

$$\frac{\partial \bar{x}_1[p_1,p_2,v(p_1,p_2,I)]}{\partial p_1} = a(a-1)p_1^{-2}I.$$

Now we can verify the Slutsky equation:

$$\frac{\partial \bar{x}_1[p_1,p_2,v(p_1,p_2,I)]}{\partial p_1} - \frac{\partial x_1^*(p_1,p_2,I)}{\partial I}x_1^* = a(a-1)p_1^{-2}I - \frac{a}{p_1}\frac{aI}{p_1} = -\frac{aI}{p_1^2}$$

$$= \frac{\partial x_1^*(p_1,p_2,I)}{\partial p_1},$$

where $a(a-1)p_1^{-2}I$ is the substitution effect and $-\frac{a}{p_1}\frac{aI}{p_1}$ the income effect. For this case, both substitution and income effects are negative. Therefore, the total change in x_1^* is negative. ∎

Example 2.5 Demand functions for a single good. Suppose that x is the good in question, and y represents all other goods. Let p be the price of x and q be the price of y. The consumer's problem is

$$\max_{x,y} u(x,y)$$
$$\text{s.t.} \quad px+qy = I.$$

Since the ordinary demand $x(p,q,I)$ is zero homogenous in prices, we can denote the demand as $x(p/q, I/q)$. In practice, q is some consumer price index, so that the demand is a function of the real price and the real income. Consumer theory tells us that virtually any functional form of $x(p/q, I/q)$ is consistent with utility maximization.[5] We can therefore arbitrarily choose a convenient functional form for $x(p/q, I/q)$ and it can be justified as a demand function derived from utility maximization. Some popular functional forms are:

1. Linear demand: $x = a + b\frac{p}{q} + c\frac{I}{q}$.
2. Logarithmic demand: $\log(x) = a + b\log\frac{p}{q} + c\log\frac{I}{q}$.
3. Semi-logarithmic demand: $\log(x) = a + b\frac{p}{q} + c\frac{I}{q}$. ∎

---

[5]Essentially, the only requirement is that the compensated own price effect is negative.

3 Properties

Example 2.6 The Linear expenditure system. Suppose that the utility function is

$$u(x) = \sum_{i=1}^{k} a_i \log(x_i - \gamma_i),$$

where $x_i > \gamma_i$. Here, γ_i is the subsistence level of consumption. Utility maximization subject to budget constraint $\sum p_i x_i = I$ immediately implies the ordinary demand:

$$x_i = \gamma_i + a_i \frac{I}{p_i} - \sum_{j=1}^{k} \gamma_j \frac{p_j}{p_i},$$

which is linear in parameters and thus can be used directly in econometric regression. ∎

Example 2.7 The Almost Ideal Demand System (AIDS) is a popular system of demand equations. It provides a first-order approximation to any arbitrary demand system, has desirable aggregation properties, and can "almost" be made consistent with classical demand theory through linear restrictions on the parameters of the system. AIDS has an expenditure function of the form:

$$\log e(p, u) = \alpha_0 + \sum_{i=1}^{n} \alpha_i \log p_i + \frac{1}{2} \sum_{i=1}^{n} \sum_{j=1}^{n} \gamma_{ij}^* \log p_i \log p_j + u \beta_0 \prod_{i=1}^{n} p_i^{\beta_i}. \quad (8)$$

Proposition 2.5 implies testable restrictions:

$$\sum_{j=1}^{n} \alpha_i = 1, \quad \sum_{j=1}^{n} \gamma_{ij} = \sum_{i=1}^{n} \gamma_{ij} = \sum_{i=1}^{n} \beta_i = 0,$$

where $\gamma_{ij} \equiv \frac{\gamma_{ij}^* + \gamma_{ji}^*}{2}$. To see this, consider the expenditure share:

$$s_i \equiv \frac{p_i \bar{x}_i}{e(p, u)} = \frac{p_i}{e(p, u)} \frac{\partial e(p, u)}{\partial p_i} = \frac{\partial \ln[e(p, u)]}{\partial \ln(p_i)}$$

$$= \alpha_i + \frac{1}{2} \sum_{k} \gamma_{ki}^* \ln(p_k) + \frac{1}{2} \sum_{j} \gamma_{ij}^* \ln(p_j) + u \beta_0 e^{\sum_k \beta_k \ln p_k} \beta_i$$

$$= \alpha_i + \sum_{j} \frac{1}{2} \left(\gamma_{ji}^* + \gamma_{ij}^* \right) \ln(p_j) + \beta_i u \beta_0 \prod_{k} p_k^{\beta_k}.$$

The adding-up condition $e(p, u) = \sum p_i \bar{x}_i$ implies $\sum s_i = 1$ and thus

$$\sum_{i=1}^{n} \alpha_i = 1, \quad \sum_{i=1}^{n} \beta_i = 0, \quad \sum_{i=1}^{n} \gamma_{ij} = 0.$$

The zero homogeneity condition $\bar{x}_i(p, u) = \bar{x}_i(\lambda p, u)$ implies $s_i(p, u) = s_i(\lambda p, u)$ and thus

$$\sum_{j=1}^{n} \gamma_{ij} = 0.$$

By definition, $\gamma_{ij} = \gamma_{ji}$, which implies

$$\sum_{i=1}^{n} \gamma_{ij} = \sum_{j=1}^{n} \gamma_{ji} = \sum_{j=1}^{n} \gamma_{ij} = 0.$$

There are also some inequality conditions that come from the concavity of the expenditure function. ∎

4 Aggregation

Suppose that there are n consumers with ordinary demand functions $x_i(p, I_i)$. The aggregate demand is

$$x(p, I_1, \ldots, I_n) = \sum_{i=1}^{n} x_i(p, I_i).$$

Since available data in reality are often on aggregate demand and aggregate income $\sum I_i$ only, we are interested in the following property:

$$\sum_{i=1}^{n} x_i(p, I_i) = x\left(p, \sum_{i=1}^{n} I_i\right), \quad \text{for } p, I_i \geq 0, \qquad (9)$$

for some function $x(p, I)$. If this holds, we can then apply our demand theory for individual consumers to aggregate variables.

Proposition 2.12 *A necessary and sufficient condition for the existence of a function $x(p, I)$ such that (9) holds is that each consumer's indirect utility function is linear in income with a common multiplier $b(p)$ for income:*

$$v_i(p, I_i) = a_i(p) + b(p) I_i. \qquad ∎$$

5 Integrability

Given a utility function, we can derive the expenditure function; and symmetrically, given a production function, we can derive the cost function. Integrability asks the opposite question: given an expenditure function or a cost function, can we respectively find the utility function or the production function? This question is important because utility and production functions are unobservable while expenditure and cost functions are observable. The issue here is how to recover unobservable variables from observable variables.

The answer is yes if the expenditure function and the cost function are differentiable and satisfy the properties in Propositions 2.5 and 1.7, respectively. This means that given sufficient differentiability of utility and production functions, the properties in Propositions 2.5 and 1.7 are sufficient and necessary conditions for a function to be an expenditure or cost function.

Here is a simple way to recover utility and production functions. Given an expenditure function, we can first find compensated demand functions using Shephard's lemma and then find the utility function by eliminating the prices. Symmetrically, given a cost function, we can first find conditional demand functions using Shephard's lemma and then find the production function by eliminating the prices. Since compensated and conditional demand functions are zero homogenous in prices, we are basically guaranteed to be able to eliminate the prices. In the following, we use two examples to illustrate the process. These two examples show the methods to recover production functions, which are also applicable to recovering utility functions.

Example 2.8 Consider a cost function $c(w, y) = w_1^a w_2^{1-a} y$. By Shephard's lemma, we have

$$x_1(w, y) = a \left(\frac{w_2}{w_1}\right)^{1-a} y, \quad x_2(w, y) = (1-a) \left(\frac{w_2}{w_1}\right)^{-a} y.$$

These two expressions can be used jointly to eliminate the price ratio, which yields:

$$y = a^{-a}(1-a)^{a-1} x_1^a x_2^{1-a}.$$

This Cobb-Douglas production function can indeed generate the cost function. ∎

The method in this example works for usual production technologies. The Leontief technology is quite special, which requires a special method. This is illustrated in the following example.

Example 2.9 Let $c(w, y) = (aw_1 + bw_2)\sqrt{y}$. By Shephard's lemma,

$$x_1(w, y) = a\sqrt{y}, \quad x_2(w, y) = b\sqrt{y}.$$

Thus, to produce y, since the elasticity of substitution is $\sigma = 0$, we need inputs $(x_1, x_2) \geq [x_1(w, y), x_2(w, y)]$, that is,

$$x_1 \geq a\sqrt{y}, \qquad x_2 \geq b\sqrt{y}.$$

Hence, given (x_1, x_2), output y satisfies

$$y \leq \left(\frac{x_1}{a}\right)^2, \qquad y \leq \left(\frac{x_2}{b}\right)^2.$$

implying that the maximum amount of output that (x_1, x_2) can produce is

$$y = \min\left\{\left(\frac{x_1}{a}\right)^2, \left(\frac{x_2}{b}\right)^2\right\},$$

which is by definition the production function. ∎

6 Revealed Preferences

In reality, not only we do not observe consumer preferences directly, but we also often do not have an expenditure function. What we do have is a set of observations $(p_t, x_t) \in \mathbb{R}^l_+ \times \mathbb{R}^l_+$, $t = 1, \ldots, T$, of consumption bundles x_t at prices p_t. The question that we would like to ask is: Does this set of observations possibly come from utility-maximizing behavior? More specifically, what is the necessary and sufficient condition under which this set of observations can be from a utility-maximizing consumer?[6]

We say that a utility function u <u>rationalizes</u> the observed data set $\{(p_t, x_t)\}_{t=1}^T$ if $x_t = x^*(p_t, p_t \cdot x_t)$ for $t = 1, \ldots, T$, where x^* is the ordinary demand given the utility function. If so, when $p_x \cdot x \geq p_x \cdot y$, we must have $u(x) \geq u(y)$. In other words, one has to spend more in order to get higher satisfaction. Conversely, by the same argument, if $u(x) > u(y)$, then we must have $p_y \cdot x \geq p_y \cdot y$. Hence,

$$p_x \cdot x \geq p_x \cdot y \quad \Rightarrow \quad u(x) \geq u(y) \tag{10}$$

$$u(x) \geq u(y) \quad \Rightarrow \quad p_y \cdot x \geq p_y \cdot y. \tag{11}$$

Since we cannot observe u, we try to use a relation such as $p_x \cdot x \geq p_x \cdot y$ to reveal the preferences. This motivates the following definition.

Denote the relation $p \cdot x \geq p \cdot y$ as $x \succsim_p y$, and say that x is <u>directly preferred</u> to y under price p. Similarly, denote $p \cdot x > p \cdot y$ as $x \succ_p y$. We say that x is <u>indirectly preferred</u> to y and write $x \succsim_I y$ if there is a sequence (p_i, x_i) for $i = 1, \ldots, n$ such that

---
[6]To rule out trivial cases, we have to require the utility function to be at least locally nonsatiated.

$$x \succsim_{p_x} x_1, \quad x_1 \succsim_{p_1} x_2, \quad \ldots, \quad x_n \succsim_{p_n} y.$$

Proposition 2.13

1. If $\{(p_i, x_i)\}$ are rationalized by u, then $x \succsim_I y \Rightarrow u(x) \geq u(y)$.
2. Both direct and indirect preferences are preference relations.
3. The direct preference is complete.
4. The direct preference implies the indirect preference: $x \succsim_{p_x} y \Rightarrow x \succsim_I y$. ∎

If the data are generated by utility maximization, then we can easily show that $x \succsim_I y$ implies $u(x) \geq u(y)$. Figure 8 shows that $x \succsim_{p_x} y$ implies $u(x) \geq u(y)$. Also, (10) and (11) can be easily seen from Fig. 8, especially the failure of the converse.

The necessary and sufficient condition for the data to be rationalized by a utility function is the following.

Generalized Axiom of Revealed Preferences (GARP). $x \succsim_I y \Rightarrow x \succsim_{p_y} y$. ∎

By (10), $x \succsim_I y$ implies $u(x) \geq u(y)$. Then, by (11), $u(x) \geq u(y)$ implies $x \succsim_{p_y} y$. Thus, GARP is necessary. Is it sufficient? The following is the fundamental theorem about revealed preferences. The proof can be found in Varian (1992, pp. 133, 134).

Theorem 2.2 (Afriat). *Let $\{(p_t, x_t)\}_{t=1}^{T}$ be a finite set of observations. Then, the following conditions are equivalent.*

(1) *The data satisfy GARP.*
(2) *There exists a locally nonsatiate utility function that rationalizes the data.*
(3) *There exists a locally nonsatiate, continuous, concave, monotonic utility function that rationalizes the data.* ∎

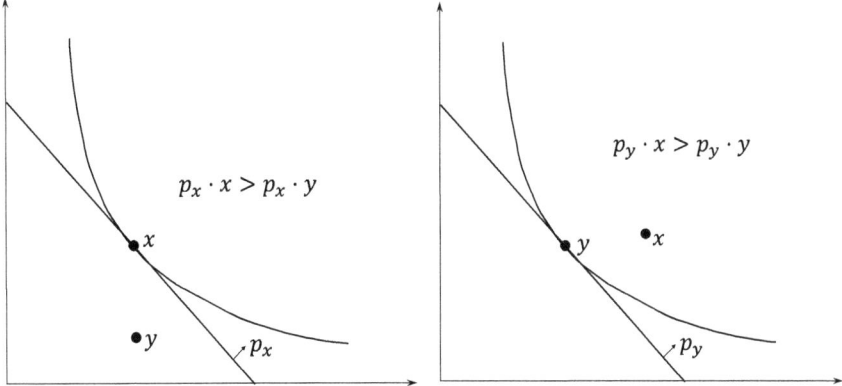

Fig. 8 Revealed preferences

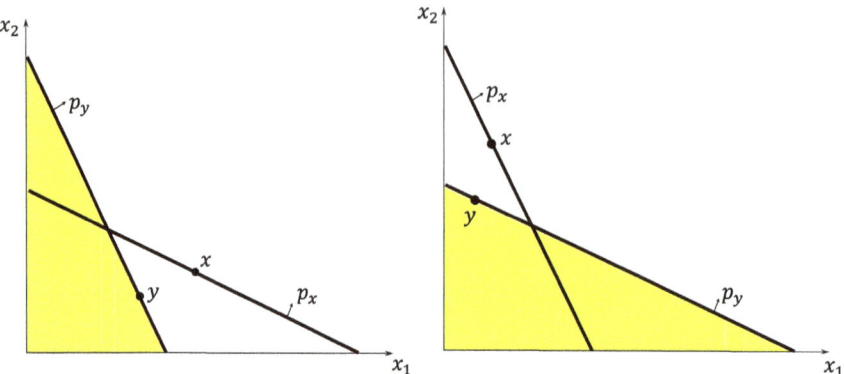

Fig. 9 Preferences under WARP

There are two more alternative conditions of revealed preferences:
Weak Axiom of Revealed Preferences (WARP). $x \succsim_{p_x} y, x \neq y \Rightarrow x \succ_{p_y} y$
(Fig. 9). ∎
Strong Axiom of Revealed Preferences (SARP). $x \succsim_I y, x \neq y \Rightarrow y \not\succsim_I x$. ∎

Proposition 2.14 *SARP implies WARP.*

Proof $x \succsim_{p_x} y$ implies $x \succsim_I y$. By SARP, we have $y \not\succsim_I x$. By completeness of the direct preference, either $y \succsim_{p_y} x$ or $y \prec_{p_y} x$ must be true. Since $y \succsim_{p_y} x$ cannot be true, we must have $x \succ_{p_y} y$. ∎

By (10) and (11), we find that both WARP and SARP are necessary. Will they be sufficient? SARP and WARP require that there be a unique demanded bundle at each budget line, while GARP allows for multiple demanded bundles. Both WARP and SARP are necessary conditions for utility-maximizing behavior in the case of single-valued demand, but only SARP is sufficient in this case. However, when there are only two goods, WARP is also sufficient in the case of single-valued demand.

Figure 10 helps us understand Afriat's theorem. For simplicity, assume that there are two data points (p_1, x_1) and (p_2, x_2) only. With two data points, GARP is the same as WARP. In the left figure, WARP is satisfied: $x_2 \succ_{p_2} x_1$ and $x_2 \succ_{p_1} x_1$. We can see that it will not be difficult to find a utility function to justify (p_1, x_1) and (p_2, x_2) as coming from utility maximization in this case. That is, the two indifferent curves in the figure can be from the same utility function. In the right figure, WARP fails: $x_2 \succ_{p_2} x_1$ and $x_2 \prec_{p_1} x_1$. We can see also that the two indifferent curves in the figure cannot be possibly from the same utility function, since indifferent curves of the same utility function can never intersect with each other.

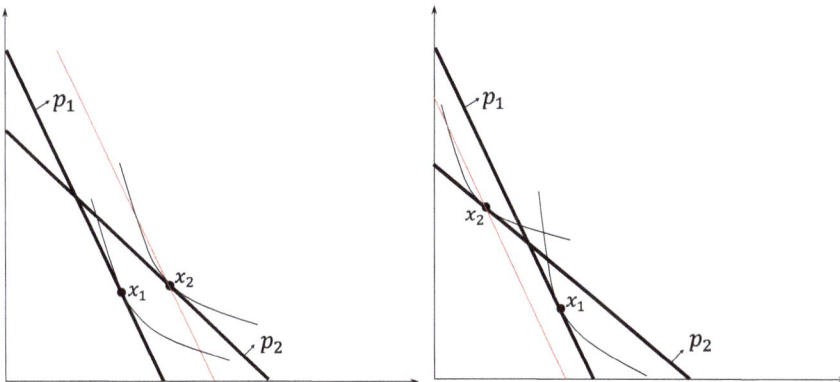

Fig. 10 Afriat's theorem

One interesting observation about Afriat's theorem is that item (2) in the theorem implies item (3). This suggests that market data cannot tell us whether the preferences are convex or monotonic. We can produce a set of data (x_t, p_t) using a non-conventional utility function. By the equivalence of items (2) and (3) in the theorem, we are guaranteed to find a nicely behaved utility function that rationalizes the data. Therefore, from the data themselves, we do not know whether the utility function that actually produces them is concave or not.

Notice that the difference between this section and the last section is that the demand points are finite. In the last section, we recover the utility function from the compensated demand, while in this section we recover the utility function from finite demand points. Given a demand function, we can recover a unique utility function that rationalizes it uniquely up to a strictly monotonic transformation. For finite demand points, there is enough freedom for many more utility functions to rationalize the points.

7 Intertemporal Analysis

This section discusses some very basic concepts of intertemporal consumption together with the idea of separation of production and consumption. Consider an economy with two periods and one consumption good. The present consumption is denoted as c_1 and the future consumption is denoted as c_2. For the main part, we assume that the spot prices of the good are $p_1 = p_2 = 1$ for the two periods, and we will later drop this assumption.

We first assume

1. the individual has endowment $(\bar{c}_1, \bar{c}_2)$;
2. there is a capital market for intertemporal borrowing and lending at interest rate $r \geq 0$.

Then, the second-period consumption cannot exceed $\bar{c}_2$ plus the savings from the first period $\bar{c}_1 - c_1$ plus the interest on saving $r(\bar{c}_1 - c_1)$: $c_2 \leq (1+r)(\bar{c}_1 - c_1) + \bar{c}_2$. Therefore, the budget constraint is

$$c_1 + \frac{1}{1+r}c_2 \leq w,$$

where $w \equiv \bar{c}_1 + \frac{1}{1+r}\bar{c}_2$ is the overall wealth. We see that the interest rate r acts as a discount rate that converts future consumption into present consumption. By this explanation, w is the total wealth in terms of present consumption good, $c_1 + \frac{1}{1+r}c_2$ is the total consumption in terms of present consumption, and the budget constraint simply says that the total consumption should be less than or equal to the total wealth.

The consumer's problem is

$$\max_{c_1,c_2} u(c_1, c_2)$$
$$\text{s.t.} \quad c_1 + \frac{1}{1+r}c_2 \leq \bar{c}_1 + \frac{1}{1+r}\bar{c}_2.$$

The solution to this problem can be illustrated by Fig. 11.

Let us then consider a consumer like Robinson Crusoe, who is completely isolated from any market. But, Crusoe can engage in production given an endowment $(\bar{c}_1, \bar{c}_2)$. Suppose that Crusoe has a production possibility set defined by $G(y_1, y_2) \leq 0$ with $G(0, 0) = 0$. Here, a production plan (y_1, y_2) means that if the firm invests an amount $-y_1 \geq 0$ of the consumption good in the first period, it will produce an amount y_2 of the consumption good in the second period, that is, y_1 and y_2 are net outputs and $G(y_1, y_2) \leq 0$ defines the limit that the firm can produce given input $-y_1$. Condition $G(0, 0) = 0$ allows the agent to produce nothing. This production setup is the same as that in our producer theory of Chap. 1 with strict monotonicity and quasi-convexity on G. In this case, we assume

1. the individual has endowment $(\bar{c}_1, \bar{c}_2)$;
2. there is a production possibility set defined by $G(y_1, y_2) \leq 0$, with $G(0, 0) = 0$, for intertemporal saving and investment.

The consumer's problem is

$$\max_{c_1,c_2,y_1,y_2} u(c_1, c_2)$$
$$\text{s.t.} \quad G(y_1, y_2) \leq 0 \qquad (12)$$
$$c_1 \leq \bar{c}_1 + y_1, c_2 \leq \bar{c}_2 + y_2.$$

In this case, savings can be made through the production process, but not through the market. The solution of this problem can be illustrated by Fig. (12).

The solution R^* is called an "autarky" solution, meaning a solution without markets.

7 Intertemporal Analysis

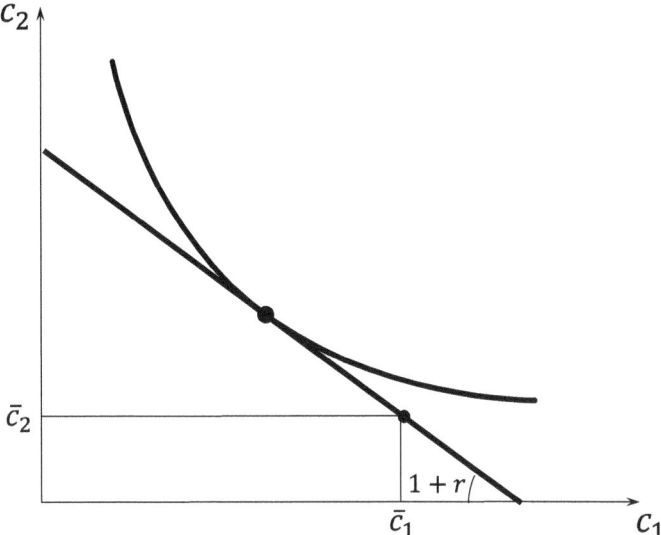

Fig. 11 Consumer's problem, given endowment and a capital market

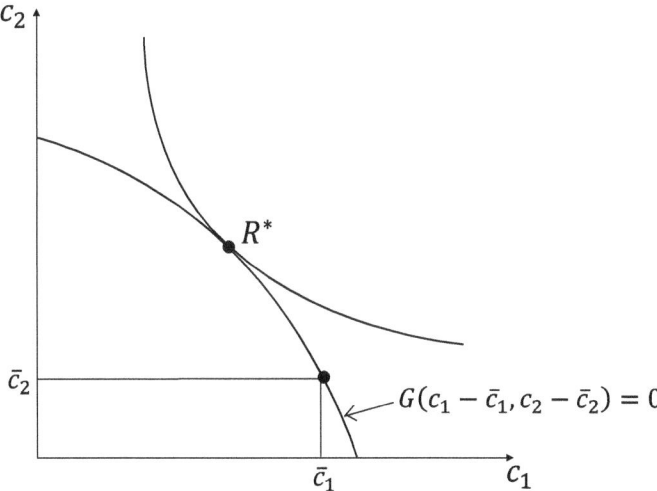

Fig. 12 Robinson Crusoe's problem, given endowment and a production frontier

Finally, let us assume that the individual not only has a productive means but also has access to a capital market for borrowing and lending. That is, we assume

1. the individual has endowment $(\bar{c}_1, \bar{c}_2)$;
2. there is a production possibility set defined by $G(y_1, y_2) \leq 0$, with $G(0,0) = 0$, for intertemporal saving and investment;

3. there is a capital market for intertemporal borrowing and lending at interest rate r.

The consumer's problem is

$$\begin{aligned}\max_{c_1,c_2,y_1,y_2} & \; u(c_1,c_2) \\ \text{s.t.} & \; G(y_1,y_2) \le 0 \\ & \; c_1 + \tfrac{1}{1+r}c_2 \le \bar{c}_1 + y_1 + \tfrac{1}{1+r}(\bar{c}_2 + y_2).\end{aligned} \tag{13}$$

To solve this problem and show it graphically, we need the following theorem.

Theorem 2.3 (The Fisher Separation Theorem). *The individual maximization problem can be divided into two problems: (1) profit maximization given the production frontier; (2) utility maximization given the endowment and the profit from the profit maximization problem. That is, the individual's problem (13) can be divided into two problems:*

$$\begin{aligned} \pi^* = \max_{y_1,y_2} & \; y_1 + \tfrac{1}{1+r}y_2 \\ \text{s.t.} & \; G(y_1,y_2) \le 0 \end{aligned} \tag{14}$$

and

$$\begin{aligned}\max_{c_1,c_2} & \; u(c_1,c_2) \\ \text{s.t.} & \; c_1 + \tfrac{1}{1+r}c_2 \le \bar{c}_1 + \tfrac{1}{1+r}\bar{c}_2 + \pi^*.\end{aligned} \tag{15}$$

Proof We can easily verify that problem (13) and dual problems (14) and (15) give the same set of first-order conditions and the same set of budget constraints. Since these conditions and constraints determine the solutions, the two sets of problems thus give the same solution. ∎

With Theorem 2.3, the solution (c_1^*, c_2^*) of problem (13) can be illustrated by Fig. 13. By the resource constraints $c_1 = y_1 + \bar{c}_1$ and $c_2 = \bar{c}_2 + y_2$, the firm's profit can be expressed in terms of consumption variables:

$$\pi = y_1 + \frac{y_2}{1+r} = c_1 - \bar{c}_1 + \frac{c_2 - \bar{c}_2}{1+r}. \tag{16}$$

Thus, the firm's problem (14) can be expressed as

$$\begin{aligned}\pi^* = \max_{c_1,c_2} & \; c_1 - \bar{c}_1 + \tfrac{c_2-\bar{c}_2}{1+r} \\ \text{s.t.} & \; G(c_1 - \bar{c}_1, c_2 - \bar{c}_2) = 0,\end{aligned}$$

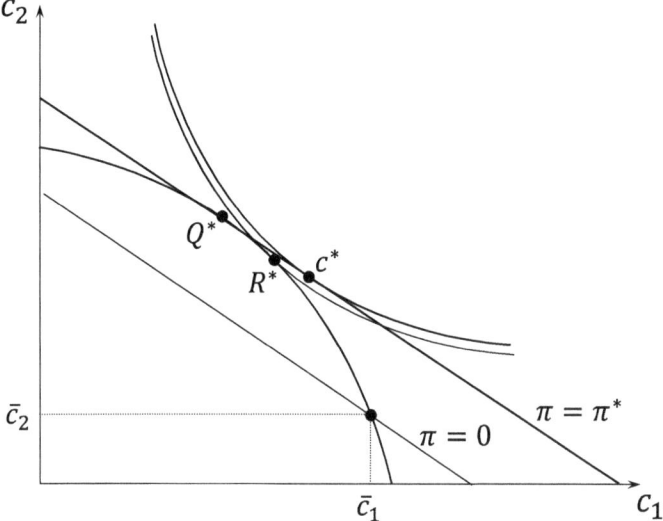

Fig. 13 Optimal consumption and investment plan

which can be illustrated in the consumption space for (c_1, c_2), as in Fig. 13. The solution point is Q^*. Given profit π^*, the consumer's problem is

$$\max_{c_1, c_2} u(c_1, c_2)$$
$$\text{s.t.} \quad c_1 + \frac{c_2}{1+r} = \pi^* + \bar{c}_1 + \frac{\bar{c}_2}{1+r}.$$

The solution point is c^* in the diagram. We can see that the consumer's budget line is the same as the isoprofit line defined in (16) with $\pi = \pi^*$.

Let us use Fig. 13 to explain the theorem. The Fisher Separation Theorem says that if c^* is the solution of the consumer's problem, then this c^* can also be obtained by the following two separate problems: (1) the individual can first consider the optimal production problem given the production frontier and pick up point Q^*, which is totally independent of the individual's preferences; (2) he can then consider the utility maximization problem given the budget constraint taking the optimal production Q^* as the endowment point for the budget, and he will pick up point c^* as his optimal consumption plan.

From Fig. 13, we see that an improvement of the individual's welfare (maximum utility) from problem (12) to problem (13), and the individual consumption improves from R^* to c^*. This improvement is because of the existence of a capital market. In Fig. 14, we can see how the individual manages production and financing in order to improve welfare.

The Separation Theorem requires a perfect capital market (economic agents can borrow and lend at the same interest rate). When the borrowing and lending rates are different, this Separation Theorem may fail. Let the lending rate be r_l and the

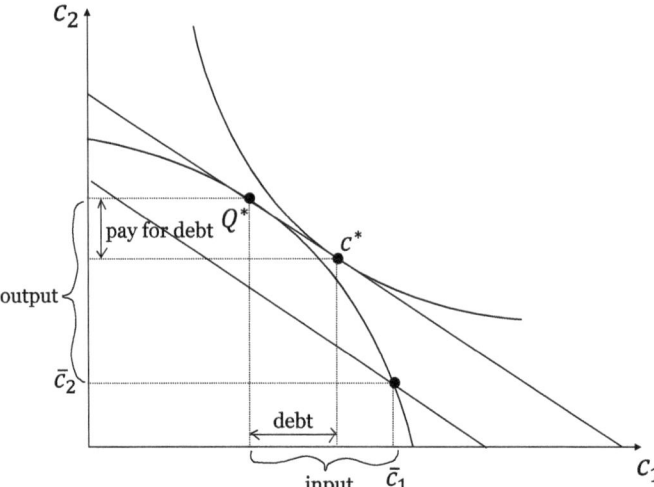

Fig. 14 Investment and consumption

borrowing rate be r_b. Suppose that $r_l < r_b$. By Fig. 15, the profit maximization problem (14) will pick point Q^*, and the utility maximization problem (15) using a budget constraint determined by the rates and endowment Q^* will pick point c'. As shown in Fig. 15, this c'. cannot be the optimal solution, since c'. is even worse than the autarky solution R^* (we know that the optimal solution must be better than R^*). This failure is because of the difference between the profit line faced by a producer and the budget line faced by a consumer. This difference is due to the fact that when the individual acts as a producer, the interest rate changes at $\bar{c}_1$, but when the individual acts as a consumer, the interest rate changes at Q^*.

We now drop the assumption of $p_1 = p_2 = 1$. Without it, the budget constraint is:

$$p_2 c_2 \leq p_2 \bar{c}_2 + (1+r) p_1 (\bar{c}_2 - c_2),$$

implying

$$c_1 + \frac{p_2}{(1+r)p_1} c_2 \leq \bar{c}_1 + \frac{p_2}{(1+r)p_1} \bar{c}_2.$$

Let $p_2/p_1 \equiv 1 + \pi$, and call π the inflation rate. Define ρ by

$$1 + \rho \equiv \frac{1+r}{1+\pi},$$

and call ρ the <u>real interest rate</u> and, accordingly, call r the <u>nominal interest rate</u>. The nominal interest rate tells us how many extra dollars we can get from savings, while the real interest rate tells us how many extra units of goods we can get. We have $\rho \approx r - \pi$. Then, the budget constraint becomes

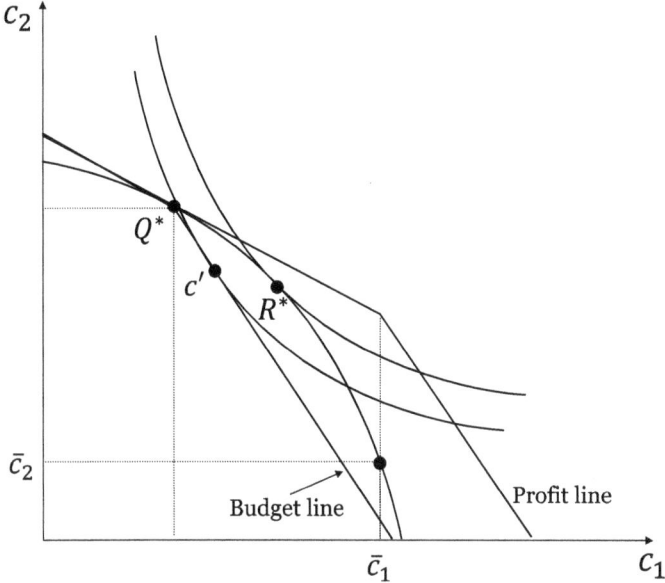

Fig. 15 Failure of separation with different rates

$$c_1 + \frac{1}{1+\rho}c_2 \leq \bar{c}_1 + \frac{1}{1+\rho}\bar{c}_2.$$

Hence, all the conclusions in this section are still valid when the prices are not constant; we only need to replace r by ρ.

The essential message from the Fisher Separation Theorem is that, under perfect markets (without distortions, transaction costs and monopolistic powers), we can assume that the firm's objective is always to maximize profits or expected profits. The owner(s) of the firm can be treated as two persons: one maximizes profit subject to technology, market and resource constraints, and the other is a typical consumer who maximizes utility subject to a budget constraint. This is actually the approach taken in the standard general equilibrium theory (in Chap. 4) in which producers and consumers have distinct objectives. However, as shown in Fig. 15, this approach fails if markets are not perfect.

Notes

Materials in this chapter are fairly standard. Many books cover them. Good references are Varian (1992), Jehle (2001), Dixit (1990), and Mas-Colell et al. (1995). In particular, a good reference for Sect. 4 is Mas-Colell et al. (1995, p. 107), and a good reference for Sect. 7 is Hirshleifer (1984, Chap. 14).

Risk Theory

1 Introduction

So far, we have assumed a world under certainty. In the real world, however, choices are often made under uncertainty. People's preferences and choices are affected by uncertainty. Risk is another dimension to economic problems, in addition to the price ratio, elasticity and substitutability. Hence, the attitude towards risk is an important issue.

Consider a game in which a coin is flipped repeatedly until the heads is up:

if on the nth time the heads is up, you get 2^n dollars and the game is over. (1)

You cannot lose in this game, but you could become very rich from this game. How much will you be willing to pay for participating in this game? The expected money value is

$$E(\text{game}) = 2 \cdot \frac{1}{2} + 2^2 \cdot \frac{1}{2^2} + \cdots + 2^n \cdot \frac{1}{2^n} + \cdots = \infty.$$

That is, you are expected to become very rich by playing this game. However, no one is willing to pay a large amount of money to play this game. That means that people do not necessarily care only about the expected money value. Common sense tells us that there is a large chance that you will not be able get much out of this game and you do not like this risk. Hence, even though you could become very rich at this game, your willingness to pay for this game is low. One explanation to this behavior is that people assign their own personal value to the prizes, which is not necessarily the monetary value of the prize. The personal value of the prize is called the utility value.

For a person with utility function $u(x)$, the personal value for a game or lottery with random payoff $\tilde{x}$ is written as $u(\tilde{x})$. If the person cares about the expected utility (we call this person an <u>expected utility maximizer</u>), then he is willing to pay amount e such that

$$E[u(\tilde{x})] = u(e),$$

where e is called the <u>certainty equivalent</u>, which is the equivalent amount in certainty for the uncertain amount $\tilde{x}$.

Suppose that a person has utility value $u(x) = \sqrt{x}$ for monetary value x Then, the expected utility of the prize for the game in (1) is

$$E[u(\text{game})] \equiv \frac{1}{2}\sqrt{2} + \frac{1}{2^2}\sqrt{2^2} + \cdots + \frac{1}{2^n}\sqrt{2^n} + \cdots$$

$$= \sum_{n=1}^{\infty} 2^{-\frac{n}{2}} = \sum_{n=1}^{\infty} \left(\frac{1}{\sqrt{2}}\right)^n = \frac{1}{\sqrt{2}-1} = 2.41.$$

In this case, $e = \$2.41^2 = \$5.83$. That is, the person is willing to pay $5.83 to play this game.

Consider another game that flips a coin once:

$$\text{if it is tails, you get \$0; and if it is heads, you get \$100.} \tag{2}$$

This game can be written as $\tilde{x} \equiv (0, 0.5; 100, 0.5)$, which means that with probability 0.5 you get 0 and with probability 0.5 you get 100. We call $\tilde{x}$ a <u>lottery</u>. The personal value over the lottery is written as $u(\tilde{x})$ or $u(0, 0.5; 100, 0.5)$. Let $u(x) = \sqrt{x}$ for monetary value x again. How much is this person willing to pay for this game? The expected utility of the lottery is

$$Eu(\tilde{x}) = 0.5u(0) + 0.5u(100) = 0.5 \times 0 + 0.5 \times \sqrt{100} = 5.$$

By $u(e) = E[u(\tilde{x})]$, we find $e = 25$, meaning that the person is willing to pay $25 to play the game in (2).

2 Expected Utility Theory

We shall imagine that the choices facing the consumer take the form of lotteries. A lottery is denoted by a vector of the form:

$$x = (x_1, p_1; x_2, p_2; \ldots; x_n, p_n),$$

for some n, where $p_i \geq 0$ and $\sum p_i = 1$. It is interpreted as: "the consumer receives prize x_i with probability p_i for $i = 1, \ldots, n$". The prize x_i may be money, bundles of goods, or even playing in additional lotteries. Many situations involving behavior under risk can be considered in this lottery framework.

In order to define the lottery space, we make three assumptions:

Axiom 10 $(x, 1; \quad y, 0) = x$.

Axiom 11 $(x, p; \quad y, 1 - p) = (y, 1 - p; \quad x, p)$.

Axiom 12 <u>Reduction of Compound Lottery (RCLA)</u>.

$$((x, p; y, 1 - p), q; y, 1 - q) = (x, pq; y, 1 - pq). \quad \blacksquare$$

Axiom 10 says that getting a prize with probability one is the same as getting the prize for certain. Axiom 11 says that the order of "x with probability p" and "y with probability $1 - p$" expressed in lottery $(x, p; \quad y, 1 - p)$ will not matter to the consumer. Axiom 12 says that the consumer's perception of a lottery depends only on the net probabilities of receiving the prizes and the sequence of events does not matter; see Fig. 1. The RCLA is equivalent to this general form: for $y_i \equiv \left(x_1, p_1^i; \ldots; x_n, p_n^i\right)$, we have

$$(y_1, q_1; \ldots; y_m, q_m) = \left(x_1, \sum_{i=1}^{m} q_i p_1^i; \ldots; x_n, \sum_{i=1}^{m} q_i p_n^i\right).$$

With RCLA, a lottery can be generally expressed as $(x, p; y, 1 - p)$. For example,

$$(x, p_1; y, p_2; z, p_3) = \left[\left(x, \frac{p_1}{1 - p_3}; y, \frac{p_2}{1 - p_3}\right), 1 - p_3; z, p_3\right].$$

Under Axioms 1–3, we can define the <u>lottery space</u> $\mathcal{L}$, in which a typical lottery can be expressed as $(x, p; y, 1 - p)$. The consumer is assumed to have a preference relation $\succsim$ on $\mathcal{L}$, and $x \succsim y$ means that x is better than y. As before, we will assume that the preference relation is a complete preference relation.

Axiom 13 The preference relation $\succsim$ is complete.

Axiom 14 (Continuity). $\{p \in [0, 1] | (x, p; y, 1 - p) \succsim z\}$ and $\{p \in [0, 1] | (x, p; \quad y, 1 - p) \precsim z\}$ are closed sets in $[0, 1]$, $\forall x, y, z \in \mathcal{L}$. $\blacksquare$

Under minor additional assumptions, Theorem 2.1 on the existence of a utility function representing a preference relation on a consumption space can be applied to show the existence of a utility function u representing this preference relation $\succsim$ on the lottery space $\mathcal{L}$. That is, we can find a function $u : \mathcal{L} \to \mathbb{R}$ such that

Fig. 1 Reduction of compound lotteries

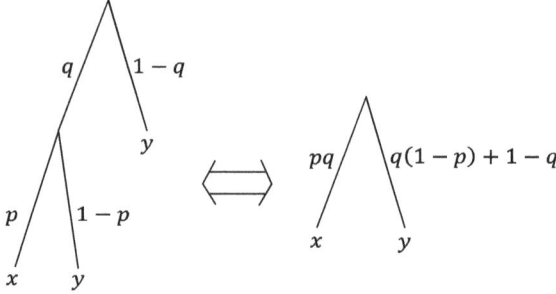

$$(x, p; y, 1-p) \succsim (w, q; z, 1-q) \quad \Leftrightarrow \quad u((x, p; y, 1-p)) \geq u((w, q; z, 1-q))$$

and

$$(x, p; y, 1-p) \succ (w, q; z, 1-q) \quad \Leftrightarrow \quad u((x, p; y, 1-p)) > u((w, q; z, 1-q)).$$

However, such a utility function is too general to be interesting. It cannot tell us much about the special behaviors of the consumer under uncertainty. We want some kind of separation between the prize and its probability in the utility function. One interesting utility representation with such a property is the so-called expected utility, namely the one that satisfies the expected utility property (EUP):

$$u((x, p; y, 1-p)) = pu(x) + (1-p)u(y),$$

for all lotteries in $\mathcal{L}$. It says that the utility of a lottery is the expected utility of its prizes. It should be emphasized that, by Theorem 2.1, the existence of a utility function is not an issue; any well-behaved preferences can be represented by a utility function. What is of interest here is the existence of a utility that satisfies the expected utility property. The key feature in an expected utility is the separation of probabilities from the valuation of prizes. For that, we need an additional axiom; see Fig. 2:

Axiom 15 (Independence). $x \sim y \quad \Rightarrow \quad (x, p; z, 1-p) \sim (y, p; z, 1-p).$ ∎

We can expand a lottery by adding an additional prize and allocating a probability to it. The independence axiom says that two expanded lotteries must be indifferent if their original lotteries are indifferent, whatever the probability assigned to the additional prize is. This is a form of separation between prizes and their probabilities. However, this axiom may be questionable. For example, if $z = x$, then getting x for certain may be better than getting x or y with uncertainty even though x and y are indifferent in preferences under certainty.

In order to avoid some technical details in the proof of existence, we will make two further assumptions.

$x \sim y \implies$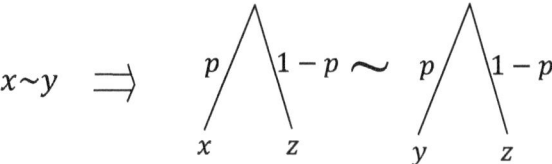

Fig. 2 Independence

Axiom 16 $\mathcal{L}$ is bounded: $\exists a, b \in \mathcal{L}$ s.t. $a \succsim x \succsim b, \forall x \in \mathcal{L}$.

Axiom 17 $(a, p; b, 1-p) \succ (a, q; b, 1-q) \iff p > q$.

Axiom 16 is purely for convenience. Axiom 17 can be derived from the other axioms. It says that if one lottery about prizes a and b is better than another, this must be because there is a higher probability of getting the best prize.

Under the above axioms, we present the following theorem about the existence of a utility representation that satisfies the expected utility property.

Theorem 3.1 (Existence of Expected Utility). *If $(\mathcal{L}, \succsim)$ satisfies Axioms 1–6, then there is a utility representation $u : \mathcal{L} \to \mathbb{R}$ that has the expected utility property.*

Proof For convenience, we will also include Axioms 7 and 8. For any $z \in \mathcal{L}$, define $u(z) \in \mathbb{R}$ to be the unique number that satisfies

$$[a, u(z); b, 1 - u(z)] \sim z.$$

That is, the utility value of a lottery is the probability of getting the best prize a. For this function $u(z)$ to be well defined, we have to ensure that such a number exists and is unique.

(1) Does such a number exist? By Axioms 5 and 7, sets A and B defined below are closed and nonempty:

$$A \equiv \{p \in [0, 1] | (a, p; b, 1-p) \succsim z\}, \quad B \equiv \{p \in [0, 1] | (a, p; b, 1-p) \precsim z\}.$$

By Axiom 13, every $p \in [0, 1]$ is in one of the above two sets (i.e. $A \cup B = [0, 1]$). By Lemma 2.1, $[0, 1]$ is connected. If $A \cap B = \emptyset$, then $A = B^c$ and $B = A^c$, which means that (by Axiom 14) A^c and B^c are nonempty open sets in $[0, 1]$, $A^c \cup B^c = B \cup A = [0, 1]$, and $A^c \cap B^c = B \cap A = \emptyset$. This contradicts with the fact that $[0, 1]$ is connected. Therefore, $A \cap B \neq \emptyset$, and for any $p_z \in A \cap B$, by the definition of A and B,

$$(a, p_z; b, 1 - p_z) \sim z.$$

(2) Is the number p_z unique? For any $p_z, p'_z \in A \cap B$, suppose that $p_z > p'_z$. Then, by Axiom 17,

$$(a, p_z; b, 1 - p_z) \succ (a, p'_z; b, 1 - p'_z).$$

This contradicts the fact that both lotteries are indifferent to z. Thus, $p_z \not\succ p'_z$. Similarly, $p_z \not\prec p'_z$. Therefore, $p_z = p'_z$.

(3) $u(z)$ has the expected utility property. In the following derivation, the number above $\sim$ indicates the corresponding axiom used. We have

$$(x, p; y, 1 - p) \overset{6}{\sim} ([a, u(x); b, 1 - u(x)], p; [a, u(y); b, 1 - u(y)], 1 - p)$$
$$\overset{3}{\sim} [a, pu(x) + (1 - p)u(y); b, 1 - pu(x) - (1 - p)u(y)].$$

Hence,

$$u[(x, p; y, 1 - p)] = pu(x) + (1 - p)u(y).$$

(4) Finally, we verify that u represents the preferences. We have

$$u(x) > u(y) \overset{8}{\Leftrightarrow} [a, u(x); b, 1 - u(x)] \succ a, u(y); b, 1 - u(y)] \Leftrightarrow x \succ y$$

and

$$u(x) = u(y) \Leftrightarrow [a, u(x); b, 1 - u(x)] = [a, u(y); b, 1 - u(y)] \overset{8}{\Leftrightarrow} x \sim y.$$

That is, u is a utility function on the lottery space. ∎

Example 3.1 Given a lottery $x \equiv (x_1, p_1; \ldots; x_n, p_n)$, where $x_1, \ldots, x_n$ are monetary values, the following two utility functions have the expected utility property:

$$u(x) \equiv \sum p_i v(x_i), \quad u(x) \equiv \sum p_i x_i,$$

where $v(x)$ is some function. A risk-averse person will have a concave function v. If the person cares only about his expected monetary value, then his preferences can be represented by the second utility function. Under the assumption of perfect capital markets, as shown by the separation theorem in Chap. 2, we can treat producers as this kind of individual. ∎

We know that any strictly monotonic transform of utility function u on the lottery space $(\mathcal{L}, \succsim)$ is also a utility function on $(\mathcal{L}, \succsim)$. But, in general, such a transformation will not preserve the expected utility property. Only when the transformation is linear will the expected utility property be preserved. Specifically, given the mapping $u : \mathcal{L} \to \mathbb{R}$, the mapping $v : \mathcal{L} \to \mathbb{R}$ is a monotonic linear transformation of u if there are two constants $a, b \in \mathbb{R}$, $a > 0$, such that

$$v(x) = au(x) + b.$$

We find that if u is an expected utility function on $(\mathcal{L}, \succsim)$, v is also an expected utility function on $(\mathcal{L}, \succsim)$. We also have the converse: any strictly monotonic transform of an expected utility that preserves the expected utility property must be a monotonic linear transform.

Proposition 3.1 (Uniqueness of Expected Utility). *The expected utility function is unique up to a monotonic linear transformation. That is, any two expected utility representations u and v of the same preferences must have a positive linear relationship $v = au + b$, where $a > 0$ and $b \in \mathbb{R}$.*

Proof Given two utility functions u and v on $(\mathcal{L}, \succsim)$, by Proposition 2.2, there must be a strictly monotonic function $f : \mathbb{Y} \to \mathbb{R}$ such that $v(\cdot) = f \circ u(\cdot)$, where $\mathbb{Y} \subset \mathbb{R}$ is the range $u(\mathcal{L})$ of u. If both u and v have the expected utility property, then, for arbitrary $x, y \in \mathcal{L}, u(x) > u(y)$, we have

$$f \circ u[(x, p; y, 1 - p)] = v[(x, p; y, 1 - p)]$$

or

$$f[pu(x) + (1-p)u(y)] = pv(x) + (1-p)v(y), \quad \forall p.$$

Let $t \equiv pu(x) + (1-p)u(y)$. Then, $p = \frac{t - u(y)}{u(x) - u(y)}$, where $p \in [0, 1]$ and $t \in [u(y), u(x)]$. Substituting this into the above equation, we have

$$\begin{aligned} f(t) &= \frac{t - u(y)}{u(x) - u(y)} v(x) - \frac{t - u(x)}{u(x) - u(y)} v(y) \\ &= \frac{v(x) - v(y)}{u(x) - u(y)} t + \frac{u(x)v(y) - u(y)v(x)}{u(x) - u(y)}. \end{aligned}$$

This means that f is a monotonic linear transformation on interval $[u(y), u(x)]$. Therefore, f is a monotonic linear transformation on the range $\mathbb{Y}$ of u. ∎

An expected utility theory has now been well established. Under fairly normal conditions, we are guaranteed to have an expected utility representation for preferences under uncertainty and this representation is unique up to a linear transformation. This utility representation is also fairly convenient to use, due to the separation of the utility value from its probability in each event.

However, the expected utility has been found to be inconsistent with actual individual behaviors in many cases. The following are two such examples.

Example 3.2 (Allais Paradox). Consider the following four lotteries with possible prizes US$5 million, US$1 million, or nothing. Each row is a lottery.

	0.1	0.1	0.8
x_1	5m	0	0
x_2	1m	1m	0
y_1	5m	0	1m
y_2	1m	1m	1m

If we look at only the first two columns, we have

$$x_1 \succ x_2 \iff y_1 \succ y_2.$$

When the third column is included, many people have the following preferences:

$$x_1 \succ x_2, \quad y_1 \prec y_2. \tag{3}$$

The preferences in (3) are not consistent with the expected utility. To see this, suppose that the consumer has an expected utility function u with $u(0) = 0$. Then, the preferences in (3) imply

$$0.1u(5m) > 0.2u(1m),$$
$$0.1u(5m) + 0.8u(1m) < u(1m).$$

These two inequalities contradict each other. Hence, the consumer's preferences cannot possibly be represented by expected utility. The fact that the third column causes a reverse of preferences violates the independence axiom. One possible explanation for the preference $y_1 \prec y_2$ is that one million US dollars is enough for life; if you take a chance to go for lottery y_1 and lose, you would kill yourself. ∎

Example 3.3 (Common Ratio Effect). Consider the following four lotteries:

$$x_1 = (3000, 1),$$
$$x_2 = (4000, 0.8; 0, 0.2),$$
$$y_1 = (3000, 0.25; 0, 0.75),$$
$$y_2 = (4000, 0.2; 0, 0.8).$$

2 Expected Utility Theory

Many people have the following preferences:

$$x_1 \succ x_2, \quad y_1 \prec y_2.$$

These preferences again cannot be represented by expected utility. With an expected utility function u with $u(0) = 0$, the preferences imply

$$u(3000) > 0.8u(4000), \quad 0.25u(3000) < 0.2u(4000).$$

These two inequalities contradict each other. ∎

These two examples show that expected utility cannot explain individual preferences under uncertainty. Consequently, an extended utility theory, called the non-expected utility theory, is developed by scrutinizing the axioms more carefully, especially the independence axiom. Such a utility representation is more flexible than the expected utility property so that various paradoxes in individual choices under uncertainty are explainable. However, non-expected utility is beyond the scope of this book.

3 Mean-Variance Utility

A utility function is a mean-variance utility function if it depends only on the mean and variance of the random choice variable. Such utility functions are popular in applications, since in practice people often focus on the mean and variance only.

Proposition 3.2 *If an agent has an exponential utility function of the form*:

$$u(x) = -e^{-rx},$$

and $\tilde{x}$ follows the normal distribution $N(\mu, \sigma^2)$, then maximizing $E[u(\tilde{x})]$ is the same as maximizing the mean-variance utility[1]:

$$v(\mu, \sigma) = \mu - \frac{r}{2}\sigma^2.$$

Proof We have

---
[1] It is from Sargent (1987b, 154–155).

$$E[u(\tilde{x})] = -\int_{-\infty}^{\infty} e^{-rx} \frac{1}{\sqrt{2\pi}\sigma} e^{-\frac{(x-\mu)^2}{2\sigma^2}} dx = -\frac{1}{2\pi\sigma} \int_{-\infty}^{\infty} e^{-\frac{x^2 + 2(r\sigma^2 - \mu)x + \mu^2}{2\sigma^2}} dx$$

$$= -\frac{1}{2\pi\sigma} \int_{-\infty}^{\infty} e^{-\frac{(x + r\sigma^2 - \mu)^2 - (r\sigma^2 - \mu)^2 + \mu^2}{2\sigma^2}} dx = -e^{-\frac{-(r\sigma^2-\mu)^2 + \mu^2}{2\sigma^2}}$$

$$= -e^{-\frac{2\mu r - r^2 \sigma^2}{2\sigma^2}}. \qquad\blacksquare$$

An alternative approach to justify a mean-variance utility function is to assume a quadratic utility function:

$$u(x) = ax + bx^2, \quad a, b \in \mathbb{R}.$$

The agent is risk averse ($u'' < 0$) if $b < 0$, or risk loving ($u'' > 0$) if $b > 0$. To guarantee $u' \geq 0$, we restrict the domain of x:

$$x \leq -\frac{a}{2b} \text{ if } b < 0; \quad x \geq -\frac{a}{2b} \text{ if } b > 0. \qquad (4)$$

This implies $a + 2bx \geq 0$, and thus $a + 2bE(\tilde{x}) \geq 0$ (Fig. 3).

Let $\mu \equiv E(\tilde{x})$ and $\sigma^2 \equiv \text{var}(\tilde{x})$. The expected utility is

$$E[u(\tilde{x})] = a\mu + b\sigma^2 + b\mu^2.$$

This expected utility again depends on the mean μ and variance σ^2 only.

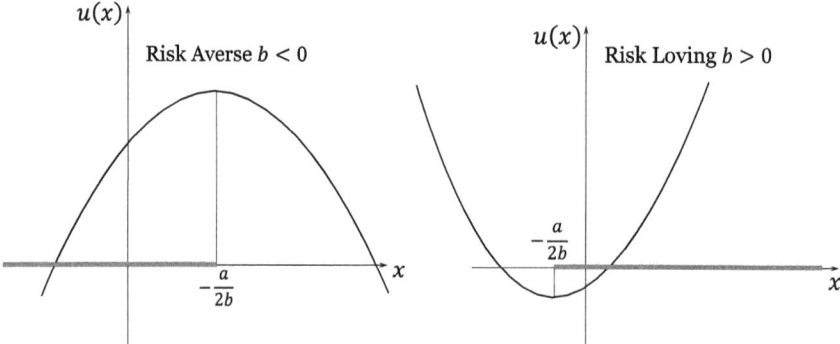

Fig. 3 The utility function

4 Measurement of Risk Aversion

In Sect. 1, we show in two examples that economic agents are risk averse. When facing risk, economic agents are not willing to pay as much as the expected value of a lottery. This risk attitude is personal, as some may be more risk averse than others. This implies different degrees of risk aversion among economic agents. In this section, we develop some measures of risk aversion.

We say that an individual is <u>risk averse</u> if his utility value of the expected income from a lottery is greater than his expected utility of the income; <u>risk neutral</u> if his utility value of the expected income is equal to his expected utility of the income; and <u>risk loving</u> if his utility value of the expected income is less than his expected utility of the income. That is, for any lottery $\tilde{x} \in \mathcal{L}$,[2]

$$\begin{aligned} u[E(x)] &> Eu(x) & \text{if} \quad \text{he is risk averse,} \\ u[E(x)] &= Eu(x) & \text{if} \quad \text{he is risk neutral,} \\ u[E(x)] &< Eu(x) & \text{if} \quad \text{he is risk loving.} \end{aligned} \quad (5)$$

We can illustrate such cases in a figure. Let $\tilde{x} = (x_l, p; \ x_h, 1-p)$. Then, the first situation in (5) is illustrated in Fig. 4, from which we can see that

$$u[E(x)] \geq Eu(x) \quad \Leftrightarrow \quad u(x) \text{ is concave.}$$

Hence,

$$\begin{aligned} & u \text{ is concave} & \text{if} \quad \text{he is risk averse,} \\ & u \text{ is linear} & \text{if} \quad \text{he is risk neutral,} \\ & u \text{ is convex} & \text{if} \quad \text{he is risk loving.} \end{aligned}$$

In Fig. 4, we can further see that the more concave u is, the bigger the difference $u[E(\tilde{x})] - E[u(\tilde{x})]$ is. That is, the more concave the expected utility function is, the more risk averse the individual is. Thus, u'' must be closely related to risk aversion and we might think that we could measure risk aversion by u''. However, this measure is not invariant to a linear transformation of the expected utility function, although the utility function represents the same preferences after a linear transformation. To address this issue, we normalize u'' by dividing it by u'; this ratio is invariant to a linear transformation. Therefore, we now have a sensible measure of risk aversion:

---

[2] In reality, a person can be risk averse under one circumstance and risk loving in another. For example, when a person buys insurance, he is risk averse; and when he buys a lottery, he is risk loving. The key in these two cases is that, when the person buys insurance, he pays to move from a risky situation to a non-risky situation; and when he buys a lottery, he pays to move from a non-risky situation to a risky situation. Our definition of risk attitudes in (5) does not include such a person. In (5), a person is either always risk averse, or risk neutral, or risk loving.

Fig. 4 A risk-averse individual

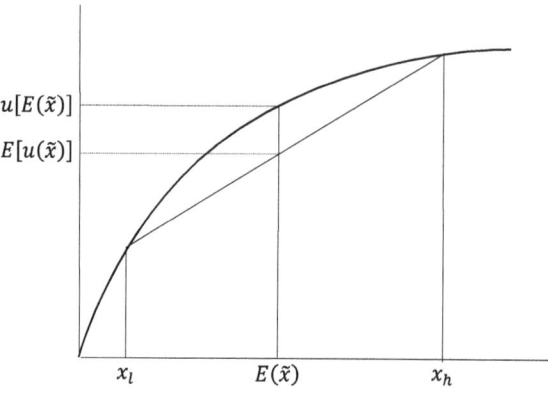

$$\text{absolute risk aversion}: \quad R_a(x) \equiv -\frac{u''(x)}{u'(x)},$$

where the negative sign makes the measure a positive number.

There are two types of fluctuations: fluctuations in the level of income $\bar{x} + \tilde{\varepsilon}$ and fluctuations in the percentage of income $\bar{x}(1 + \tilde{\varepsilon})$, where $\bar{x}$ is the mean of income and $\tilde{\varepsilon}$ represents fluctuations. The absolute risk aversion deals with the first case. The second case is dealt with by the so-called relative risk aversion, which is defined as

$$\text{relative risk aversion}: \quad R_r(x) \equiv -\frac{xu''(x)}{u'(x)}.$$

To justify the above two definitions more rigorously, for any lottery $\tilde{x}$, let $\bar{x} \equiv E(\tilde{x})$. Define risk premium π_a by

$$u(\bar{x} - \pi_a) = E[u(\tilde{x})]. \tag{6}$$

This risk premium is the amount that the individual is willing to pay for avoiding the risk; see Fig. 5. Here, $\bar{x} - \pi_a$ and $\tilde{x}$ are indifferent in the expected utility, but one has risk while the other does not.

We now find an approximated solution of (6) for π_a. Define $\tilde{\varepsilon}$ by $\tilde{x} \equiv \bar{x} + \tilde{\varepsilon}$. We have $E(\tilde{\varepsilon}) = 0$. By Taylor's expansion,

the left-hand side of (3.6) $\approx u(\bar{x}) - u'(\bar{x})\pi_a$;
the right-hand side of (3.6) $= E[u(\bar{x} + \tilde{\varepsilon})] \approx E\left[u(\bar{x}) + u'(\bar{x})\tilde{\varepsilon} + \frac{1}{2}u''(\bar{x})\tilde{\varepsilon}^2\right]$.

Equalizing the above two formulae implies an approximated solution of π_a:

4 Measurement of Risk Aversion

Fig. 5 The risk premium

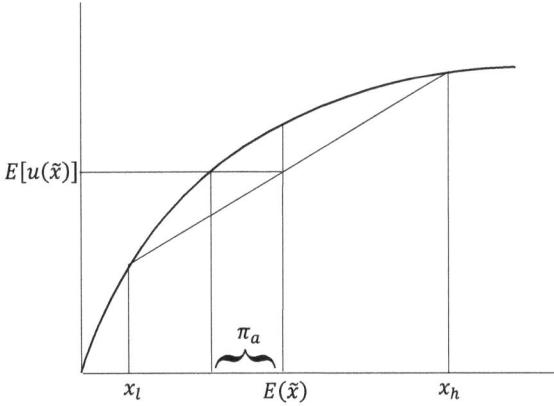

$$\pi_a \approx \frac{1}{2}\sigma^2 R_a(\bar{x}),$$

where $\sigma^2 \equiv var(\tilde{\varepsilon})$. This approximation becomes exact when $\sigma \to 0$. In this formula, σ^2 has nothing to do with individual preferences while $R_a(\bar{x})$ depends on preferences. This formula shows that $R_a(\bar{x})$ is the multiplier of how much the individual is willing to pay for avoiding the risk σ^2.

Similarly, define the <u>relative risk premium</u> π_r by

$$u[\bar{x}(1-\pi_r)] = E[u(\tilde{x})],$$

and define $\tilde{\varepsilon}$ by $\tilde{x} = \bar{x}(1+\tilde{\varepsilon})$. By the same method, we find

$$\pi_r \approx \frac{1}{2}\sigma_\varepsilon^2 R_r(\bar{x}).$$

We have therefore justified the definitions of the two risk aversion measures. Absolute risk aversion measures the aversion for the risk contained in the absolute magnitude of a lottery, while relative risk aversion measures the aversion for the risk contained in the relative magnitude of a lottery.

5 Stochastic Dominance

Besides risk aversion, we now introduce another important risk attitude: stochastic dominance. We denote F_ξ and f_ξ, respectively, as the accumulated distribution function and the density function of random variable ξ. Assume that all the distributions in this section have finite supports and the supports are bounded by interval $[\underline{x}, \bar{x}]$, where $\underline{x}, \bar{x} \in \mathbb{R}$ are finite numbers.

If $F_{\tilde{x}}(t) \leq F_{\tilde{y}}(t)$ for any $t \in \mathbb{R}$, we say that $\tilde{x}$ dominates $\tilde{y}$, or $F_{\tilde{x}}$ dominates $F_{\tilde{y}}$, by first-order stochastic dominance (FOSD), denoted as $\tilde{x} \succsim_F \tilde{y}$ or $F_{\tilde{x}} \succsim_F F_{\tilde{y}}$. It means that, for any $t \in \mathbb{R}$, $\tilde{x}$ is more likely than $\tilde{y}$, to have a value larger than t. FOSD focuses on the expected return of an asset.

If $\int_{-\infty}^{t} F_{\tilde{x}}(\tau) d\tau \leq \int_{-\infty}^{t} F_{\tilde{y}}(\tau) d\tau$ for any $t \in \mathbb{R}$, we say that $\tilde{x}$ dominates $\tilde{y}$, or $F_{\tilde{x}}$ dominates $F_{\tilde{y}}$, by second-order stochastic dominance (SOSD), denoted as $\tilde{x} \succsim_S \tilde{y}$ or $F_{\tilde{x}} \succsim_S F_{\tilde{y}}$. SOSD focuses on the risk of an asset.

FOSD is based on the idea of a higher probability of getting a higher value. SOSD is based on the idea of relative riskiness or dispersion.

We first have two results that relate the concepts of stochastic dominance to preferences. It turns out that the two concepts closely relate to the monotonicity and concavity of a utility function.

Theorem 3.2 (Hanoch-Levy 1969). *$\tilde{x}$ dominates $\tilde{y}$ by FOSD iff $E[u(\tilde{x})] \geq E[u(\tilde{y})]$ for any increasing u.*

Proof We have[3]

$$\int_{-\infty}^{+\infty} u(t) d\big[F_x(t) - F_y(t)\big] = u(t)\big[F_x(t) - F_y(t)\big]\Big|_{-\infty}^{+\infty} - \int_{-\infty}^{+\infty} u'(t)\big[F_x(t) - F_y(t)\big] dt$$

$$= -\int_{-\infty}^{+\infty} u'(t)\big[F_x(t) - F_y(t)\big] dt.$$

Then, the increasingness of u tells us whether or not we have $Eu(\tilde{x}) \geq Eu(\tilde{y})$. ∎

One immediate implication is that if $\tilde{x}$ dominates $\tilde{y}$ by FOSD, then $E(\tilde{x}) \geq E(\tilde{y})$. However, the converse is obviously not true.

Theorem 3.3 *Suppose that $\tilde{x}$ and $\tilde{y}$ have the same mean. Then, $\tilde{x}$ dominates $\tilde{y}$ by SOSD iff $E[u(\tilde{x})] \geq E[u(\tilde{y})]$ for any concave u.*

Proof We have

---

[3]In the proofs of this section, we assume sufficient differentiability for u; this is for convenience. Also, we use $-\infty$ and ∞ in places where $\underline{x}$ and $\bar{x}$ should be used for mathematical rigor.

5 Stochastic Dominance

$$\int_{-\infty}^{+\infty} u(t)d[F_x(t) - F_y(t)] = -\int_{-\infty}^{+\infty} u'(t)[F_x(t) - F_y(t)]dt$$

$$= -u'(t)\int_{-\infty}^{t} [F_x(z) - F_y(z)]dz \Big|_{-\infty}^{+\infty}$$

$$+ \int_{-\infty}^{+\infty} u''(t)\int_{-\infty}^{t} [F_x(z) - F_y(z)]dzdt$$

$$= -u'(\infty)\int_{-\infty}^{\infty} [F_x(z) - F_y(z)]dz$$

$$+ \int_{-\infty}^{+\infty} u''(t)\int_{-\infty}^{t} [F_x(z) - F_y(z)]dzdt.$$

We also have

$$\int_{-\infty}^{\infty} [F_x(z) - F_y(z)]dz = z[F_x(z) - F_y(z)]\Big|_{-\infty}^{+\infty} - \int_{-\infty}^{+\infty} zd[F_x(z) - F_y(z)]$$

$$= -[E(x) - E(y)] = 0.$$

Hence,

$$\int_{-\infty}^{+\infty} u(t)d[F_x(t) - F_y(t)] = \int_{-\infty}^{+\infty} u''(t)\int_{-\infty}^{t} [F_x(z) - F_y(z)]dzdt.$$

Therefore, the concavity of u determines whether or not we have $Eu(\tilde{x}) \geq Eu(\tilde{y})$. ∎

The rest of this section takes a different approach to understanding the concepts of stochastic dominance.

Example 3.4 Consider a two-stage lottery that offers $\tilde{x}$ in the first stage and offers $\tilde{\varepsilon}$ in the second stage. The final lottery is $\tilde{y} = \tilde{x} \oplus \tilde{\varepsilon}$, where, for each realization x of $\tilde{x}$ in the first period, $\tilde{\varepsilon}$ follows a distribution $H_{\tilde{\varepsilon}}(\cdot|x)$. Here, the symbol $\oplus$ means a sum of the two lotteries in two stages; more precisely, $\tilde{y}$ is a two-stage lottery for which the value of $\tilde{y}$ is a sum of realized values of $\tilde{x}$ and $\tilde{\varepsilon}$ but the distribution of the second-stage lottery $\tilde{\varepsilon}$ is conditional on a realization of the first-stage lottery $\tilde{x}$. Assume $H_{\tilde{\varepsilon}}(0|x) = 0$ for any x, meaning that $\tilde{\varepsilon}$ is a positive shock with probability 1. Thus, for any increasing function u we have

$$\int_{-\infty}^{+\infty} u(y)dF_{\tilde{y}}(y) = \int_{-\infty}^{+\infty}\int_{-\infty}^{+\infty} u(x+\varepsilon)dH_{\tilde{\varepsilon}}(\varepsilon|x)dF_{\tilde{x}}(x)$$

$$= \int_{-\infty}^{+\infty}\int_{0}^{\infty} u(x+\varepsilon)dH_{\tilde{\varepsilon}}(\varepsilon|x)dF_{\tilde{x}}(x)$$

$$\geq \int_{-\infty}^{+\infty}\int_{0}^{\infty} u(x)dH_{\tilde{\varepsilon}}(\varepsilon|x)dF_{\tilde{x}}(x)$$

$$= \int_{-\infty}^{+\infty} u(x) \int_{0}^{\infty} dH_{\tilde{\varepsilon}}(\varepsilon|x)dF_{\tilde{x}}(x)$$

$$= \int_{-\infty}^{+\infty} u(x)dF_{\tilde{x}}(x).$$

Hence, $\tilde{y}$ dominates $\tilde{x}$ by FOSD. That is, given a random variable $\tilde{x}$ we can construct another random variable $\tilde{y}$ such that $\tilde{y} \succsim_F \tilde{x}$, where $\tilde{y}$ is obtained from $\tilde{x}$ by adding a positive shock $\tilde{\varepsilon}$. Conversely, given a random variable $\tilde{y}$ we can similarly construct another random variable $\tilde{x}$ such that $\tilde{y} \succsim_F \tilde{x}$.

Notice that risk does not play a role in this case, which reflects the fact that FOSD is concerned with a potential return only. However, adding a shock will add more risks. In particular, when this shock does not contribute much to an increase in expected return, risk will then become a crucial issue. This is what the next example will discuss. ∎

Example 3.5 Consider a two-stage lottery that offers $\tilde{x}$ in the first stage and offers $\tilde{\varepsilon}$ in the second stage. The final lottery is $\tilde{y} = \tilde{x} \oplus \tilde{\varepsilon}$, where $\tilde{\varepsilon}$ is another random variable with $E(\tilde{\varepsilon}|x) = 0$ for each realization x of $\tilde{x}$ in the first period, and $\tilde{\varepsilon}$ follows a distribution $H_{\tilde{\varepsilon}}(\cdot|x)$. We have

$$E(y) = \int_{-\infty}^{+\infty}\int_{-\infty}^{+\infty} (x+\varepsilon)dH_{\tilde{\varepsilon}}(\varepsilon|x)dF_{\tilde{x}}(x) = \int_{-\infty}^{+\infty} xdF_{\tilde{x}}(x) = E(x).$$

We say in this case that $F_{\tilde{y}}$ is a mean-preserving spread of $F_{\tilde{x}}$, which means that $\tilde{y}$ has the same mean as $\tilde{x}$, but $\tilde{y}$ is more random (with a larger variance) than $\tilde{x}$ is. Then, for any concave function u,

$$\int_{-\infty}^{+\infty} u(y) dF_{\tilde{y}}(y) = \int_{-\infty}^{+\infty} \int_{-\infty}^{+\infty} u(x+\varepsilon) dH_{\tilde{\varepsilon}}(\varepsilon|x) dF_{\tilde{x}}(x)$$

$$\leq \int_{-\infty}^{+\infty} u\left[\int_{-\infty}^{+\infty} (x+\varepsilon) dH_{\tilde{\varepsilon}}(\varepsilon|x)\right] dF_{\tilde{x}}(x) = \int_{-\infty}^{+\infty} u(x) dF_x(x).$$

Hence, $\tilde{x}$ dominates $\tilde{y}$ by SOSD. That is, given a random variable $\tilde{x}$ we can construct another random variable $\tilde{y}$ with the same mean such that $\tilde{x} \succsim_S \tilde{y}$. Conversely, given a random variable $\tilde{y}$ we can similarly construct another random variable $\tilde{x}$ with the same mean such that $\tilde{x} \succsim_S \tilde{y}$. The basic idea here is that $\tilde{x}$ dominates $\tilde{y}$ by SOSD if $\tilde{y}$ is $\tilde{x}$ plus some noise. ∎

The conclusion from Example 3.5 is summarized in the following theorem.

Theorem 3.4 *For any two random variables $\tilde{x}$ and $\tilde{y}$ with the same mean, $\tilde{y}$ is dominated by $\tilde{x}$ by SOSD iff $\tilde{y}$ is a mean-preserving spread of $\tilde{x}$*

Example 3.6 For two distribution functions F and G, we say that G constitutes an elementary increase in risk from F if G is generated from F by taking all the mass of F in an interval $[a, b]$ and assigning it to the endpoints a and b in such a manner that the mean is preserved. This is illustrated in the following figure. Since the two triangle areas indicated in Fig. 6 are equal in size, we have $\int_{-\infty}^{+\infty} F(x)dx = \int_{-\infty}^{+\infty} G(x)dx$ Since

$$0 = \int_{-\infty}^{+\infty} [F(x) - G(x)] dx = -\int_{-\infty}^{+\infty} x d[F(x) - G(x)],$$

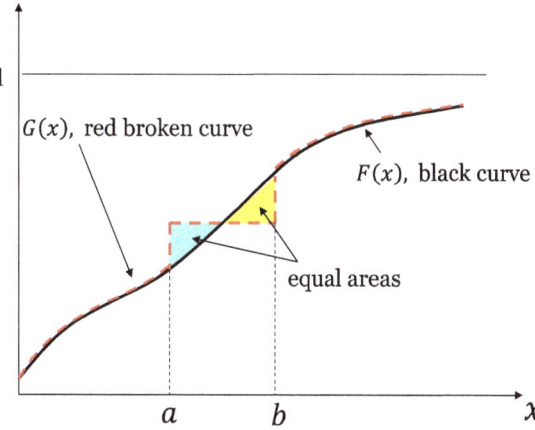

Fig. 6 An elementary increase in risk

the two random variables defined by F and G have equal means. We can easily verify that F dominates G by SOSD by comparing $\int_{-\infty}^{t} F(\tau)d\tau$ with $\int_{-\infty}^{t} G(\tau)d\tau$ for any t. Thus, an elementary increase in risk is a mean-preserving spread. ∎

With well-established expected utility theory, the rest of this chapter studies individual behavior under risk.

6 Demand for Insurance

When facing a potential loss, what is an individual's demand for insurance? Consider an individual who has wealth w but may lose an amount l with probability p. That is, he faces a lottery $(w - L, p; w, 1 - p)$. This individual can buy insurance for a cost qz from a competitive insurance company, which pays him z dollars if the bad event happens. Here, q is called the <u>price of insurance</u>. Given this price, the consumer decides an insurance amount z. The consequences are shown below:

Probability:	p	$1-p$
Before insurance:	$w - L$	w
After insurance:	$w - L + z - qz$	$w - qz$

What is the optimal amount q^* for the individual? The consumer's problem is:

$$\max_{z \geq 0} \; pu(w - L + z - qz) + (1-p)u(w - qz).$$

The FOC is

$$pu'[w - L + (1-q)z^*](1-q) - (1-p)u'(w - qz^*)q = 0. \tag{7}$$

This equation determines the demand for insurance $z^* = z^*(q)$.

The expected profit per capita for the company is: $qz - pz = (q - p)z$. In a competitive market, the expected profit will be driven to zero, implying $q = p$. Substituting this into (7), we get

$$u'[w - L + (1-p)z^*] = u'(w - pz^*).$$

If the consumer is strictly risk averse so that $u'' < 0$, then $w - L + (1-p)z^* = w - pz^*$, implying $z^* = L$. That is, the consumer will completely insure himself against the loss L.

This problem can be shown graphically. Let

$$I_1 \equiv w - qz, \quad I_2 \equiv w - L + (1-q)z, \tag{8}$$

6 Demand for Insurance

Fig. 7 The insurance problem

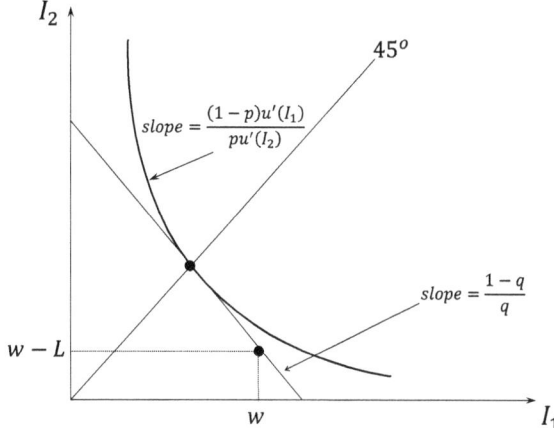

which are the income levels in the two events. By eliminating q from the two expressions, we have the budget constraint: $(1-q)I_1 + qI_2 = w - qL$. Then, the consumer's problem becomes

$$\max_{I_1, I_2 \geq 0} (1-p)u(I_1) + pu(I_2)$$
$$\text{s.t. } (1-q)I_1 + qI_2 = w - qL.$$

This problem is illustrated in Fig. 7.

The slope of the budget line is $\frac{1-q}{q}$ and the slope of the indifference curve at (I_1, I_2) is $MRS \equiv \frac{(1-p)u'(I_1)}{pu'(I_2)}$. At the optimal point (I_1^*, I_2^*), the two slopes must be equal:

$$\frac{(1-p)u'(I_1^*)}{pu'(I_2^*)} = \frac{1-q}{q}. \quad (9)$$

In a competitive insurance market, zero profit implies $q = p$. Hence, (9) implies $u'(I_1^*) = u'(I_2^*)$ or $I_1^* = I_2^*$, which is on the 45°-line. A point on the 45°-line offers the same income in any event. This situation is called <u>full insurance</u>.

In summary, when the insurance market is competitive, the consumer will demand full insurance. However, this conclusion holds only when the market has complete information. We will revisit this issue later when the market has incomplete information.

7 Demand for Risky Assets

Since the absolute risk aversion R_a measures an individual's attitude towards fluctuations in the *level* of income, we would expect a risk-averse individual to become less risk averse to this kind of uncertainty when his income rises. That is, we expect $R_a(x)$ to decrease as x increases. Since the relative risk aversion R_r measures an individual's attitude towards fluctuations in a *percentage* of income, we would expect a risk-averse individual to become more risk averse to this kind uncertainty when his income rises. That is, we expect $R_r(x)$ to increase as x increases. Hence, we assume a decreasing $R_a(x)$ and an increasing $R_r(x)$ in this section, and we want to find out what they imply to an individual's portfolio choices.

Suppose that an economic agent has initial wealth A. He can invest an amount m in money and the rest $a = A - m$ in a risky asset. The net return from the money is zero, and the net return from the risky asset is $\tilde{x}$, which is a random variable and can be negative. He consumes nothing. Then, his final wealth is $\tilde{y} = m + (1+\tilde{x})a$. As an expected utility maximizer who cares about the expected utility value of future income, his problem is:

$$v(A) \equiv \max_{m, a \geq 0} Eu[m + (1+\tilde{x})a]$$
$$\text{s.t. } m + a = A.$$

This is equivalent to:

$$v(A) \equiv \max_{a \in [0, A]} Eu(A + a\tilde{x}).$$

For $a \in [0, A]$, define

$$W(a) \equiv E[u(A + a\tilde{x})].$$

For an interior solution $a^* \in (0, A)$, the FOC is:

$$W'(a^*) = 0 \quad \text{or} \quad E[u'(A + a^*\tilde{x})\tilde{x}] = 0. \tag{10}$$

Since $W'(a) \equiv E[u'(A + a\tilde{x})\tilde{x}]$ is a continuous function, it cannot change its sign without taking zero value. This means that either

$$W'(a) > 0, \; \forall a \in (0, A); \text{ or } W'(a) < 0, \; \forall a \in (0, A); \text{ or } \exists a_0$$
$$\in (0, A) \text{ s.t. } W'(a_0) = 0.$$

We know that if $W'(a) > 0$ for all $a \in (0, A)$, then $a^* = A$; if $W'(a) < 0$ for all $a \in (0, A)$, then $a^* = 0$. Assuming that u is concave, we have $W''(a) = E[u''(A + a\tilde{x})\tilde{x}^2] \leq 0$. That is, W is concave, which means that any $a_0 \in (0, A)$ satisfying $W'(a_0) = 0$ is an optimal solution. Therefore, there exists a solution a^*

such that $W'(a^*) \geq 0$ if $a^* = A$; $W'(a^*) \leq 0$ if $a^* = 0$; and $W'(a^*) = 0$ if $a^* \in (0, A)$.

Since $W''(a) \equiv E[u''(A+a\tilde{x})\tilde{x}^2]$, if $\tilde{x}$ is not equal to zero with probability 1, the strict concavity of $u(x)$ will guarantee the strict concavity of $W(a)$ and therefore the uniqueness of a^*. We denote this optimal solution as $a^* = a^*(A)$.

Taking a derivative w.r.t. A on the FOC (10), we have

$$E\left[u''(A+a^*\tilde{x})\tilde{x}\left(1+\frac{da^*}{dA}\tilde{x}\right)\right] = 0.$$

Thus,

$$\frac{da^*}{dA} = -\frac{E[u''(A+a^*\tilde{x})\tilde{x}]}{E[u''(A+a^*\tilde{x})\tilde{x}^2]}. \tag{11}$$

The following proposition shows that, for a risk-averse individual, a decreasing absolute risk aversion $R_a(x)$ will imply an increasing $a^*(A)$, meaning that the individual will invest more in the risky asset if his initial income increases.

Proposition 3.3 $u' > 0$, $u'' < 0$, $R_a \downarrow \;\Rightarrow\; a^* \uparrow$.

Proof Decreasing R_a implies

$$R_a(A+a^*x) \lessgtr R_a(A) \quad \text{if} \quad x \gtrless 0,$$

or

$$-\frac{u''(A+a^*x)}{u'(A+a^*x)} \lessgtr R_a(A) \quad \text{if} \quad x \gtrless 0,$$

implying

$$u''(A+a^*x) \gtrless -R_a(A)u'(A+a^*x) \quad \text{if} \quad x \gtrless 0,$$

implying

$$u''(A+a^*x)x \geq -R_a(A)u'(A+a^*x)x, \quad \forall x,$$

implying

$$E[u''(A+a^*\tilde{x})\tilde{x}] \geq -R_a(A)E[u'(A+a^*\tilde{x})\tilde{x}].$$

Using (10), we have $E[u''(A+a^*\tilde{x})\tilde{x}] \geq 0$. Using (11) and $u'' < 0$, we therefore have

$$\frac{da^*}{dA} = -\frac{E[u''(A+a^*\tilde{x})\tilde{x}]}{E[u''(A+a^*\tilde{x})\tilde{x}^2]} \geq 0. \quad \blacksquare$$

The <u>income elasticity</u> of demand for money is defined by $\eta_m \equiv \frac{A}{m}\frac{dm}{dA}$. Money is a luxury good if $\eta_m > 1$.

Proposition 3.4 $u' > 0$, $u'' < 0$, $R_r \uparrow \;\Rightarrow\; \eta_m > 1$.

Proof We have $\tilde{y} = m + (1+\tilde{x})a^* = A + a^*\tilde{x}$ and $A = m + a^*$. Using (11),

$$\begin{aligned}
\eta_m - 1 &= \frac{A}{m}\frac{dm}{dA} - 1 = \frac{A}{m}1 - \frac{da^*}{dA} - 1 = \frac{a^*}{m} - \frac{A}{m}\frac{da^*}{dA} \\
&= \frac{a^*}{m} + \frac{A}{m}\frac{E[u''(\tilde{y})\tilde{x}]}{E[u''(\tilde{y})\tilde{x}^2]} = \frac{E[u''(\tilde{y})\tilde{x}(a^*\tilde{x}+A)]}{mE[u''(\tilde{y})\tilde{x}^2]} = \frac{E[u''(\tilde{y})\tilde{x}\tilde{y}]}{mE[u''(\tilde{y})\tilde{x}^2]}.
\end{aligned} \qquad (12)$$

Increasing R_r implies that

$$R_r(A+a^*x) \gtreqless R_r(A) \quad \text{if} \quad x \gtreqless 0$$

or

$$-\frac{yu''(y)}{u'(y)} \gtreqless R_r(A) \quad \text{if} \quad x \gtreqless 0,$$

implying

$$yu''(y) \lesseqgtr -R_r(A)u'(y) \quad \text{if} \quad x \gtreqless 0,$$

implying

$$u''(y)xy \leq -R_r(A)u'(y)x, \quad \forall x,$$

implying

$$E[u''(\tilde{y})\tilde{x}\tilde{y}] \leq -R_r(A)E[u'(\tilde{y})\tilde{x}] = 0.$$

where (10) implies $E[u'(\tilde{y})\tilde{x}] = 0$. From $u'' < 0$ and (12), we therefore have

$$\eta_m - 1 = \frac{E[u''(\tilde{y})\tilde{x}\tilde{y}]}{mE[u''(\tilde{y})\tilde{x}^2]} \geq 0. \quad \blacksquare$$

8 Portfolio Analysis

Assume that an economic agent invests in two risky assets $\tilde{x}$ and $\tilde{y}$ with the proportion $\lambda \in [0, 1]$ in $\tilde{x}$ and the rest in $\tilde{y}$. His income is $\tilde{z} = \lambda\tilde{x} + (1-\lambda)\tilde{y}$. We call λ the portfolio [or call $(\lambda, 1-\lambda)$ the portfolio]. The agent's problem is

$$\max_{\lambda \in [0,1]} Eu(\tilde{z})$$
$$\text{s.t. } \tilde{z} = \lambda\tilde{x} + (1-\lambda)\tilde{y}.$$

Assume that the utility function is quadratic:

$$u(z) = az + bz^2, \quad a, b \in \mathbb{R}.$$

The agent is risk averse ($u'' < 0$) if $b < 0$, or risk loving ($u'' > 0$) if $b > 0$. We can transform the consumer's problem of choosing λ into a problem of choosing (μ, σ). Let $\mu \equiv E(\tilde{z})$ and $\sigma^2 \equiv \text{var}(\tilde{z})$. Then, the expected utility is

$$U(\mu, \sigma) \equiv E[u(\tilde{z})] = a\mu + b\sigma^2 + b\mu^2.$$

Let

$$\mu_x \equiv E(\tilde{x}), \quad \mu_y \equiv E(\tilde{y}), \quad \sigma_x^2 \equiv \text{var}(\tilde{x}), \quad \sigma_y^2 \equiv \text{var}(\tilde{y}), \quad \rho \equiv \frac{\text{cov}(\tilde{x},\tilde{y})}{\sigma_x \sigma_y}.$$

We now need to express the budget condition in terms of (μ, σ). Taking an expectation and variance on the budget condition $\tilde{z} = \lambda\tilde{x} + (1-\lambda)\tilde{y}$ yields

$$\mu = \lambda\mu_x + (1-\lambda)\mu_y, \quad \sigma^2 = \lambda^2\sigma_x^2 + (1-\lambda)^2\sigma_y^2 + 2\lambda(1-\lambda)\rho\sigma_x\sigma_y. \quad (13)$$

They can be combined into one condition by eliminating λ:

$$\sigma^2 = \frac{\mu - \mu_y}{\mu_x - \mu_y}^2 \left(\sigma_x^2 + \sigma_y^2 - 2\rho\sigma_x\sigma_y\right) + 2\frac{\mu - \mu_y}{\mu_x - \mu_y}\left(\rho\sigma_x\sigma_y - \sigma_y^2\right) + \sigma_y^2. \quad (14)$$

We call this the portfolio curve. We can thus treat the two assets $\tilde{x}$ and $\tilde{y}$ as two consumption bundles $X \equiv (\mu_x, \sigma_x)$ and $Y \equiv (\mu_y, \sigma_y)$ available in the market and the individual is to choose $Z \equiv (\mu, \sigma)$. The consumer's problem becomes:

$$\max_{\mu, \sigma} U(\mu, \sigma) = a\mu + b\sigma^2 + b\mu^2$$
$$\text{s.t. the portfolio curve.}$$

The consumer now chooses (μ, σ) rather than λ.

Because of the simplicity of the utility function, we can use a graph to analyze the solution. For this, we need to draw the indifference curves and the portfolio curve on a diagram. First, the indifference curve with utility level C is

$$a\mu + b\sigma^2 + b\mu^2 = C. \tag{15}$$

Suppose that this indifference curve determines a function $\mu = \mu(\sigma)$. By taking the derivative w.r.t. σ, we have

$$\frac{d\mu}{d\sigma} = -\frac{2b\sigma}{a+2b\mu} \gtreqless 0 \Leftrightarrow b \lesseqgtr 0. \tag{16}$$

In the space of (σ, μ), as in Fig. 8, this means that the indifference curve is increasing in the risk-averse case and decreasing in the risk-loving case. We also have

$$\frac{d^2\mu}{d\sigma^2} = \frac{d}{d\sigma}\frac{d\mu}{d\sigma} = \frac{d}{d\sigma}\frac{2b\sigma}{a+2b\mu} = \frac{-2b(a+2b\mu)+4b^2\sigma\frac{d\mu}{d\sigma}}{(a+2b\mu)^2}.$$

Therefore,

$$\frac{d^2\mu}{d\sigma^2} \gtreqless 0 \Leftrightarrow b \lesseqgtr 0.$$

This means that the indifference curve is convex in the risk-averse case and concave in the risk-loving case. We also know, in either case, that

$$\left.\frac{d\mu}{d\sigma}\right|_{\sigma=0} = 0.$$

This together with the concavity and the monotonicity of the indifference curves give us the shapes of the indifference curves in Fig. 8.

We also need to find which direction the indifference curve shifts when the utility level C is increased. For this, we have

$$\frac{\partial U(\mu,\sigma)}{\partial \sigma^2} = b, \quad \frac{\partial U(\mu,\sigma)}{\partial \mu} = a + 2b\mu \geq 0, \tag{17}$$

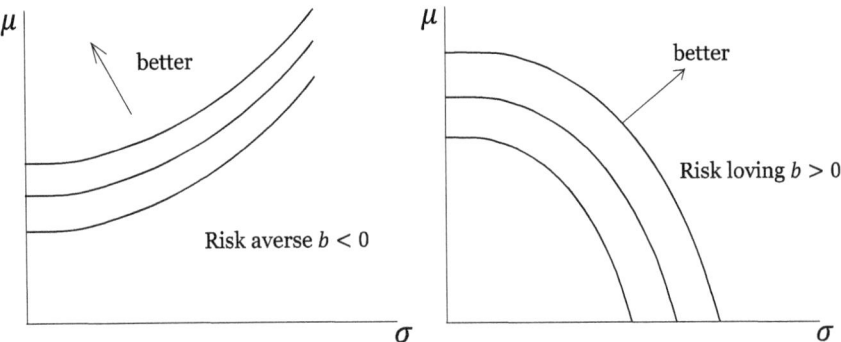

Fig. 8 The indifference curves

where "$a + 2b\mu \geq 0$" is because of the assumption in (4). Expressions in (17) tell us the directions of increasing utility for the indifference curves as shown in Fig. 8.

Second, let us now draw the portfolio curve. From elementary geometry, we know that Eq. (14) in the space for (μ, σ) is a hyperbola if μ and σ are allowed to take negative values. There are two branches in this hyperbola. We will show that our portfolio curve is one of the branches. We know from (13) that, for any $\rho_1 < \rho_2$, the portfolio curve for ρ_1 is on the left of that for ρ_2. Furthermore, the right-most portfolio curve is for $\rho = 1$ and is a straight line connecting X and Y:

$$\sigma = \frac{\mu - \mu_y}{\mu_x - \mu_y}\left(\sigma_x - \sigma_y\right) + \sigma_y.$$

Therefore, the portfolio curve for any given ρ must be the branch that is bent towards the μ axis (i.e., $\frac{\partial^2 \sigma}{\partial \mu^2} \geq 0$ on the curve). Besides that, for the left-most portfolio curve with $\rho = -1$, by (13), we have

$$\sigma^2 = [\lambda \sigma_x - (1-\lambda)\sigma_y]^2 \quad \text{or} \quad \sigma = \lambda \sigma_x - (1-\lambda)\sigma_y.$$

This portfolio curve is straight except at $\sigma = 0$; it takes $\sigma = 0$ if $\lambda = \frac{\sigma_y}{\sigma_x + \sigma_y}$. These imply the shape of the portfolio curves in Fig. 9.

With this graphic illustration, we can now easily combine Figs. 7 and 8 to analyze portfolio behaviors of an investor under various conditions. For example, from Fig. 9, we can see that a risk-averse person tends to invest in both assets while a risk-loving person will always invest all his money in one asset only (Fig. 10).

Fig. 9 The portfolio curves

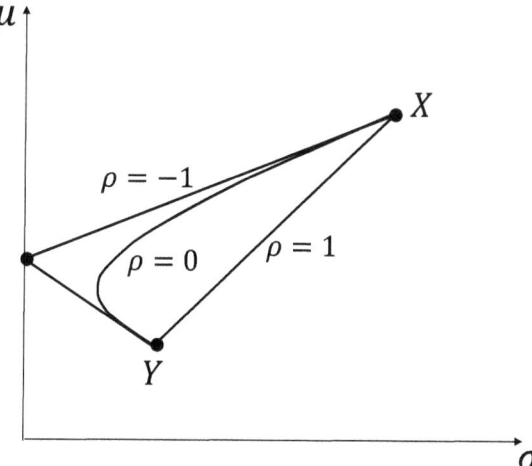

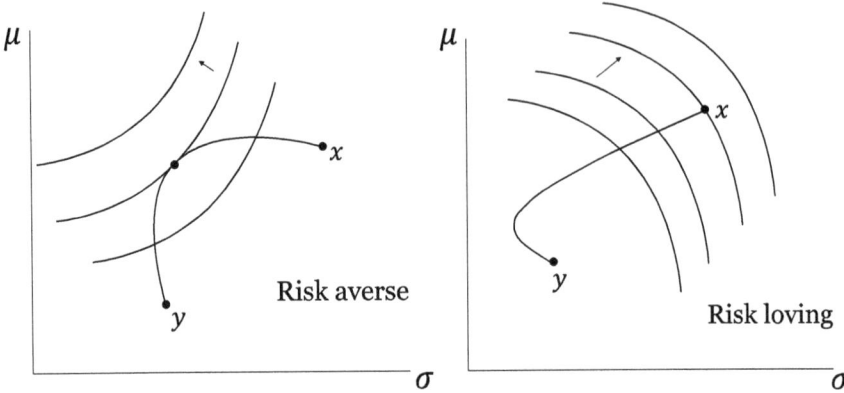

Fig. 10 Portfolio decisions

Proposition 3.5 *A risk-averse person tends to invest in both assets, while a risk-loving person will always invest in one asset only.*

Notes
Good references for Sect. 6 are Campbell (1987) and Laffont (1995), who offer detailed discussions on insurance problems under various conditions. A good reference for Sect. 7 is Huang and Litzenberger (1988, 20–25). A good reference for Sect. 8 is Huang and Litzenberger (1988, Chap. 3). A good reference for Sect. 5 is Mas-Colell et al. (1995, 194–199).

General Equilibrium Theory 4

This chapter deals with three subjects: general equilibrium, Pareto optimality, and welfare properties.

1 The General Equilibrium Concept

Previous chapters assume exogenous prices, that is, consumers and producers take prices as given and choose their best actions based on their own budget/resource constraints. In this chapter, all prices are endogenous and they adjust to clear all markets. An equilibrium is reached when all markets are clear and when no one wants to change anymore.

This equilibrium concept is applicable to an individual market, an industry, a sector, or the whole economy. Given an economic model, we say that we have a general equilibrium if there is an equilibrium in each of the individual markets and there is an overall equilibrium in the whole economy.[1] A general equilibrium takes into account the functioning of individual markets as well as the interactions among all markets. An overall equilibrium for the whole economy is necessary since, although some prices can be determined by individual markets, there are always some prices that must be determined by the whole economy. It is the interactions of markets that affect the determination of these prices. An equilibrium in an individual market is called a partial equilibrium. The crucial difference between a partial equilibrium and a general equilibrium is that in a partial equilibrium some prices are given while in a general equilibrium all prices are endogenously determined within the model.

---

[1] A necessary condition for the whole economy to be in equilibrium is that each individual market is in equilibrium. Hence, we can simply define a general equilibrium to be an equilibrium for the whole economy. However, there is a purpose to mention individual markets here.

The concept of *goods* considered here is very broad. Services, such as banking services, are taken to be just another kind of good. Goods can be distinguished by time, location, and the state of the world. This means that the value of a good is not only determined by its physical characteristics, but also by the location where it is delivered and consumed, by the time when it is produced and consumed, and by the environment in which it is made available. For example, a good consumed today and the same good consumed in the future are considered as two different goods. By this approach, a dynamic problem can be dealt with in a one-period model.

General equilibrium theory applies to the Arrow-Debreu world, in which a few ideal conditions on the economic environment must be satisfied. The markets are complete if there is a market for each good; more precisely, if any affordable consumption bundle is available in the markets. The markets are perfect if there is no friction and no distortion such as transaction costs, taxes, etc. Perfect competition requires all economic agents, including consumers and producers, to take prices as given. There are no consumption externalities if each consumer is concerned with his own consumption only (private consumption). There are no production externalities if each firm's production function depends on its own inputs only. Information is symmetric if all economic agents have the same information. Information is complete if all economic agents' preferences and payoffs are common knowledge. An economy is the so-called Arrow-Debreu world if information is symmetric and complete, there are no externalities, and markets are complete, perfect, and competitive.[2]

2 GE in a Pure Exchange Economy

We first establish general equilibrium theory in a pure exchange economy in which there is no production.

A pure exchange economy consists of

(1) k commodities $j = 1, 2, \ldots, k$;
(2) n consumers $i = 1, 2, \ldots, n$;
(3) each consumer i has an endowment $w_i \in \mathbb{R}_+^k$ of goods, a consumption space $\mathbb{R}_+^k$, and a utility function $u_i : \mathbb{R}_+^k \to \mathbb{R}$ representing his preferences $\succsim_i$.

Definition 4.1 Any $x = (x_1, x_2, \ldots, x_n)$, $x_i \in \mathbb{R}_+^k$ for agent i, is an allocation. ■

Definition 4.2 An allocation $x = (x_1, x_2, \ldots, x_n)$ is feasible if $\sum_{i=1}^n x_i \leq \sum_{i=1}^n w_i$. ■

---

[2]See Campbell (1988) for such a definition.

2 GE in a Pure Exchange Economy

Denote x_i^j as consumption of person i for good j. If there are only two goods $j = 1, 2$ and two agents $i = A, B$, then the economy can be conveniently illustrated by the <u>Edgeworth box</u>; see Fig. 1.
The Edgeworth box consists of points defined by

$$x_A^1 + x_B^1 = w_A^1 + w_B^1, \quad x_A^2 + x_B^2 = w_A^2 + w_B^2, \quad x_i^j \geq 0, \quad i = A, B; \quad j = 1, 2.$$

The Edgeworth box has two distinct features:

1. Each point represents a feasible allocation.
2. Given a price vector, the budget line for one person happens to be the budget line for the other person. The budget line for person A is

$$p_1 x_A^1 + p_2 x_A^2 = p_1 w_A^1 + p_2 w_A^2.$$

By the feasibility conditions, this becomes

$$p_1 \left(w_A^1 + w_B^1 - x_B^1 \right) + p_2 \left(w_A^2 + w_B^2 - x_B^2 \right) = p_1 w_A^1 + p_2 w_A^2,$$

implying

$$p_1 x_B^1 + p_2 x_B^2 = p_1 w_B^1 + p_2 w_B^2,$$

which is the budget line for person B.

Because of these useful features, the Edgeworth box is often used to illustrate equilibria in a two-person two-good pure exchange economy.

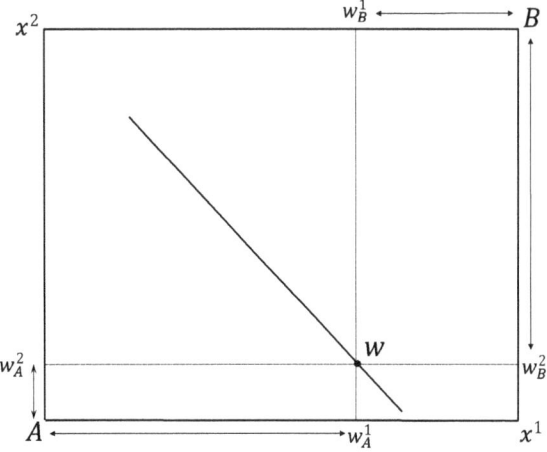

Fig. 1 The Edgeworth box

Suppose that for each good, there is a market for economic agents to buy and sell the good. The vector of market prices for the k goods is $p = (p_1, p_2, \ldots, p_k)$. Each consumer takes prices as given and chooses his most preferred bundle of goods from his consumption set; that is, each consumer i acts as if he is solving the following problem:

$$\max_{x_i \geq 0} u_i(x_i) \tag{1}$$
$$\text{s.t. } p \cdot x_i \leq p \cdot w_i.$$

The optimal solution to this problem is the Marshallian demand function $x_i(p, p \cdot w_i)$. Under the assumption of strict convexity of preferences and by Proposition 2.3, the demand function is a well-defined continuous function. The equilibrium price p^* is a solution of the following inequalities:

$$\sum_{i=1}^{n} x_i(p^*, p^* \cdot w_i) \leq \sum_{i=1}^{n} w_i. \tag{2}$$

Since the ordinary demand $x_i^*(p, p \cdot w_i)$ is zero homogeneous in prices, it means that if p^* is an equilibrium price vector, then λp^* for any $\lambda > 0$ is also an equilibrium price vector. This means that the equilibrium price vector is not unique. In fact, under normal conditions, one of the prices can be completely free and the rest of the $k-1$ prices are uniquely determined; alternatively speaking, the $k-1$ price ratios are uniquely determined.

It will be shown later that if all goods are desirable, then the inequalities in (2) become equalities, that is, the equilibrium price clears all markets. Equilibrium allocation x^* is the vector of the quantities demanded at the equilibrium prices: $x_i^* = x_i(p^*, p^* \cdot w_i)$, $\forall i$. And the pair (p^*, x^*) is called the general equilibrium. That is, (p^*, x^*) is a general equilibrium if it satisfies (1) and (2) jointly. The following is the formal definition.

Definition 4.3 (x^*, p^*) is a general equilibrium (GE) if

(1) For each i, x_i^* solves consumer i's utility maximization problem:

$$\max_{x_i \geq 0} u_i(x_i)$$
$$\text{s.t. } p^* \cdot x_i \leq p^* \cdot w_i.$$

2 GE in a Pure Exchange Economy

(2) x^* is feasible:
$$\sum_{i=1}^{n} x_i^* \leq \sum_{i=1}^{n} w_i. \blacksquare$$

We now try to illustrate GE in the Edgeworth box. Given endowment w_i, the function $\varphi_i(p) \equiv x_i(p, p \cdot w_i)$ defines the so-called <u>offer curve</u>. We can draw the offer curves in the Edgeworth box; see Fig. 2. The general equilibrium occurs at the intersection point of the two offer curves. Each point on a consumer's offer curve is the best consumption bundle for the consumer given the price vector. Only at the intersection point of the offer curves do the two consumers choose the same consumption bundle, which means that the allocation is feasible. It is a general equilibrium since the two conditions in Definition 4.3 are satisfied.

We are going to show the existence of the general equilibrium. For that, we first need to introduce a fixed-point theorem and two lemmas. Let the aggregate <u>excess demand function</u> be:

$$z(p) \equiv \sum_{i=1}^{n} x_i(p, p \cdot w_i) - \sum_{i=1}^{n} w_i.$$

Then, the equilibrium price vector p^* is defined by $z(p^*) \leq 0$. Given the demand functions, the existence of an equilibrium is equivalent to the existence of an equilibrium price vector, and the equilibrium allocation is the demand at the equilibrium prices. For this reason, we often simply call the equilibrium price vector p^* the equilibrium.

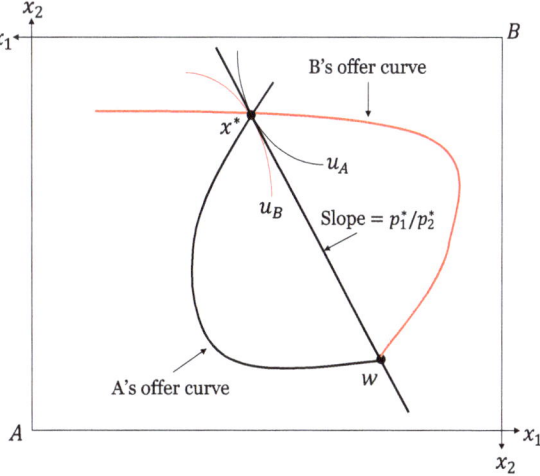

Fig. 2 The general equilibrium in the Edgeworth box

Proposition 4.1 (Walras's Law). *If preferences are strictly monotonic, then, for any price p, we have $p \cdot z(p) = 0$; i.e., the value of the excess demand is always zero.*

Proof By strict monotonicity of preferences, more is always better. Hence, the budget equation must be binding for the optimal consumption: $p \cdot x_i(p, p \cdot w_i) = p \cdot w_i$, $\forall i$. We thus have

$$p \cdot z(p) = \sum \left[p \cdot x_i(p, p \cdot w_i) - p \cdot w_i \right] = 0. \qquad \blacksquare$$

Walras's Law actually says something quite obvious: if each individual balances his budget, then the aggregate spending balances as well. An important implication of Walras's Law is this: if, at some $p \gg 0$, $k-1$ markets clear, then the remaining market must also clear. Technically, this means that condition $z(p^*) = 0$ only implies $k-1$ independent equations; the other equation is satisfied automatically. Since the demand function is zero homogenous, only price ratios $p_1/p_k, p_2/p_k, \ldots, p_{k-1}/p_k$ need to be determined. Hence, the $k-1$ equations are just enough to determine p^*.

The following fixed-point theorem can be found in Smart (1980).

Lemma 4.1 (Brouwer). *For any nonempty compact convex set $A \subset \mathbb{R}^n$ and any continuous mapping $f : A \to A$, there is an $x \in A$ such that $x = f(x)$.*

Lemma 4.2 *Let $\Delta^{k-1} \equiv \{p \in \mathbb{R}_+^k \,|\, \sum p_i = 1\}$. If a mapping $f : \Delta^{k-1} \to \mathbb{R}^k$ is continuous and satisfies $p \cdot f(p) = 0, \forall p \in \Delta^{k-1}$, then there is a $p^* \in \Delta^{k-1}$ such that $f(p^*) \leq 0$.*

Proof Define $g : \Delta^{k-1} \to \Delta^{k-1}$ by

$$g(p) = \frac{1}{\sum_i \left[\max\{0, f_i(p)\} + p_i \right]} \left[\max\{0, f(p)\} + p \right],$$

where

$$\max\left\{0, f(p)\right\} \equiv \begin{pmatrix} \max\{0, f_1(p)\} \\ \vdots \\ \max\{0, f_k(p)\} \end{pmatrix}.$$

By Brouwer's fixed-point theorem, there is a p^* such that $g(p^*) = p^*$. Together with the fact that $\sum p_i^* = 1$, this implies

$$\max\{0, f(p^*)\} = p^* \sum_i \max\{0, f_i(p^*)\}.$$

Since $p^* \cdot f(p^*) = 0$, by multiplying $f(p^*)$ on both sides of the above equation, we have

$$\sum f_i(p^*) \max\{0, f_i(p^*)\} = 0. \tag{3}$$

Since $f_i(p^*) \max\{0, f_i(p^*)\} \geq 0$ for any i, (3) implies $f_i(p^*) \max\{0, f_i(p^*)\} = 0$ for any i. Thus, $f(p^*) \leq 0$. ∎

Theorem 4.1 (Existence of Equilibrium). *If preferences are strictly convex, strictly monotonic, and continuous, then an equilibrium exists.*

Proof By Proposition 2.3, $z : \mathbb{R}_+^k \to \mathbb{R}^k$ is continuous. By Proposition 4.1, z satisfies Walras's law. Hence, by Lemma 4.2, $z : \Delta^{k-1} \to \mathbb{R}^k$ has an equilibrium. ∎

A good is said to be desirable if $p_j = 0$ implies $z^j(p) > 0$.

Proposition 4.2 (Market Clearing). *Suppose that preferences are strictly monotonic. If good j is desirable, then, in the general equilibrium, we have $p_j^* > 0$ and the market for that good must clear: $z^j(p^*) = 0$.*

Proof By Walras's Law, $p^* \cdot z(p^*) = 0$. Since $z(p^*) \leq 0$, we must have $p_j^* z^j(p^*) = 0$ for all j. Then, by desirability, we cannot have $p_j^* = 0$, implying $z^j(p^*) = 0$. ∎

Proposition 4.2 tells us that if preferences are strictly monotonic and if all goods are desirable, then, in general equilibrium, all markets clear.

Theorem 4.2 (Uniqueness of Equilibrium). *If preferences are strictly monotonic, all demand functions are differentiable, and all goods are desirable and substitutes in excess demand ($\partial z^i / \partial p_j > 0$ for all $i \neq j$), then the equilibrium price vector is unique up to a positive multiplier (if p^* and p' are equilibrium prices, then there is $\lambda > 0$ such that $p' = \lambda p^*$). That is, equilibrium price ratios $(p_1^*/p_k^*, \ldots, p_{k-1}^*/p_k^*)$ are unique.*

Proof For any equilibrium price vector p^*, since z is zero homogeneous, any price $p' = \lambda p^*$, $\lambda > 0$, is also an equilibrium price.

Conversely, given any two equilibrium price vectors p' and p^*, if $p' \neq \lambda p^*$ for some $\lambda > 0$, then we find a contradiction. By desirability, $p^* \gg 0$ and $z(p^*) = 0$. Let $m \equiv \max_j \{p_j'/p_j^*\}$. Suppose $m = p_1'/p_1^*$. By zero homogeneity, we have $z(mp^*) = 0$ and thus $z^1(mp^*) = 0$. By the definition of m, $p_j' \leq mp_j^*$, $\forall j$, and $p_1' = mp_1^*$. Since p' and p^* are not proportional, $\exists l$ such that $p_l' < mp_l^*$. We now lower each price mp_j^* to p_j'. Since at least one price, p_l, goes down strictly and p_1 is constant, by gross substitutability and $z^1(mp^*) = 0$, we have $z^1(p') < 0$. This contradicts the fact that p' is an equilibrium price vector and $p_1' \neq 0$. ∎

Example 4.1 There are two goods and two consumers with:

$$u_1(x,y) = xy, \quad w_1 = (10, 20);$$
$$u_2(x,y) = x^2 y, \quad w_2 = (20, 5).$$

Let us find the equilibrium. The ordinary demand functions are:

$$x_1 = \tfrac{I_1}{2p_x}, \quad y_1 = \tfrac{I_1}{2p_y}, \quad x_2 = \tfrac{2I_2}{3p_x}, \quad y_2 = \tfrac{I_2}{3p_y},$$

where the incomes are $I_1 = 10p_x + 20p_y$ and $I_2 = 20p_x + 5p_y$. We have

$$\frac{\partial(x_1 + x_2 - 30)}{\partial p_y} > 0, \quad \frac{\partial(y_1 + y_2 - 25)}{\partial p_x} > 0,$$

implying that the two goods are gross substitutes in excess demand. By Theorem 4.2, the equilibrium price is unique. As explained in Theorem 4.2, only the ratio of the prices is determined by market clearing conditions. Let p represent the price ratio: $p \equiv p_x/p_y$, or equivalently, $p \equiv p_x$ and $p_y = 1$. Then,

$$x_1 = 5 + \tfrac{10}{p}, \quad x_2 = \tfrac{40}{3} + \tfrac{10}{3p}.$$

The market clearing condition for good x is:

$$x_1 + x_2 = \bar{x}_1 + \bar{x}_2,$$

where $\bar{x}_1 \equiv 10$ and $\bar{x}_2 \equiv 20$ are the endowments of good x. This condition implies that $p^* = \tfrac{8}{7}$; see Fig. 3.

By Walras's law, the market for good y automatically clears. Therefore, a typical equilibrium price vector is $\left(p_x^*, p_y^*\right) = \lambda(8/7, 1)$, $\lambda > 0$. The equilibrium price is unique up to a positive multiplier. By substituting the price ratio into the demand functions, we have the equilibrium allocation: $x_1^* = 13.75$, $x_2^* = 16.25$, $y_1^* = \tfrac{110}{7}$, $y_2^* = \tfrac{65}{7}$. ∎

Example 4.2 There are two goods and two consumers with:

$$u_1(x,y) = \min\{x,y\}, \quad w_1 = (40, 0);$$
$$u_2(x,y) = \min\{x,y\}, \quad w_2 = (0, 20).$$

Let us find the equilibria. Different from the last example, because of the specialty of the utility functions, we are going to use the Edgeworth box to find the equilibria; see Fig. 4. Notice that the two goods in this case are gross complements and thus the uniqueness theorem is not applicable.

Let $p \equiv p_x/p_y$, which is the slope of the budget line going through the endowment point W. We can see that, when $0 < p \leq \infty$, demands from the two consumers are

Fig. 3 A unique equilibrium

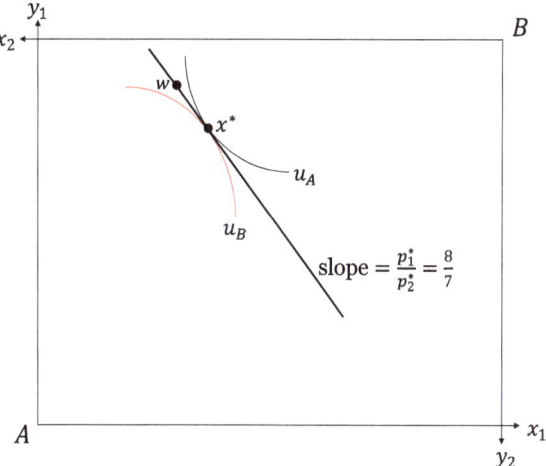

Fig. 4 General equilibria

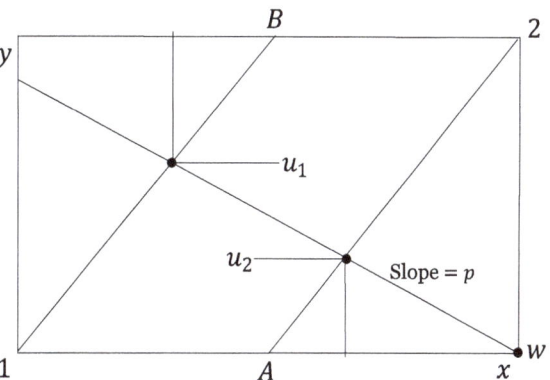

not the same point in the Edgeworth box. Only when $p = 0$ can demands from the two consumers be the same point. Therefore, equilibria are those with price ratio $p^* = 0$ and allocations on the interval $1A$. ∎

3 Pareto Optimality

Is such an equilibrium desirable for the society as a whole? Since a society includes many consumers, in what sense is an equilibrium a good solution for the society? Should we maximize a joint utility function? It turns out that Pareto optimality is a minimum group optimization criterion in the sense that any sensible criterion for group optimization will have it as a necessary condition.

Definition 4.4 A feasible allocation x is <u>Pareto optimal (PO)</u> or <u>weakly Pareto optimal</u> if there is no other feasible allocation x' such that $x'_i \succ_i x_i$, $\forall i$. That is, we can no longer make everyone better off. ∎

Definition 4.5 A feasible allocation x is <u>strongly Pareto optimal</u> if there is no other feasible allocation x' such that

(1) $x'_i \succsim_i x_i$, $\forall i$;
(2) there is a i_0 s.t. $x'_{i_0} \succ_{i_0} x_{i_0}$.

That is, we can no longer make anyone better off without hurting others. ∎

These definitions of Pareto optimality are just like the definition of technological efficiency in producer theory, except that Pareto optimality is defined for preferences $\succ_i$, while technological efficiency is defined for vector inequality $>$.

Example 4.3 Suppose that there is only one good and two agents with strictly increasing preferences. Individual 1's consumption is $x_1 \in \mathbb{R}$ and individual 2's consumption is $x_2 \in \mathbb{R}$. The allocation is a vector $x = (x_1, x_2) \in \mathbb{R}^2$. The feasible set of allocations is the shaded area in Fig. 5. Then, the boundary $ABCD$ is the set of weakly PO allocations and the BC boundary is the set of strongly PO allocations. ∎

Example 4.4 Suppose that there are two goods and two agents with utility functions $u_1(x, y) = xy$ and $u_2(x, y) = 1$. These two utility functions are continuous but not strictly monotonic. Because of the failure of monotonicity, weakly PO allocations may not be strongly PO. As shown in Fig. 6, for these two individuals, all the points in the Edgeworth box are weakly Pareto optimal, but only point 2 is strongly Pareto optimal. ∎

Fig. 5 One good and two agents

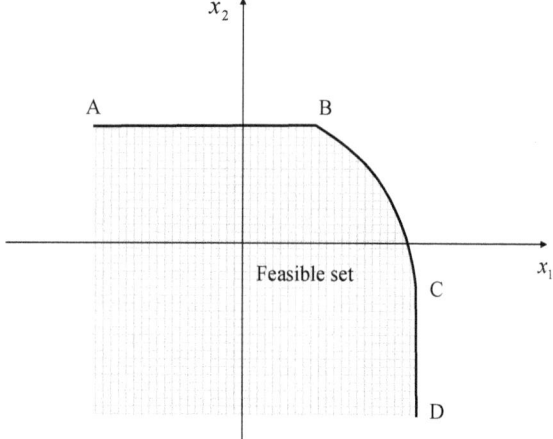

Fig. 6 Weak Pareto optimality $\not\Rightarrow$ strong Pareto optimality

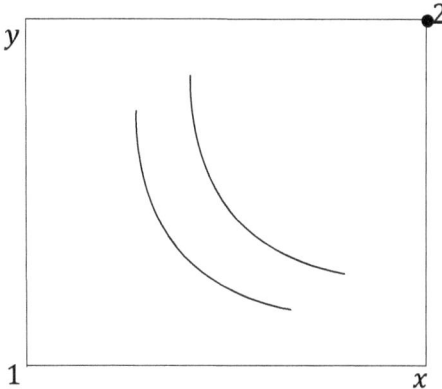

Proposition 4.3 *A strongly Pareto optimal allocation is weakly Pareto optimal. Conversely, if preferences for all the consumers are continuous and strictly monotonic, then a weakly Pareto optimal allocation is strongly Pareto optimal.*

Proof To show "weak PO => strong PO," we use a counter proof and show "not strong PO => not weak PO" instead. Given an allocation $\bar{x} = (\bar{x}_1, \ldots, \bar{x}_n)$, if it is not strong PO, there exists a feasible allocation $x = (x_1, \ldots, x_n)$ that makes at least one person, say $i = 1$, better off without hurting others. By continuity, we can take a little bit of the goods away, say $\varepsilon > 0$, from x_1 and yet maintain $x_1 - \varepsilon \succ_1 \bar{x}_1$. We can equally distribute ε to the rest of the population. By strict monotonicity, we have $x_i + \frac{1}{n-1}\varepsilon \succ x_i \succsim \bar{x}_i$. That is, we find a feasible allocation

$$\left(x_1 - \varepsilon, x_2 + \frac{1}{n-1}\varepsilon, \ldots, x_n + \frac{1}{n-1}\varepsilon\right)$$

that makes everyone strictly better off than $\bar{x}$. Thus, $\bar{x}$ is not weakly PO. ∎

Agent i's marginal rate of substitution between two goods l and h is:

$$MRS_i^{lh}(x_i) \equiv \frac{\partial u_i(x_i)}{\partial x_i^l} \bigg/ \frac{\partial u_i(x_i)}{\partial x_i^h} \equiv \text{pay in good } h \text{ for one more unit of good } l.$$

The MRS_i^{lh} is the slope of the indifference curve. It measures the substitutability between two goods.

Proposition 4.4 *Suppose that u_i is differentiable, quasi-concave, and $u_i'(x) > 0$, for all i. Then, a feasible allocation x is Pareto optimal iff*

$$MRS_1^{lh}(x_1) = MRS_2^{lh}(x_2) = \cdots = MRS_n^{lh}(x_n), \quad \forall l, h. \tag{4}$$

Proof Suppose $\bar{x}$ is a Pareto optimal allocation. By the definition of Pareto optimality, $\bar{x}$ is the solution of the following problem:

$$\max_{(x_1,\ldots,x_n)} u_1(x_1)$$
$$\text{s.t. } u_i(x_i) \geq u_i(\bar{x}_i), \quad i \geq 2 \tag{5}$$
$$\sum x_i \leq \sum w_i.$$

The Lagrange function is

$$\mathcal{L} \equiv u_1(x_1) + \sum_{i=2}^{n} \lambda_i [u_i(x_i) - u_i(\bar{x}_i)] + \mu \cdot \left(\sum_{i=1}^{n} w_i - \sum_{i=1}^{n} x_i \right).$$

The FOCs are

$$\frac{\partial u_1(\bar{x}_1)}{\partial x_1} = \mu, \quad \lambda_i \frac{\partial u_i(\bar{x}_i)}{\partial x_i} = \mu, \quad \forall i \geq 2,$$

implying

$$\lambda_i \frac{\partial u_i(\bar{x}_i)}{\partial x_i} = \frac{\partial u_1(\bar{x}_1)}{\partial x_1}, \quad \forall i \geq 2,$$

implying

$$MRS_i^{lh}(\bar{x}_i) = MRS_1^{lh}(\bar{x}_1), \quad \forall i \geq 2. \tag{6}$$

Conversely, given (6), let

$$\mu \equiv \frac{\partial u_1(\bar{x}_1)}{\partial x_1}, \quad \lambda_i \equiv \mu^1 / \frac{\partial u_i(\bar{x}_i)}{\partial x_i^1}.$$

Then, by (6), for each l,

$$\lambda_i \frac{\partial u_i(\bar{x}_i)}{\partial x_i^l} = \lambda_i \frac{\partial u_i(\bar{x}_i)}{\partial x_i^1} MRS_1^{l1}(\bar{x}_1) = \mu^1 MRS_1^{l1}(\bar{x}_1) = \frac{\partial u_1(\bar{x}_1)}{\partial x_1^l} = \mu^l.$$

That is,

$$\lambda_i \frac{\partial u_i(\bar{x}_i)}{\partial x_i} = \mu.$$

Thus, (6) implies the FOCs. We have therefore proven the equivalence between the FOCs and (6). By the quasi-concavity of all u_i, the FOCs are necessary and sufficient for the solution of problem (5). ∎

Proposition 4.4 can be easily illustrated by Fig. 7. First, if two indifference curves at an allocation point are tangents, we can see from the left figure of Fig. 7 that we can no longer find another feasible allocation that makes everyone strictly

3 Pareto Optimality

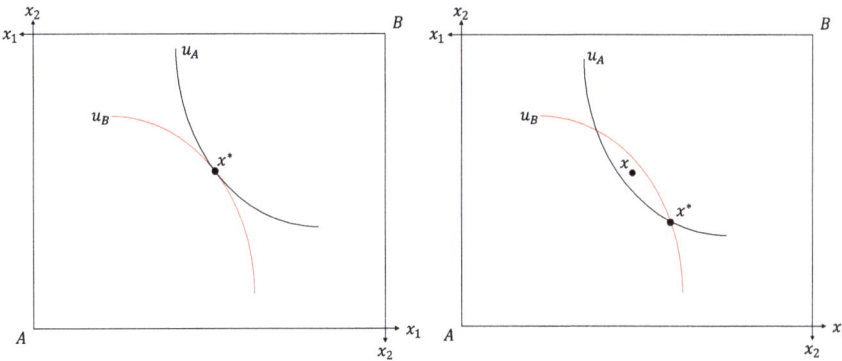

Fig. 7 PO allocations

better off. Thus, the allocation x^* is PO. Conversely, if the two indifference curves intersect at x^*, as shown in the right figure of Fig. 7, then we can find another allocation x that makes everybody strictly better off.

The <u>contract curve</u> is the set of all Pareto optimal allocations. Figure 8 shows such a curve.

Example 4.5 For the consumers in Example 4.1, we now find the contract curve. The feasible allocations satisfy

$$x_1 + x_2 = 30, \quad y_1 + y_2 = 25.$$

The PO allocations are determined by

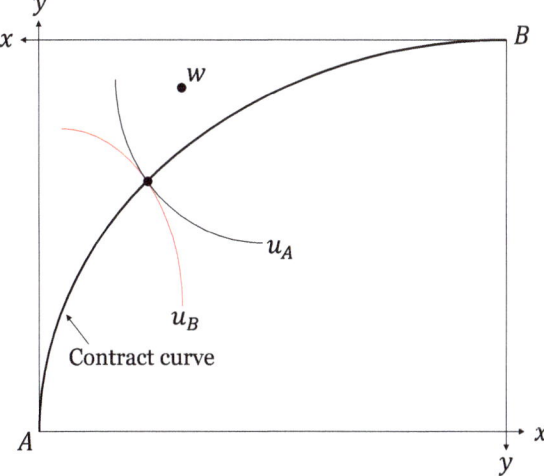

Fig. 8 The contract curve

$$MRS_1^{xy} = MRS_2^{xy} \Rightarrow \frac{y_1}{x_1} = \frac{2x_2y_2}{(x_2)^2} \Rightarrow \frac{y_1}{x_1} = \frac{2(25-y_1)}{30-x_1} y_1 = \frac{50x_1}{30+x_1}.$$

Thus, $y_1 = \frac{50x_1}{30+x_1}$ for $x_1 \in [0, 30]$ defines the so-called contract curve, as shown in Fig. 8. We have

$$\frac{dy_1}{dx_1} = \frac{1500}{30+x_1} > 0, \quad \frac{d^2y_1}{dx_1^2} = -\frac{3000}{(30+x_1)^2} < 0.$$

That is, the contract curve is increasing and concave. ∎

Example 4.6 For the consumers in Example 4.2, we now find the contract curve. Since the utility functions are not differentiable, to find PO allocations, we use a graphical means. In Fig. 9, point a is a weakly PO point if the interior areas of A and B (better-off areas) have no common points. Point a is a strongly PO point if the interior area of A and the indifference curve defined by u_1 have no common points and also the interior area of B and the indifference curve defined by u_2 have no common points. We thus find that the shaded area in Fig. 10 is the set of PO allocations. Furthermore, the weakly PO allocations are all strongly PO. ∎

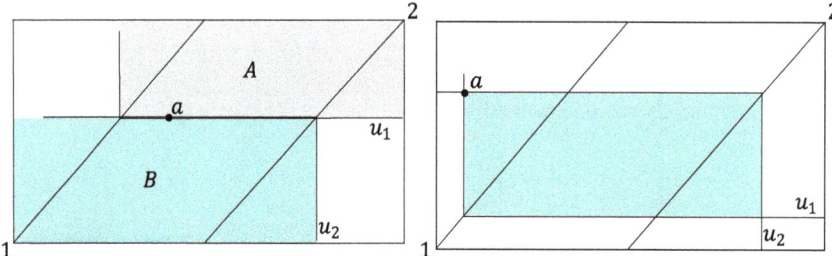

Fig. 9 Find PO points using indifference curves

Fig. 10 The Pareto optimal set

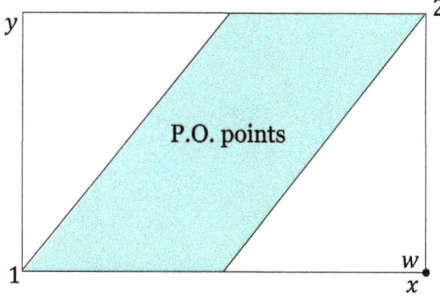

4 Welfare Theorems

We now present two welfare theorems on Pareto optimality of general equilibria.

Theorem 4.3 (First Welfare Theorem). *If (x^*, p^*) is a general equilibrium, then x^* is Pareto optimal.*[3]

Proof Suppose not. Let x' be a feasible allocation that all agents strictly prefer x' to x^*. Then, since x_i^* is the best choice among the consumption bundles inside the budget set, x_i' must be outside the budget set:

$$p^* \cdot x_i' > p^* \cdot w_i, \quad i = 1, 2, \ldots, n \quad \Rightarrow \quad p^* \cdot \sum x_i' > p^* \cdot \sum w_i.$$

But, by $\sum x_i' \leq \sum w_i$, we have

$$p^* \cdot \sum x_i' \leq p^* \cdot \sum w_i.$$

This is a contradiction. ∎

The First Welfare Theorem indicates that the general equilibrium in Fig. 3 must be on the contract curve in Fig. 8. This situation is shown in Fig. 11. Similarly, the GEs in Fig. 4 must be in the region of PO points in Fig. 10. This situation is shown in Fig. 12.

The First Welfare Theorem tells us that a general equilibrium must be Pareto optimal. The next theorem shows the converse.

Theorem 4.4 (Second Welfare Theorem). *Suppose that preferences are continuous, strictly monotonic, and strictly convex. Then, any Pareto optimal allocation x^* can become a general equilibrium allocation by a redistribution of endowments.*

Proof Given wealth levels $w_i \equiv x_i^*, \forall i$, by Theorem 4.1, there exists a GE (x', p'). Since x_i^* satisfies the budget constraint $p' \cdot x_i \leq p' \cdot w_i$ for each consumer, we must have $x_i' \succsim_i x_i^*, \forall i$. If there were j such that $x_j' \succ_j x_j^*$, it would contradict with the Pareto optimality of x^* (strong PO is the same as weak PO in this case). Therefore, $x_i' \sim_i x_i^*, \forall i$. Thus, since x_i' provides maximum utility on the budget line, so does x_i^*. Also, x^* is obviously feasible. Hence, (p', x^*) is a general equilibrium. ∎

The second welfare theorem can be illustrated by Fig. 13. For any point x^* on the contract curve, if the indifference curves are convex and continuous, we can then find a straight line with a positive slope going through point x^* and separating the two indifference curves. The slope of this straight line is the equilibrium price ratio. The original endowment point w may not happen to be on this straight line. We can arbitrarily pick a point w_0 on this straight line as the new endowment point. This straight line becomes a budget line for the two consumers. By this, x^* becomes

---

[3]If preferences are strictly monotonic, a general equilibrium allocation is strongly PO.

Fig. 11 GE on the contract curve

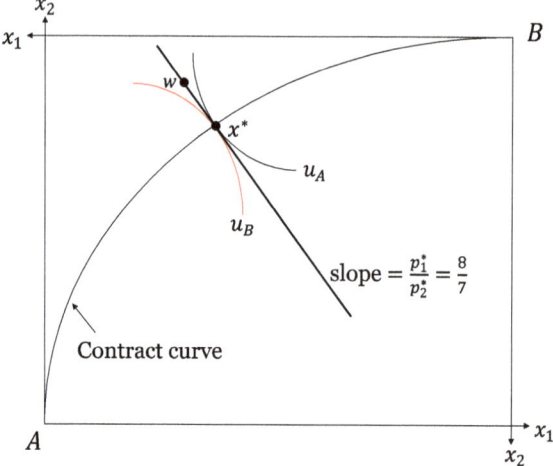

Fig. 12 GEs in the contract region

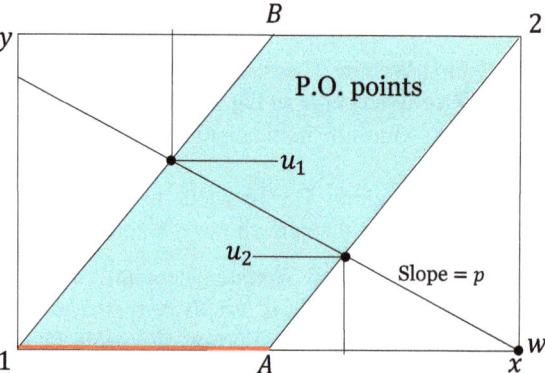

Fig. 13 The second welfare theorem

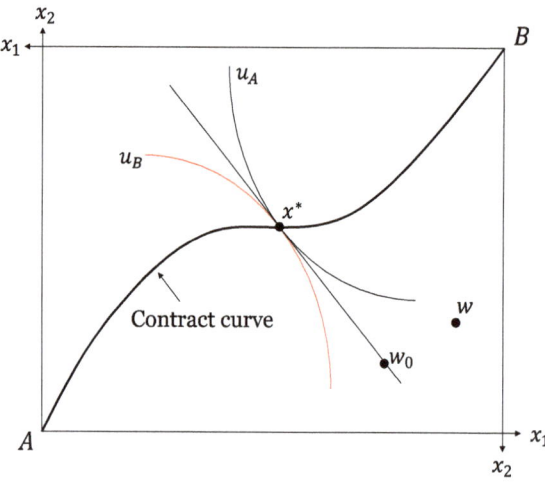

4 Welfare Theorems

an equilibrium allocation. Here, changing the endowment point from w to w_0 is what we mean by a redistribution of endowments.

Why is a redistribution of endowments sometimes necessary? Imagine a PO allocation $x_1^* = x_2^* = (1, 1)$ for both agents. If the original endowments are $w_1 = (0, 0)$ and $w_2 = (2, 2)$, we will not be able to find a price vector that supports (x_1^*, x_2^*), since the second agent has nothing to gain by trading with the first agent in the markets. Hence, in order for (x_1^*, x_2^*) to become a GE allocation, we need to have a reallocation of endowments.

Why is convexity of preferences necessary? In Fig. 14, we have two indifference curves, one of which has an unusual shape. This unusual shape means that the consumer's preferences are not convex. As a result, we can see from the figure that an equilibrium price does not exist no matter where the endowment point is—we cannot find a straight line that separates the two indifference curves. This example shows the necessity of convexity of preferences.

Pareto optimality is only concerned with the optimality of an allocation, but it has nothing to say about social welfare. There are generally many Pareto optimal allocations. Even if we agree that we should have a Pareto optimal allocation, we still do not know which Pareto optimal allocation we should have.

To deal with the issue, we assume that there exists a <u>social welfare function</u> $W : \mathbb{R}^n \to \mathbb{R}$ that aggregates individual utility values $u_1, u_2, \ldots, u_n$ to a social welfare value $W(u_1, u_2, \ldots, u_n)$. In other words, for each distribution $(u_1, u_2, \ldots, u_n)$ of individual utilities, there is a social welfare value. As it should be, we will assume that W is strictly increasing. An allocation x^* is said to be <u>socially optimal (SO)</u> if it solves:

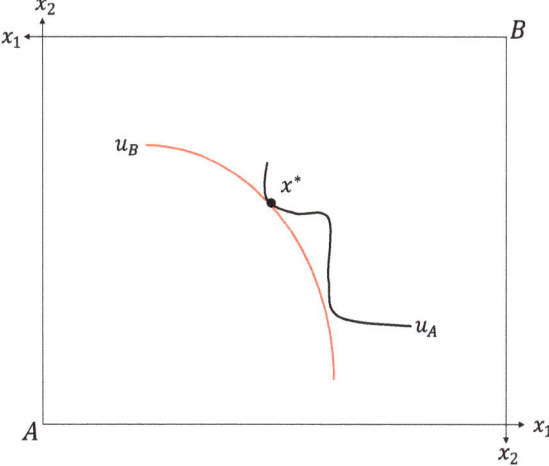

Fig. 14 Nonconvexity of preferences

$$\max_{(x_1,\ldots,x_n)} \quad W[u_1(x_1), u_2(x_2), \ldots, u_n(x_n)]$$
$$\text{s.t.} \quad \sum_{i=1}^n x_i \le \sum_{i=1}^n w_i.$$

That is, a socially optimal allocation maximizes social welfare subject to the resource constraint. Similar to the consideration of the relation between Pareto optimality and the general equilibrium, we now investigate the relation between Pareto optimality and social optimality. We will have two results parallel to the first and second welfare theorems.

Proposition 4.5 *If x^* is socially optimal, then x^* is strongly Pareto optimal.*

Proof Suppose not. Then there would be some feasible allocation x' such that $u_i(x'_i) \ge u_i(x^*_i), \forall i$; and $u_j(x'_j) > u_j(x^*_j)$ for some j. By strict monotonicity of W, this contradicts the fact that x^* maximizes the social welfare function. ∎

Proposition 4.5 comes directly from the monotonicity of the social welfare function. The converse of Proposition 4.5 holds under certain conditions.

Proposition 4.6 *Suppose that preferences are continuous, strictly monotonic, and strictly convex. Then, for any Pareto optimal allocation x^* with $x^*_i \gg 0, \forall i$, there are weights $a_i > 0, i = 1, \ldots, n$, such that x^* solves*

$$\max_{(x_1,\ldots,x_n)} \quad \sum a_i u_i(x_i)$$
$$\text{s.t.} \quad \sum x_i \le \sum x^*_i. \quad (7)$$

That is, x^* is socially optimal given the social welfare function $W = \sum a_i u_i(x_i)$.

Proof By the second welfare theorem, there is a price vector p^* such that (x^*, p^*) is a general equilibrium with endowments $(x^*_1, x^*_2, \ldots, x^*_n)$. This implies that x^*_i solves

$$v_i(p^*, p^* \cdot x^*_i) \equiv \max u_i(x_i)$$
$$\text{s.t.} \quad p^* \cdot x_i \le p^* \cdot x^*_i.$$

By the Lagrange theorem, there exist $\lambda_i, i = 1, \ldots, n$, such that

$$u'_i(x^*_i) = \lambda_i p^*, \quad i = 1, 2, \ldots, n. \quad (8)$$

Since u_i is strictly increasing by assumption, (8) implies that $\lambda_i > 0$ for all i. The Kuhn-Tucker conditions further $p^* \cdot x_i = p^* \cdot x^*_i$ for all i. Consider

$$\max_{(x_1,\ldots,x_n)} \quad \sum \frac{1}{\lambda_i} u_i(x_i)$$
$$\text{s.t.} \quad \sum x_i \le \sum x^*_i \quad (9)$$

4 Welfare Theorems

and take

$$\mathcal{L} \equiv \sum \frac{1}{\lambda_i} u_i(x_i) + \mu \cdot \left(\sum x_i^* - \sum x_i \right).$$

By (8), if we take $\mu = p^*$, (x^*, μ) satisfies the FOCs of maximizing $\mathcal{L}$ and the Kuhn-Tucker condition. Therefore, since u_i is quasi-concave, by Wang (2008, Theorem 3.9) or Wang (2015, Theorem 3.11), x^* solves (9). ∎

Figure 15 offers a graphic illustration of Proposition 4.6. Suppose that there are two consumers but one good. Problem (7) can be transformed into the following problem. The curve AB is the contract curve. For any PO point on the curve AB, we can find positive weights a_1 and a_2 such that the PO point is the maximum of $a_1 u_1 + a_2 u_2$ subject to the feasibility condition, as presented in the following problem.

$$\max_{(u_1,\ldots,u_n)} \sum a_i u_i$$

$$\text{s.t.} \quad (u_1,\ldots,u_n) \in \left\{ (u_1(x_1),\ldots,u_n(x_n)) \,\Big|\, \sum x_i \leq \sum x_i^* \right\}.$$

Proposition 4.6 suggests that any PO allocation is socially optimal. However, social optimality is conditional on a social welfare function. A socially optimal allocation with a given social welfare function may not be socially optimal with a different social welfare function. Proposition 4.6 indicates that a PO allocation is socially optimal only with a linear social welfare function and special weights. We know by the Envelope Theorem that λ_i is the marginal utility of income:

$$\lambda_i = \frac{\partial v_i(p^*, p^* \cdot x_i^*)}{\partial I},$$

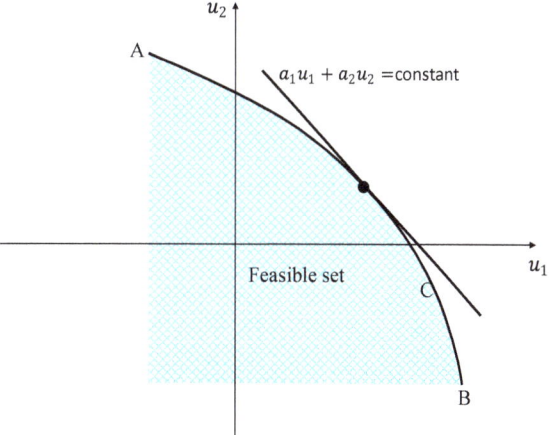

Fig. 15 One good and two agents

i.e., the weights $\{a_i\}$ are the reciprocals of the marginal utilities of income. This means that if an economic agent has a large income, his weight in the social welfare function will be large since his marginal utility of income is small. Hence, the competitive market "favors" individuals with large incomes.

We have now covered all the main ideas on general equilibrium. Here is a brief summary of the relationships among general equilibria, Pareto optimal allocations, and socially optimal allocations:

(1) GE allocations and SO allocations are always Pareto optimal.
(2) PO allocations are GE allocations under convexity of preferences and endowment redistribution.
(3) PO allocations are SO under convexity of preferences and a linear social welfare function with special weights.

The rest of this chapter is to extend these results to more general economies.

5 General Equilibrium with Production

In this section, we allow the economy to have a production sector.

Besides n consumers with utility functions and endowments $(u_1, w_1), \ldots, (u_n, w_n)$, there are m firms $j = 1, \ldots, m$, with production possibilities defined by functions $G_j, j = 1, \ldots, m$. These G_j satisfy Assumptions 1 and 2 in Chap. 1. The firms are assumed to be owned by the n consumers, with δ_{ij} being the share of firm j owned by agent i, $0 \leq \delta_{ij} \leq 1$, and $\sum_i \delta_{ij} = 1, \forall j$. Let $x_i \in \mathbb{R}^k_+$ be consumer i's consumption bundle, $y_j \in \mathbb{R}^k$ be firm j's net output bundle, and $p \in \mathbb{R}^k_+$ be a price vector.

5.1 General Equilibrium with Production

Like the pure exchange economy in Sect. 2, we have parallel definitions for the various concepts.

Definition 4.6 An allocation (x, y), where $x \geq 0$, is <u>feasible</u> if

$$\sum_{i=1}^n x_i \leq \sum_{i=1}^n w_i + \sum_{j=1}^m y_j; \quad G_j(y_j) \leq 0, \quad \forall j.$$

5 General Equilibrium with Production

Definition 4.7 (x^*, y^*, p^*) is a <u>general equilibrium</u> if

(1) For each i, x_i^* solves consumer i's utility maximization problem:

$$\max_{x_i \geq 0} u_i(x_i)$$

$$\text{s.t.} \quad p^* \cdot x_i \leq p^* \cdot w_i + \sum_{j=1}^{m} \delta_{ij} p^* \cdot y_j^*.$$

(2) For each j, y_j^* solves firm j's profit maximization problem:

$$\max_{y_j} p^* \cdot y_j$$

$$\text{s.t.} \quad G_j(y_j) \leq 0.$$

(3) (x^*, y^*) is feasible:

$$\sum_{i=1}^{n} x_i^* \leq \sum_{i=1}^{n} w_i + \sum_{j=1}^{m} y_j^*. \quad \blacksquare$$

Notice here that we have supposed that the consumer's problem can be separated into profit maximization and utility maximization, as indicated by the Fisher separation theorem in Chap. 2.

We can illustrate the equilibrium in an Edgeworth box. Suppose that there is only one firm with net output vector y. Since $y = \sum x_i - \sum w_i$ by feasibility, letting $w \equiv \sum w_i$ and $x \equiv \sum x_i$, the firm's problem can be equivalently written as

$$\pi + p \cdot w = \max_{x} p \cdot x$$

$$\text{s.t.} \quad G(x - w) \leq 0.$$

If the optimal solution from this problem is $\hat{x}$, then this $\hat{x}$ can be allocated to individuals. This profit maximization problem can be shown in the consumption space for (x^1, x^2) in Fig. 16. Since consumption is $x = w + y$, condition $G(y) \leq 0$ becomes $G(x - w) \leq 0$. Note that if $G(0) = 0$, the endowment point is on the production frontier (PPF).

Example 4.7 We add a firm to the economy in Example 4.1. The firm inputs x to produce y by production function $y = \sqrt{x}$. The two consumers share the firm equally. Let $p = p_x$ and $p_y = 1$. We first consider the firm's profit maximization problem[4]:

---

[4]We here have $G(-x, y) \equiv y - \sqrt{x}$. Thus, the PPF is defined by $G(x - w_x, y - w_y) \equiv y - w_y - \sqrt{w_x - x} = 0$.

Fig. 16 General equilibrium

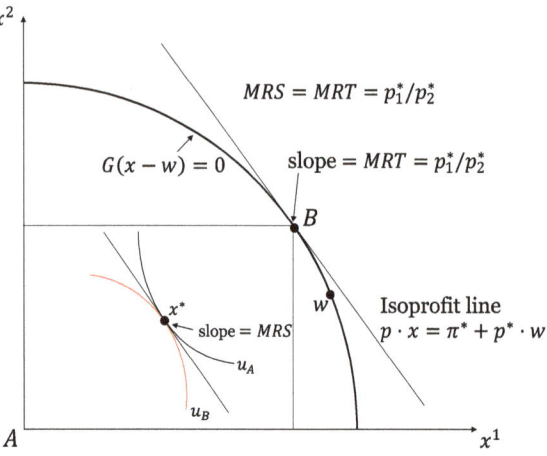

$$\pi = \max_{0 \le x \le 30} \sqrt{x} - px.$$

The solution is

$$x_0 = \frac{1}{4p^2}, \quad y_0 = \frac{1}{2p}, \quad \pi = \frac{1}{4p}.$$

We next consider the consumers' utility maximization problems. Given incomes I_1 and I_2, the individual demand functions are:

$$x_1 = \frac{I_1}{2p}, \quad y_1 = \frac{I_1}{2}, \quad x_2 = \frac{2I_2}{3p}, \quad y_2 = \frac{I_2}{3}.$$

We have

$$I_1 = 10p + 20 + \frac{\pi}{2} = 10p + 20 + \frac{1}{8p}, \quad I_2 = 20p + 5 + \frac{\pi}{2} = 20p + 5 + \frac{1}{8p}.$$

Hence,

$$x_1 = 5 + \frac{10}{p} + \frac{1}{16p^2}, \quad x_2 = \frac{40}{3} + \frac{10}{3p} + \frac{1}{12p^2}.$$

The feasible allocations are

$$x_0 + x_1 + x_2 = 30, \quad y_1 + y_2 = 25 + y_0.$$

The first feasibility condition implies

5 General Equilibrium with Production

$$30 = 5 + \frac{10}{p} + \frac{1}{16p^2} + \frac{40}{3} + \frac{10}{3p} + \frac{1}{12p^2} + \frac{1}{4p^2},$$

implying

$$\frac{19}{16}\left(\frac{1}{p}\right)^2 + 40\frac{1}{p} - 35 = 0,$$

implying

$$p^* \approx 1.17.$$

Using this price, we further find the equilibrium allocation:

$$x_0^* \approx 0.18, \quad x_1^* \approx 13.58, \quad x_2^* \approx 16.24. \quad \blacksquare$$

Example 4.8 Add a firm to the economy in Example 4.2. The firm inputs x to produce y by production function $y = \sqrt{x}$. The two consumers share the firm equally. Let $p = p_x$ and $p_y = 1$. We first consider the firm's profit maximization problem:

$$\pi = \max_{0 \leq x \leq 40} \sqrt{x} - px,$$

which implies

$$x_0 = \tfrac{1}{4p^2}, \quad y_0 = \tfrac{1}{2p}, \quad \pi = \tfrac{1}{4p}.$$

We next consider the consumers' utility maximization problems. The budget equation is $px_i + y_i = I_i$. With the special utility function u_i, each individual must consume at point (x_i, y_i) where $x_i = y_i$. By the budget condition, the individual demand functions are:

$$x_i = y_i = \frac{I_i}{1+p}.$$

We have

$$I_1 = 40p + \tfrac{1}{8p}, \quad I_2 = 20 + \tfrac{1}{8p}.$$

Hence,

$$x_1 = y_1 = \tfrac{1}{1+p}\left(40p + \tfrac{1}{8p}\right), \quad x_2 = y_2 = \tfrac{1}{1+p}\left(20 + \tfrac{1}{8p}\right).$$

The feasibility conditions are

$$x_0 + x_1 + x_2 = 40, \quad y_1 + y_2 = 20 + y_0.$$

The first feasibility condition implies

$$40 = \frac{1}{1+p}\left(40p + \frac{1}{8p}\right) + \frac{1}{1+p}\left(20 + \frac{1}{8p}\right) + \frac{1}{4p^2},$$

which implies

$$p^* = 1/8.$$

Using this price, we can further calculate the equilibrium allocation. ∎

5.2 Efficiency of General Equilibrium

Definition 4.8 A feasible allocation (x, y) is <u>Pareto optimal (PO)</u> if there is no feasible allocation (x', y') such that $x'_i \succ_i x_i, \forall i$. ∎

For a point $y = (y^1, y^2)$ on the production frontier, the slope of the production frontier at this point is the marginal rate of transformation (MRT):

$$MRT(y) = \frac{G_{y^1}(y)}{G_{y^2}(y)}.$$

With two consumers A and B and one firm, we find that, under normal conditions (in Proposition 4.7), a feasible allocation (x_A, x_B, y) is Pareto efficient iff

$$MRS_A(x_A) = MRS_B(x_B) = MRT(y). \tag{10}$$

In other words,

(1) the two indifference curves at $x = (x_A, x_B)$ are tangents of each other, and
(2) the slope of the two indifference curves at x is the same as the slope of the production frontier at y.

This is intuitive. Given net output y, Pareto efficiency implies that the two individuals will divide the total supply $w + y$ of goods so that $MRS_A(x_A) = MRS_B(x_B)$, where w is the total endowment. Further, if

$$MRT(y) > MRS_A(x_A), \tag{11}$$

then we can reduce the output of x^1 by one unit in exchange for an increase of output in x^2 by $MRT(y)$ units. Consumer A is willing to sacrifice one unit of x^1 for an $MRS_A(x_A)$ unit increase in x^2. Thus, such a change in production plan will at least make A better off without hurting B. Thus, the situation in (11) could not happen to a Pareto efficient allocation. Therefore, Pareto efficiency must imply condition (10). Conversely, if (10) holds for an allocation (x_A, x_B, y), then, first, given y, by $MRS_A(x_A) = MRS_B(x_B)$, no change can improve both; second, given x_B, by $MRT(y) = MRS_A(x_A)$, no change can improve both A's welfare and the firm's profitability; finally, given x_A, by $MRT(y) = MRS_B(x_B)$, no change can improve both B's welfare and the firm's profitability. Thus, such an allocation must be Pareto efficient (Fig. 17).

We can easily understand $MRT(y) = MRS_A(x_A)$ with Fig. 18. If $MRT(y) \neq MRS_A(x_A)$, we consider the curve defined by $G(x_A + x_B^* - w) = 0$. This curve will intersect with the u_A-indifference curve at x^*. Then, we can easily find another feasible point that improves consumer A's welfare while consumer B keeps x_B^*.

In general, we have the following result.

Proposition 4.7 *Suppose that u_i is differentiable, quasi-concave, and $u_i'(x) > 0$, for all i, and G_j is differentiable, quasi-convex and $G_j'(y_j) > 0$. Then, a feasible allocation (x, y) is Pareto optimal iff.*

$$MRS_1^{lh}(x_1) = \cdots = MRS_n^{lh}(x_n) = MRT_1^{lh}(y_1) = \cdots = MRT_m^{lh}(y_m), \quad \forall l, h. \quad (12)$$

Example 4.9 We add a firm to the economy in Example 4.1. The firm inputs x to produce y by production function $y = \sqrt{x}$. The two consumers share the firm equally. Let us find the PO allocations. We use index $i = 0$ for the firm. By (12), the condition for PO is

$$\frac{y_1}{x_1} = \frac{2x_2 y_2}{(x_2)^2} = \frac{1}{2\sqrt{x_0}},$$

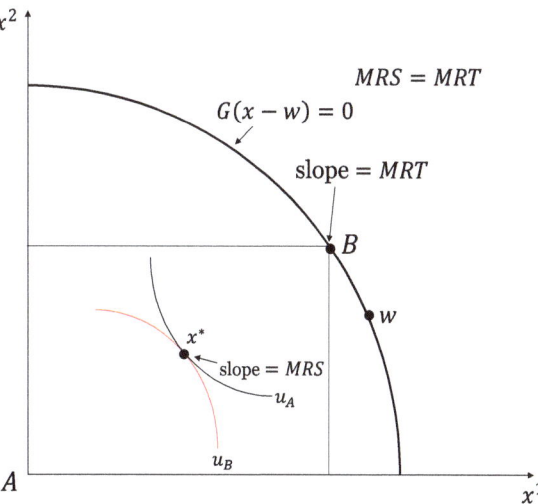

Fig. 17 Pareto efficiency

Fig. 18 A Pareto inefficient allocation

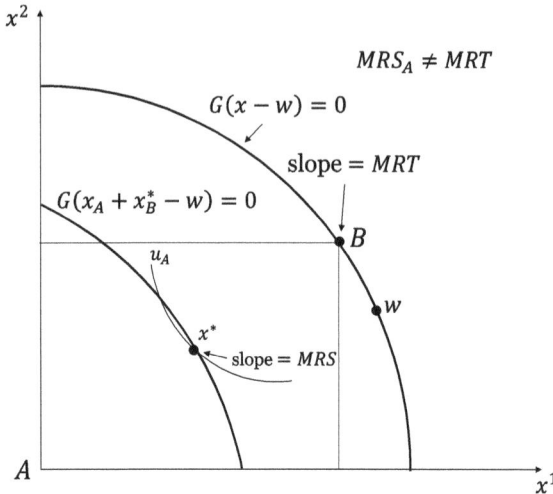

implying

$$y_1 = \frac{x_1}{2\sqrt{x_0}}, \quad \frac{y_1}{x_1} = \frac{2(25 + y_0 - y_1)}{30 - x_0 - x_1}. \tag{13}$$

The second equation of (13) implies

$$y_1 = \frac{50 + 2y_0 x_1}{30 - x_0 + x_1}. \tag{14}$$

Then, using the first equation of (13) and (14) becomes

$$\frac{1}{2\sqrt{x_0}} = \frac{50 + 2\sqrt{x_0}}{30 - x_0 + x_1},$$

implying

$$x_1 = 5x_0 + 100\sqrt{x_0} - 30. \tag{15}$$

Thus, given any x_0, we can get y_0 from $y_0 = \sqrt{x_0}$, x_1 from (15), y_1 from (14), and x_2 and y_2 from the feasibility constraints.

Such a solution is a PO allocation (Fig. 19). Since x_0 is totally free, all points on the PPF belong to PO allocations; but, by (15), for each (x_0, y_0), there is only one point (x_1, y_1) belong to the PO allocation. ∎

Example 4.10 We add a firm to the economy in Example 4.2. The firm inputs x to produce y by production function $y = \sqrt{x}$. The two consumers share the firm equally. Let us find the PO allocations. Given any x_0, if the Edgeworth box defined by $x_0 + x_1 + x_2 = 40$ and $y_1 + y_2 = 20 + \sqrt{x_0}$ is not a square box, then at any point

(x_1, y_1) in the Edgeworth box, one of the consumers will be willing to give up some units of x or y to the firm for free, as shown in Fig. 20. Such an allocation cannot be PO.

Thus, the only possible PO allocation must have a square Edgeworth box, implying

$$40 - x_0 = 20 + y_0,$$

which determines a unique x_0 satisfying

$$20 = x_0 + \sqrt{x_0},$$

implying

$$(\sqrt{x_0} - 4)(\sqrt{x_0} + 5) = 0,$$

implying $x_0 = 16$ and $y_0 = 4$. Thus, the PO allocations $\{(x_0, y_0), (x_1, y_1), (x_2, y_2)\}$ are those with $x_0 = 16$, $y_0 = 4$ and $\{(x_1, y_1), (x_2, y_2)\}$ on the 45°-line of the 24 × 24 Edgeworth box; see Fig. 21. ∎

Similar to the discussions in Sect. 4, we also have first and second welfare theorems. The proofs are also similar to the proofs of Theorems 4.3 and 4.4.

Theorem 4.5 (First Welfare Theorem). *If (x^*, y^*, p^*) is a general equilibrium, then (x^*, y^*) is Pareto optimal.* ∎

To see this result, consider a simple case with two consumers, two goods and one firm. By utility maximization, we have

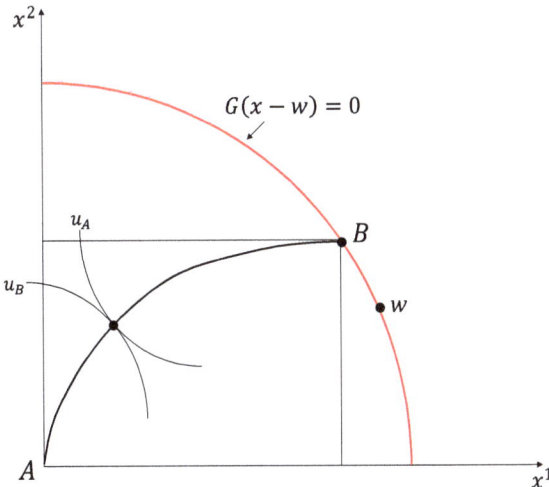

Fig. 19 Pareto optimal allocations

Fig. 20 Pareto optimal allocations

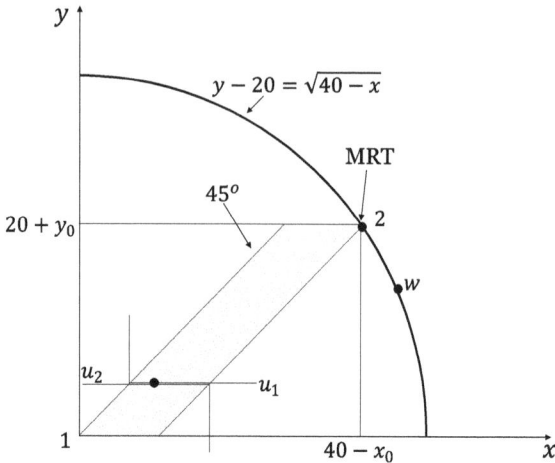

Fig. 21 Pareto optimal allocations

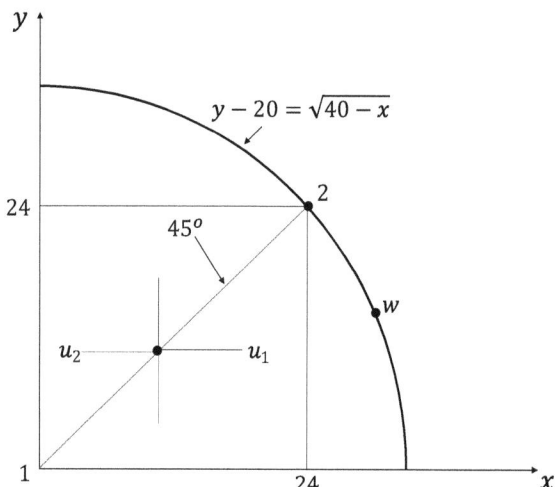

$$MRS_A(x_A) = MRS_B(x_B) = p_1/p_2.$$

Profit maximization implies

$$MRT(y) = p_1/p_2.$$

Hence, in a general equilibrium,

$$MRS_A(x_A) = MRS_B(x_B) = MRT(y) = p_1/p_2. \qquad (16)$$

5 General Equilibrium with Production

Then, according to Proposition 4.7, the equilibrium allocation is Pareto efficient.

How about the converse of Theorem 4.5? That is, is any Pareto efficient allocation a general equilibrium allocation? The answer is yes. Given a Pareto efficiency allocation (x^*, y^*) as shown in Fig. 17, we can draw a straight line through x^* that is tangent to the two indifference curves at x^*. We can choose prices p_1^* and p_2^* such that

$$p_1^*/p_2^* = MRT(y^*) = MRS^i(x^{i*}).$$

The slope of the straight line is p_1^*/p_2^*. Thus, as shown in Fig. 13, given this price vector p^*, each individual maximizes his utility x^* and the firm maximizes its profit at y^*. Finally, we assign profit shares (δ_A, δ_B) and an endowment point (w_A, w_B) such that $p^* \cdot x_i^* = p^* \cdot w_i + \delta_i p^* \cdot y^*$ for $i = A$ and B.[5] Thus, we have shown that any Pareto efficient allocation can be a general equilibrium allocation with a proper distribution of endowments and profit shares. The following is a general result.

Theorem 4.6 (Second Welfare Theorem). *If preferences are continuous, strictly monotonic, and strictly convex, then any Pareto optimal allocation (x^*, y^*) is a general equilibrium allocation with a proper distribution of profit shares and endowments.* ∎

The two results about social optimality in Sect. 4 also hold in this case.

Keep in mind that the number of firms is fixed in the Arrow-Debreu world. Due to this, even though the firms take market prices as given, positive profits are possible in equilibrium. Hence, calling a general equilibrium a general equilibrium is actually a bit misleading. In Chap. 6, the number of firms will be endogenously determined in a long-run equilibrium in a competitive industry, in which case competitive firms will have zero profit in the long run.

6 General Equilibrium with Uncertainty

We now further allow uncertainty in a competitive economy. We will treat the same good in different states of nature as different goods. We label a good in different states by index s. Let $\mathbb{S}$ be the set of possible states of nature or realizations of random events. There are S different possible states: $\mathbb{S} \equiv \{1, 2, \ldots, S\}$. Each state s occurs with probability α^s.

There are k goods, and these goods are different due to their differences in physical descriptions, delivery dates, and locations, but not in terms of the state of nature in which a transaction is made. Given $s \in \mathbb{S}$, consumer i's consumption bundle is a vector $x_i^s \in \mathbb{R}_+^k$. Including different realizations of the state, consumer

---
[5]There are two additional conditions in this assignment: $x_A^* + x_B^* = y^* + w_A + w_B$ and $\delta_A + \delta_B = 1$. We have five equations for six variables. This situation is the same as that in Fig. 13, where the choice of w_0 has one degree of freedom.

i's consumption bundle is $x_i \equiv (x_i^1, x_i^2, \ldots, x_i^S) \in \mathbb{R}_+^{kS}$, from which consumer i gains utility $u_i(x_i)$.

There are n individuals indexed by $i = 1, 2, \ldots, n$. If state s is realized, then individual i will have endowment $w_i^s \in \mathbb{R}_+^k$. Hence, $w_i \equiv (w_i^1, w_i^2, \ldots, w_i^S) \in \mathbb{R}_+^{kS}$ is individual i's endowment profile. Each individual i has a utility function u_i over his consumption space $\mathbb{R}_+^{kS}$. When the utility function u_i has the expected utility property, his utility function is

$$u_i(x_i) = \sum_{s=1}^{S} \alpha^s u_i(x_i^s).$$

There are m firms $j = 1, 2, \ldots, m$, with production possibilities defined by $G_j : \mathbb{R}^{kS} \to \mathbb{R}$, $j = 1, 2, \ldots, m$. The condition for firm j's production possibilities is $G_j(y_j^1, y_j^2, \ldots, y_j^S) \leq 0$.[6] The functions G_j satisfy Assumptions 1 and 2 in Chap 1. The firms are owned by n individuals, with δ_{ij} being the share of firm j owned by agent i, $0 \leq \delta_{ij} \leq 1$, and $\sum_i \delta_{ij} = 1$, $\forall j$. Let $y_j^s \in \mathbb{R}^k$ be firm j's net output bundle in state s, $y_j \equiv (y_j^1, y_j^2, \ldots, y_j^S) \in \mathbb{R}^{kS}$ be firm j's net output bundle, and $p = (p^1, p^2, \ldots, p^S) \in \mathbb{R}_+^{kS}$ be the price vector.

In this economy, contracts are made before the events occur. In other words, consumers buy contingent claims to goods in advance, and firms sell contingent claims to their output in advance with deliveries contingent on the realization of the state of nature. Shareholders (the consumers) receive their share of the profits from contingent sales before the state of nature is realized. This economy is called the contingent contracts economy (Fig. 22).

The markets are markets for contracts or contingent claims. The prices are prices of contingent claims. Market tradings occur ex ante; no trading occurs ex post. After uncertainty is resolved (a state is realized), production, delivery and consumption of goods are made ex post. All contracts are enforceable and ex-post renegotiation is not allowed.

Definition 4.9 An allocation (x, y), where $x \geq 0$, is feasible if

$$\sum_{i=1}^{n} x_i^s \leq \sum_{i=1}^{n} w_i^s + \sum_{j=1}^{m} y_j^s, \quad \forall s; \quad G_j(y_j) \leq 0, \quad \forall j. \quad \blacksquare$$

Definition 4.10 A feasible allocation (x, y) is Pareto optimal (PO) if there is no feasible allocation (x', y') such that $x_i' \succ_i x_i$, $\forall i$. $\blacksquare$

---

[6] We can also assume production possibilities to be dependent on the state. At state s, firm j's production possibilities are defined by $G_j^s : \mathbb{R}^k \to \mathbb{R}$, and the conditions for production possibilities are $G_j^1(y_j^1) \leq 0, \ldots, G_j^S(y_j^S) \leq 0$, instead of $G_j(y_j^1, y_j^2, \ldots, y_j^S) \leq 0$.

6 General Equilibrium with Uncertainty

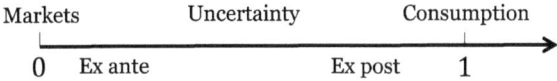

Fig. 22 Contingent contracts markets

Definition 4.11 (x^*, y^*, p^*) is a <u>general equilibrium</u> if

(1) For each i, $x_i^* \in \mathbb{R}_+^{kS}$ solves consumer i's utility maximization problem:

$$\max_{x_i \geq 0} u_i(x_i)$$

$$\text{s.t.} \sum_{s=1}^{S} p^{*s} \cdot x_i^s \leq \sum_{s=1}^{S} p^{*s} \cdot w_i^s + \sum_{s=1}^{S}\sum_{j=1}^{m} \delta_{ij} p^{*s} \cdot y_j^{*s}.$$

(2) For each j, $y_j^* \in \mathbb{R}^{kS}$ solves firm j's profit maximization problem[7]:

$$\max_{y_j} \sum_{s=1}^{S} p^{*s} \cdot y_j^s$$

$$\text{s.t.} \ G_j\left(y_j^1, y_j^2, \ldots, y_j^S\right) \leq 0.$$

(3) (x^*, y^*) is feasible:

$$\sum_{i=1}^{n} x_i^{*s} \leq \sum_{i=1}^{n} w_i^s + \sum_{j=1}^{m} y_j^{*s}, \quad \forall s. \ \blacksquare$$

The three conditions in Definition 4.11 are actually the same as those in Definition 4.7. Hence, this model is actually a special case of that in the last section. This model is more specific in that it specifically labels the difference in the state of nature. The previous model allows this difference but mixes it with other differences in goods. Hence, the first and second welfare theorems still hold; so do the theorems on social optimality.

This contingent contracts economy requires $k \times S$ markets. This system is not informationally efficient since there is a much simpler way of establishing the same equilibrium, in which economic agents need much less information.

In the next chapter, by the introduction of S contingent securities markets, together with k ex post goods markets, the number of markets is reduced from $k \times S$ to $k + S$. A <u>contingent security</u> is a financial asset that pays one unit of money if a specified state is realized; otherwise it pays nothing. We can show that the same equilibrium allocation can be achieved in an economy with k goods markets and S contingent securities markets.

---

[7] The profits here are profits from contingent contracts; as are the consumer expenditures and incomes.

Notes
Materials in this chapter are fairly standard. Many books cover them. Good references for this chapter are Varian (1992), Jehle (2001), Dixit (1990), and Mas-Colell et al. (1995). In particular, good references for Sect. 5 are Campbell (1987, 40–47) and Varian (1992, Chap. 18).

Part II
Micro-Foundation of Markets

Micro-foundation of Finance

5

In this chapter, we look at financial markets. With complete and perfect markets, the models in this chapter are within the Arrow-Debreu world. However, to problems under uncertainty, the approach taken in this chapter is distinctly different from that in Chap. 4. Specifically, we introduce security markets into the Arrow-Debreu world. Financial securities allow economic agents to allocate their portfolios across time and across states of nature. We first see how securities can be used to obtain the same efficient equilibrium allocations as that in the contingent contracts economy of Chap. 4. We then try to price these securities under various circumstances.

1 Security Markets

1.1 Contingent Markets

We have S uncertain states $\{1, 2, \ldots, S\}$ and C commodities $\{1, 2, \ldots, C\}$. The consumption space is $\mathbb{C} \equiv \mathbb{R}^{S \times C}$, with S states and C commodities. There are I agents $\{1, 2, \ldots, I\}$ with endowment $w_i \in \mathbb{C}$ and utility function $u_i : \mathbb{C} \to \mathbb{R}$. There are no firms. Denote

$$\begin{aligned}
x_i &= \left(\left(x_{i1}^1, \ldots, x_{iC}^1 \right), \ldots, \left(x_{i1}^S, \ldots, x_{iC}^S \right) \right) = \left(x_i^1, \ldots, x_i^S \right), \\
p &= \left(\left(p_1^1, \ldots, p_C^1 \right), \ldots, \left(p_1^S, \ldots, p_C^S \right) \right) = \left(p^1, \ldots, p^S \right), \\
w_i &= \left(\left(w_{i1}^1, \ldots, w_{iC}^1 \right), \ldots, \left(w_{i1}^S, \ldots, w_{iC}^S \right) \right) = \left(w_i^1, \ldots, w_i^S \right), \\
x &= (x_1, \ldots, x_I), \\
w &= (w_1, \ldots, w_I).
\end{aligned}$$

Here, p_c^s is the price of commodity c in state s, x_{ic}^s is the consumption of; commodity c by agent i in state s, and w_{ic}^s is the endowment of commodity c to agent i in state s. Thus, p^s is the price vector of the contingent market for commodities x^s.

© Springer Nature Singapore Pte Ltd. 2018
S. Wang, *Microeconomic Theory*, Springer Texts in Business and Economics,
https://doi.org/10.1007/978-981-13-0041-7_5

Consider a market for contingent contracts. Each contract delivers a particular commodity contingent on a particular state. There are a total of $S \times C$ such contracts. Contracts are settled before the true state is revealed; then, contracted deliveries occur and consumption ensues. In this market, the commodity prices p_c^s and consumption demand x_{ic}^s are formed and determined before the actual commodities are produced; they are all determined in the contingent contracts.

A <u>contingent market equilibrium (CME)</u> is a pair (x,p), where x_i solves

$$\begin{array}{ll} \max\limits_{x_i \in \mathbb{C}} & u_i(x_i) \\ \text{s.t.} & p \cdot x_i \leq p \cdot w_i, \end{array} \tag{1}$$

for all i, and x is feasible:

$$\sum_{i=1}^{I} x_i^s = \sum_{i=1}^{I} w_i^s, \quad \forall s.$$

This is a competitive equilibrium in the pure exchange economy defined in Chap. 4. The welfare theorems in Chap. 4 state that a competitive equilibrium is Pareto optimal and, conversely, that a Pareto optimal allocation is a competitive equilibrium under some regularity conditions and with a reallocation of endowments.

1.2 Security Markets

A security market is an effective and much simpler alternative to the contingent market. Any CME can be implemented by a security market equilibrium (SME) in a complete market. As explained by Campbell (1987), the contingent market is informationally inefficient in the sense that the same equilibrium in a contingent market can be implemented in a security market using only $S+C$ markets rather than $S \times C$ markets.

Suppose that there are N securities that are traded at prices $q \equiv (q_1, \ldots, q_N) \in \mathbb{R}_+^N$ before the uncertainty is resolved. The security dividend vector for each state is $d^s \equiv (d_1^s, \ldots, d_N^s)$, where d_n^s is the dividend paid by the nth security in state s. Denote the dividend matrix as

$$d = \begin{pmatrix} d^1 \\ \vdots \\ d^S \end{pmatrix} = \begin{pmatrix} d_1^1 & \cdots & d_N^1 \\ \vdots & & \vdots \\ d_1^S & \cdots & d_N^S \end{pmatrix} \in \mathbb{R}_+^{S \times N}.$$

Spot markets for commodities are opened after uncertainty is resolved, with prices

$$P = \left((P_1^1, \ldots, P_C^1), \ldots, (P_1^S, \ldots, P_C^S)\right) = (P^1, \ldots, P^S),$$

1 Security Markets

where P_c^s is the price of good c in state s in the spot market. Agent i's plan is $(\theta_i, x_i) \in \mathbb{R}^N \times \mathbb{R}^{S \times C}$, where θ^i is a portfolio of securities (θ_{in} is the chosen number of the n th security). Here, p_c^s is the ex-ante price and P_c^s is the ex-post price of a commodity. Also, $\{d_n^s\}$ are all in the same unit of measure, which is the same as the unit of measure for P. This unit of measure can be money or one of the available goods. Similarly, p can be in the unit of money or in the unit of one of the goods. In general, p and P can be in different units of measure, but q and p are in the same unit of measure.

If $d^1, \ldots, d^S$ are linearly independent, the securities are called <u>independent securities</u> and the financial market is said to be <u>complete</u>. Here, a complete market actually requires two things: the number N of securities is larger or equal to the number S of states and among these N securities there are S independent securities.[1] Securities allow consumers to move resources across states. A complete market means that there are enough linearly independent securities that allow consumers to move to any point in the consumption space. For example, suppose that there are two states, 1 and 2, and a consumer wishes to consume at the point (c_1^*, c_2^*). Given two securities with dividend vectors $d^1 = (d_1^1, d_2^1)$ and $d^2 = (d_1^2, d_2^2)$ and with a proportion λ of the consumer's income I in d^1, consumption point $I[\lambda d^1 + (1 - \lambda) d^2]$ is available to the consumer. If d^1 and d^2 are linearly independent and (c_1, c_2) is affordable, there must exist a λ^* such that

$$(c_1^*, c_2^*) = I[\lambda^* d^1 + (1 - \lambda^*) d^2].$$

If (c_1^*, c_2^*) is Pareto optimal (PO), then this complete market is efficient.

Assume that the security market opens before uncertainty resolves; after the uncertainty resolves, the spot market opens. Before the spot market opens, individuals do not have endowments. A <u>security market equilibrium (SME)</u> is $[(\theta_1, x_1), \ldots, (\theta_I, x_I), (q, P)]$, where (θ_i, x_i) solves

$$\begin{aligned}
\max_{(\theta_i, x_i)} \quad & u_i(x_i) \\
\text{s.t.} \quad & q \cdot \theta_i \leq 0, \\
& P^s \cdot x_i^s \leq P^s \cdot w_i^s + \theta_i \cdot d^s, \quad \forall s,
\end{aligned} \tag{2}$$

for all i, and $[(\theta_1, x_1), \ldots, (\theta_I, x_I)]$ is feasible:

$$\sum_{i=1}^I \theta_i = 0, \quad \sum_{i=1}^I x_i^s = \sum_{i=1}^I w_i^s, \quad \forall s.$$

---

[1] This is implied by the fact that, for any matrix, the rank of its column vectors is the same as the rank of its row vectors.

Here, $q \cdot \theta_i \leq 0$ is the budget constraint for securities.[2] Each individual is endowed with zero units of securities; thus, the total expenditure on securities is less than or equal to the value of the endowed securities. We need this constraint to determine the prices of the securities. We can treat each security just like a good.

Proposition 5.1 *Suppose that utility functions are quasi-concave. Given a CME (x,p) and an arbitrary non-singular dividend matrix $d \in \mathbb{R}^{S \times S}$, let $N = S$, for any i, s and n, take*

$$q = (p_1^1, \ldots, p_1^S)d, \quad P^s = p^s/p_1^s,$$
$$\theta_i = (P^1 \cdot (x_i^1 - w_i^1), \ldots, P^S \cdot (x_i^S - w_i^S))(d^{-1})^T, \quad (3)$$

where superscript T represents the transpose of a vector or a matrix. Then, $[(\theta, x), (q, P)]$ is an SME.

Proof Since the utility functions are quasi-concave, we only need to verify that the optimal allocation of a CME must satisfy the equations (FOCs, budget constraints and feasibility constraints) that determine an SME.

The Lagrangian function for the security market problem (2) is

$$L = u_i(x_i) - \gamma_i q \cdot \theta_i + \sum_s \mu_i^s (P^s \cdot w_i^s + \theta_i \cdot d^s - P^s \cdot x_i^s),$$

where γ_i and μ_i^s are the Lagrangian multipliers. The FOCs are

$$\frac{\partial u_i(x_i)}{\partial x_i^s} = \mu_i^s P^s, \quad \gamma_i q = \sum_s \mu_i^s d^s. \quad (4)$$

We need to find the Lagrangian multipliers (γ, μ), prices (q, P) and a portfolio θ such that, for the CME allocation x, the pair (θ, x) satisfies the equations in (4) and the budget and feasibility constraints in (2).

The FOC for (1) is

$$p = u_i'(x_i)/\lambda_i, \quad (5)$$

where λ_i is the marginal utility of income. Besides (3), we also take

$$\mu_i^s = \lambda_i p_1^s, \quad \gamma_i = \lambda_i. \quad (6)$$

---

[2] Suppose that there are two assets and one asset is money and the other is bonds. If I issue 10 $100 bonds, I get $1000 but lose 10 bond units. My budget for this transaction is ($1000 × 1) − ($1000 × 1) = 0. In this example, q_m = $1000, q_b = $100, and θ_b = 10, where m indicates money and b indicates bonds.

Then, the first equation in (4) is the same as (5):

$$\frac{\partial u_i(x_i)}{\partial x_i^s} = \lambda_i p_1^s P^s = \lambda_i p^s.$$

The second equation in (4) is satisfied by:

$$\gamma_i q = \lambda_i q = \lambda_i (p_1^1, \ldots, p_1^S) d = \lambda_i \sum_s p_1^s d^s = \sum_s \mu_i^s d^s.$$

Also, by the budget condition in (1), the first budget constraint in (2) is satisfied by:

$$q \cdot \theta_i = (p_1^1, \ldots, p_1^S) d\theta_i^T = \sum_s p_1^s P^s \cdot (x_i^s - w_i^s) = \sum_s p^s \cdot (x_i^s - w_i^s)$$
$$= p \cdot (x_i - w_i) \leq 0.$$

By the definition of θ_i in (3), we have

$$\theta_i \left((d^1)^T, \ldots, (d^S)^T \right) = \theta_i d^T = \left(P^1 \cdot (x_i^1 - w_i^1), \ldots, P^S \cdot (x_i^S - w_i^S) \right).$$

Hence, the second budget constraint in (2) is satisfied by:

$$\theta_i \cdot d^s = P^s \cdot (x_i^s - w_i^s), \quad \text{for all } s.$$

The first feasibility condition is satisfied by

$$\sum_i \theta_i = \left(\sum_i P^1 \cdot (x_i^1 - w_i^1)(d^{-1})^T, \ldots, \sum_i P^S \cdot (x_i^S - w_i^S)(d^{-1})^T \right)$$
$$= \left(P^1 \cdot \left(\sum_i x_i^1 - \sum_i w_i^1 \right), \ldots, P^S \cdot \left(\sum_i x_i^S - \sum_i w_i^S \right) \right) (d^{-1})^T = 0.$$

Finally, the second feasibility condition is already satisfied by the fact that x is a CME allocation. The proof is complete. ∎

Proposition 5.1 immediately implies the following simpler version.

Corollary 5.1 *Suppose that utility functions are quasi-concave. Given a CME (x, p), let $N = S$, $d = I_S$, where I_S is the S-dimension identity matrix, and, for any i, s and n, take*

$$q_n = p_1^n, \quad P^s = p^s / p_1^s, \quad \theta_i^s = P^s \cdot (x_i^s - w_i^s). \tag{7}$$

Then, $[(\theta, x), (q, P)]$ is an SME. ∎

Proposition 5.1 means that an allocation x in a CME can always be implemented in an SME. We have a few remarks on this result.

First, P^s in (6) can be defined as $\frac{1}{\alpha^s}p^s$ for any arbitrary constant $\alpha^s > 0$. Since P in Proposition 5.1 is in units of the first good, we have $P^s = \frac{p^s}{\alpha^s}/\frac{p_1^s}{\alpha^s}$. This arbitrary α^s allows us to use any other good as the unit of measure.

Second, we can choose a different set of (γ, μ, q, P) in (6), which will imply a different set of expressions in (3). For example,

- In Duffie (1988, p. 4), by taking $d = I_S$, $\mu_i^s = \lambda_i$ and $\gamma_i = \lambda_i$ for all s and i, then

$$q^s \equiv 1, \quad P \equiv p, \quad \theta_i^s \equiv p^s \cdot \left(x_i^s - w_i^s\right), \quad \forall i, s.$$

- If we take $d = I_S$, $\mu_i^s = \lambda_i \alpha^s$ and $\gamma_i = \lambda_i$ for an arbitrary $\alpha^s > 0$, we have $p^s = \alpha^s P^s$ and $q = \sum_s \alpha^s d^s$.

There are many possible versions of SME. Our version can be conveniently used to derive asset pricing formulas in the rest of this chapter.

Third, the converse of Proposition 5.1 holds. That is, any SME under a complete market can be supported by a CME. However, when the complete market assumption is dropped, one can easily construct a counter example in which this equivalence of the two markets fails. This result is summarized by the following theorem.

Theorem 5.1 *If the market is complete, then any CME can be supported by an SME, and vice versa. Furthermore, the completeness of the market is a necessary and sufficient condition to guarantee the equivalence of the two types of equilibrium.* ∎

2 Static Asset Pricing

Consider a two-period model with *one commodity* and S different states of nature in the second period. The consumption space is $\mathbb{C} = \mathbb{R}^{S+1}$, where $(x^0, x^1, \ldots, x^S) \in \mathbb{C}$ represents x^0 units of consumption in the first period and x^s units in state s of the second period, $s = 1, 2, \ldots, S$. Suppose that preferences are given by an expected utility:

$$u_i\left(x^0, x^1, \ldots, x^S\right) = v_i\left(x^0\right) + \beta \sum_{s=1}^{S} \alpha^s v_i(x^s), \tag{8}$$

where α^s is the probability of state s occurring and β is the time discount factor. Then, the FOC for (1) is

$$p = u_i'(x_i)/\lambda_i, \tag{9}$$

which is equivalent to

$$p^k/p^l = MRS_i^{kl}(x_i), \quad \forall k,l.$$

That is,

$$\frac{p^s}{p^0} = \alpha^s \beta \frac{v_i'(x_i^s)}{v_i'(x_i^0)}, \quad \forall s. \tag{10}$$

Together with the budget and resource constraints, this determines a CME.

Suppose now that the market is complete with dividend matrix $d \in \mathbb{R}^{S \times C}$. By Proposition 5.1, we can convert the above CME into an SME in which the spot price is $P^s = 1$ for each $s = 0, 1, \ldots, S$,[3] and in which the market value of the nth security is

$$q_n = \sum_{s=1}^{S} p^s d_n^s.$$

Assuming that $p^0 = 1$,[4] by (10), for each agent i,

$$q_n = \beta \sum_{s=1}^{S} \alpha^s \frac{v_i'(x_i^s)}{v_i'(x_i^0)} d_n^s. \tag{11}$$

That is, the market value q_n of a security is the expected future dividends $\tilde{d}_n$ discounted by the marginal rate of substitution $MRS_i^{10}(x_i^0, \tilde{x}_i)$ between two periods. Since $\tilde{x}_i = x_i^s$ and $\tilde{d}_n = d_n^s$ with probability α^s, (11) can be written as[5]

$$q_n = \beta E\left[\frac{v_i'(\tilde{x}_i)}{v_i'(x_i^0)} \tilde{d}_n\right]. \tag{12}$$

This pricing formula (12) is a direct implication of Proposition 5.1 and it is the basis for all available equilibrium asset pricing models, whether in discrete-time or continuous-time settings.

---

[3] This is implied from (3). Intuitively, since the equilibrium price vector is unique up to a positive multiplier and since there is only one good, the price can be simply 1.
[4] This means that the good x_0 is used as the numeraire for p and q. As we know, for CME prices, one of them can be set to 1.
[5] The MRS $v_i'(\tilde{x}^i)/v_i'(x_0)$ is there because of risk aversion. Without it, the price of the security will be the discounted expected income.

3 Representative Agent Pricing

The pricing formulae in the last section are agent-specific. We now derive the corresponding formulae for a representative agent.

For the model outlined in Sect. 2, the pricing formula (12) depends on the agent. This means that it can only be applied to a single-agent economy. We intend to extend the pricing formula (12) to a multi-agent economy by the construction of a representative agent in a given equilibrium $(x_1, \ldots, x_I, p)$. We know that (12) is derived from (8) and (9). If we can extend (8) and (9) to a multi-agent economy, then (12) is naturally extendable to a multi-agent economy.

Define a utility function for the <u>representative agent</u> $u_g : \mathbb{R}^C \to \mathbb{R}$ as

$$u_g(y) \equiv \max_{y_1, \ldots, y_I} \sum_{i=1}^{I} \gamma_i u_i(y_i) \quad \text{s.t.} \quad y_1 + \cdots + y_I \leq y, \tag{13}$$

for some $\gamma \in \mathbb{R}^I_{++}$. Here, $u_g(y)$ is the social welfare maximum function used in the welfare theorems of the Arrow-Debreu world. We first generalize (9) to a multi-agent economy.

Proposition 5.2 *For any given CME $(x_1, \ldots, x_I, p)$, if $\gamma_i = 1/\lambda_i$, where λ_i is the marginal utility of income defined in (1)[6] and $w \equiv \sum_i w_i$, then $(x_1, \ldots, x_I)$ solves (13) for $y = w$ and*

$$p = u'_g(w). \tag{14}$$

Proof By Proposition 4.6, since $w = \sum x_i$, $(x_1, \ldots, x_I)$ is a solution of (13) for $y = w$ and weights $\gamma_i = 1/\lambda_i$, the Lagrangian function for (13) is

$$\mathcal{L} \equiv \sum_{i=1}^{I} \gamma_i u_i(y_i) + \lambda(w - y_1 - \cdots - y_I).$$

The FOCs of (13) are

$$\gamma_i u'_i(x_i) = \lambda, \quad \forall i.$$

The Envelope Theorem implies

$$u'_g(w) = \lambda.$$

Together with (9), we have

---

[6] Let $v_i(p, I_i) \equiv \max_{x_i \in C}\{u_i(x_i) | p \cdot x_i \leq I_i\}$. Then, $\lambda_i \equiv \partial v_i(p, I_i)/\partial I_i$.

3 Representative Agent Pricing

$$p = u'_i(x_i)/\lambda_i = \gamma_i u'_i(x_i) = \lambda = u'_g(w). \blacksquare$$

For the utility functions in (8), we can similarly define a utility function $v_g : \mathbb{R} \to \mathbb{R}$ for the representative agent by

$$v_g(c) \equiv \max_{y_1,\ldots,y_I} \sum_{i=1}^{I} \gamma_i v_i(c_i) \quad \text{s.t.} \quad c_1 + \cdots + c_I \leq c, \quad c \in \mathbb{R}.$$

Then,

$$u_g(y^0, y^1, \ldots, y^S)$$
$$= \max_{\substack{x_1^s,\ldots,x_I^s \\ s=0,1,\ldots,S}} \left\{ \sum_{i=1}^{I} \gamma_i \left[v_i(x_i^0) + \beta \sum_{s=1}^{S} \alpha^s v_i(x_i^s) \right] \middle| \sum_{i=1}^{I} x_i^s \leq y^s, \quad s = 0, 1, \ldots, S \right\}$$
$$= \max_{x_1^0,\ldots,x_I^0} \left\{ \sum_{i=1}^{I} \gamma_i v_i(x_i^0) \middle| \sum_{i=1}^{I} x_i^0 \leq y^0 \right\} + \beta \sum_{s=1}^{S} \alpha^s \max_{x_1^s,\ldots,x_I^s} \left\{ \sum_{i=1}^{I} \gamma_i v_i(x_i^s) \middle| \sum_{i=1}^{I} x_i^s \leq y^s \right\}$$
$$= v_g(y^0) + \beta \sum_{s=1}^{S} \alpha^s v_g(y^s).$$

That is, the representative agent has a similar utility function as in (8):

$$u_g(y^0, y^1, \ldots, y^S) = v_g(y^0) + \beta \sum_{s=1}^{S} \alpha^s v_g(y^s), \quad y^s \in \mathbb{R}. \tag{15}$$

We can now extend (12) to a multi-agent economy.

Theorem 5.2 *If the market is complete, for a given equilibrium $(x_1, \ldots, x_I, p)$, if one chooses $\gamma_i = 1/\lambda_i$, then*

$$q_n = \beta E \left[\frac{v'_g(\tilde{w})}{v'_g(w^0)} d_n \right], \tag{16}$$

where $w^0 \equiv \sum_i w_i^0$ and $\tilde{w} \equiv \sum_i \tilde{w}_i$ are the aggregate endowments in the two periods. Further, the completeness of the market is generically necessary.

Proof With (15) replacing (8) and (14) replacing (9), we can immediately use (16) to replace (12). $\blacksquare$

The generic necessity of a complete market means that, if the financial market is not complete, the representative-agent pricing formula does not hold, except in pathological or extremely special cases.

4 The Capital Asset Pricing Model

Consider a set Y of random variables (or lotteries) with finite variance on some probability space. Each $y \in Y$ corresponds to the random payoff of some security. As in Sect. 2, there is only *one* commodity, i.e., $C = 1$. The choice space for agents is $\mathbb{C} = \mathrm{span}(Y)$. Each random variable in $\mathbb{C}$ is an <u>asset</u>. The total endowment $M \equiv \sum_{i=1}^{I} w_i$ is called the <u>market asset</u>. Denote 1 as some riskless security paying value 1 in all states. For simplicity, assume also that $\mathbb{C}$ is finite-dimensional. Each agent i's utility function is assumed to be <u>strictly</u> variance averse, meaning that

$$u_i(x) > u_i(y) \quad \text{whenever} \quad E(x) = E(y) \quad \text{and} \quad \mathrm{var}(x) < \mathrm{var}(y).$$

By Theorem 3.3 in Chap. 3, under expected utility, any concave utility function is variance averse.

Suppose that $(x_1, \ldots, x_I, p)$ is a CME for this economy. Each x_i can be viewed as a random variable. A linear functional L on $\mathbb{C}$ can be written as $L(x) = p \cdot x$.[7] Assume that $p \cdot 1$ and $p \cdot M$ are not zero. We can show that p can be represented by a unique $\pi \in \mathbb{C}$ via the formula:

$$p \cdot x = E(\pi x), \quad \forall x \in \mathbb{C}. \tag{17}$$

π is called the <u>state price</u> of the asset since it is based on the state, while p is an ex-ante price. For discrete time, we can easily find this π from Sect. 2, as shown in the following example. For continuous time, we need the so-called Riez representation theorem to define this π.

Example 5.1 Assume $I = 1$ and $N = 1$. Given a CME $(x^{0*}, x^{1*}, \ldots, x^{S*}, p)$ for the model in Sect. 2, by (10), define

$$\pi^s \equiv \beta \frac{v'(x^{s*})}{v'(x^{0*})}.$$

Then, for any dividend bundle $d = (d^1, \ldots, d^S) \in \mathbb{R}^S$, we have

$$p \cdot d = \sum_{s=1}^{S} p^s d^s = \sum_{s=1}^{S} \alpha^s \pi^s d^s = E(\pi d). \blacksquare$$

---

[7]If there are only finite states $s = 1, \ldots, S$, then we write p as a vector $p = (p^1, \ldots, p^S)$. For any $x = (x^1, \ldots, x^S)$, we have $p \cdot x = \sum_s p^s x^s$.

Proposition 5.3 *For a CME* $(x_1, \ldots, x_I, p)$, *let* π *be the state price of p. Then,*

$$x_i = \alpha_i + \beta_i \pi, \tag{18}$$

where $\beta_i \equiv \frac{cov(x_i, \pi)}{var(\pi)}$ *and* $\alpha_i \equiv E(x_i) - \beta_i E(\pi)$.

Proof For the CME $(x_1, \ldots, x_I, p)$, consider the least-squares regression model:

$$x_i = \alpha_i + \beta_i \pi + \varepsilon.$$

By the definition of α_i and β_i in the proposition, we have $E(\varepsilon) = \text{cov}(\varepsilon, \pi) = 0$.[8] For asset $\hat{x}_i = \alpha_i 1 + \beta_i \pi$, we have $x_i = \hat{x}_i + \varepsilon$. By (17), $p \cdot \varepsilon = E(\pi \varepsilon) = \text{cov}(\varepsilon, \pi) = 0$. Thus, $p \cdot \hat{x}_i = p \cdot x_i$, which means that $\hat{x}_i$ is affordable for agent i. By $\hat{x}_i = \alpha_i 1 + \beta_i \pi$, we have $\text{cov}(\hat{x}_i, \varepsilon) = 0$. Hence, we have

$$E(x_i) = E(\hat{x}_i) \quad \text{and} \quad \text{var}(x_i) = \text{var}(\hat{x}_i) + \text{var}(\varepsilon) > \text{var}(\hat{x}_i),$$

unless ε is zero. Therefore, $u_i(\hat{x}_i) > u_i(x_i)$ unless ε is zero. Since x_i is optimal for agent i, inequality $u_i(\hat{x}_i) > u_i(x_i)$ cannot be true, implying $\varepsilon = 0$ The proof is complete. ∎

Theorem 5.3 *Given a CME price p, and assuming that* $\text{var}(M) \neq 0$, *for any asset x, we have the* <u>Capital Asset Pricing Model (CAPM)</u>:

$$\bar{R}_x - R_1 = \beta_x(\bar{R}_M - R_1), \tag{19}$$

where $R_x \equiv \frac{x}{p \cdot x}$, $\bar{R}_x \equiv E(R_x)$, *and* β_x *is the* <u>beta</u> *defined by*

$$\beta_x \equiv \frac{cov(R_x, R_M)}{var(R_M)}.$$

Proof (18) implies

$$M = \sum_{i=1}^{I} w_i = \sum_{i=1}^{I} x_i = \alpha + \beta \pi,$$

where $\alpha \equiv \sum \alpha_i$ and $\beta \equiv \sum \beta_i$. Since $\text{var}(M) \neq 0$, we have $\beta \neq 0$. Then, for any asset x,

$$p \cdot x = E(\pi x) = E\left(\frac{M - \alpha}{\beta} x\right) = \frac{E(M - \alpha)}{\beta} E(x) + \frac{1}{\beta} \text{cov}(x, M),$$

implying

---
[8] Conversely, these two conditions actually imply α_i and β_i as defined in the proposition.

$$\frac{\beta}{E(M-\alpha)} = \bar{R}_x + \frac{p \cdot M}{E(M-\alpha)} cov(R_x, R_M). \tag{20}$$

Applying (20) to the two special assets 1 and M, we find that

$$\frac{\beta}{E(M-\alpha)} = R_1,$$
$$\frac{\beta}{E(M-\alpha)} = \bar{R}_M + \frac{p \cdot M}{E(M-\alpha)} var(R_M),$$

which can be used to eliminate the two parameters α and β in (20) and change it to

$$R_1 = \bar{R}_x + \frac{R_1 - \bar{R}_M}{var(R_M)} cov(R_x, R_M).$$

This then implies the CAPM. ∎

Equation (19) means that the expected excess return on any asset is the beta times the expected excess return of the market asset. One interesting feature is that, from (19), any asset whose return is uncorrelated with the market return is expected to have the riskfree rate of return.

5 Dynamic Asset Pricing

We now extend our discussions of pricing models to the infinite horizon.

5.1 The Euler Equation

Consider a multiperiod model with one representative agent and one good. Given a probability space, let the consumption choice space be

$$\mathbb{C} \equiv \{c = (c_0, c_1, \ldots) | c_i \text{ is a bounded random variable}, c_i \in \mathbb{R}\}.$$

Given a random process $\{X_t\}_{t=0}^{\infty}$, the state at time t is a realization x_t of X_t. Assume that $\{X_t\}$ is a first-order Markov process, in the sense that X_{t+1} has a conditional distribution of the form $F(\cdot|X_t)$ at time t. A security is defined by a dividend sequence in $\mathbb{C}$. There are N securities with dividend vector $d(X_t) \in \mathbb{R}^N$ whose ith component is the payoff of the ith security at state X_t. Let the prices of the securities be $q(X_t) \in \mathbb{R}_+^N$ whose ith component is the price of the ith security in state X_t. In a recursive model with a state variable being a first-order Markov process, such equilibrium security prices exist under mild conditions. We take the security prices to be *ex dividend*, so that purchasing a portfolio $\theta \in \mathbb{R}^N$ of securities in state X_t

5 Dynamic Asset Pricing

requires an investment of $\theta \cdot q(X_t)$ and yields a market value of $\theta \cdot [q(X_{t+1}) + d(X_{t+1})]$. The informational restrictions are that, for any time t, both c_t and θ_t must depend only on observations of $X_0, X_1, \ldots, X_t$, or in technical terms, that there is a function f_t such that

$$(c_t, \theta_t) = f_t(X_0, X_1, \ldots, X_t), \quad t = 0, 1, 2, \ldots$$

Given $W_0 \geq 0$ and the plan (θ, c), the <u>wealth process</u> $W = \{W_0, W_1, \ldots\}$ of the agent is defined by

$$W_t = \theta_{t-1} \cdot [q(X_t) + d(X_t)], \quad t = 1, 2, \ldots$$

Notice that there is no income endowment, i.e. no human capital. For simplicity, we require $c_t \geq 0$ and <u>no short sales</u> $\theta_t \geq 0$, for all t. The supply of shares is 1 for each security.

The representative agent has a utility function $u : \mathbb{R} \to \mathbb{R}$ and his problem is:

$$V(x_0, w_0) \equiv \max_{\{c_t\} \in \mathbb{C}, \{\theta_t\}} E_0 \sum_{t=0}^{\infty} \beta^t u(c_t)$$

$$\text{s.t.} \quad c_t + \theta_t \cdot q(X_t) \leq W_t, \quad \text{for all } t$$
$$W_t = \theta_{t-1} \cdot [q(X_t) + d(X_t)], \quad \text{for all } t$$
$$\text{given } X_0 = x_0, W_0 = w_0.$$

The Bellman Equation for this problem is (see the Appendix):

$$V(X_t, W_t) = \max_{(c_t, \theta_t) \in \mathbb{R}_+ \times \mathbb{R}_+^N} u(c_t) + \beta E_t V(X_{t+1}, W_{t+1})$$

$$\text{s.t.} \quad c_t + \theta_t \cdot q(X_t) \leq W_t,$$
$$W_t = \theta_{t-1} \cdot [q(X_t) + d(X_t)],$$

where $E_t(\cdot) = E(\cdot | X_t)$. The problem can be simplified to

$$V(X_t, W_t) = \max_{\theta_t} u[W_t - \theta_t \cdot q(X_t)] + \beta E_t V\{X_{t+1}, \theta_t \cdot [q(X_{t+1}) + d(X_{t+1})]\}.$$

By the Envelope Theorem,

$$\frac{\partial V(X_t, W_t)}{\partial w} = u'[W_t - \theta_t^* \cdot q(X_t)].$$

The FOC is

$$-u'[W_t - \theta_t^* \cdot q(X_t)]q(X_t) + \beta E_t\left\{\frac{\partial V(X_{t+1}, W_{t+1})}{\partial w}[q(X_{t+1}) + d(X_{t+1})]\right\} = 0.$$

These two equations imply

$$q(X_t) = \beta E_t\left\{\frac{u'[W_{t+1} - \theta_{t+1}^* \cdot q(X_{t+1})]}{u'[W_t - \theta_t^* \cdot q(X_t)]}[q(X_{t+1}) + d(X_{t+1})]\right\},$$

or

$$q(X_t) = \beta E_t\left\{\frac{u'(c_{t+1}^*)}{u'(c_t^*)}[q(X_{t+1}) + d(X_{t+1})]\right\}. \tag{21}$$

A triple (θ^*, c^*, q^*) is an <u>equilibrium</u> if (θ^*, c^*) is an optimal plan given prices q^* and dividends d, and if the markets clear:

$$c_t^* = \sum_{n=1}^{N} d_n(X_t), \quad \theta_t^* = 1, \quad t = 0, 1, 2, \ldots$$

In equilibrium, (21) becomes

$$q*(X_t) = \beta E_t\left\{\frac{u'[\sum_n d_n(X_{t+1})]}{u'[\sum_n d_n(X_t)]}[q^*(X_{t+1}) + d(X_{t+1})]\right\}. \tag{22}$$

This is the multiperiod version of the pricing formula (12). It is called the <u>Euler Equation</u>. From this, we can further derive the explicit solution for q^*. Using (22) recursively, by the fact that $E_t E_{t+i} = E_t$ for $i \geq 1$ and assuming the transversality condition (as explained in the Appendix):

$$\lim_{i\to\infty} \beta^i E_t\left\{\frac{u'[\sum_n d_n(X_{t+i})]}{u'[\sum_n d_n(X_t)]}q^*(X_{t+i})\right\} = 0,$$

we find the solution:

$$q^*(X_t) = E_t\left\{\sum_{i=1}^{\infty} \beta^i \frac{u'[\sum_n d_n(X_{t+i})]}{u'[\sum_n d_n(X_t)]}d(X_{t+i})\right\}.$$

5.2 Dynamic CAPM

Formula (22) shows that the current price of a security that pays $q(X_{t+1}) + d(X_{t+1})$ in the next period will cost $q(X_t)$. To extend this to any portfolio x of securities, for each state $X_t = s$, define a linear operator p by

$$p^s \cdot x \equiv \beta E_s \left\{ \frac{u'\left[\sum_n d_n(X_{t+1})\right]}{u'\left[\sum_n d_n(s)\right]} x \right\}. \tag{23}$$

By (12), p^s is the market price for the securities that pay x at $t+1$. Let[9]

$$R_C^s \equiv \frac{\sum_n d_n(X_{t+1})}{\sum_n d_n(s)}, \quad R_x^s \equiv \frac{x}{p^s \cdot x}, \quad \beta_x^s \equiv \frac{\text{cov}_s(R_x, R_C)}{\text{var}_s(R_C)}.$$

In other words, β_x^s is the <u>conditional beta</u> of x relative to the aggregate consumption, analogous with the static CAPM, where the market portfolio is in fact the aggregate consumption since the model is static. Assuming for illustration purposes that u is quadratic: $u(c) = Ac^2 + B$, where A and B are two positive constants. Let $\bar{R}_x^s \equiv E_s(R_x^s)$. Then, (23) becomes

$$p^s \cdot x = \beta E_s(R_C^s x),$$

implying

$$1 = \beta E_s(R_C^s R_x^s).$$

For convenience, let us drop the s temporarily. Then,

$$1/\beta = \text{cov}_s(R_x, R_C) + \bar{R}_C \bar{R}_x. \tag{24}$$

Applying it to $x = C$ and $x = 1$, we have

$$1/\beta = \text{var}_s(R_C) + \bar{R}_C^2, \quad 1/\beta = \bar{R}_C R_1.$$

Using them on (24) yields

$$\frac{\bar{R}_C R_1 - \bar{R}_C \bar{R}_x}{\bar{R}_C R_1 - \bar{R}_C^2} = \beta_x.$$

---

[9] $\sum_n d_n(X_{t+1})$ is the same as M in (19).

That is,

$$\bar{R}_x^s - R_1^s = \beta_x^s(\bar{R}_C^s - R_1^s), \quad i = 1, 2, \ldots S.$$

This is the so-called <u>Consumption-Based Capital Asset Pricing Model</u>. The interesting feature is that this CAPM has no cross-period component. That is, its components are based on the current period only.

In a continuous-time model, one can extend this Consumption-Based CAPM to non-quadratic utility functions. Under regularity conditions, the increment of a differentiable function can be approximated by the first two terms of its Taylor series expansion (which is a quadratic function); this approximation becomes exact in expected value as the time increment shrinks to zero when uncertainty is generated by a Brownian Motion.

6 Continuous-Time Stochastic Programming

6.1 Continuous-Time Random Variables

For simplicity of notation, we will deal with real-valued variables in this section. A generalization to vector-valued variables is straightforward.

The continuous-time counterpart of discrete-time accumulated white noise is the standard Brownian motion. Formally, the <u>standard Brownian motion</u> is a stochastic process $B \equiv \{B_t \in \mathbb{R} | t \in (0, \infty)\}$ on some probability space satisfying

(a) For any $s \geq 0$ and $t > s$, $B_t - B_s$ is normally distributed with zero mean and variance $t - s$.
(b) For any $0 \leq t_0 < t_1 < \cdots < t_l < \infty$, random variables $B_{t_0}, B_{t_1} - B_{t_0}, \ldots, B_{t_l} - B_{t_{l-1}}$ are statistically independent.
(c) $B_0 = 0$ with probability 1.

The three conditions basically say that (1) $B_0 = 0$ and (2) for any partition $0 \leq t_0 < t_1 < \cdots < t_l < \infty$, $\{B_{t_k} - B_{t_{k-1}}\}_{k=0}^{k=l}$ are white noises, where $t_{-1} = 0$.

Brownian motion is very popular partly because of the following convenient rule:

$$(dt)^2 = 0, \quad dt \cdot dB_t = 0, \quad (dB_t)^2 = dt. \tag{25}$$

Since $E(\Delta B_t^2) = \text{var}(\Delta B_t) = \Delta t$, we have $\Delta B_t \approx \sqrt{\Delta t}$ and $\Delta t \cdot \Delta B_t \approx (\Delta t)^{3/2}$, which justify the formulas in (25), where a term $(\Delta t)^b$ with $b > 1$ is treated as zero.

A process X is called a <u>diffusion process</u> if it follows the following stochastic differential equation,

6 Continuous-Time Stochastic Programming

$$dX_t = \mu(X_t)dt + \sigma(X_t)dB_t, \quad \text{for } t \geq 0, \tag{26}$$

where $\mu : \mathbb{R} \to \mathbb{R}$ and $\sigma : \mathbb{R} \to \mathbb{R}_+$ are bounded and Lipschitz. We may heuristically treat $\mu(X_t)dt$ and $\sigma^2(X_t)dt$ as the instantaneous mean and variance of dX_t. The following is a very useful result for the differentiation of continuous-time random variables.

Lemma 5.1 (Ito's Formula) *If f is twice continuously differentiable and X is a diffusion process, then*

$$df(X_t) = f'(X_t)dX_t + \frac{1}{2}f''(X_t)(dX_t)^2,$$

or

$$df(X_t) = \mathcal{D}f(X_t)dt + f'(X_t)\sigma(X_t)dB_t,$$

where

$$\mathcal{D}f(x) \equiv f'(x)\mu(x) + \frac{1}{2}f''(x)\sigma^2(x). \quad \blacksquare$$

A more general version of Ito's formula is

$$df(t, X_t) = f'_t(t, X_t)dt + f'_x(t, X_t)dX_t + \frac{1}{2}f''_x(t, X_t)(dX_t)^2,$$

or if X_t is a vector,

$$df(t, X_t) = f'_t(t, X_t)dt + f'_x(t, X_t) \cdot dX_t + \frac{1}{2}(dX_t)^T f''_x(t, X_t)dX_t.$$

If f' is bounded, then $\int_0^t f'(X_s)\sigma(X_s)dB_s$ is a <u>martingale</u> defined by the condition:

$$E_0 \left[\int_0^t f'(X_s)\sigma(X_s)dB_s \right] = 0.$$

It then follows that

$$\lim_{t \to 0} E_0 \left[\frac{f(X_t) - f(X_0)}{t} \right] = \mathcal{D}f(X_0).$$

That is, the expected derivative of f at any point x is $\mathcal{D}f(x)$.

6.2 Continuous-Time Stochastic Programming

Consider a single-agent economy with uncertainty generated by a Brownian motion B. There are two securities: a risky security and a riskless security. The risky security has (ex-dividend) price S_t, which is a diffusion process:

$$dS_t/S_t = \mu dt + \sigma dB_t, \quad t \geq 0. \tag{27}$$

This security pays dividend $\delta S_t dt$ at any time t, where μ and δ are strictly positive constants. We may think heuristically of $\mu + \delta$ as the instantaneous expected return and σ^2 as the instantaneous variance of the return. The riskless security has a price that is always 1, and it pays dividends at a constant rate $r \geq 0$ (the interest rate). Here, we assume that $\mu + \delta > r$, otherwise no one would hold the risky security. Let θ_t be the proportion of total wealth invested at time t in the risky security and c_t be the consumption at time t. Then, the wealth process W follows

$$dW_t = \theta_t W_t \left(\delta dt + \frac{dS_t}{S_t} \right) + (1 - \theta_t) W_t r dt - c_t dt,$$

which can be written as

$$dW_t = [\theta_t W_t (\mu + \delta - r) + rW_t - c_t]dt + \theta_t W_t \sigma dB_t. \tag{28}$$

Hence, the individual's problem is

$$V(w) \equiv \max_{(\theta, c)} E_0 \int_0^\infty u(c_t) e^{-\rho t} dt$$

$$\text{s.t.} \quad dW_t = [\theta_t W_t (\mu + \delta - r) + rW_t - c_t]dt + \theta_t W_t \sigma dB_t, \tag{29}$$

$$\text{given } W_0 = w,$$

where $\rho > 0$ is a discount rate, and u is strictly increasing, differentiable, and strictly concave.

The optimal solution (θ_t, c_t) should depend on the information available at time t. Since the wealth W_t contains all relevant information at any time t, we may limit the choice of the optimal consumption and portfolio to the case with

$$\theta_t^* = A(W_t), \quad c_t^* = C(W_t),$$

where A and C are some measurable functions. Then, W is a diffusion process:

$$dW_t = [A(W_t) W_t (\mu + \delta - r) + rW_t - C(W_t)]dt + A(W_t) W_t \sigma dB_t, \tag{30}$$

and

$$V(w) = E_0 \int_0^\infty e^{-\rho t} u[C(W_t)] dt.$$

For any $\varepsilon > 0$, we can break the expression into two parts:

$$V(w) = E_0 \int_0^\varepsilon e^{-\rho t} u[C(W_t)] dt + e^{-\rho \varepsilon} E_0 E_\varepsilon \int_\varepsilon^\infty e^{-\rho(t-\varepsilon)} u[C(W_t)] dt$$

$$= E_0 \int_0^\varepsilon e^{-\rho t} u[C(W_t)] dt + e^{-\rho \varepsilon} E_0 [V(W_\varepsilon)].$$

Notice here that we have assumed that $C(W_t)$ is a time-consistent solution. Then,

$$\frac{1 - e^{-\rho \varepsilon}}{\varepsilon} V(w) = E_0 \left[\frac{1}{\varepsilon} \int_0^\varepsilon e^{-\rho t} u[C(W_t)] dt \right] + e^{-\rho \varepsilon} E_0 \left[\frac{V(X_\varepsilon) - V(w)}{\varepsilon} \right].$$

Take limits as $\varepsilon \to 0$ and use Ito's Lemma to arrive at

$$\rho V(w) = u[C(w)] + \mathcal{D}V(w), \tag{31}$$

where

$$\mathcal{D}V(x) = V'(x)\mu(x) + \frac{1}{2} V''(x) \sigma^2(x),$$

where $\mu(w)$ and $\sigma(w)$ are defined in (30), i.e.,

$$\mu(w) = A(w)w(\mu + \delta - r) + rw - C(w), \quad \sigma(w) = A(w) w \sigma.$$

Then, by (31), the consumer's problem of maximizing $V(w)$ becomes the following Bellman equation:

$$\max_{A(w), C(w)} \mathcal{D}V(w) + u[C(w)]. \tag{32}$$

The first-order conditions are

$$(\mu + \delta - r) w V'(w) + V''(w) A(w) w^2 \sigma^2 = 0, \quad u'[C(w)] = V'(w),$$

which gives us the optimal solution:

$$A(w) = -\frac{V'(w)}{V''(w)w}\frac{\mu+\delta-r}{\sigma^2}, \quad C(w) = (u')^{-1}[V'(w)]. \tag{33}$$

One difficulty with this solution is that function $V(w)$ is not explicitly given yet and we have to use (31) to determine $V(w)$. That is, we need to substitute $A(w)$ and $C(w)$ into Eq. (31) to determine $V(w)$. After that, we return to (33) to find the final solution for $A(w)$ and $C(w)$.

Example 2 If $u(c_t) = c_t^\alpha$, $\alpha \in (0,1)$, then we must have $V(w) = kw^\alpha$ for some constant k. To verify this, for initial wealth λw, by inspecting the budget constraint (28), we know that $[\lambda C(w), A(w)]$ satisfies the budget constraint. Also, $\lambda C(w)$ maximizes the utility, and

$$V(\lambda w) = \max\ E_0 \int_0^\infty e^{-\rho t}[\lambda C(W_t)]^\alpha dt = \lambda^\alpha V(w).$$

By taking $\lambda = 1/w$, we have $V(1) = V(w)/w^\alpha$. Thus, $V(w) = kw^\alpha$ for some k. Then, by (33),

$$A(w) = \frac{\mu+\delta-r}{(1-\alpha)\sigma^2}, \quad C(w) = k^{\frac{1}{\alpha-1}}w.$$

Hence, it is optimal to consume a fixed fraction of wealth and to hold fixed fractions of wealth in assets. We now need to determine k. By (31), we have

$$\rho k w^\alpha = k^{\frac{\alpha}{\alpha-1}}w^\alpha + \alpha k w^{\alpha-1}\mu(w) + \frac{1}{2}k\alpha(\alpha-1)w^{\alpha-2}\sigma^2(w),$$

where, by (30),

$$\mu(w) = \frac{\mu+\delta-r}{(1-\alpha)\sigma^2}w(\mu+\delta-r) + rw - k^{\frac{1}{\alpha-1}}w = \left[\frac{(\mu+\delta-r)^2}{(1-\alpha)\sigma^2} + r - k^{\frac{1}{\alpha-1}}\right]w,$$

$$\sigma(w) = \frac{\mu+\delta-r}{(1-\alpha)\sigma^2}w\sigma = \frac{\mu+\delta-r}{(1-\alpha)\sigma}w,$$

implying

$$\rho k w^\alpha = k^{\frac{\alpha}{\alpha-1}}w^\alpha + \alpha k w^\alpha \left[\frac{(\mu+\delta-r)^2}{(1-\alpha)\sigma^2} + r - k^{\frac{1}{\alpha-1}}\right] + \frac{1}{2}k\alpha(\alpha-1)w^\alpha\left[\frac{\mu+\delta-r}{(1-\alpha)\sigma}\right]^2,$$

implying

$$k^{\frac{1}{\alpha-1}} = \frac{\rho - \alpha r}{1-\alpha} - \frac{\alpha(\mu + \delta - r)^2}{2(1-\alpha)^2 \sigma^2}.$$

Thus, the solution is

$$C(w) = \left[\frac{\rho - r\alpha}{1-\alpha} - \frac{\alpha\mu + \delta - r^2}{2(1-\alpha)^2 \sigma^2}\right] w, \quad A(w) = \frac{\mu + \delta - r}{(1-\alpha)\sigma^2}. \quad \blacksquare$$

7 The Black-Scholes Pricing Formula

The classical example of pricing a derivative security is the Black-Scholes Option Pricing Formula. Consider a single-agent economy with uncertainty generated by a Brownian motion B_t. There are two independent securities: a stock and a bond. The stock has (ex-dividend) price S_t, which is a diffusion process:

$$dS_t/S_t = \mu dt + \sigma dB_t, \tag{34}$$

and pays dividend $\delta S_t dt$ at any time t, where σ and δ are strictly positive constants. Here, μ is the expected return per unit of time and σ is the standard deviation of the stock return per unit of time. Assume that there is no dividend, $\delta = 0$, in this section. The market value of the riskless security (the bond) is $R_t = R_0 e^{rt}$ for some $R_0 > 0$, which means that

$$dR_t = rR_t dt.$$

A <u>derivative security</u> is a security whose value depends on the values of other more basic underlying securities. We want to evaluate a derivative security that pays a lump sum of $g(S_T)$ at a future time T, where g is a sufficiently well-behaved function to justify the derivation in this section. Suppose that the value of the derivative security at any time $t \in [0, T]$ is $C(S_t, t)$.

Instead of holding one unit of the derivative with wealth $C(S_t, t)$ at time t, the investor can alternatively put his wealth in a stock-bond combination. Suppose that an investor decides to hold a portfolio (a_t, b_t) of the stock and the bond at time t. Then,

$$C(S_t, t) = a_t S_t + b_t R_t. \tag{35}$$

Assume that the stock and the bond are independent securities so that a return from any other asset can be achieved by a portfolio of the stock and the bond. In this complete market, a no-arbitrage condition is

$$dC(S_t, t) = a_t dS_t + b_t dR_t. \tag{36}$$

We can see that the following portfolio satisfies the feasibility condition (35):

$$a_t = C_s(S_t, t), \quad b_t = \frac{C(S_t, t) - C_s(S_t, t)S_t}{R_t}. \tag{37}$$

By Ito's Lemma,

$$dC(S_t, t) = C_t dt + C_s dS_t + \frac{1}{2} C_{ss}(dS_t)^2 = \left(C_t + \frac{1}{2} C_{ss} \sigma^2 S_t^2\right) dt + C_s dS_t. \tag{38}$$

By the definition of (a_t, b_t) in (37),

$$a_t dS_t + b_t dR_t = C_s dS_t + r(C - C_s S_t) dt.$$

Therefore, (36) and (38) imply

$$C_t + \frac{1}{2} C_{ss} \sigma^2 S_t^2 = r(C - C_s S_t), \quad t \in [0, T].$$

This means that C must satisfy the following partial differential equation:

$$C_t(s, t) + rsC_s(s, t) + \frac{1}{2} \sigma^2 s^2 C_{ss}(s, t) - rC(s, t) = 0, \quad \forall (s, t) \in (0, \infty) \times (0, T). \tag{39}$$

The boundary condition is

$$C(s, T) = g(s), \quad s \geq 0.$$

The solution is shown in the following theorem, which can be easily verified.

Theorem 5.4 (Derivative Pricing) *Equation (39) has the solution:*

$$C(s, t) = E\left[e^{-r(T-t)} g(se^{x_t})\right], \tag{40}$$

where

$$x_t \sim N\left[(T-t)\left(r - \frac{\sigma^2}{2}\right), \sigma\sqrt{T-t}\right]. \quad \blacksquare$$

Equation (40) can be solved numerically by standard Monte Carlo simulation and variance reduction methods.

7 The Black-Scholes Pricing Formula

As an example, let us apply (40) to the original Black-Scholes Option Pricing Formula. We evaluate a European call derivative. A European call option gives the holder the right to buy the underlying asset on a certain date for a certain price. The price is called the exercise price and the date is called the exercise date or maturity. Let

$$K = \text{the exercise price},$$
$$T = \text{the maturity},$$
$$C_0 = \text{the initial price of the option}$$
$$C_T = \text{the price of the option at maturity}$$
$$S_0 = \text{the initial price of the underlying stock}$$
$$S_T = \text{the price of the underlying stock at maturity}$$

Since the option is exercised only if $S_T \geq K$, and in that case the net gain to the option holder is $S_T - K$, the value of the option at maturity is

$$C_T = (S_T - K)^+ \equiv \max(0, S_T - K). \tag{41}$$

In this case, we have $g(s) = (s - K)^+$. Using Theorem 5.4, the option price C_0 is found in the following proposition.

Proposition 5.4 (The Black-Scholes Formula). *The initial price of the call option is*

$$C_0 = S_0 \Phi(\gamma) - \frac{K}{(1+r)^T} \Phi\left(\gamma - \sigma\sqrt{T}\right), \tag{42}$$

where Φ is the cumulative standard normal distribution function and

$$\gamma = \frac{1}{\sigma\sqrt{T}} \log\left[\frac{S_0(1+r)^T}{K}\right] + \frac{1}{2}\sigma\sqrt{T}. \blacksquare$$

Notes The materials in this chapter are quite standard and appear in many books. We have covered only very basic and elementary financial theory, much of which is on asset pricing. The two best books on standard financial theory are Duffie (1988) and Duffie (1992).

Micro-foundation of Industry

6

This chapter introduces the standard equilibrium theory of industry. We focus mainly on the output market and later cover briefly the input market.

All the firms are assumed to maximize profits. Given a firm's revenue function $R(y)$ and cost function $c(y)$ for a level of output y, the optimal level of output y is determined by the following well-known MR = MC equation:

$$MR(y^*) = MC(y^*), \tag{1}$$

where $MR(y)$ and $MC(y)$ are the marginal revenue and marginal cost functions, respectively. This condition is applicable to any kind of firm. The optimality of condition (1) can be easily seen in Fig. 1.

Revenue $R(y)$ is derived from market conditions in the output market, while cost $c(y)$ is derived from market conditions in the input market. When we focus on the output market, we will derive $R(y)$ and take $c(y)$ as given; when we focus on the input market, we will derive $c(y)$ and take $R(y)$ as given.

1 A Competitive Output Market

A (perfectly) competitive industry is an industry in which

- There are *many small firms (small in market shares)*: firms are independent of each other in decision making.
- Each sells an *identical product*: each firm faces a horizontal demand curve at the market price and therefore takes the price as given.
- There is *free entry*: zero profit is made in the long run.

Fig. 1 Optimal output

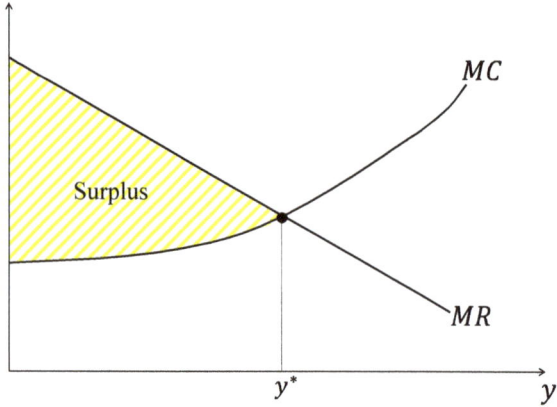

Under these conditions, no single firm can exert a significant influence on the market price of a good. Firms in such a market behave as price takers. If there is a positive profit, other firms can enter the industry, imitate the incumbent firms, and take a share of the profit. On the other hand, if there is a loss, some firms will leave the industry. Hence, in the long run, the profit is zero for all firms in such an industry.

A competitive firm is a firm that takes the market prices as given. Some caution should be exercised here. The firm faces two markets: the output market and the input market. A firm may be competitive in the output market but not so in the input market. In this section, we deal with competitive firms in the output market. Note also that the competitiveness of a firm is not an additional assumption on the firm. We only assume that the firm maximizes its profit or expected profit. The competitiveness is implied/induced by the market conditions.

Since there are many firms in the market producing the same product, any downward deviation from the market price will attract all the consumers in the market, and any upward deviation from the market price will drive away all the consumers. Therefore, any competitive firm faces a demand curve that is perfectly elastic at the prevailing market price, as shown in Fig. 2.

Hence, given a market price p, by the first-order and second-order conditions, firm i will supply a quantity y_i satisfying

$$p = c'_i(y_i), \quad c''_i(y_i) > 0. \qquad (2)$$

These two conditions correspond to the increasing part of the MC curve. However, these two conditions in (2) only guarantee the maximum profit; they do not guarantee profitability. Since the firm is free to exit, when the price is below its average variable cost, the firm will stop production or exit. Since the fixed cost is $FC = c(0)$, the variable cost is $VC(y) = c(y) - c(0)$ and the average variable cost is $AVC = VC(y)/y$. This implies the supply curve shown in Fig. 3.

1 A Competitive Output Market

Fig. 2 A firm's demand curve

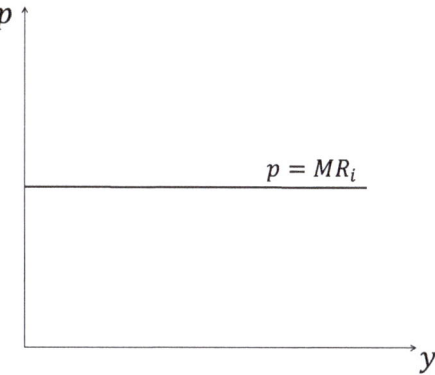

Fig. 3 A firm's supply curve

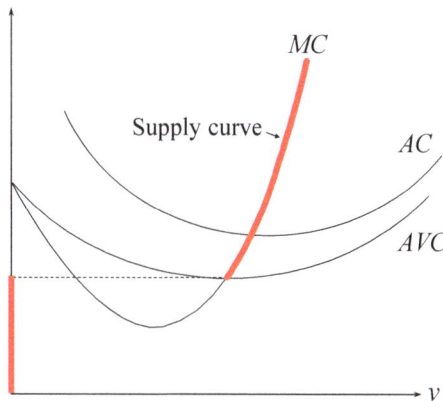

Given a price p, the industry supply $y(p)$ is the sum of individual firms' supply functions $y_1(p), \ldots, y_m(p)$ at price p:

$$y(p) \equiv \sum_{i=1}^{m} y_i(p).$$

Since $y(p)$ is the sum of $y_1(p), \ldots, y_m(p)$ at each given price p, the curve of $y(p)$ in Fig. 4 is the horizontal sum of the curves of $y_1(p), \ldots, y_m(p)$. That is, the industry supply curve is the horizontal sum of all individual firms' supply curves. Note that when the price equals the minimum AVC (assuming that all firms have the same minimum AVC), some firms choose to exit while others choose to stay so that the industry supply curve is horizontal at the minimum AVC.

The industry demand is the total quantity demanded by consumers for the product, which is typically a downward-sloping curve according to the law of demand.

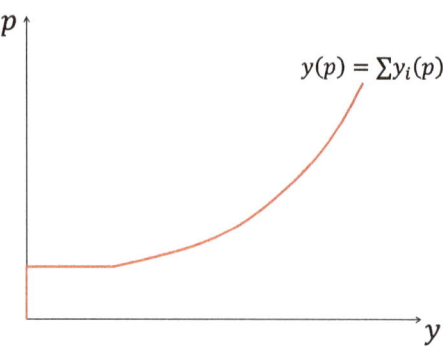

Fig. 4 The industry supply curve

We have now derived firms' demand and supply curves and the industry's demand and supply curves. We have a few remarks to our derivation:

- While each firm faces a horizontal demand curve as shown in Fig. 2, the industry faces a typical downward-sloping market demand.
- We have focused on the output market. The behavior of firms in the input market is assumed away by the given cost function $c_i(y)$.
- The supply curves in both the short run and the long run are shown in Fig. 3, except that, in the long run, $AVC = AC$. By the LeChatelier Principle, the only difference between the long run and the short run supply curves is that the long-run supply curve is more flat. On the other hand, the demand curves are assumed to be independent of time.

While all firms take the price as given, the price is determined in equilibrium. The <u>equilibrium price</u> is the price that equalizes the industry demand and the industry supply, as shown on the right side of Fig. 5. The left side of Fig. 5 shows that the firm will keep producing as long as $p \geq AVC_i$, otherwise it will shut down.

When the time horizon comes into play, we have short-run and long-run equilibria. The above definition of equilibrium is for a <u>short-run equilibrium</u>. In the short run, the number of firms is fixed. However, in the long run, the number of firms may vary. Entry and exit may then affect the industry supply curve, which may result in changes of the short-run equilibrium price over time. In this case, when the entry and exit stop (or more precisely, when there is no potential entry and exit to come), we say that the industry reaches a <u>long-run equilibrium</u> and the resulting price is called the long-run equilibrium price, as shown in Fig. 6. In the long run, the number of firms is determined in equilibrium. This equilibrium number of firms is typically determined by the zero-profit condition, as shown on the left side of Fig. 6.

Within each firm, some inputs cannot be varied in the short run. Capital inputs, such as equipment and buildings, may not be adjustable in the short run. In this case, a short-run equilibrium is dependent on given fixed amounts of capital inputs. In the long run, all inputs are variable and thus they can all be chosen optimally.

1 A Competitive Output Market

Fig. 5 Short-run equilibrium

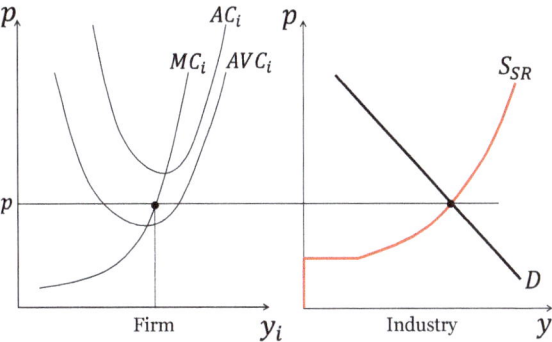

Fig. 6 Long-run equilibrium

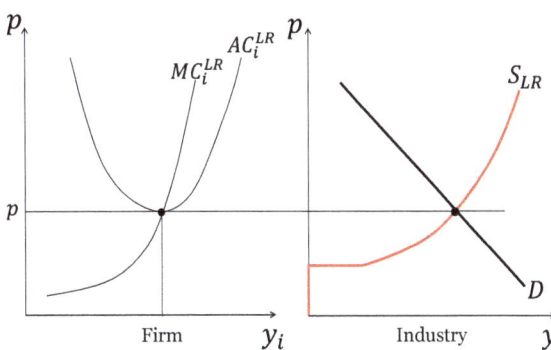

We have now established the equilibrium theory for a competitive industry. In the following, we show a few examples of how to use this theory to analyze an industry.

Example 6.1 Given a long-run equilibrium (p_0, y_0), suppose that there is a demand shock, which moves demand from D to D' in Fig. 7. What is the new long-run equilibrium?

As Fig. 7 indicates, starting from the original industry equilibrium (p_0, y_0), the shock brings the industry to a short-run equilibrium (p_1, y_1) and eventually leads to a long-run equilibrium (p_0, y_2). Here, the industry supply curve shifts to the left since some firms leave the industry due to negative long-run profit. The firms stop leaving when the price returns to the original level. Hence, in the short run, the shock causes a temporary drop in price; in the long run, the price remains the same, but there is a permanent reduction in quantity. ∎

Example 6.2 Suppose that the industry demand is

$$y^d(p) = a - bp,$$

where a and b are positive constants. Suppose that the firms have identical cost functions and their cost function is

Fig. 7 A decrease in demand

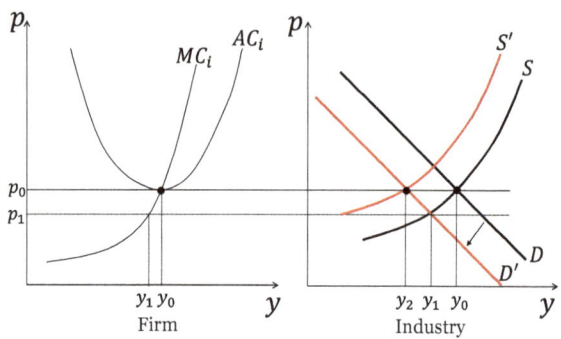

$$c_i(y_i) = y_i^2 + 1.$$

Since $MC_i = 2y_i$, the supply curve for each firm is

$$p = 2y_i.$$

If there are m firms in the industry, the industry supply curve is the horizontal sum of the MC_i's:

$$y^s = \sum y_i = \sum \frac{1}{2}p = \frac{1}{2}mp.$$

By equalizing the industry supply and demand, we find the equilibrium price:

$$p^* = \frac{2a}{2b+m}. \qquad (3)$$

We can see that the equilibrium price is lower if the number of firms increases.

In the short run, at price p^*, a firm may have a positive profit or may have a loss. From $\min AVC_i(y_i)$, we find the break-even price for each firm $p_{\min}^{SR} = 0$. Since we always have $p^* \geq p_{\min}^{SR}$, all the existing firms are profitable and will produce in the short run.[1] The short-run equilibrium price is $p^{SR} = p^*$.

In the long run, from $\min AC_i(y_i)$, we find that the break-even price for each firm is $p_{\min}^{LR} = 2$. This means that a firm will stay or enter the industry if the price is above 2 and exit if the price is below 2 in the long run. When new firms are entering the industry, they will drive down the price, as shown in (3). This process will continue until a negative profit is expected. The long-run equilibrium price is the smallest price that satisfies

---

[1] Even though a firm may decide to produce in the short run (when the firm is taking a short-term view), it may decide to leave an industry in the long run (when the firm is taking a long-term view).

1 A Competitive Output Market

$$p = \frac{2a}{2b+m} \quad \text{and} \quad p \geq 2.$$

The largest m^*, satisfying the above two conditions, i.e., the equilibrium number of firms, can be easily solved from the two conditions:

$$m^* = [a - 2b],$$

where $[x]$ represents the largest integer that is less than or equal to x. The long-run equilibrium price is then

$$p^{LR} = \frac{2a}{2b + [a-2b]}.$$

As shown in Fig. 8, the firms in the industry will generally make a positive profit even though the expected profit for a newcomer is negative. In fact, the firms in the industry will make a positive profit in the long run if and only if $a - 2b$ is not an integer. ∎

Example 6.3 Let $p^d(y)$ be the inverse industry demand curve and $p^s(y)$ be the inverse industry supply curve.[2] Suppose that the industry is originally in an equilibrium: $p^d(y^*) = p^s(y^*)$. Now, the government imposes a sales tax of t dollars per unit of output on the producers. Originally, at output level y, the firms need to be paid $p^s(y)$ for an additional unit of output on the margin. After the tax is imposed, since the firms are paying t per unit of output to the government, the firms now need to be paid $p^s(y) + t$ for an additional unit of output at y. This means that the after-tax industry supply curve is

$$p = p^s(y) + t.$$

The new equilibrium condition is therefore $p^d(y) = p^s(y) + t$.

If the time horizon is introduced, the above p^s serves as the short-run industry supply curve. We know that the long-run supply curve will be more flat. For simplicity, assume that the long-run MC curve (i.e., the long-run supply curve) for the industry is constant. Then, the effects of the tax can be easily seen in Fig. 9.

---

[2] A demand function is typically written as $y = f(p)$, meaning that the quantity demanded y is dependent on the price. We often write $y^d = f(p)$ to indicate it as a demand function. When this function f is strictly decreasing, we can also express the demand relationship by $p = f^{-1}(y)$ or $p = g(y)$, where g is the inverse function. We call this the <u>inverse demand function</u>. Again, we often write $p^d = g(y)$ to indicate that it is a demand function.

Fig. 8 Short-run and long-run equilibria

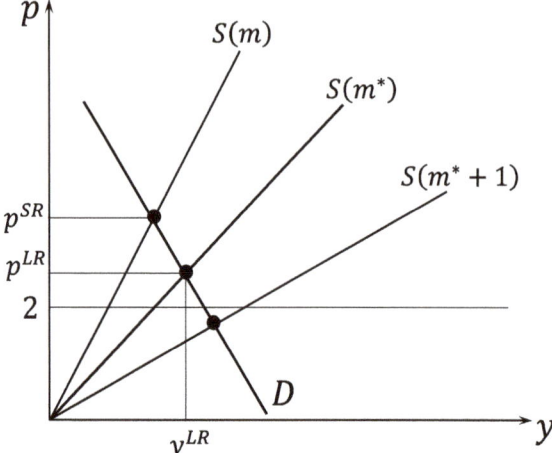

Fig. 9 The effects of a sales tax

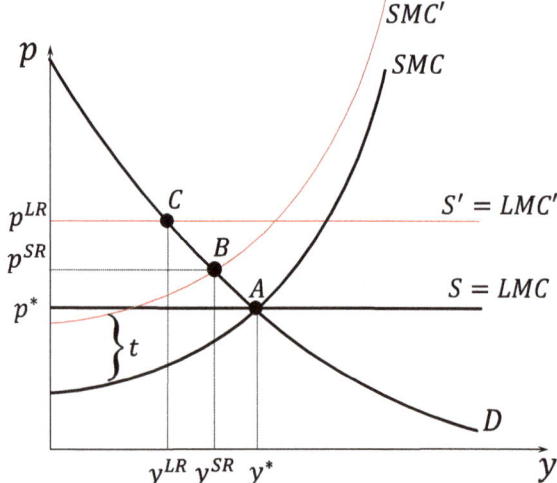

The original equilibrium is at A. With the tax, the short-run equilibrium is now at B, where the consumers and the industry share the burden of taxes. The short-run price is increased less than t. The long-run equilibrium is at C, at which point the consumers bear the tax alone. The long-run price is increased exactly by t: $p^{LR} = p^* + t$. This is expected: with zero profit for the firms in the long run, the tax burden has to be passed on to the consumers. ∎

2 A Monopoly

A monopoly is an industry in which

- There is only *one firm*: the firm is the industry.
- The firm faces a *downward-sloping demand curve*: it is able to manipulate both price and quantity.
- There is *no entry*: it is possible for the firm to maintain a positive profit in the long run.

In theory, there are three types of monopolies:

1. A single-price monopoly, which charges the same price for each and every unit of its output.
2. A price-discriminating monopoly, which charges different prices for the same good to different people or for different quantities demanded.
3. If the monopoly does not know each consumer's willingness to pay, we have monopoly under asymmetric information.

2.1 A Single-Price Monopoly

The market demand is defined by a function $p^d = p(y)$, indicating the price is dependent on the quantity demanded. A downward-sloping demand curve means that $p'(y) < 0$. The revenue is $R(y) \equiv p(y)y$. By (1), the FOC is

$$p(y^*) + p'(y^*)y^* = c'(y^*). \tag{4}$$

As shown in Fig. 10, when the firm adds one more unit of output, since it charges the same price all units, it gets paid $p(y)$ for that unit, but since the price has been lowered for all units, it loses $|p'(y)|y$ on all previous units. Hence, $MR(y) = p(y) + p'(y)y$, as indicated in Fig. 10.
Since $p'(y) < 0$, we have $MR(y) < p(y)$, meaning that the MR curve is always below the demand curve except at $y = 0$.

The monopoly's problem is shown in Fig. 11. The monopoly is to produce y^*, determined by condition (4). To sell this quantity y^*, since the monopoly charges one price for all units, the highest price is p^*, as shown in Fig. 11. Hence, a single-price monopoly's solution is to produce y^* and charge p^*, as shown in Fig. 11.

We have a few observations of this solution. First, we have

$$MR(y) = p(y)\left(1 + \frac{y}{p}\frac{dp}{dy}\right) = p(y)\left(1 - \frac{1}{\eta}\right),$$

Fig. 10 Marginal revenue

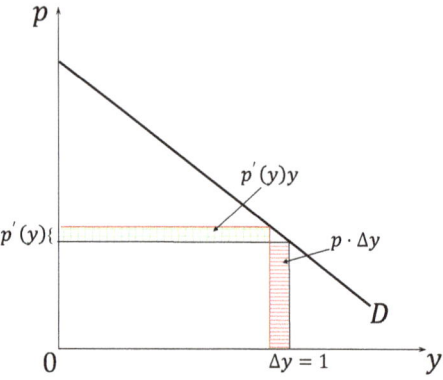

Fig. 11 A single-price monopoly's problem

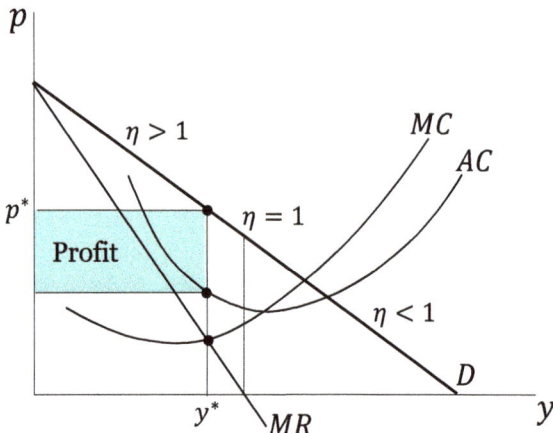

where η is the price elasticity of demand. With $MC(y) \geq 0$, the profit maximization condition $MR(y^*) = MC(y^*)$ implies that a single-price monopoly will always produce in the elastic range ($\eta \geq 1$) of the demand curve.

Second, as shown in Fig. 11, there is no guarantee that a monopoly will make a positive profit. A monopoly may make zero profit or even a loss.

Third, unlike a competitive firm, a monopoly has no supply curve. The monopoly picks a combination of output and price using its MR and MC curves. Therefore, the monopoly only supplies at one single point.

Example 6.4 Consider a simple demand function: $p^d = A - ay$, where A and a are positive constants. In this case, the MR curve is half way between the demand curve and the vertical axis. The solution is shown in Fig. 12.

Fig. 12 A single-price monopoly's problem with linear demand

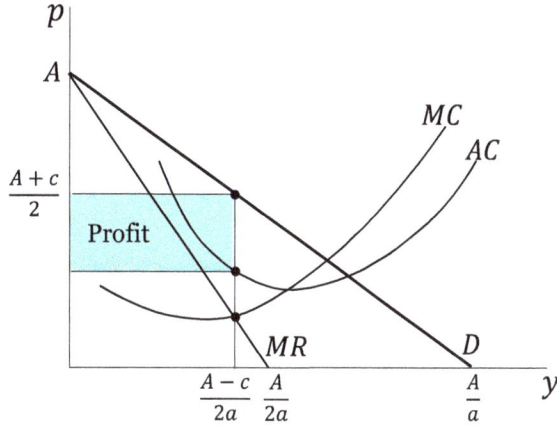

2.2 A Price-Discriminating Monopoly

Price discrimination is the practice of charging some customers a higher price than others for an identical good or of charging an individual customer a higher price on a purchase of a few units of the good than on a purchase of many units of the good. Some people may pay less with price discrimination, but others pay more. But, the firm can make more profit by price discrimination.

Price discrimination can be practiced in varying degrees. Perfect price discrimination occurs when a firm charges a different price for each unit sold. Given an output y, the left side of Fig. 13 shows that a monopoly will do better by charging multiple prices p_1 and p_2 than a single price p_2 for the total quantity y. The right side of the figure shows that for a given quantity y, by charging a different price for each different unit, the monopoly achieves the maximum revenue for selling that amount y. That is, perfect discrimination implies the highest profit. The monopoly can alternatively charge a fee equal to the area abp and charge a single price for the quantity y.

The right side of Fig. 13 shows that the revenue is the area $aby0$, suggesting that $MR(y) = p(y)$.[3] In other words, the MR curve is the same as the demand curve. By (1), the profit is maximized when marginal revenue equals marginal cost. Therefore, a perfectly price-discriminating monopoly will produce its output at the point where the marginal cost curve intersects with the demand curve, as shown in Fig. 14.

We have a few observations of the solution. First, the monopoly price is generally higher than the competitive price. Second, the monopoly output is generally less than the competitive quantity. Third, the more perfectly the monopoly can price discriminate, the closer its output is to the competitive output.

---

[3]Rigorously, the revenue function is $R(y) = \int_0^y p(x)dx$, implying $MR(y) = p(y)$ for any $y \geq 0$.

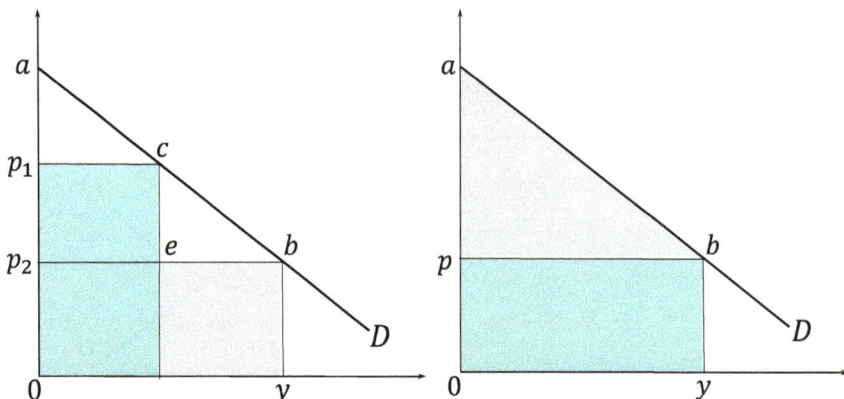

Fig. 13 Price discrimination

Fig. 14 A perfect price-discriminating monopoly

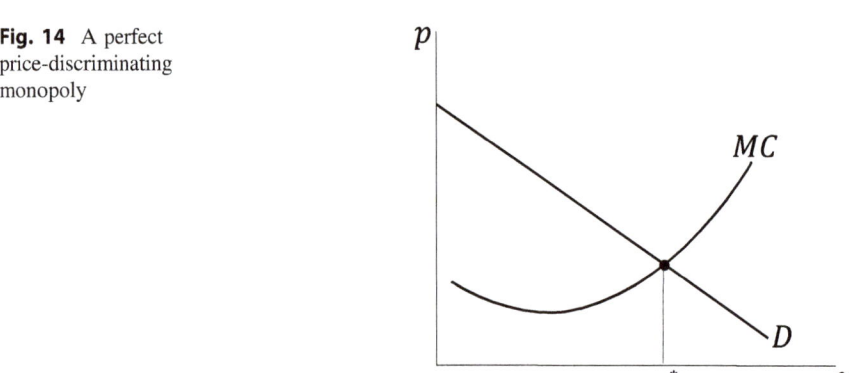

2.3 Monopoly Pricing Under Asymmetric Information

We have so far assumed complete information (or symmetric information): demand and supply functions are all public knowledge. Price discrimination becomes difficult when the monopolist does not know the demand curve. Suppose now that the monopolist does not know the demand curve of each consumer although it can practice price discrimination. More specifically, the monopolist knows all existing types of consumers, but it does not know which consumer belongs to which type. The information is asymmetric since the consumers know their own types but the monopolist does not. What should the monopoly do?

This problem belongs to the field of mechanism design, which is formally covered in Chap. 9. In this section, we give a simple example to illustrate the solution. Suppose that there are two consumers with the demand curves indicated in Fig. 15, where consumer 2's demand curve is above consumer 1's demand curve, with maximum demand $\bar{x}_1$ and $\bar{x}_2$, respectively, at zero price. Assume zero cost. We have labeled the areas as A, B, C, D and E. As shown in Fig. 15a, with perfect price

2 A Monopoly

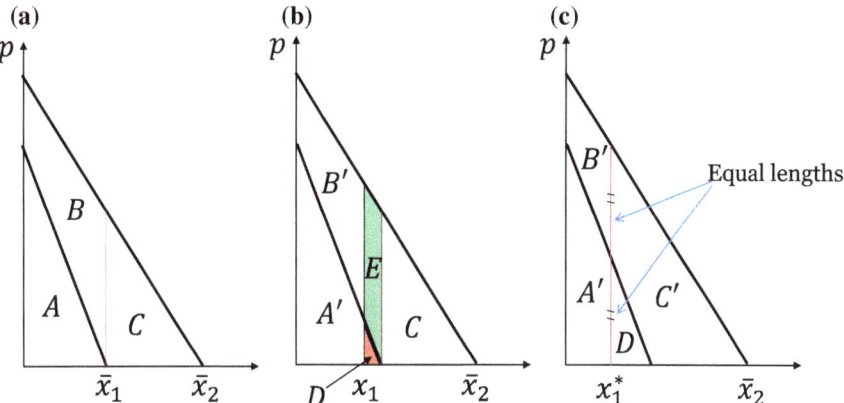

Fig. 15 Monopoly price under incomplete information

discrimination, the monopolist would like to sell $\bar{x}_1$ to consumer 1 for price A with package $(\bar{x}_1, A)$ and sell $\bar{x}_2$ to consumer 2 for price $A+B+C$ (the price here is the total charge rather than a unit price) with package $(\bar{x}_2, A+B+C)$. However, since the monopolist cannot distinguish the consumers, if these two packages are offered to the market, consumer 1 will pick $(\bar{x}_1, A)$, but consumer 2 will also pick $(\bar{x}_1, A)$, by which he retains a surplus of B instead of no surplus for the second package. The profit for the monopolist in this case is $\pi = 2A$.

To avoid the above problem, the monopolist can make two alternative offers: $(\bar{x}_1, A)$ and $(\bar{x}_2, A+C)$. If so, both consumers will choose the package intended for them. In particular, consumer 2 will choose $(\bar{x}_2, A+C)$ since the other yields the same surplus B. This strategy yields a higher profit for the monopolist. The profit is now $\pi = 2A + C$.

Comparing this profit with the previous one, we realize that profit improves if the monopoly designs a set of offers so that each consumer will pick what is intended for him. Such a set of offers is said to be incentive compatible. In the following, we impose the condition of incentive compatibility and try to find the profit-maximizing solution under this condition.

The above solution may not the best to the monopolist. For example, as shown in Fig. 15b, the monopolist can offer a lower x_1. The monopolist now offers: (x_1, A') and $(\bar{x}_2, A' + D + C + E)$. Each consumer will still choose the package that is intended for him. The monopolist's profit is increased since she loses the small red area but gains the larger green area. The profit is $\pi = 2A + C + E - D$, which is better than before as long as $E > D$.

Hence, the condition for profit maximization is

$$\frac{\partial \pi}{\partial x_1} = \frac{\partial E}{\partial x_1} - \frac{\partial D}{\partial x_1} = 0,$$

i.e., the optimal x_1^* is the one where the marginal benefit from one more unit of reduction in x_1 equals the marginal loss. As shown in Fig. 15c, the optimal offers are (x_1^*, A') and $(\bar{x}_2, A' + D + C')$, which maximize the monopolist's profit. Consumer 1 gets zero surplus and consumer 2 gets surplus B'. The profit is $\pi^* = 2A + C' - D$.

Is the optimal solution efficient? Under complete information, social welfare is

$$2A + B + C = 2A + B' + C'.$$

Under incomplete information, with the optimal solution, social welfare is

$$\pi^* + B' = 2A + C' - D + B'.$$

Hence, the optimal solution under incomplete information is inefficient, and the loss due to incomplete information is the amount of D.

We will revisit this problem again in Chap. 11 using a more rigorous approach.

3 Allocative Efficiency

Suppose that the inverse demand function is $p = p^d(x)$. For consumption level x, suppose that the price is fixed at p. Then, the <u>consumer surplus</u> at quantity x is

$$CS(x) \equiv \int_0^x \left[p^d(t) - p\right] dt, \tag{5}$$

as shown on the left side of Fig. 16. We can break this into two parts. $p^d(t)$ is the maximum price that the consumer is willing to pay for one more unit of the good when the consumer already has t units of the good. Thus, the consumer surplus is the difference between the total amount that the consumer is willing to pay for x and what the consumer actually pays:

$$CS(x) = \int_0^x p^d(t) dt - px.$$

Note that $CS(x)$ is the surplus in terms of "money" value (as opposed to utility value). See Varian (1992, 160–163) for the equivalence between this definition and a surplus in welfare when the utility function is quasi-linear.

Producer surplus can also be similarly defined. Let $MC(y)$ be the firm's marginal cost at output level y. Suppose that the price at which the firm sells that quantity y is fixed at p. Then, the <u>producer surplus</u> at quantity y is

3 Allocative Efficiency

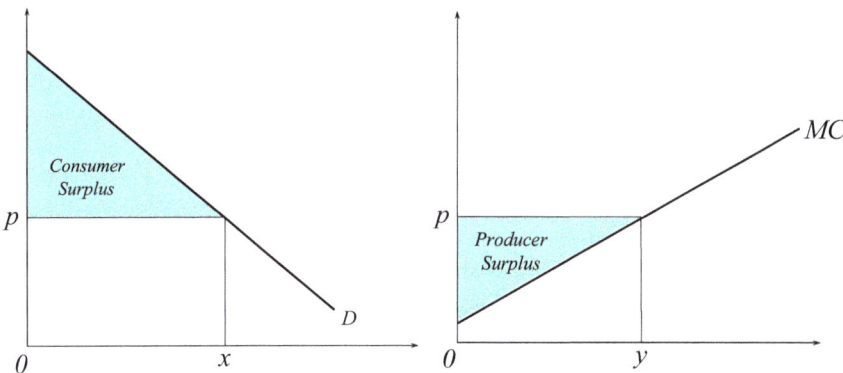

Fig. 16 Consumer surplus and producer surplus

$$PS(y) \equiv \int_0^y [p - MC(t)]dt, \qquad (6)$$

as shown on the right side of Fig. 16. It is the gain without taking into account the fixed cost. We have

$$c(y) - c(0) = \int_0^y MC(t)dt,$$

where $c(y) - c(0)$ is the variable cost and $c(0)$ is the fixed cost. Hence, the profit is

$$\pi(y) = PS(y) - c(0).$$

The social welfare is defined as the sum of all participants' surpluses. In particular, if only consumers and producers are involved in a trade, the social welfare is:

Social Welfare = Consumer Surplus + Producer Surplus.

A market equilibrium reaches allocative efficiency when the social welfare is maximized.[4]

Using this definition, we can now compare the allocative efficiency of a monopoly with that of a competitive industry. Under perfect competition, market equilibrium occurs where the industry supply curve and the industry demand curve

---

[4]In our definition, we implicitly assume that there are no external costs and external benefits. In this case, the demand curve is the marginal social benefit (MSB) curve and the marginal cost curve is the marginal social cost (MSC) curve. The social optimal point is where the MSB and MSC curves intersect.

intersect, as shown by point E in Fig. 17a. Now, suppose that the industry is taken over by a single firm. Assume that there are no changes in production techniques so that the new combined firm has the same cost structure as the cost structures of the separate firms. If it is a single-price monopoly, it will produce quantity M and charge price P_M in Fig. 17b. We see that the monopoly produces less and charges a higher price. If it is a perfectly price-discriminating monopoly, it will produce quantity C and charge prices along the demand curve. The highest price will be P_A for the first unit sold and the lowest price will be P_C for the last unit sold.

In a competitive equilibrium, as shown in Fig. 17a, the producer surplus is the area $P_C EF$, the consumer surplus is $P_A E P_C$, and the social welfare is the largest possible area $P_A EF$. Hence, a competitive industry is allocatively efficient.

In a single-price monopoly, as shown in Fig. 17b, the consumer surplus drops to area $P_A H P_M$, the producer surplus increases to area $P_M H G F$, and the social welfare drops to area $P_A H G F$. The loss in social welfare is the area HEG, which is called the <u>deadweight loss</u> because no one gets it. The size of the deadweight loss measures the loss in efficiency. The monopoly captures some of the consumer surplus by charging a higher price. But it also eliminates some producer surplus and consumer surplus by producing at an inefficient output level. The social welfare is not the maximum. Hence, a single-price monopoly is allocatively inefficient.

However, there is no deadweight loss if the monopoly practices perfect price discrimination. We see that this monopoly will produce at level C, which means that the monopoly is as efficient as a competitive industry. Thus, a perfectly price-discriminating monopoly is allocatively efficient. But in this case, the monopoly takes all the consumer surplus away. That is, with a perfectly price discriminating monopoly, the consumer surplus is zero and the producer surplus is equal to the total social welfare, which is the area $P_A EF$.

Further, as shown in Fig. 15, the profit for a monopoly under incomplete information is $\pi = 2A + C + E - D$, and the consumer surplus is B'. If the monopoly has complete information, its profit will be $\pi = 2A + B + C$, and the consumer surplus is 0. Hence, the loss of social welfare due to incomplete

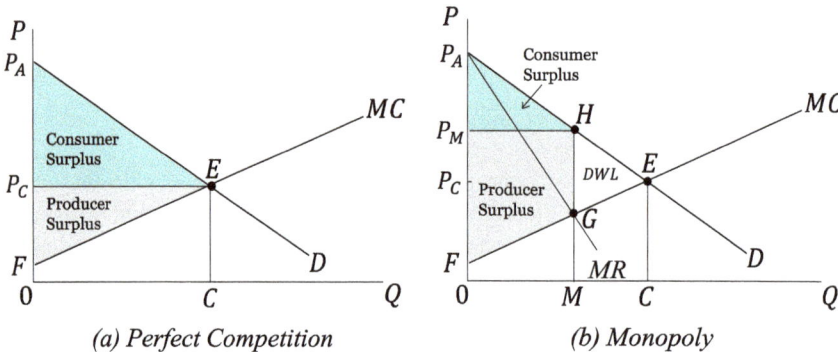

Fig. 17 Welfare implications

information is the size of D. Therefore, a monopoly under incomplete information is allocatively inefficient.

However, our simple-minded analysis may not include all gains and losses. Since monopoly generally implies a positive profit, there is competition for monopoly rights. If there is no barrier to this opportunity, in equilibrium, the resources used up in such an activity may equal the monopoly's profit. That is, competition for monopoly rights may result in the use of resources to acquire those rights. Those resources may be equal in value to the monopoly's potential profit. As a consequence,

Cost of Monopoly to a Society = Dead Weight Loss + Monopoly's Economic Profit.

On the other hand, there are gains from a monopoly such as economies of scale, economies of scope, and the incentive to innovate. This is especially true in the case of natural monopolies.

Example 6.5 Let us now consider the welfare implications of Example 6.3.

After a tax is imposed, in the short run, the consumer surplus drops from App^* to Bpp^{SR} on the left side of Fig. 18, and the producer surplus drops from Ap^*F to ECF. The tax revenue is $Bp^{SR}CE$. The tax is shared by both the consumers and the producers. If the tax revenue is used as efficiently as it would be in the hands of the consumers and producers, then the net loss (the deadweight loss) is BEA.

In the long run, the consumer surplus drops from App^* to Cpp^{LR} on the right side of Fig. 18. The tax revenue is paid by the consumers only. If the tax revenue is used as efficiently as it would be in the hands of the consumers, then the dead weight loss is ACG.

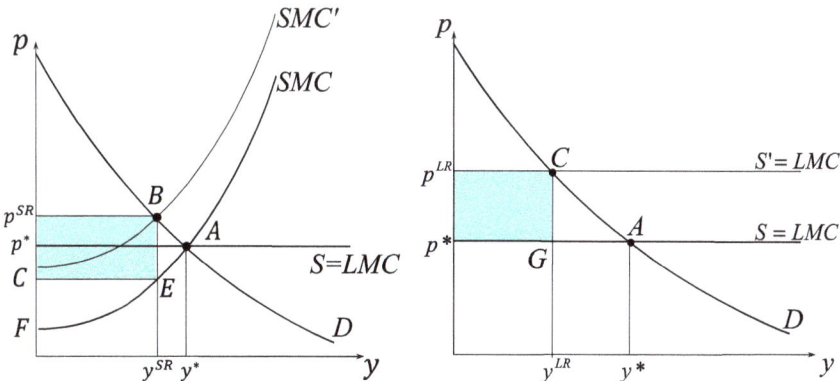

Fig. 18 Welfare implications of a sales tax

In both the long run and the short run, the tax creates a loss of efficiency. This is not surprising since the industry was competitive and hence efficient before the tax was levied. A price distortion in the industry caused by the tax can only reduce its efficiency. ∎

4 Monopolistic Competition

A <u>monopolistically competitive industry</u> is an industry in which

- There are *many firms*: firms are independent of each other in their decisions.
- *Product differentiation* exists: each firm faces a downward-sloping demand curve and therefore can choose its own price and quantity.
- There is *free entry*: zero profit is made in the long run.

An important difference between monopoly and monopolistic competition is free entry. Because of free entry, profit cannot persist in the long run. Apart from this, each firm behaves just like a monopoly.

Suppose that firm i faces inverse demand $p_i = p_i(y_1, y_2, \ldots, y_n)$. The demand not only depends on firm i's own output y_i but also on other firms' outputs. For $y = (y_1, \ldots, y_n)$, denote

$$y_{-i} \equiv (y_1, \ldots, y_{i-1}, y_{i+1}, \ldots, y_n), \quad y = (y_i, y_{-i}).$$

Since there are many firms in the industry, we can assume that each firm takes other firms' actions as given when it makes choices. That is, each firm follows a <u>Nash strategy</u> in output, where the firm chooses its optimal output given other firms' choices of output. Therefore, taking y_{-i} as given, firm i's problem is

$$\max_{y_i} \; \pi_i \equiv p_i(y_i, y_{-i}) y_i - c_i(y_i).$$

Then the FOC for firm i is

$$p_i(y_i^*, y_{-i}) + \frac{\partial p_i(y_i^*, y_{-i})}{\partial y_i} y_i^* = c_i'(y_i^*).$$

An equilibrium $(y_1^*, \ldots, y_n^*)$ is the point where every firm maximizes its own profits while taking other firms' choices as given. That is, the equilibrium is determined by:

$$p_i(y_i^*, y_{-i}^*) + \frac{\partial p_i(y_i^*, y_{-i}^*)}{\partial y_i} y_i^* = c_i'(y_i^*), \quad i = 1, \ldots, n. \tag{7}$$

4 Monopolistic Competition

When a time horizon comes into play, the equilibrium defined by (7) is only a short-run equilibrium. In the short run, the number of firms is given and fixed; no firms enter into or exit from the industry.

In the short-run equilibrium, the firms may make positive or negative profits. In the long run, when exit and entry are feasible, some firms may enter and some firms may exit until the profit is zero for every firm in the industry. This means that, in addition to condition (7), the number of firms in the industry will be determined by the following zero-profit conditions:

$$p_i(y_i^*, y_{-i}^*) y_i^* = c_i(y_i^*), \quad i = 1, 2, \ldots, n. \tag{8}$$

The short-run and long-run equilibria are illustrated in Fig. 19. In the short run, since no firm can enter, each firm acts just like a monopoly as shown in Fig. 19a. With profit being earned in the short run (a loss is also possible), new firms enter the industry and take some of the market away from the existing firms in the long run. As they do so, the demand curve for each firm starts to shift to the left. As the demand curve continues to shift to the left, each firm's profit gradually falls. In the long run equilibrium, as shown in Fig. 19b, every firm makes zero profit.

We have a few observations of the solution. First, in the long-run equilibrium, as shown in Fig. 19b, the AC curve is tangent to the demand curve at the optimal point for each firm. This can be rigorously proven. By (8), we know that firm i's AC curve cuts or touches its demand curve at the optimal point. By (7), we have

$$\frac{\partial p_i(y^*)}{\partial y_i} = \frac{MC_i(y^*)}{y_i^*} - \frac{p_i(y^*)}{y_i^*} = \frac{MC_i(y^*)}{y_i^*} - \frac{AC(y^*)}{y_i^*} = \frac{d}{dy_i}\left[\frac{c_i(y_i, y_{-i}^*)}{y_i}\right]_{y_i=y_i^*}.$$

That is, firm i's AC curve and demand curve have the same slope at the optimal point. Therefore, these two curves are tangent at the optimal point.

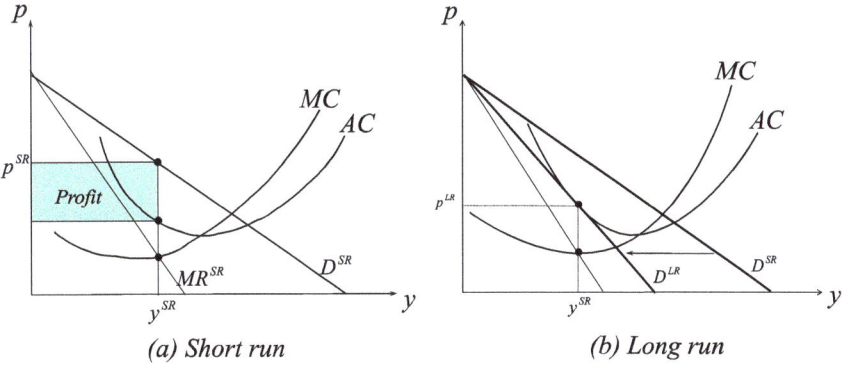

Fig. 19 Monopolistic competition

Second, in the long run, each firm in monopolistic competition always has excess capacity in that its output is lower than that at which the average total cost is minimum. This result arises from the fact that the firm faces a downward-sloping demand curve. Product differentiation causes a downward-sloping demand curve and thus excess capacity.

Third, allocative efficiency is achieved when the demand curve intersects with the MC curve. Hence, monopolistic competition is allocatively inefficient. Even though there is zero profit in the long-run equilibrium, a monopolistically competitive industry produces a level at which price exceeds marginal cost.

Finally, besides what we can see from a graphic analysis, there may be other gains and losses. Monopolistically competitive firms may attempt to differentiate the consumer's perceptions of the product, principally by advertising, which incurs costs. On the other hand, the loss in allocative efficiency in monopolistic competition has to be weighed against the gain from greater product variety and product innovation. Monopolistically competitive firms are constantly seeking out new products that will provide them with a competitive edge, even if only temporarily.

5 Oligopoly

Oligopoly is an industry in which

- There are a *small number of firms*: firms depend on each other.
- There is a single product: firms jointly face a downward-sloping industry demand curve.
- There is *no entry*: long-run positive profits are possible.

Different from a monopolistically competitive industry, in an industry with a small number of firms, the firms realize that they are dependent on each other. Since they jointly face a common demand curve, each firm's sales depends not only on its own price but also on other firms' prices. If one firm lowers its own price, its own sales increase, but the sales of the other firms in the industry decrease. In such a situation, other firms will most likely lower their prices too. Hence, before deciding to cut its own price, each firm may try to predict how other firms will react and attempt to calculate the effects of those reactions on its own profit. To analyze the behavior of the firms in such a market, game theory is needed.

Game theory analyzes strategic interactions. Strategic interaction occurs when a firm takes into account the expected behavior of other firms and their mutual recognition of interdependence. What kind of situation in such a game can be reasonably called an equilibrium or solution? A simple concept of equilibrium is called the Nash equilibrium. A player is said to play a Nash strategy if he makes a choice by assuming that others will stay where they are. A Nash equilibrium is when no one wants to change by assuming others will not change. In other words, a Nash equilibrium is when everyone is playing an optimal Nash strategy. Firms may

5 Oligopoly

play Nash strategies in quantities or in prices. The <u>Cournot equilibrium</u> is a Nash equilibrium when both firms play Nash strategies in quantities. The <u>Bertrand equilibrium</u> is a Nash equilibrium when both firms play Nash strategies in prices.

A Nash game is a game in which all players move simultaneously. A game can also be played dynamically, in which some players move ahead of others. In an oligopoly, we call the firm that moves first the leader and the firms that moves second the followers. A <u>Stackelberg equilibrium</u> is when each follower makes his choice to maximize his own payoff using a Nash strategy and the leader makes her choice to maximize her own payoff by taking into account the followers' reactions. We may think of a Nash game as simply a one-period game in which all players take actions simultaneously and of a Stackelberg game as a two-period game in which the leader moves in the first period and the followers move in the second period.

We will focus on a simple case of oligopoly called duopoly. <u>Duopoly</u> is an oligopoly in which there are only two firms.

5.1 Bertrand Equilibrium

Consider a duopoly. The two firms compete on prices, i.e., they play Nash strategies using their prices as strategic variables. Let $p \equiv (p_1, p_2)$ be the price vector and $x(p)$ be the market demand for the product. Assume that the two firms have the same cost function and the cost function has a constant marginal cost $c > 0$. The two firms simultaneously offer their prices p_1 and p_2. Sales for firm i are then given by

$$x_i(p) = \begin{cases} x(p) & \text{if } p_i < p_j, \\ x(p)/2 & \text{if } p_i = p_j, \\ 0 & \text{if } p_i > p_j. \end{cases}$$

The profit is

$$\pi_i = (p_i - c)x_i(p).$$

An equilibrium is said to be <u>stable</u> if, after a deviation from the equilibrium, the system is able to move back to the original equilibrium by itself. Figure 21 illustrates such a situation.

Proposition 6.1 (Bertrand). *In a duopoly with a constant marginal cost c, there is a unique stable Bertrand equilibrium (p_1^*, p_2^*), in which $p_1^* = p_2^* = c$.*

Proof Suppose that firm 2 offers price $p_2 > c$. Then, if firm 1 offers the same price $p_1 = p_2$, they share the market half-half, by which firm 1's profit is

$$\pi_1' = \frac{1}{2}(p_2 - c)x(p_2).$$

Fig. 20 Converging to a Bertrand equilibrium

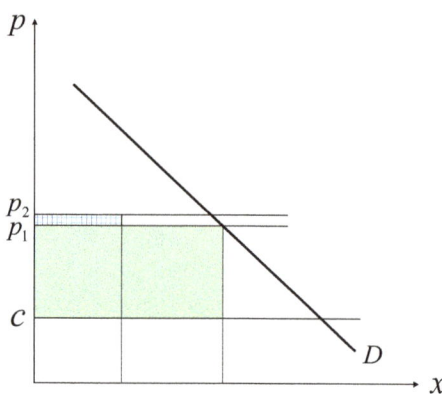

However, if firm 1 chooses a price slightly lower than p_2, say $p_1 = p_2 - \varepsilon$, firm 1 takes over the whole market and firm 1's profit becomes (see Fig. 20):

$$\pi_1'' = (p_2 - c - \varepsilon)x(p_2 - \varepsilon).$$

Since $\pi_1'' \to (p_2 - c)x(p_2)$ as $\varepsilon \to 0$, we have $\pi_1'' > \pi_1'$ when ε is small enough. Therefore, as long as $p_2 > c$, firm 1 will offer a price lower than p_2. Symmetrically, as long as $p_1 > c$, firm 2 will offer a price lower than p_1. Therefore, it is not a Nash equilibrium if $p_1 > c$ or $p_2 > c$.

Finally, we can easily verify that the situation when $p_1 = p_2 = c$ is a Nash equilibrium. Also, by the way that we derive the equilibrium, it must be stable. ■

We can similarly show that when there are n identical firms with a constant marginal cost c, the Bertrand equilibrium is still the outcome when $p_1^* = \cdots = p_n^* = c$.

The striking implication of the Bertrand equilibrium is that a competitive outcome can be obtained with two firms only. In fact, price competition in this case results in a horizontal demand curve for each firm.

However, this solution is problematic by common sense. For example, in Fig. 20, given p_2, when firm 1 thinks about lowering its price slightly below p_2, it should realize that its action will cause the other firm to react, which is likely to lower p_2 below p_1. If so, firm 1's action leads to zero profit eventually. Hence, firm 1 is better setting its price $p_1 = p_2$. That is, if firm 1 takes into account a possible reaction from firm 2 when it considers lowering its price, firm 1 may not choose to do so. If so, the two parties may set a price above c and stays there. This is an equilibrium (not a Nash equilibrium) since a possible reaction from the other party serves as a deterrent for a deviation. See the concept of reactive equilibrium in Chap. 7.[5]

---

[5]The reactive equilibrium is defined by Riley (1979). Wilson (1977) defines the anticipated equilibrium, which is very similar to the reactive equilibrium.

5.2 Cournot Equilibrium

Consider the duopoly again, but the two firms now play Nash strategies in quantity. In a Cournot game, taking y_2 as given, firm 1's problem is

$$\max_{y_1} p(y_1 + y_2)y_1 - c_1(y_1); \tag{9}$$

and, taking y_1 as given, firm 2's problem is

$$\max_{y_2} p(y_1 + y_2)y_2 - c_2(y_2).$$

The FOCs are

$$\begin{aligned} p'(\hat{y}_1 + y_2)\hat{y}_1 + p(\hat{y}_1 + y_2) &= c'_1(\hat{y}_1), \\ p'(y_1 + \hat{y}_2)\hat{y}_2 + p(y_1 + \hat{y}_2) &= c'_2(\hat{y}_2). \end{aligned} \tag{10}$$

Hence, the Cournot equilibrium (y_1^*, y_2^*) is determined by the two FOCs:

$$p'(y_1^* + y_2^*)y_1^* + p(y_1^* + y_2^*) = c'_1(y_1^*), \quad p'(y_1^* + y_2^*)y_2^* + p(y_1^* + y_2^*) = c'_2(y_2^*).$$

We can also look at this equilibrium from another angle. The FOC for firm 1 determines a <u>reaction function</u> of firm 1:

$$\hat{y}_1 = f_1(y_2),$$

and the FOC for firm 2 determines another reaction function of firm 2:

$$\hat{y}_2 = f_2(y_1).$$

A firm's reaction function is implied by the firm's Nash strategy. The Cournot equilibrium is where the two reaction curves intersect, as shown in Fig. 21.
As shown in Fig. 21, the Nash equilibrium is stable if f_1 is steeper than f_2 at the intersection point, as shown on the left of Fig. 21. If f_2 is steeper than f_1, it would be unstable, as shown on the right of Fig. 21. The left diagram means that the stability condition is

$$\left| \frac{\partial f_2}{\partial y_1} \right| < \left| \frac{\partial y_2}{\partial f_1} \right| \quad \text{or} \quad \left| \frac{\partial f_2}{\partial y_1} \frac{\partial f_1}{\partial y_2} \right| < 1.$$

Of course, the intersection points may not be unique, implying the possibility of multiple equilibria.

Proposition 6.2 (Cournot). *In a Cournot equilibrium with constant marginal cost c for both firms, we have*

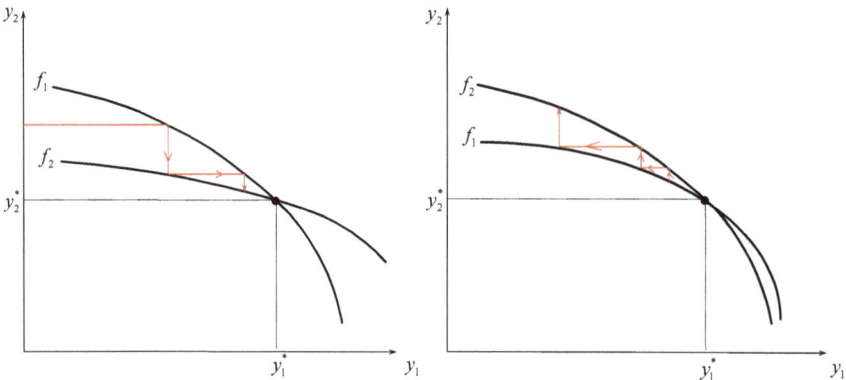

Fig. 21 The stability of Cournot equilibria

$$p^c \leq p^* \leq p^m,$$

where p^* is the market price in the Cournot equilibrium, p^c is the competitive price, and p^m is the monopoly price (see Fig. 22).

Proof By (10), we have

$$p(y_1^* + y_2^*) + p'(y_1^* + y_2^*)y_i^* = c, \tag{11}$$

where y_i^* is firm i's output in the Cournot equilibrium. Since $p'(y) \leq 0$ for any $y \geq 0$, we have $p(y_1^* + y_2^*) \geq c = p^c$.

Let y^m be the monopoly output. Given outputs (y_1^*, y_2^*), if $y^m > y_1^* + y_2^*$, then firm 1 can increase its output from y_1^* to $y_1' = y^m - y_2^*$, while firm 2 keeps at y_2^*. Then, the joint profit for the two firms is the monopoly profit π^m, which cannot be less than the joint profit π^* in the Cournot equilibrium. Also, when the total output is increased, the price must fall, implying that firm 2 must be worse off. With the joint profit larger than before and with firm 2 losing profit, firm 1's profit must increase by deviating from y_1^* to y_1'. Thus, (y_1^*, y_2^*) could not be a Nash equilibrium, which is a contradiction. Therefore, we must have $y^m \leq y_1^* + y_2^*$. ∎

In general, with n identical firms, condition (11) becomes

$$p(y^*) + p'(y^*)\frac{y^*}{n} = c, \tag{12}$$

where y^* is the joint output in a Cournot equilibrium. In one extreme, when $n = 1$, (12) gives us the monopoly solution y^m. In the other extreme, when $n \to \infty$, since y^* is always less than the competitive solution y^c, we have $\frac{1}{n}y^* \to 0$ so that (12) gives us the competitive solution y^c.

Fig. 22 A Cournot equilibrium

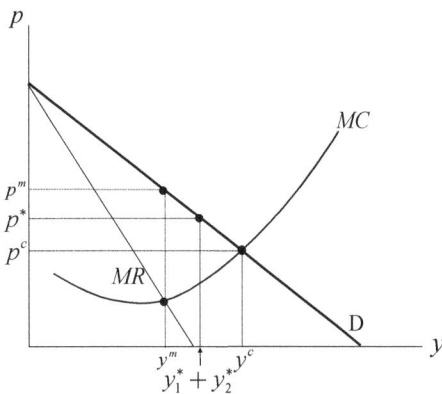

Most firms in reality seem to use their prices, not their quantities, as their strategy variables, yet the reality tends to produce an outcome that is close to the Cournot solution. That is, the Cournot model gives the right answer for the wrong reason. One possible explanation for this is capacity constraints. We can think of a quantity choice in the Cournot model as a long-run choice of capacity, with the price being an outcome of short-run price competition.

What is the outcome in a model in which firms first choose their capacity levels and then compete on prices? Kreps and Scheinkman (1983) show that the unique subgame perfect Nash equilibrium in a two-stage model with short-run capacity constraints is the Cournot outcome. In other words, the competition of price in the Bertrand model can be thought as the second-stage price competition. We can thus think of the Cournot quantity competition as capturing long-run competition through a capacity choice, with price competition occurring in the short run given the chosen level of capacity. See Wang and Zhu (2001) for such a model.

Example 6.6 Consider two identical firms with $c_i(y_i) \equiv cy_i$ and with industry demand $p = a - y$, where $a > c > 0$. Then, the firms' profits are

$$\pi_i \equiv (a - y_1 - y_2)y_i - cy_i, \quad i = 1, 2. \tag{13}$$

If both play Nash strategies in quantity, the FOC for firm i is

$$a - c - 2y_i - y_j = 0. \tag{14}$$

By symmetry, with $y_1^N = y_2^N$ in equilibrium, we find the Nash equilibrium solution:

$$y_1^N = y_2^N = \frac{a-c}{3}, \quad \pi_1^N = \pi_2^N = \frac{1}{9}(a-c)^2. \quad \blacksquare$$

5.3 Stackelberg Equilibrium

The Stackelberg equilibrium is a two-period solution. By applying the dynamic optimality principle, the equilibrium can be solved using the backward-solving method.

Suppose that firm 1 is the leader and firm 2 is the follower, which means that firm 1 moves first by taking into account firm 2's possible reactions and firm 2 follows based on its own reaction function. Firm 2's reaction function is $\hat{y}_2 = f_2(y_1)$ determined by (10). Then, firm 1's problem is

$$\max_{y_1} p[y_1 + f_2(y_1)]y_1 - c_1(y_1). \tag{15}$$

Notice that this is different from (9) in that (9) treats y_2 as a constant. Let y_1^{**} be a solution of (15), then firm 2's optimal choice is $y_2^{**} \equiv f_2(y_1^{**})$ and the Stackelberg equilibrium is (y_1^{**}, y_2^{**}).

Example 6.7 Consider the firms in Example 6.6. Suppose now that firm 1 is the leader and firm 2 is the follower. By (14), firm 2's reaction function is $\hat{y}_2 = \frac{a-c-y_1}{2}$. Thus, firm 1's problem is

$$\max_{y_1} \left(a - c - y_1 - \frac{a-c-y_1}{2} \right) y_1,$$

implying the Stackelberg equilibrium solution:

$$y_1^S = \frac{a-c}{2}, \quad y_2^S = \frac{a-c}{4}, \quad \pi_1^S = \frac{(a-c)^2}{8}, \quad \pi_1^S = \frac{(a-c)^2}{16}.$$

In this case, the leader has a larger output share and a higher profit. But, this is not a general conclusion. In some cases, a firm is better off to be a follower rather than a leader. ∎

5.4 Cooperative Equilibrium

The Nash game and the Stackelberg game are <u>noncooperative games</u> in which the two players do not cooperate in any way to enhance their common objectives. A <u>cooperative game</u> is a game in which the players have a common objective even though they may act in their own best interests.[6] In the duopoly case, if the two firms agree to cooperate with each other and to maximize their total profit, they will first consider the cost minimization problem:

---

[6] The fact that each player acts in his own best interest does not automatically mean that the game is noncooperative. For example, the players may work as a team for the total surplus of a project, but they may compete for or react to their shares of the surplus in their own best interests.

5 Oligopoly

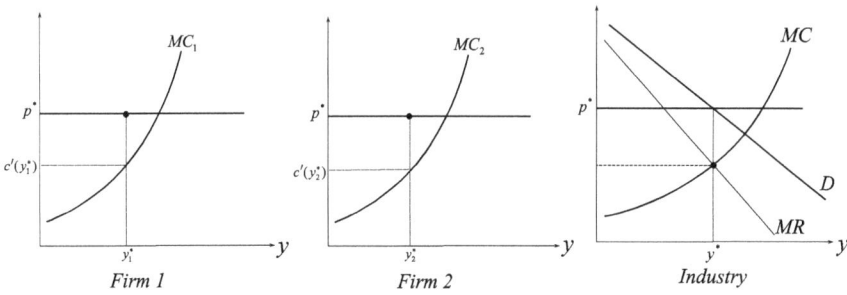

Fig. 23 The cooperative solution of a Duopoly

$$c(y) \equiv \min_{y_1, y_2 \geq 0} c_1(y_1) + c_2(y_2)$$
$$\text{s.t.} \quad y = y_1 + y_2,$$

and then the profit maximization problem:

$$\max_{y \geq 0} p(y)y - c(y).$$

The FOCs of the two problems for the cooperative equilibrium are

$$c'(y^*) = c'_1(y_1^*) = c'_2(y_2^*), \quad p'(y^*)y^* + p(y^*) = c'(y^*), \quad (16)$$

where $y^* = y_1^* + y_2^*$. This problem is the same as the problem of a single monopoly with two plants. The solution is shown in Fig. 23, where the MC curve is the horizontal sum of MC_1 and MC_2.

The difficulty in a cooperative game is how to share the benefit among the players. Although the cooperative equilibrium tells us how much each firm should produce, it does not tell us how the two firms should share the benefit. In practice, the two firms typically bargain to divide the benefit.

The cooperative equilibrium is prone to cheating. In Fig. 23, we can see that the price is higher than the marginal costs for individual firms, implying that given the price each firm can do better by expanding its own output. Hence, when the output is not observable or enforceable, this equilibrium is prone to cheating.

Example 6.8 Reconsider the firms in Example 6.6 again. If both choose cooperation, they will maximize

$$\pi \equiv \pi_1 + \pi_2 = (a - y)y - cy.$$

Assuming equal shares, we find the solution:

$$y_1^C = y_2^C = \frac{a-c}{4}, \quad \pi_1^C = \pi_2^C = \frac{1}{8}(a-c)^2. \quad \blacksquare$$

5.5 Competition Versus Cooperation

In the duopoly, we have seen three possible games that the two firms may play: Nash, Stackelberg, and cooperation. Which game will they play? This question is like the well-known story of the prisoners' dilemma.

The <u>prisoners' dilemma</u> goes as follows. Two prisoners who will be in jail for the next two years are suspected of having committed another crime. Each prisoner is told that if they both confess to that crime, they will be convicted for the crime and each will stay an additional year in jail; if one prisoner confesses to the crime and the accomplice does not, they will be convicted of the crime and the confessor will stay one year less in jail and his accomplice will spend another eight years in jail; and if neither of them confesses, they will not be convicted of the crime. This is a two-person game. Each player has two strategies: confess or deny. There are four outcomes and their payoffs are listed in the following table, where, for example, $(-10, -1)$ means that prisoner 1 loses 10 years of freedom and prisoner 2 loses 1 year of freedom (Fig. 24).

The Nash equilibrium of this game is that both choose to confess. Furthermore, this equilibrium is a stronger type of equilibria called a dominant strategy equilibria. A <u>dominant strategy</u> is a strategy that is the best strategy regardless of actions taken by the other players. A <u>dominant strategy equilibrium</u> occurs when every player is taking a dominant strategy in equilibrium. The Nash equilibrium for the prisoner's dilemma is a dominant strategy equilibrium.

Let us now apply this story to the duopoly case. Consider the firms in Example 6.6. They have two strategies: produce at the cooperative output or cheat by playing Nash strategies on quantity. For the two firms in Example 6.6, if one firm, say firm 1, produces at the cooperative quantity,

$$y_1^C = \frac{a-c}{4},$$

		Prisoner 2	
		Confess	Deny
Prisoner 1	Confess	-3, -3	-1, -10
	Deny	-10, -1	-2, -2

Fig. 24 The prisoners' dilemma

and the other firm, firm 2, cheats by playing a Nash strategy. Firm 2 will maximize

$$\pi_2^{NC} \equiv (a - y_1^C - y_2)y_2 - cy_2,$$

and produce at

$$y_2^{NC} = \frac{3(a-c)}{8}.$$

The profits in this case are

$$\pi_1^{NC} = \frac{3}{32}(a-c)^2, \quad \pi_2^{NC} = \frac{9}{64}(a-c)^2,$$

where subscript N stands for the Nash strategy and C stands for the cooperative strategy. The profits in other situations can be found in Examples 6.6 and 6.8. The profits are put into the following matrix box, where the entries are the profits without the constant term $(a-c)^2$.

It is just like the prisoners' dilemma. The Nash equilibrium for this game is (Nash, Nash), i.e., both play the Nash strategy. This Nash equilibrium is also a dominant strategy equilibrium. Hence, this result predicts that the cooperative solution is not sustainable and the two firms will end up in the Nash equilibrium.

5.6 Cooperation in a Repeated Game

For the prisoners or the firms in the duopoly, the Nash equilibrium is not a Pareto optimal outcome. Isn't there some way by which the better solution can be achieved?

The duopoly game is played once. But if the duopolists play the game more than once, they might find some way to cooperate. For example, if a game is played repeatedly, called a <u>repeated game</u>, one player will be able to penalize the other player for 'bad' behavior by a trigger strategy. If one player cheats once, the other will then refuse to cooperate from now on. This form of punishment is referred to as the <u>trigger strategy</u>. By this trigger strategy, a player may find that the one-time gain from cheating may not cover the future losses. If each player takes this into account, none of them may cheat. Thus, a cooperative solution may be sustainable under the trigger strategy.

Example 6.9 For the game in Fig. 25, suppose that the two firms are initially in a cooperative equilibrium and the discount rate of time preference is $\delta \in (0, 1)$. With the trigger strategy, the net gain from cheating is

$$\left(\frac{9}{64} - \frac{1}{8}\right) + \sum_{t=1}^{\infty} \delta^t \left(\frac{1}{9} - \frac{1}{8}\right) = \frac{1}{64} - \frac{\delta}{72(1-\delta)}.$$

		Firm 2	
		Nash	Coop
Firm 1	Nash	1/9, 1/9	9/64, 3/32
	Coop	3/32, 9/64	1/8, 1/8

Fig. 25 The Duopoly's game on games

Hence, the cooperative solution is sustainable if $\delta \geq 9/17$, but the Nash equilibrium will be the outcome if $\delta < 9/17$. ∎

6 Production Differentiation

Even if firms have the same product, price differentiation may result if firms are located in different locations and consumers have to pay transportation costs to purchase the product. Here, the distance between two firms may not be the physical distance. The distance may represent product differentiation of the same product due to differences in color, design, style, and consumer perception.

Consider two identical firms, firms 1 and 2, located at 0 and 1 respectively on interval $[0, 1]$ selling the same product with a constant marginal cost c. Consumers are located uniformly on $[0, 1]$. The total size of the consumers is normalized to 1 and each consumer buys one and only one unit of the good.[7] The total cost of buying the product from firm i is $p_i + td$, where p_i is the price charged by firm i, d is the distance between the consumer and the firm, and t is a positive constant representing transportation cost. See Fig. 26.

Let $\hat{z}$ be the location of the consumer, called the neutral consumer, who is indifferent between the two firms, where $\hat{z}$ satisfies

$$p_1 + t\hat{z} = p_2 + t(1 - \hat{z}),$$

implying

$$\hat{z} = \frac{t + p_2 - p_1}{2t}.$$

We can see that consumers on the left of $\hat{z}$ will go to firm 1 and the rest will go to firm 2. Then, the demand for firm 1's product is $\hat{z}$, assuming $\hat{z} \in [0, 1]$. Thus, given p_2, firm 1's problem is

---

[7]We ignore the possibility that a consumer may not buy if the costs from both firms are too high. See Mas-Collel et al. (1995, p. 396) for a more general case in which this possibility is considered.

6 Production Differentiation

Fig. 26 Location differentiation

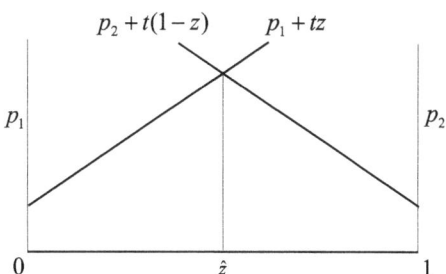

$$\max_{p_1}(p_1 - c)(t + p_2 - p_1)\frac{1}{2t}.$$

The FOC is

$$t + p_2 - 2p_1 + c = 0.$$

Given the symmetry of the model, we look for a symmetric equilibrium. With $p_1^* = p_2^*$ in equilibrium, the above equation immediately implies the equilibrium solution:

$$p_1^* = p_2^* = c + t. \qquad (17)$$

This model offers a different solution from that in Proposition 6.1. One natural question is: why not lower down the price further as in the case in Proposition 6.1? The explanation is that when a firm lowers its price, it will attract some consumers but not all the consumers. Due to the transportation cost, some consumers will not change their choices. Hence, a firm has to weigh between attracting more consumers and a lower price on all its sales. Consequently, the equilibrium prices in this model are higher than that in Proposition 6.1, although both are in Bertrand equilibrium.

7 Location Equilibrium

Have you ever wondered why some shops selling similar products gather in a neighborhood or even in a single building? For example, bank branches from different banks often gather in a neighborhood. Most computer shops in Hong Kong gather in several buildings throughout Hong Kong; each building houses many dozens of computer shops selling similar products. Why don't they spread out? By spreading out, they can charge a higher price for the same product, as shown in (17).

In the previous section, the locations are fixed. If firms can change locations, as they understand the effect of location, they may locate themselves in the best locations. How will firms locate themselves in equilibrium?

Consider the same model as in the last section, except that locations are part of the firms' choices. Now, the two firms play a two-stage game, in which the firms decide locations in the first stage and then play a Bertrand game in prices in the second stage.

Suppose that firm i sits at location x_i and charges price p_i. Given prices (p_1, p_2) and locations (x_1, x_2) of the two firms, assuming $x_1 \leq x_2$,[8] a consumer $z \in [0, 1]$ will go to firm 1 iff

$$p_1 + t|z - x_1| < p_2 + t|x_2 - z|.$$

Consider first the case with $x_1 < x_2$ (as opposed to $x_1 = x_2$). If the neutral consumer is at $[0, x_1]$, then everyone at $[0, x_1]$ will be neutral. This is because, given the prices, when a consumer moves along $[0, x_1]$, his preference will not change.[9] Symmetrically, if the neutral consumer is at $[x_2, 1]$, then everyone at $[x_2, 1]$ will be neutral. Thus, if there is a Nash equilibrium with $x_1 < x_2$, a neutral consumer can be assumed to be at $[x_1, x_2]$ (Fig. 27).

Denote the neutral consumer at $[x_1, x_2]$ as $\hat{z}$. For this consumer, we must have

$$p_1 + t(\hat{z} - x_1) = p_2 + t(x_2 - \hat{z}),$$

implying

$$\hat{z} = \frac{p_2 - p_1 + t(x_2 + x_1)}{2t} = \frac{p_2 - p_1}{2t} + \frac{x_2 + x_1}{2}.$$

All consumers in $[0, \hat{z})$ will go to firm 1 and the rest will go to firm 2. Hence, the profits are

$$\pi_1 = (p_1 - c)\hat{z} = (p_1 - c)\left(\frac{p_2 - p_1}{2t} + \frac{x_2 + x_1}{2}\right),$$
$$\pi_2 = (p_2 - c)(1 - \hat{z}) = (p_2 - c)\left(1 - \frac{p_2 - p_1}{2t} - \frac{x_2 + x_1}{2}\right).$$

We solve the game backwards. Given the locations x_1 and x_2, consider the Bertrand equilibrium in the second stage. The FOC for firm 1 is

$$\frac{p_2 - p_1}{2t} + \frac{x_2 + x_1}{2} - \frac{p_1 - c}{2t} = 0,$$

implying firm 1's reaction function:

$$\hat{p}_1 = \frac{p_2 + c + t(x_1 + x_2)}{2}. \tag{18}$$

---

[8]There is no loss of generality with this assumption as we can always name the firm on the left to be firm 1.
[9]This is because, for $x \in (0, x_1)$, if $p_2 + t(x_2 - x) = p_1 + t(x_1 - x)$, then, for any $\varepsilon \in \mathbb{R}$, $p_2 + t(x_2 - x + \varepsilon) = p_1 + t(x_1 - x + \varepsilon)$.

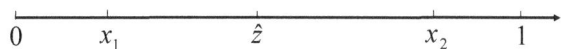

Fig. 27 Location equilibrium

Symmetrically, firm 2's reaction function is

$$\hat{p}_2 = \frac{p_1 + c + t(1 - x_1 + 1 - x_2)}{2}. \tag{19}$$

Substituting (19) into (18) yields the equilibrium prices:

$$p_1^* = c + \frac{2}{3}t + \frac{1}{3}t(x_1 + x_2), \quad p_2^* = c + \frac{4}{3}t - \frac{1}{3}t(x_1 + x_2). \tag{20}$$

The prices are generally different. When $x_1 = 0$ and $x_2 = 1$, as expected, these prices are the same as those in (17).

We now consider the problem in the second stage. Given the prices in (20), the neutral consumer is

$$\hat{z} = \frac{1}{3} + \frac{x_2 + x_1}{6}.$$

Then, the profits are

$$\pi_1^* = (p_1^* - c)\hat{z} = t\left(\frac{2}{3} + \frac{x_1 + x_2}{3}\right)\left(\frac{2}{3} + \frac{x_2 + x_1}{6}\right),$$
$$\pi_2^* = (p_2^* - c)(1 - \hat{z}) = t\left(\frac{4}{3} - \frac{x_1 + x_2}{3}\right)\left(\frac{1}{3} - \frac{x_2 + x_1}{6}\right).$$

We find

$$\frac{\partial \pi_1^*}{\partial x_1} = \frac{\partial \pi_1^*}{\partial (x_1 + x_2)} = t\left[\frac{1}{3}\left(\frac{2}{3} + \frac{x_2 + x_1}{6}\right) + \frac{1}{6}\left(\frac{2}{3} + \frac{x_1 + x_2}{3}\right)\right] > 0,$$

$$\frac{\partial \pi_2^*}{\partial x_2} = \frac{\partial \pi_2^*}{\partial (x_1 + x_2)} = -t\left[\frac{1}{3}\left(\frac{1}{3} - \frac{x_2 + x_1}{6}\right) + \frac{1}{6}\left(\frac{4}{3} - \frac{x_1 + x_2}{3}\right)\right] < 0.$$

Therefore, given x_2 and given the assumption that $x_1 < x_2$, firm 1 will move as close to x_2 as it can; symmetrically, given x_1, firm 2 will also move as close to firm 1 as it can. Thus, there is no equilibrium satisfying $x_1 < x_2$. Hence, the only possible equilibrium is where $x_1 = x_2$. In other words, the two firms must locate at the same location in equilibrium.

If $x_1^* = x_2^* > 1/2$, one firm can move slightly to the left without changing its price, and this firm will capture all the consumers on the left although it loses all the consumers on the right. This firm will be better off since the size of consumers on the left of x_i^* is larger than that on the right of x_i^*. This contradicts the definition of Nash equilibrium. The situation is the same if $x_1^* = x_2^* < 1/2$. Therefore, in equilibrium, both firms must sit in the middle:

$$x_1^* = x_2^* = 1/2.$$

When the two firms do sit together, the price behavior derived above no longer applies. In fact, once they both sit together, Proposition 6.1 applies and the Bertrand equilibrium must be $p_1^* = p_2^* = c$.

We have a few remarks about the solution. First, if there are more than two identical firms, there does not exist a Nash equilibrium. The reason is intuitive. For example, if there are three firms, it is still the case that the only possible equilibrium is the situation in which all three firms sit in the middle. But this situation cannot be an equilibrium since one of them can capture half of the market, instead of one third, by moving slightly to one side.

Second, what will happen if n identical firms are located on a circle? What will be the equilibrium positions? It seems that the Nash equilibrium must be the situation in which all firms sit around the circle at an equal distance from each other. For example, if there are two firms, they must have the same price in the Nash equilibrium. If so, no matter how they locate themselves, they will get the same market share. However, if they sit opposite to each other, the consumers will have the lowest overall transportation cost. If so, the profit must be the highest. Thus, the Nash equilibrium is one in which the two firms sit opposite to each other.[10]

Third, in reality, consumers are distributed on a flat surface, rather than on a line or a circle.[11] What will happen? If there are no buildings to block a path, the behavior of the two firms should be just like the two firms on a line. That is, the two firms should sit together in the middle of the region in equilibrium. In this case, on a surface, a Nash equilibrium exists for an arbitrary n.

Finally, the non-existence of the equilibrium may be due to the naivety of the Nash equilibrium concept—each player acts on the assumption that others will not react to his move. Under the reactive equilibrium concept (see Footnote 5), a player will not make a move if others' reactions can cause him to lose in the end. Under this equilibrium concept, the situation where all firms sit in the middle is an equilibrium. In fact, any situation in which all the firms sit together is a reactive equilibrium.

8 Entry Barriers

A monopolist industry can be a result of entry barriers and entry costs. In this section, we use a simple model to discuss the effect of entry cost on the formation of an industry.

---

[10]In this case, we need to include a participation constraint: the consumer is willing to buy a unit from firm i only if $p_i + td \geq v$, where is the consumer's utility value from consumption of the product. See Mas-Colell et al. (1995, p. 399).

[11]There are two standard models of spatial competition in the literature: the linear city model pioneered by Hotelling (1929), and the circular city model pioneered by Lerner and Singer (1939), developed by Vickrey (1964) and made popular by Salop (1979).

8 Entry Barriers

Consider an industry in which

Stage 1. All potential firms simultaneously decide to be in or out. If a firm decides to be in, it pays a setup cost $K > 0$. This K is a sunk cost.

Stage 2. All firms that have entered play a Cournot game.

We use simple cost and demand functions:

$$c_i(y) = cy, \quad p^d(y) = a - y, \tag{21}$$

where a and c are two constants with $a > c \geq 0$. In stage 2, suppose that there are n firms in the industry. Then, each firm considers the following problem

$$\pi_i = \max_{y_i \geq 0} \left(a - \sum_{j=1}^{n} y_j - c \right) y_i.$$

The reaction function is

$$\hat{y}_i = \frac{1}{2} \left(a - c - \sum_{j \neq i}^{n} y_j \right).$$

In the symmetric equilibrium with $y_1^* = \cdots = y_n^*$, the above reaction function implies

$$y_i^* = \frac{a-c}{n+1},$$

which implies a profit of

$$\pi_i^* = (a - c - ny_i^*)y_i^* = \left(\frac{a-c}{n+1} \right)^2.$$

The total output ny_i^* approaches the competitive output $a - c$ when $n \to \infty$.

How many firms will the industry have in equilibrium? Firms will continue to enter the industry until the expected profit is zero:

$$\left(\frac{a-c}{n+1} \right)^2 - K = 0,$$

which implies the equilibrium number of firms[12]:

---

[12] If an integer is needed, the equilibrium number of firms is $\hat{n} = [n^*]$.

$$n^* = \frac{a-c}{\sqrt{K}} - 1.$$

As K decreases, n^* increases and the total output increases. Indeed, as $K \to 0$, we have $n^* \to \infty$. That is, a reduction in entry cost will increase the number of firms in equilibrium.

On the other hand, an increase in the intensity of competition may reduce the number of firms in equilibrium. See Question 6.5 in the problem set for the result of the Bertrand competition in the second stage, by which the equilibrium number of firms is only 1: $n^* = 1$. Hence, more intense competition in stage 2 lowers the equilibrium level of competition in the market.

Let y_n denote the output of a firm in an n-firm industry. All the firms are identical with the same cost function $c(y_i)$. Then, social welfare is

$$W(n) = \int_0^{ny_n} p^d(y)dy - nc(y_n) - nK.$$

Instead of a zero-profit condition, let n^s be the socially optimal number of firms. For the specific functions defined in (21), with $y_n = \frac{a-c}{n+1}$, the socially optimal number of firm, determined by $W'(n^s) = 0$, is

$$n^s = \frac{(a-c)^{2/3}}{K^{1/3}} - 1. \tag{22}$$

Therefore,

$$n^* + 1 = (n^s + 1)^{3/2},$$

implying $n^* > n^s$. That is, there are too many firms in the industry. This is due to the loss of social welfare K for each entry. Without it, the socially optimal number of firms is infinity.

The following proposition from Mankiw and Whinston (1986) is a general result on entry bias. For the specific functions defined in (21), the three conditions in the proposition are all satisfied.

Proposition 6.3 *Suppose $p'(y) < 0$ and $c''(y) \geq 0$. Let y_n be the symmetric equilibrium output. Assume*

(1) *ny_n is increasing in n.*
(2) *y_n is decreasing in n.*
(3) *$p(ny_n) \geq c'(y_n)$ for all n.*

Then, $n^* \geq n^s - 1$. ∎

9 Strategic Deterrence Against Potential Entrants

Incumbent firms in an industry often make strategic investments to deter potential entrants. These investments include investments in cost reduction, capacity, and new-product development.

Consider a two-stage duopoly model in which firm 1 is the incumbent and firm 2 is a new entrant:

Stage 1. Firm 1 has the option to make a strategic investment $k > 0$.
Stage 2. If firm 2 enters the industry, firms 1 and 2 play a Cournot game, choosing strategies $y_1, y_2 \in \mathbb{R}$, respectively, resulting in profits $\pi_1(y_1, y_2, k)$ and $\pi_2(y_1, y_2)$, respectively.

Let the reaction functions be

$$\hat{y}_1 = \hat{y}_1(y_2, k), \quad \hat{y}_2 = \hat{y}_2(y_1).$$

Suppose that there is a unique Nash equilibrium $[y_1^*(k), y_2^*(k)]$ in stage 2 satisfying the local stability condition:

$$\left| \frac{\partial \hat{y}_1}{\partial y_2} \frac{\partial \hat{y}_2}{\partial y_1} \right| < 1. \tag{23}$$

Suppose also that

$$\frac{\partial \pi_1(y_1, y_2, k)}{\partial y_2} < 0, \quad \frac{\partial \pi_2(y_1, y_2)}{\partial y_1} < 0. \tag{24}$$

These two are fairly natural assumptions. By differentiating the equilibrium condition $y_1^* = \hat{y}_1 [\hat{y}_2(y_1^*), k]$ w.r.t. k, we find that

$$\frac{dy_1^*(k)}{dk} = \frac{\frac{\partial \hat{y}_1}{\partial k}}{1 - \frac{\partial \hat{y}_1}{\partial y_2} \frac{\partial \hat{y}_2}{\partial y_1}}.$$

We also have

$$\frac{d\pi_2[y_1^*(k), y_2^*(k)]}{dk} = \frac{\partial \pi_2[y_1^*(k), y_2^*(k)]}{\partial y_1} \frac{\partial y_1^*(k)}{\partial k} = \frac{\partial \pi_2[y_1^*(k), y_2^*(k)]}{\partial y_1} \frac{\frac{\partial \hat{y}_1}{\partial k}}{1 - \frac{\partial \hat{y}_1}{\partial y_2} \frac{\partial \hat{y}_2}{\partial y_1}},$$

where we have used the Envelope Theorem or the FOC $\partial \pi_2/\partial y_2 = 0$. Hence, to discourage entry, by (23) and (24), we need

$$\frac{\partial \hat{y}_1}{\partial k} > 0, \tag{25}$$

implying a negative effect of k on π_2, by which increasing k can discourage entry. Here, condition (25) means that k is a productive investment for firm 1. If there is a k such that $\pi_2[y_1^*(k), y_2^*(k)] < 0$, then entry can be prevented.

However, investing in k may be costly for firm 1. We also need to look at the effect of a larger k on firm 1's profit. Similarly, using the Envelope Theorem and equation $y_2^* = \hat{y}_2[\hat{y}_1(y_2^*, k)]$, we find that

$$\frac{dy_2^*(k)}{dk} = \frac{\frac{\partial \hat{y}_1}{\partial k}\frac{\partial \hat{y}_2}{\partial y_1}}{1 - \frac{\partial \hat{y}_1}{\partial y_2}\frac{\partial \hat{y}_2}{\partial y_1}},$$

and

$$\frac{d\pi_1[y_1^*(k), y_2^*(k), k]}{dk} = \frac{\partial \pi_1[y_1^*(k), y_2^*(k), k]}{\partial k} + \frac{\partial \pi_1[y_1^*(k), y_2^*(k), k]}{\partial y_2} \frac{\frac{\partial \hat{y}_1}{\partial k}\frac{\partial \hat{y}_2}{\partial y_1}}{1 - \frac{\partial \hat{y}_1}{\partial y_2}\frac{\partial \hat{y}_2}{\partial y_1}}. \tag{26}$$

The first term on the right side of (26) is the <u>direct effect</u>, and the second term is the <u>strategic effect</u>, which depends on the strategic response of firm 2. There are two possible responses:

$$\text{Strategic substitutes}: \frac{\partial \hat{y}_2}{\partial y_1} < 0,$$
$$\text{Strategic complements}: \frac{\partial \hat{y}_2}{\partial y_1} > 0.$$

Under condition (25), if y_2 is a strategic substitute of y_1, the strategic effect is positive. If the direct effect is positive or not very negative, firm 1 will expand k without any hesitation to prevent entry.

Example 6.10 Suppose that firm 1 is the incumbent and firm 2 is a new entrant. Let

$$C_1(y) = c_1(k)y, \quad C_2(y) = c_2 y, \quad p^d(y) = a - y,$$

where a and c_2 are constants with $a > c_2 \geq 0$, $c_1(k) \geq 0$ and $c_1'(k) < 0$. The profit functions are

$$\pi_1(y_1, y_2, k) = [a - y_1 - y_2 - c_1(k)]y_1 - k,$$
$$\pi_2(y_1, y_2) = (a - y_1 - y_2 - c_2)y_2 - K,$$

where K is the entry cost. Conditions in (24) are satisfied. If the second-stage game is a Cournot game, the reaction functions are

Fig. 28 Equilibrium under strategic deterrence

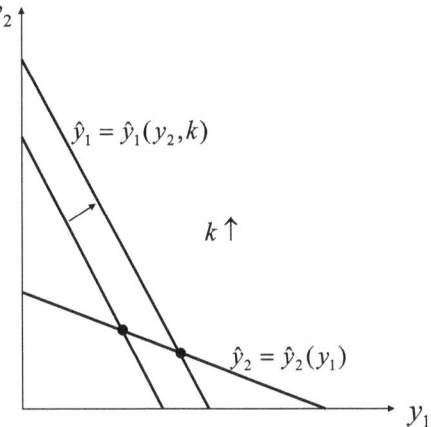

$$\hat{y}_1 = \frac{1}{2}[a - c_1(k) - y_2], \quad \hat{y}_2 = \frac{1}{2}(a - c_2 - y_1).$$

Condition (25) is satisfied, implying $\partial \pi_2^*/\partial k < 0$. Also, since y_1 and y_2 are strategic substitutes, the strategic effect on π_1 is positive. The solution is shown in the following figure. When k increases, firm 1's reaction curve shifts out, implying $\frac{\partial y_1^*}{\partial k} > 0$, which in turn implies $\frac{d\pi_2^*}{dk} < 0$ (Fig. 28).

In fact, the equilibrium outputs are

$$y_1^* = \frac{a - 2c_1(k) + c_2}{3}, \quad y_2^* = \frac{a - 2c_2 + c_1(k)}{3},$$

and the profits are

$$\pi_1^* = \frac{1}{9}[a - 2c_1(k) + c_2]^2 - k, \quad \pi_2^* = \frac{1}{9}[a - 2c_2 + c_1(k)]^2 - K.$$

We can easily see that $\frac{\partial \pi_2^*}{\partial k} < 0$, implying that an increase in k reduces the threat of entry. We can also have $\frac{\partial \pi_1^*}{\partial k} > 0$ if $c_1'(k)$ is negative enough, i.e., if k is very effective in cost reduction. ∎

10 Competitive Input Markets

We now turn to input markets. The input markets are for factors of production, which can be divided into four broad categories: labor, capital, land and raw materials. The owners of factors of production receive incomes from the firms that use those factors as inputs. These incomes, which are the opportunity costs to firms of using those input factors, are payments made for raw materials, wages paid for labor, rental rates paid for capital, and rent paid for land. In the short run, usually labor and raw materials are variable inputs while capital and land are fixed inputs.

Given revenue function $R(y)$ and cost function $C(y)$ for output $y \in \mathbb{R}_+$, as well as production function $y = f(x)$, define their counterparts for input $x \in \mathbb{R}_+^n$:

$$R_I(x) \equiv R[f(x)], \quad C_I(x) \equiv C[f(x)].$$

The <u>marginal revenue product</u> is the increase in revenue resulting from one more unit of input: $MRP(x) \equiv R_I'(x)$, and the <u>marginal cost product</u> is the increase in cost resulting from one more unit of input: $MCP(x) \equiv C_I'(x)$, where

$$R_I'(x) = \left(\frac{\partial R_I(x)}{\partial x_1}, \ldots, \frac{\partial R_I(x)}{\partial x_n}\right), \quad C_I'(x) = \left(\frac{\partial C_I(x)}{\partial x_1}, \ldots, \frac{\partial C_I(x)}{\partial x_n}\right).$$

Then,

$$MRP(x) = MR(y) \cdot MP(x), \quad MCP(x) = MC(y) \cdot MP(x),$$

where $MP(x) \equiv f'(x)$ is the marginal product of input. By (1), we find that maximum profit means that the marginal revenue product (MRP) equals the marginal cost (MCP):

$$MRP(x^*) = MCP(x^*). \tag{27}$$

Just like (1), this condition is applicable to any type of firm in the input markets.

10.1 Demand and Supply

In competitive input markets, input prices are determined in input markets by demand and supply, much the same way as output prices are determined in goods markets. A profit-maximizing firm that is a competitive company in the input markets takes the market prices for inputs as given. This means that $MCP(x) = w$, where $w \in \mathbb{R}_+^n$ is the given market price vector for inputs $x \in \mathbb{R}_+^n$. By (27), the optimal inputs x^* are determined by $MRP(x^*) = w$. This means that a competitive firm's demand curves for inputs are the MRP curves:

$$w^d = MRP(x). \tag{28}$$

This is shown in Fig. 29 for the case of a single input $x \in \mathbb{R}_+$.

In a competitive input market, the industry demand curve for a factor of production is the horizontal sum of all the firms' demand curves for that factor. The industry supply curve is taken as given and is typically a upward-sloping supply curve. The price and quantity traded for the factor of production are determined in

Fig. 29 The optimal problem for a competitive buyer

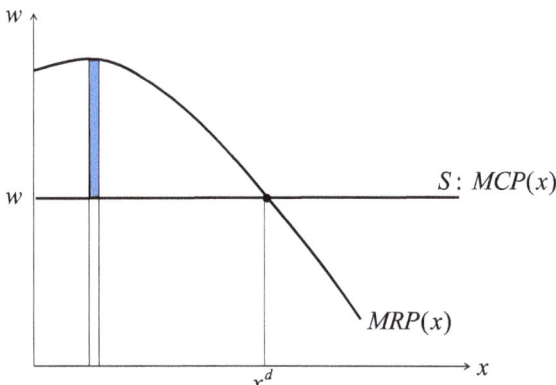

equilibrium by the intersection of the demand and supply curves. If time horizon is taken into account, this is a short-run equilibrium. In a long-run equilibrium, the number of firms is endogenously determined in equilibrium by the zero-profit condition.

In the rest of this chapter, for simplicity of notation, especially in figures, we will assume a single input. All results hold for multiple inputs $x \in \mathbb{R}^n_+$.

10.2 Equilibrium and Welfare

Economic rent is the income received by the supplier over and above the amount required to induce him to offer the input. Transfer earnings is the income required to induce the supply of the input. These two together yield the total income of an input:

$$\text{Total income of an input} = \text{Economic Rent} + \text{Transfer Earnings}.$$

In Fig. 30, the income of the factor is the square area, which is split into two parts: the transfer earnings and the economic rent. The supply curve shows the minimum price at which an additional factor unit is willingly supplied. If the supplier receives only the minimum amount required to induce him to supply each unit of the factor, he will be paid a different price for each unit; those prices will trace the supply curve and the earnings received will be the transfer earnings. The economic rent is similar to the consumer surplus, which is the difference between what the supplier is willing to accept and what is actually paid to him.

In two extreme cases, when the supply is perfectly inelastic (e.g., land) and the supply curve is vertical, the entire income is the economic rent, as shown in Fig. 31; when the supply is perfectly elastic (e.g., unskilled labor) and the supply curve is horizontal, the entire income is the transfer earnings.

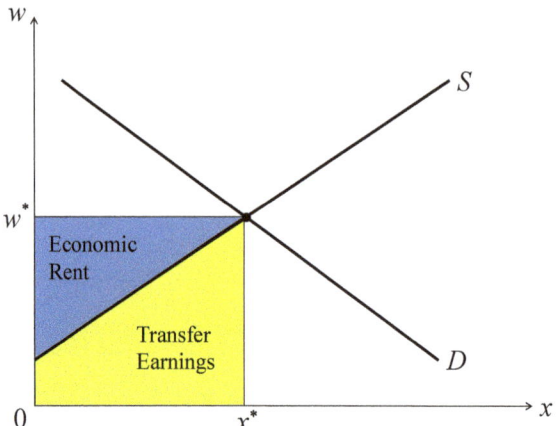

Fig. 30 Economic rent and transfer earnings

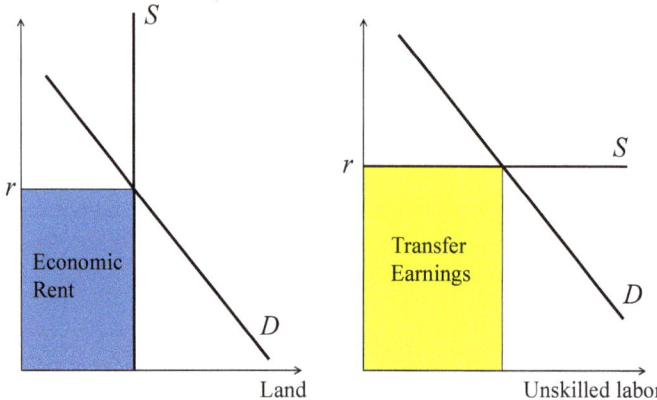

Fig. 31 Land and unskilled labor

Example 6.11 (Housing Market). An equilibrium in the housing market is shown in Fig. 32. The left chart represents the newly built units, which are flows of demand and supply. The right chart represents the stock of demand and supply, which determines the price or the rental rate. The housing stock is a balance between newly built units and demolished units. The housing stock is fixed in the short run. When new units exceed demolished units, the housing stock will increase over time, and vice versa. For each equilibrium flow in the new housing supply on the left chart, there is a balance between newly built units and demolished units, which determines a corresponding housing stock on the right chart. With this framework, we present three special cases in the following.

10 Competitive Input Markets

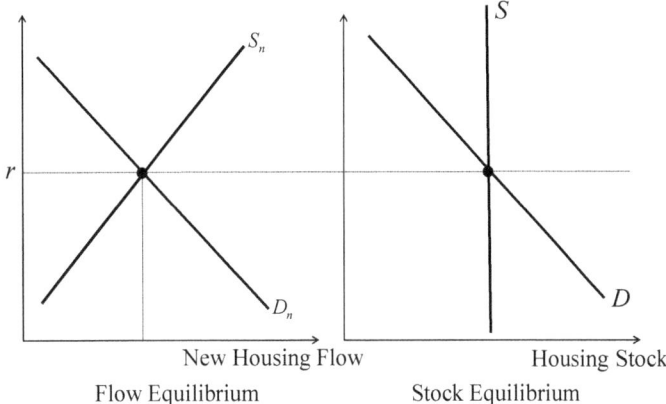

Fig. 32 Short-run versus long-run equilibria in the housing market

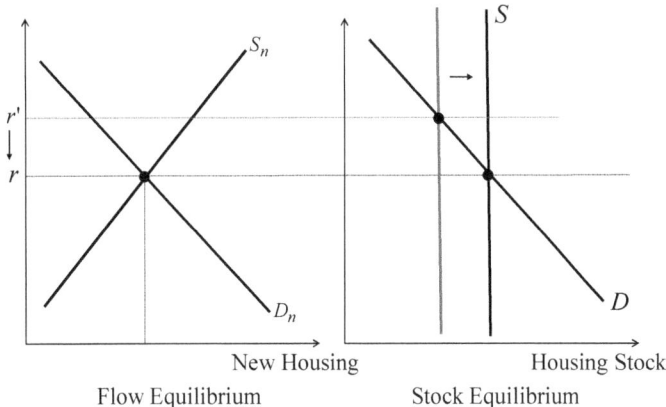

Fig. 33 An earthquake

(1) An earthquake destroys a chunk of housing stock, which pushes up the rental rate immediately from r to r'. The short-run supply of new housing units increases and it is larger than the original equilibrium supply of new housing, which leads to an accumulation of the housing stock over time. In the end, the housing stock will return to its original level together with the rental rate, as shown in Fig. 33.

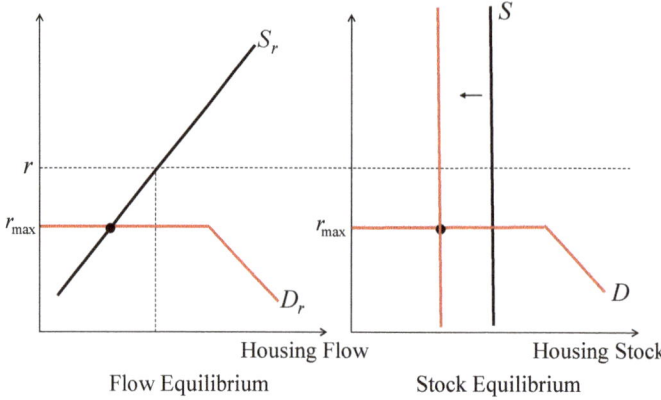

Fig. 34 Rental ceiling

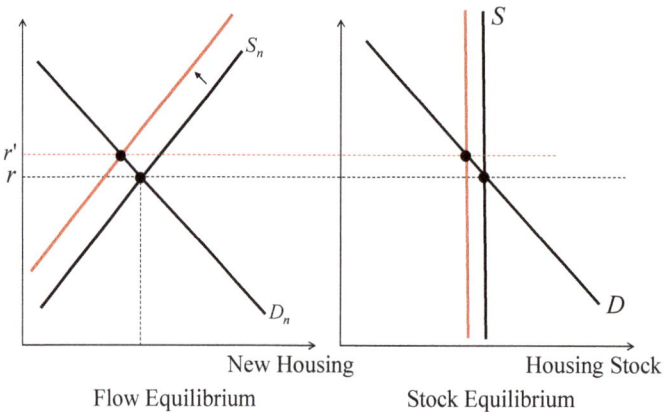

Fig. 35 Restricting land supply

(2) Suppose that the government imposes a rental ceiling r_{max} that is lower than the market rate r. With the policy, the demand curves are changed, as shown in Fig. 34.[13] It reduces the flow of new apartments. The housing stock thus decreases over time. In the long-run, the ceiling rate will be the market rate, but the housing stock is reduced as a result.[14]

---

[13]Since a demand curve represents the maximum price (that the demander is willing to accept), a price ceiling is a ceiling on the demand curve. Since a supply curve represents the minimum price (that the supplier is willing to accept), a price floor is a floor on the supply curve.

[14]Since the rental rate is low, the landlords may have less incentive to maintain housing quality, which may reduce demand further (a leftward shift of the demand curves). If this reduction is large enough, it may cause an even lower equilibrium housing stock.

(3) The Hong Kong Government has a policy to restrict the supply of land, i.e., it pushes S_n up. It will lead to a reduction of housing stock and a higher rental rate r' (Fig. 35).

11 A Monopsony

If there is only one firm that demands a factor of production in the input market, then this firm is called a <u>monopsony</u>. When the firm is a monopsony in the input market, the price of input will be determined by the supply curve, i.e., the price w of the input will depend on input x, i.e., $w = w(x)$. Assuming a single-price monopsony (the firm pays a single price for all its units of input, just like a single-price monopoly), the MCP is

$$MCP(x) = \frac{d}{dx}\{w(x)x\} = w(x) + xw'(x). \tag{29}$$

The formula (29) can be understood by analysis of Fig. 36. When you buy one more unit of x, you pay $w(x)$ for that unit; but, in addition, since the price for all units is increased by $w'(x)$, you also pay $xw'(x)$ for the existing units x. Hence, the total cost for that additional unit is $w(x) + xw'(x)$.

The monopsony's problem is

$$\max_x R(x) - w(x)x.$$

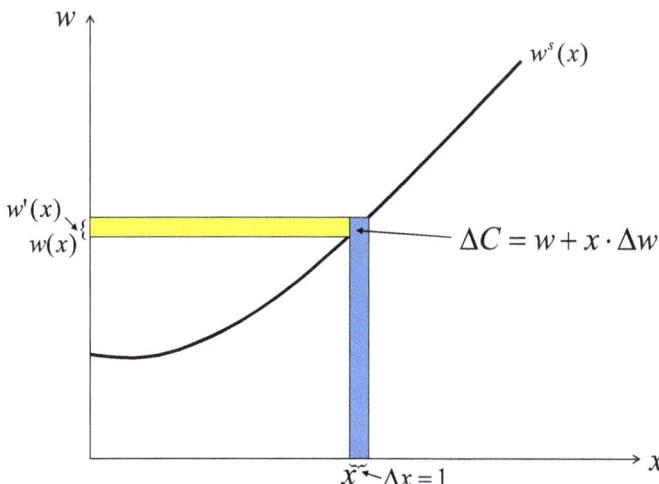

Fig. 36 Marginal cost product

Fig. 37 Monopsony's optimization problem

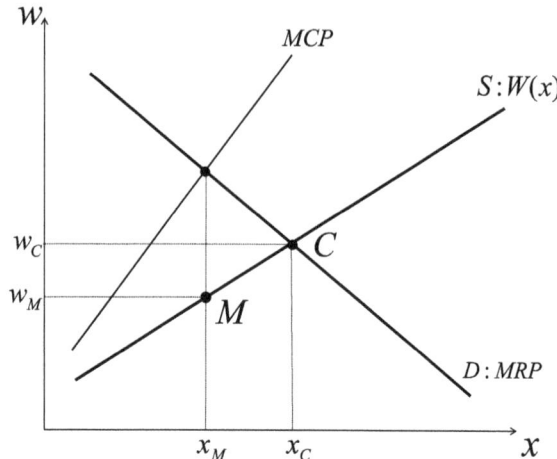

The FOC is:

$$R'(x^*) = w(x^*) + w'(x^*)x^*.$$

This can be written as

$$R'(x^*) = w(x^*)\left(1 + \frac{1}{\varepsilon}\right),$$

where ε is the price elasticity of supply. As expected, when ε goes to infinity, the behavior of a monopsony approaches that of a competitive firm in the input market.

This equilibrium is illustrated in Fig. 37,[15] where $MRP(x) \equiv R'(x)$ is the marginal revenue of input x, $MCP(x) \equiv [w(x) \cdot x]'$ is the marginal cost of input x, and S is the supply curve defined by $w = w(x)$. The condition $MRP(x^*) = MCP(x^*)$ determines the optimal x^*. Once x^* is determined, the monopsony picks the lowest possible price w_M to purchase the input. As shown in the figure, the factor price and quantity in the monopsony equilibrium M are lower than those in the competitive equilibrium C.

Both the monopoly and monopsony equilibria can be thought of as Stackelberg equilibria in which the firm is the leader and the consumers or the suppliers of inputs are the followers.

Example 6.12 (Minimum Wage). Suppose that the government imposes a minimum wage $w_{\min}$ on the labor market. This means that the part of the supply curve where laborers are willing to accept a lower wage becomes $w = w_{\min}$, as shown in Fig. 38. That is, the minimum wage condition replaces the original upward-sloping

---

[15]We have drawn the MCP curve for a linear supply curve $w(x) = a + bx$, which gives us some idea about the position of the MCP curve relative to the supply curve.

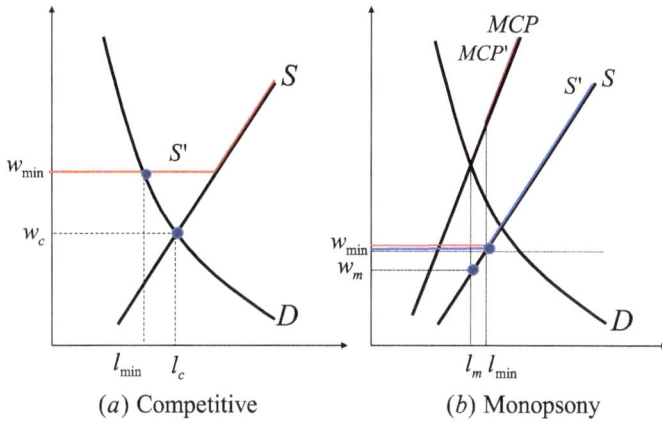

Fig. 38 Minimum wage

labor supply curve $w = w^s(l)$ by the horizontal supply curve $w = w_{\min}$ for the lower part of the original supply curve.

The impact of this minimum wage on employment depends on the nature of the labor market. If the labor market is competitive, as shown in Fig. 38a, the equilibrium employment of labor is reduced from l_c to $l_{\min}$. However, if the labor market is dominated by a monopsonist, as shown in Fig. 38b, the employment will be increased from l_m to $l_{\min}$. In both cases, the equilibrium wage rate will be the minimum wage, but the implication on employment is completely different.

In the case of monopsony, the new MCP' is the broken curve that takes value w_m when $l \leq l_{\min}$ but coincides with MCP when $l > l_{\min}$ in Fig. 38b and we can see that it does not intersect with the demand curve D, which means that the FOC will not be satisfied and the solution must be a corner solution. At level $l_{\min}$, there are two possible corner solutions and the lower one must be the one (the upper one is above the firm's willingness to pay). We can also look the monopsony's surplus as shown in Fig. 39 to see where the monopsony maximizes its profit/surplus.

Notice the differences of impact of a government policy Figs. 33 and 38. A demand curve represents a price ceiling, while a supply curve represents a price floor. Hence, when a policy represents a new price ceiling, it is imposed on the demand curve; when a policy represents a new price floor, it is imposed on the supply curve.

Fig. 39 Monopsony's problem under minimum wage law

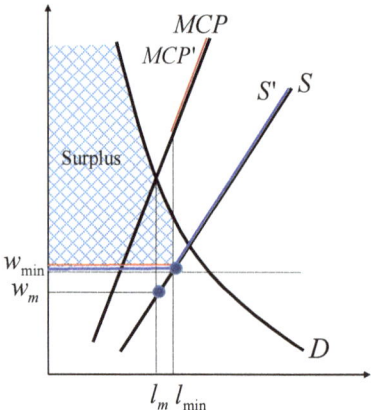

12 Vertical Relationships

So far, firms' relations have been horizontal: firms interact with each other for the same good in the same market (input or output market). This section considers an example in which two firms interact with each other vertically: one firm is the supplier of the other.

Consider a situation in which there is a upstream firm that produces output x with cost $c(x)$ and there is downstream firm that inputs x to produce output y for revenue $R(x)$. The upstream firm is a monopolist in its output market (the downstream firm's input market). The downstream firm is a competitive firm in its input market. The nature of the upstream firm in its input market is captured in its cost function $c(x)$; the nature of the downstream firm in its output market is captured in its revenue function $R(x)$.

Consider a simple case, as shown in Fig. 40, with

$$R(x) = (a - bx)x, \quad c(x) = cx.$$

12.1 Independent Firms

Suppose that the two firms are independent. Since the downstream firm is a competitive firm in the input market, it will take the input price w as given. Its problem is

$$\max_x (a - bx)x - wx,$$

implying

Fig. 40 Vertical relationship

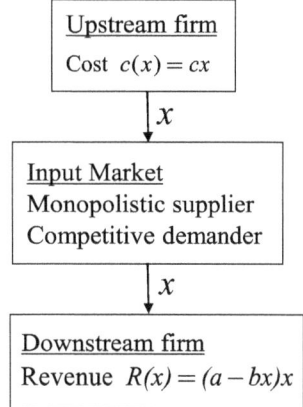

$$\hat{x} = \frac{a-w}{2b},$$

which implies a demand function for x:

$$w^d = a - 2bx.$$

Then, the upstream monopolist's problem is

$$\max_x (a - 2bx)x - cx,$$

which implies

$$x^* = \frac{a-c}{4b}. \tag{30}$$

12.2 An Integrated Firm

Suppose now that the two firms merge into one firm. The integrated firm has revenue function $R(x) = (a - bx)x$ and cost function $c(x) = cx$. The integrated firm's problem is

$$\max_x (a - bx)x - cx,$$

implying

$$\bar{x} = \frac{a-c}{2b}. \tag{31}$$

From (30) and (31), we can see that the integrated firm produces double of the independent firms.

12.3 Explanation

The reason is as follows. Given revenue function $R_D(x)$ for the downstream firm, its problem is

$$\max_x R_D(x) - wx,$$

implying

$$MR_D(x) = w.$$

This is the demand function for the upstream firm. Then, the problem of the upstream firm is

$$\max_x MR_D(x)x - c(x).$$

That is, the revenue function for the upstream firm is $R_U(x) = MR_D(x)x$. The FOC is

$$MR_U(x^*) = MC_U(x^*), \qquad (32)$$

where $MC_U(x) \equiv c'(x)$, which is the marginal cost for the upstream firm. In Fig. 41, the upstream firm faces demand MR_D and has marginal revenue MR_U and marginal cost MC_U. Since it is a monopoly, according to $MR_U(x^*) = MC_U(x^*)$, the upstream firm sells output x^* at price w^*. The downstream firm, on the other hand, has marginal revenue MR_D and marginal cost MC_D. Since it is a competitive firm in the input market, its marginal cost is fixed at w^*: $MC_D(x) = w^*$. Again, according to $MR_D(x^*) = MC_D(x^*)$, the downstream firm buys input x^* at price w^*.

On the other hand, the problem for the integrated firm is

$$\max_x R_D(x) - c(x),$$

which implies

$$MR_D(\bar{x}) = MC_U(\bar{x}). \qquad (33)$$

As shown in Fig. 42, by the condition $MR_D(\bar{x}) = MC_U(\bar{x})$, the integrated firm inputs at $\bar{x}$.

12 Vertical Relationships

Fig. 41 Independent firms

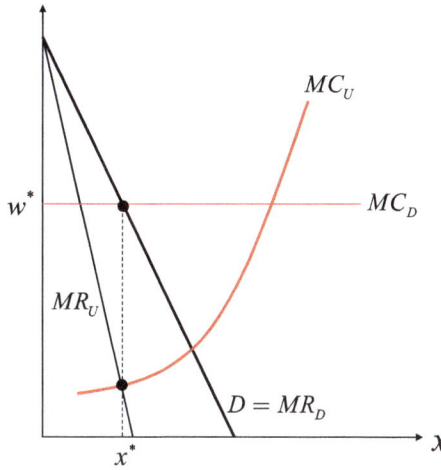

Fig. 42 Integrated firm

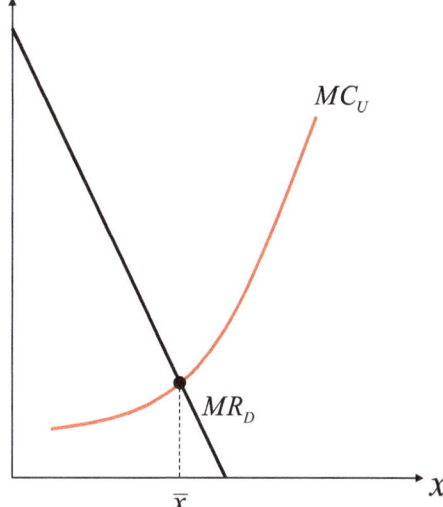

From Fig. 41, since MR_U is about half way below MR_D, x^* will be about half of $\bar{x}$. In general, we can show that an integrated firm will always produce more than an upstream-downstream pair of independent firms. With an upstream-downstream pair, the upstream monopolist cuts its input and raises its price above its MC; the downstream competitive firm, as a price taker, passively follows the upstream monopolist's choices.

We can similarly consider other alternative situations.

If both are competitive in their own markets, the solution with separate firms is the same that under integration.

If the upstream firm is competitive in its output market and the downstream firm is a monopsony in its input market. In this case, the upstream firm's problem is

$$\max_x wx - c(x),$$

which implies the inverse supply function:

$$w = c'(x).$$

Then, the downstream firm's problem is

$$\max_x R(x) - c'(x)x,$$

which implies

$$R'(x) = c''(x)x + c'(x).$$

This equation determines the optimal solution x^*. We can also use backward induction. Given a supply function $w(x)$, the downstream firm's problem is

$$\max_x R(x) - w(x)x,$$

which implies

$$R'(x) = w'(x)x + w(x). \tag{34}$$

The upstream firm's problem is

$$\max_x wx - c(x),$$

which implies the inverse supply function:

$$w = c'(x).$$

That is, the supply function is $w(x) = c'(x)$. Substituting this into (34) yields

$$R'(x) = c''(x)x + c'(x).$$

In summary, if the downstream firm is competitive, it offers a demand curve for the other firm to manipulate, as shown in Fig. 41; if the upstream is competitive, it offers a supply curve for the other firm to manipulate, as shown in Fig. 43. If both firms are competitive, one offers a demand curve and other offers a supply curve; the intersection point is the solution. If both firms are monopolistic in their own

Fig. 43 Independent firms

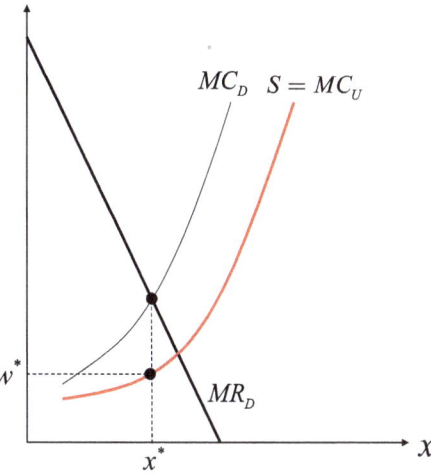

Fig. 44 Integrated firm

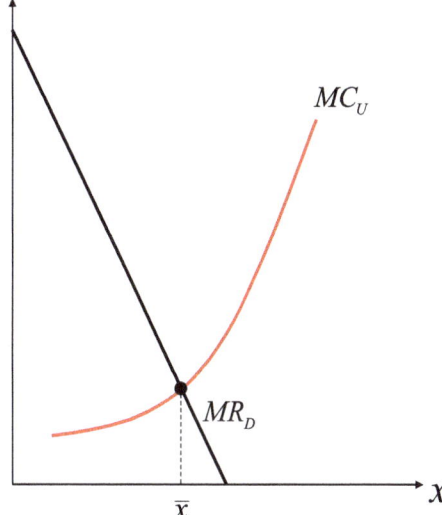

market, no one is to offer a demand or supply curve for the other to manipulate; if so, the two firms have to bargain to settle for a trade (Fig. 44).

Notes
Good references are Varian (1992, Chaps. 13–16) and Mas-Colell et al. (1995, Chap. 12).

Part III
Game Theory

Imperfect Information Games 7

In the last chapter, we saw that game theory is a powerful tool in dealing with the economic problems, especially when there are a small number of economic agents with conflicts of interest. Besides the issue of externalities, game theory is particularly useful for economic problems under imperfect and incomplete information.

There are two kinds of information issues: imperfect information and incomplete information relating two kinds of information: knowledge of actions taken by other players and knowledge of other players' payoffs. If each player perfectly knows other players' actions, it is a game of perfect information; otherwise it is a game of imperfect information. If all players' payoffs are common knowledge, it is a game of complete information; otherwise, it is a game of incomplete information.

This chapter focuses on games of imperfect information, including games of perfect information as a special. We will present several popular equilibrium concepts.

1 Two Game Forms

Games come in various sorts. Many can only be roughly described in words. However, many relatively simple games can be clearly defined by the two standard forms: the extensive form and the normal form. This section defines these two forms of games.

1.1 The Extensive Form

The extensive form relies on a game tree. We first present two examples, by which we gain some basic understanding of a game tree.

Example 7.1 (Matching Pennies under Perfect Information). There are two players P1 and P2, each holding a penny. P1 puts his penny down first. Then, after seeing P1's choice, P2 puts her penny down. If the sides (heads or tails) of the two pennies match, P1 pays $1 to P2; otherwise P2 pays $1 to P1.

This game is presented in the game tree in Fig. 1. The game starts at the initial decision node and ends at a terminal node. The terminal nodes are assigned payoffs and the non-terminal nodes are decision points. The branches represent possible moves/actions of each player. ■

Example 7.2 (Matching Pennies under Imperfect Information). This game is just like the previous game except that when P1 puts his penny down, he keeps it covered. Hence, when P2 moves, she does not know what P1 has chosen.

This game is presented in the game tree on the left of Fig. 2. In the game tree, P2's two decision nodes are contained in an information set. This means that, when it is time for P2 to make a choice, she does not know which decision node she is at, since she does not know which choice P1 has made. The only thing that P2 knows is that the game has arrived at her information set and she has to make a choice without knowing which decision node she is at. Hence, P2's decision will be dependent on the knowledge of her own information set. In order to indicate payoffs, we draw two identical copies of P2's actions, as indicated in the game tree on the right.

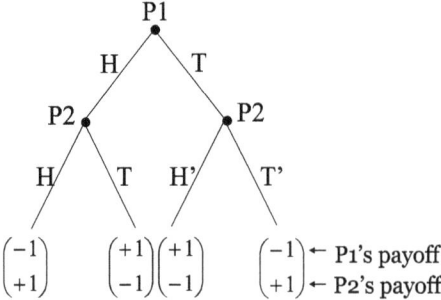

Fig. 1 Matching Pennies under perfect information

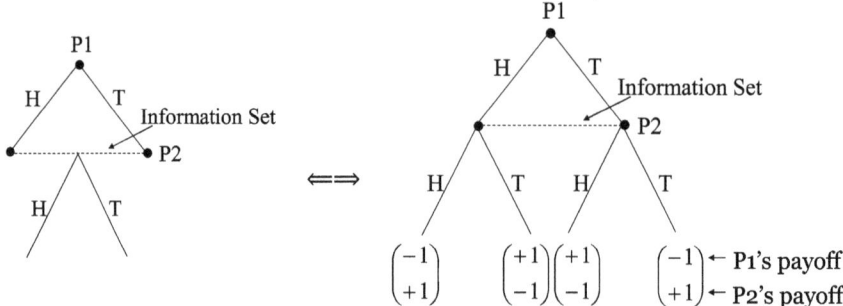

Fig. 2 Matching Pennies under imperfect information

1 Two Game Forms

Note that the game tree is the same if the two players move simultaneously. As long as they cannot observe each other's choices, the timing of a move is irrelevant. ∎

In Example 7.1, P2's possible actions $\{H, T\}$ following H are different from the possible actions $\{H', T'\}$ following T. However, in Example 7.2, the possible actions $\{H, T\}$ following H are the same as those following T. This is because the strategies are dependent on information sets rather than on decision nodes. This will be clear in the following definition.

A player's <u>information set</u> is a set of this player's decision nodes that are indistinguishable to other players. Information sets cause difficulties for players to act or react. The Nash approach to this problem is to let each player choose an optimal strategy given other players' strategies. This strategy is called a <u>Nash strategy</u> or reaction function. More specifically, if a player assumes what other players will do at all information sets (even though some players may never do so), then this player knows what she should do. The Bayesian approach is to assign beliefs over the nodes in all information sets. By this, an information problem becomes a problem under uncertainty. We will present these approaches in Sects. 2 and 3.

With observations of some game trees in the above examples, we now formally define a game tree.

Definition 7.1 A game in extensive form consists of the following items:

1. *Sets.* A set of nodes $\mathbb{X}$, a set of possible actions $\mathbb{A}$, a set of players $\mathbb{N} = \{1, \ldots, n\}$, and a collection of information sets $\mathbb{H}$.[1]
2. *Sequence.* A game starts from a single node. Except the initial node, each node follows from a single immediate predecessor node. The set of terminal nodes is $\mathbb{T}$. All other nodes in $\mathbb{X}$ are called decision nodes.
3. *Information Structure.* Each decision node belongs to one and only one information set. Denote $H(x)$ as the information set that contains node x. When $H(x)$ is a singleton, i.e., $H(x) = \{x\}$, we often refer to $H(x)$ as node x. Each information set is followed by a few branches. Each branch represents a possible action taken by the player who is to make a decision upon observing that information set, i.e., when the play reaches that information set. For $H \in \mathbb{H}$, let $A(H) = \{$all the branches following $H\}$. If it is player i's turn to make a move at an information set H, we call H player i's information set. Each information set belongs to one and only one player, including a special player called Nature.
4. *Nature.* Sometimes Nature is included. Let H_0 be the information set where Nature makes a move. A function $\rho : A(H_0) \to [0, 1]$ assigns probabilities to actions at information set H_0. Nature is like a player in the model, except that it does not have a payoff function and it does not optimize its choices.[2]

---
[1]We can actually allow infinite steps, infinite possible actions, and infinite players.
[2]In game theory, if Nature is involved in a game, Nature typically moves first. In particular, in a game of incomplete information, Nature always moves first.

5. *Payoffs.* A collection of payoff functions $\mathcal{U} = \{u_1(\cdot), \ldots, u_n(\cdot)\}$ assigns utilities to the players at each terminal node, $u_i : \mathbb{T} \to \mathbb{R}$.

Thus, a game in extensive form is specified by the collection

$$\Gamma_E = \{\mathbb{X}, \mathbb{A}, \mathbb{N}, \mathbb{H}, \mathcal{U}, \rho(\cdot), A(\cdot), H(\cdot)\}. \quad \blacksquare$$

Figure 3 shows a typical game tree.

A game is a <u>game of perfect information</u> if every information set is a singleton set; otherwise, it is a <u>game of imperfect information</u>. The structure of a game is assumed to be <u>common knowledge</u>, in the sense that all players know the structure of the game, know that their rivals know it, know their rivals know that they know it, and so on. We will only consider games of <u>perfect recall</u>, meaning that a player does not forget what she once knew, including her own actions.

1.2 The Normal Form

A strategy is a complete contingency plan that specifies how a player will act in every possible distinguishable circumstance. The set of circumstances for a player is his set of information sets, with each information set representing a different distinguishable circumstance in which he may need to make a move. Thus, a player's strategy amounts to a complete specification of how he plans to move at each of his information set.

Definition 7.2 Let $\mathbb{H}_i$ denote the collection of player i's information sets, $\mathbb{A}$ the set of possible actions in the game, and $A(H) \subset \mathbb{A}$ the set of possible actions at information set H. A (pure) <u>strategy</u> for player i is a function $s_i : \mathbb{H}_i \to \mathbb{A}$ such that $s_i(H) \in A(H)$ for all $H \in \mathbb{H}_i$. Denote $\mathbb{S}_i$ as the strategy space of player i, which contains all possible strategies of player i. $\quad \blacksquare$

If $\mathbb{H}_i = \{H_{1i}, H_{2i}, \ldots, H_{m_i i}\}$ is player i's collection of information sets, a strategy s_i of player i specifies an action $a_{ki} \in A(H_{ki})$ under each information set H_{ki} in $\mathbb{H}_i$. We can denote s_i as

$$s_i = <a_{1i}, a_{2i}, \ldots, a_{m_i i}>,$$

where s_i lists the planned action at each and every information set of player i. With many possible combinations of actions the player can take, this player has many possible strategies.

Example 7.3 (Matching Pennies under Perfect Information). Given his information set, player 1 has two strategies:

$$s_{11} = H, \quad s_{21} = T.$$

1 Two Game Forms

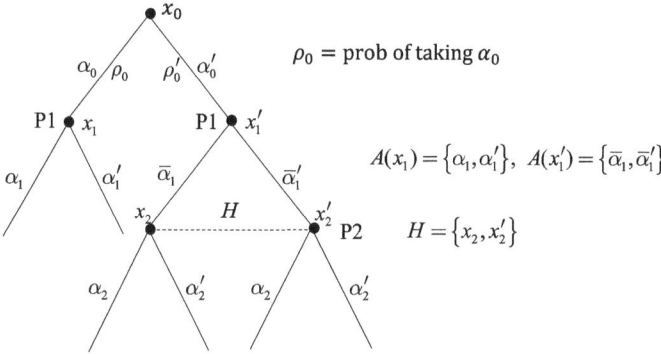

Fig. 3 A game tree

Player 2 has two information sets. Denote a typical strategy of Player 2 as $s_2 = \langle a_2^l, a_2^r \rangle$, where a_2^l is an action if the information set on the left is reached and a_2^r is an action if the information set on the right is reached. Then, player 2 has four possible strategies:

$$s_{12} = H, H', \quad s_{22} = \langle H, T' \rangle, \quad s_{32} = \langle T, H' \rangle, \quad s_{42} = \langle T, T' \rangle.$$

Their strategy spaces are $\mathbb{S}_1 = \{s_{11}, s_{21}\}$ and $\mathbb{S}_2 = \{s_{12}, s_{22}, s_{32}, s_{42}\}$. ■

Example 7.4 (Matching Pennies under Imperfect Information). Each player has one information set. Each player i has two strategies $s_{1i} = H$ and $s_{2i} = T$. Thus, $\mathbb{S}_i = \{H, T\}$ for both players. ■

Denote a profile of strategies as $s = (s_1, \ldots, s_n)$, where $s_i \in \mathbb{S}_i$ is a strategy from player i. The normal form of a game is a specification of strategies and their associated payoffs.

Definition 7.3 For a game with n players, the normal form of a game specifies for each player i a set of strategies $\mathbb{S}_i$ and a payoff function $u_i(s_1, \ldots, s_n)$. Thus, a game in normal form is specified by the collection $\Gamma_N = [\mathbb{N}, \{\mathbb{S}_i\}, \{u_i\}]$. ■

Example 7.5 (Matching Pennies under Perfect Information). The strategy sets are defined in Example 7.3. The payoff functions are

$$u_1(s_1, s_2) = \begin{cases} +1 & \text{if } s_1 = s_{11}, \quad s_2 = s_{32} \quad \text{or} \quad s_{42} \\ -1 & \text{if } s_1 = s_{11}, \quad s_2 = s_{12} \quad \text{or} \quad s_{22} \\ +1 & \text{if } s_1 = s_{21}, \quad s_2 = s_{12} \quad \text{or} \quad s_{32} \\ -1 & \text{if } s_1 = s_{21}, \quad s_2 = s_{22} \quad \text{or} \quad s_{42} \end{cases}$$

$$u_2(s_1, s_2) = -u_1(s_1, s_2).$$

A more convenient way to present this game is the game box below:

P1\P2	s_{12}	s_{22}	s_{32}	s_{42}
s_{11}	-1, +1	-1, +1	+1, -1	+1, -1
s_{21}	+1, -1	-1, +1	+1, -1	-1, +1

Example 7.6 (Matching Pennies under Imperfect Information). Given the strategies in Example 7.4, the normal form is

P1\P2	s_{12}	s_{22}
s_{11}	-1, +1	+1, -1
s_{21}	+1, -1	-1, +1

Starting from a verbal description, a game is usually initially presented by a game tree or the extensive form. The extensive form is a complete and precise representation of a game. The normal form is a reduced form of the extensive form. In the normal form, all players are assumed to simultaneously pick their strategies from their own strategy sets. Dynamic features and sequencing of actions in the original extensive form are ignored in the normal form.

Each extensive-form game implies a unique normal form; the converse is however not true. For example, in Example 7.5, the normal form can also be derived from the extensive-form game in Fig. 4. Because of the condensed representation, the normal form generally omits some of the details present in the extensive form. Depending on the equilibrium concept, for some equilibrium concepts, these two forms are equivalent; but for others, these two forms are not equivalent. In particular, for Nash equilibria, these two forms are equivalent.

1.3 Mixed Strategy

Instead of following a certain strategy for sure, we now allow players to follow a strategy with a certain probability. We call the probabilistic distribution of a player's strategies a mixed strategy and call the original strategies pure strategies.

Definition 7.4 Given a normal-form game Γ_N, for $\mathbb{S}_i = \{s_{1i}, \ldots, s_{n_i i}\}$, we denote a mixed strategy as $\sigma_i = (\sigma_{1i}, \ldots, \sigma_{n_i i})$, where σ_{ki} is the probability that s_{ki} is taken. That is, a mixed strategy σ_i is a probability distribution over the pure strategies in $\mathbb{S}_i$. Denote the mixed extension of $\mathbb{S}_i$ as

$$\Delta(\mathbb{S}_i) = \left\{ (\sigma_{1i}, \ldots, \sigma_{n_i i}) \geq 0 \,\Big|\, \sum_{k=1}^{n_i} \sigma_{ki} = 1 \right\}.$$

1 Two Game Forms

Fig. 4 An alternative game tree

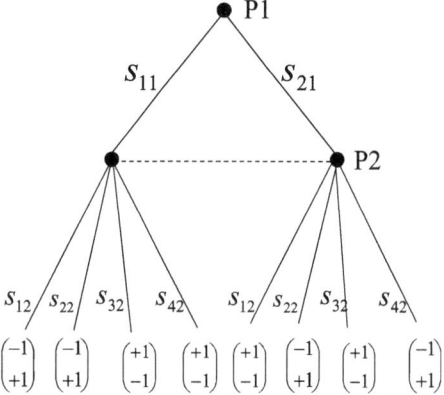

We sometimes denote σ_{ki} as $\sigma_i(s_{ki})$, i.e., $\sigma_i(s_{ki})$ is the probability that the mixed strategy σ_i assigns to the pure strategy $s_{ki} \in \mathbb{S}_i$. ∎

A pure strategy is a special mixed strategy. For example, $s_{1i} = (1, 0, \ldots, 0)$. Due to this, we can write a mixed strategy $\sigma_i = (\sigma_{1i}, \ldots, \sigma_{n_i i})$ as

$$\sigma_i = \sigma_{1i} s_{1i} + \sigma_{2i} s_{2i} + \cdots + \sigma_{n_i i} s_{n_i i}.$$

This representation is more clear.

We denote $(s_1, \ldots, s_n)$ as a profile of pure strategies, where $s_i \in \mathbb{S}_i$ is a pure strategy from player i. Similarly, we denote $(\sigma_1, \ldots, \sigma_n)$ as a profile of mixed strategies, where $\sigma_i \in \Delta(\mathbb{S}_i)$ is a mixed strategy from player i.

There is another way that a player can randomize. Rather than randomizing over the potentially a very large set of pure strategies in $\mathbb{S}_i$, she could randomize separately over the possible actions at each of her information sets $H \in \mathbb{H}_i$. This is called a behavior strategy.

Definition 7.5 Given an extensive-form game Γ_E, a <u>behavior strategy</u> for player i specifies, for every information set $H \in \mathbb{H}_i$, a probability distribution $\sigma_H = (\sigma_{1H}, \ldots, \sigma_{kH})$ over the actions in $A(H) = \{a_{1H}, \ldots, a_{kH}\}$, with $\sigma_{jH} \geq 0$ and $\sum_{j=1}^{k} \sigma_{jH} = 1$. ∎

If $\mathbb{H}_i = \{H_{1i}, H_{2i}, \ldots, H_{m_i i}\}$ is player i's collection of information sets, a behavior strategy σ_i of player i specifies a probability distribution $\sigma_{H_{ki}} = (\sigma_{1H_{ki}}, \sigma_{2H_{ki}}, \ldots)$ over the actions in $A(H_{ki})$ for each information set H_{ki} in $\mathbb{H}_i$. We can denote σ_i as

$$\sigma_i = <\sigma_{H_{1i}}, \sigma_{H_{2i}}, \ldots, \sigma_{H_{m_i i}}>,$$

where σ_i lists the probability distributions over all player i's information sets.

A mixed strategy σ_i and a behavior strategy σ'_i of player i are said to be outcome-equivalent if given any strategy profile σ_{-i} of the other players and for each terminal node $t \in \mathbb{T}$, the probability of reaching t under (σ_i, σ_{-i}) is the same as that under (σ'_i, σ_{-i}), i.e., $P(t|\sigma_i, \sigma_{-i}) = P(t|\sigma'_i, \sigma_{-i})$.

Proposition 7.1 (Kuhn). *Mixed and behavior strategies are outcome-equivalent for finite extensive-form games with perfect recall.*[3] ∎

Since the two types of randomization are equivalent for games with perfect recall, we typically use behavior strategies for extensive-form games and mixed strategies for normal-form games. In fact, for convenience, we will often refer to behavior strategies as mixed strategies.

We also need to extend payoffs in a normal-form game to include payoffs over mixed strategies. We introduce some notation first. Denote

$$s_{-i} = (s_1, \ldots, s_{i-1}, s_{i+1}, \ldots, s_n), \quad s = (s_i, s_{-i}),$$
$$\sigma_{-i} = (\sigma_1, \ldots, \sigma_{i-1}, \sigma_{i+1}, \ldots, \sigma_n), \quad \sigma = (\sigma_i, \sigma_{-i}),$$
$$\mathbb{S}_{-i} = \mathbb{S}_1 \times \cdots \times \mathbb{S}_{i-1} \times \mathbb{S}_{i+1} \times \cdots \times \mathbb{S}_n, \quad \mathbb{S} = \mathbb{S}_i \times \mathbb{S}_{-i}.$$

We have defined payoffs for pure strategies, but we have not defined payoffs for mixed strategies. We use expected utility to define payoffs for mixed strategies. Given a profile of mixed strategies (σ_i, σ_{-i}), the payoff is

$$u_i(\sigma_i, \sigma_{-i}) = \sum_{k_i=1}^{n_i} \sigma_{k_i i} u_i(s_{k_i i}, \sigma_{-i}) = \sum_{k_1=1}^{n_1} \cdots \sum_{k_n=1}^{n_n} \sigma_{k_1 1} \cdots \sigma_{k_n n} u_i(s_{k_1 1}, \ldots, s_{k_n n})$$
$$= \sum_{s_i \in \mathbb{S}_i, i \in \mathbb{N}} \sigma_1(s_1) \cdots \sigma_n(s_n) u_i(s_1, \ldots, s_n) \quad (1)$$

where $\sigma_{k_i i}$ is the probability that player i follows his strategy $s_{k_i i} \in \mathbb{S}_i$. For example, for two players $i = 1, 2$, with strategy sets $\mathbb{S}_1 = \{s_{11}, s_{21}\}$ and $\mathbb{S}_2 = \{s_{12}, s_{22}\}$, we have

$$u_1(\sigma_1, \sigma_2) = \sigma_{11} u_1(s_{11}, \sigma_2) + \sigma_{21} u_1(s_{21}, \sigma_2)$$
$$= \sigma_{11}[\sigma_{12} u_1(s_{11}, s_{12}) + \sigma_{22} u_1(s_{11}, s_{22})] + \sigma_{21}[\sigma_{12} u_1(s_{21}, s_{12}) + \sigma_{22} u_1(s_{21}, s_{22})]$$
$$= \sum_{k_1=1}^{2} \sum_{k_2=1}^{2} \sigma_{k_1 1} \sigma_{k_2 2} u_1(s_{k_1 1}, s_{k_2 2}) = \sum_{s_i \in \mathbb{S}_i, i=1,2} \sigma_1(s_1) \sigma_2(s_2) u_1(s_1, s_2).$$

We will now introduce equilibrium concepts to the two forms of game. We consider the normal form first in Sect. 2 and then the extensive form in Sect. 3.

---

[3]Kuhn's Theorem has been extended in the literature to infinite games with infinite actions and horizon. However, perfect recall is always required for the equivalence result to hold.

2 Equilibria in Normal-Form Games

In the normal form, the players are assumed to choose their strategies simultaneously. That is, a normal-form game is a simultaneous-move game. We introduce four equilibrium concepts: the Nash equilibrium, the dominant-strategy equilibrium, the trembling-hand Nash equilibrium, and the reactive equilibrium.

2.1 Nash Equilibrium

Nash treats all games as simultaneous-move games, in which the players choose their strategies (contingent plans) simultaneously. Nash further assumes that when each player is choosing her strategy she assumes that other players keep their strategies unchanged.

Definition 7.6 A mixed strategy profile $\sigma^* = (\sigma_1^*, \ldots, \sigma_n^*)$ is a <u>Nash equilibrium</u> (NE) of $\Gamma_N = [\mathbb{N}, \{\Delta(\mathbb{S}_i)\}, \{u_i\}]$ if, for every $i \in \mathbb{N}$, we have $u_i(\sigma_i^*, \sigma_{-i}^*) \geq u_i(\sigma_i, \sigma_{-i}^*)$ for all $\sigma_i \in \Delta(\mathbb{S}_i)$. ∎

If an NE consists of pure strategies only, we call it a <u>pure-strategy NE</u>; otherwise it is called a mixed-strategy NE. The following two propositions offer ways to solve for NEs.

Proposition 7.2 A strategy profile $\sigma^* = (\sigma_1^*, \ldots, \sigma_n^*)$ is a Nash equilibrium in $\Gamma_N = [\mathbb{N}, \{\Delta(\mathbb{S}_i)\}, \{u_i\}]$ if and only if, for each $i \in \mathbb{N}$,[4]

(i) $u_i(s_{ki}, \sigma_{-i}^*) = u_i(s_{ji}, \sigma_{-i}^*)$ if $\sigma_i^*(s_{ki}) > 0$ and $\sigma_i^*(s_{ji}) > 0$, (2)

(ii) $u_i(s_{ki}, \sigma_{-i}^*) \geq u_i(s_{ji}, \sigma_{-i}^*)$ if $\sigma_i^*(s_{ki}) > 0$ and $\sigma_i^*(s_{ji}) = 0$. (3)

Proof Necessity. If $(\sigma_1^*, \ldots, \sigma_n^*)$ is a NE, σ_i^* is the solution of the following problem:

$$\max_{\sigma_i = (\sigma_{1i}, \ldots, \sigma_{n_i i}) \geq 0} \sum_{k=1}^{n_i} \sigma_{ki} u_i(s_{ki}, \sigma_{-i}^*)$$

$$\text{s.t.} \quad \sum_{k=1}^{n_i} \sigma_{ki} = 1.$$

By introducing the Lagrangian multiplier λ, the problem is equivalent to

---

[4]Condition (2) is usually enough to determine a mixed strategy NE. However, there are cases in which one player strictly prefers one strategy no matter what the other player does; in this case, (2) never happens. That is, one player plays a strategy with probability 1 in equilibrium. In this case, we need (3) to identify the equilibrium.

$$\max_{\sigma_i=(\sigma_{1i},\ldots,\sigma_{n_i i})\geq 0} \sum \sigma_{ki} u_i(s_{ki},\sigma_{-i}^*) + \lambda\left(1-\sum \sigma_{ki}\right)$$
$$= \sum \sigma_{ki}\left[u_i(s_{ki},\sigma_{-i}^*) - \lambda\right] + \lambda.$$

Since this objective function is linear in σ_{ki}, we immediately have the following three necessary conditions: for all $i \in \mathbb{N}$,

(1) $u_i(s_{ki},\sigma_{-i}^*) = \lambda$, for all k with $0 < \sigma_{ki}^* < 1$,
(2) $u_i(s_{ki},\sigma_{-i}^*) \leq \lambda$, for all k with $\sigma_{ki}^* = 0$,
(3) $u_i(s_{ki},\sigma_{-i}^*) \geq \lambda$, for all k with $\sigma_{ki}^* = 1$.

These three conditions immediately imply (2)–(3).[5] Notice that if $\sigma_{ki}^* = 1$ for some k, (2) never happens.

Sufficiency. Given conditions (2) and (3), if σ^* is not a Nash equilibrium, then there is some player i who has a strategy σ_i' with $u_i(\sigma_i',\sigma_{-i}^*) > u_i(\sigma_i^*,\sigma_{-i}^*)$. If so, there must exist a pure strategy $s_{ki} \in \mathbb{S}_i$ for which $u_i(s_{ki},\sigma_{-i}^*) > u_i(\sigma_i^*,\sigma_{-i}^*)$. Since by (2) $u_i(\sigma_i^*,\sigma_{-i}^*) = u_i(s_{ji},\sigma_{-i}^*)$ for all $s_{ji} \in \mathbb{S}_i$ with $\sigma_i^*(s_{ji}) > 0$, we find that

$$u_i(s_{ki},\sigma_{-i}^*) > u_i(s_{ji},\sigma_{-i}^*). \tag{4}$$

Note that we have at least one $s_{ji} \in \mathbb{S}_i$ with $\sigma_i^*(s_{ji}) > 0$. If $\sigma_i^*(s_{ki}) > 0$, (4) contradicts (2); if $\sigma_i^*(s_{ki}) = 0$, (4) contradicts (3). Either way, we have a contradiction. Thus, σ^* must be a Nash equilibrium. ∎

Proposition 7.3 *Given any σ_{-i}, if $u_i(s_i^*,\sigma_{-i}) \geq u_i(s_i,\sigma_{-i})$ for all $s_i \in \mathbb{S}_i$, then $u_i(s_i^*,\sigma_{-i}) \geq u_i(\sigma_i,\sigma_{-i})$ for all $\sigma_i \in \Delta(\mathbb{S}_i)$.*

Proof If not, there is a mixed strategy σ_i such that $u_i(s_i^*,\sigma_{-i}^*) < u_i(\sigma_i,\sigma_{-i}^*)$. If so, there must exist $s_i \in \mathbb{S}_i$ such that $u_i(s_i^*,\sigma_{-i}^*) < u_i(s_i,\sigma_{-i}^*)$. This is a contradiction. ∎

In summary, to find a pure-strategy NE, we use Proposition 7.3, by which each player's NE strategy is a best strategy among her pure strategies. To find a mixed-strategy NE (when at least one player takes a mixed strategy), we use the two conditions in Proposition 7.2.

Example 7.7 In the following game, among the pure strategies in $\mathbb{S}_1 = \{L_1, R_1\}$ and $\mathbb{S}_2 = \{L_2, R_2\}$, there are obviously two pure-strategy NEs: (L_1, L_2) and (R_1, R_2). By Proposition 7.3, they are also NEs in mixed-strategy spaces $\Delta(\mathbb{S}_1)$ and $\Delta(\mathbb{S}_2)$.

---

[5]We can also prove the necessity in the following way. If either (2) or (3) does not hold, then there are strategies $s_{ki}, s_{ji} \in \mathbb{S}_i$ with $\sigma_i(s_{ki}) > 0$ such that $u_i(s_{ki},\sigma_{-i}^*) < u_i(s_{ji},\sigma_{-i}^*)$. If so, player i would do better by playing s_{ji} whenever he is supposed to play s_{ki} in the original NE, implying that the original equilibrium is not an equilibrium.

2 Equilibria in Normal-Form Games

P1\P2	L_2	R_2
L_1	2, 3	-2, 2
R_1	-2, 3	4, 5

To find a mixed-strategy NE, assume that P1 mixes his two strategies. By (2), P1 must be indifferent between his two pure strategies. Hence, given $\sigma_2 = (r, 1-r)$ where $r \in [0, 1]$, we have

$$2r - 2(1-r) = -2r + 4(1-r),$$

implying $r^* = 3/5$. Conversely, since P2 mixes her two strategies (with $r^* = 3/5$), P2 must be indifferent between her two pure strategies. Hence, given $\sigma_1 = (\rho, 1-\rho)$ where $\rho \in [0, 1]$, we have

$$3\rho + 3(1-\rho) = 2\rho + 5(1-\rho),$$

implying $\rho^* = 2/3$. This result is consistent with our initial assumption that P1 mixes his two strategies. Hence, we have a mixed-strategy NE:

$$NE3: \quad \sigma_1^* = \left(\frac{2}{3}, \frac{1}{3}\right), \quad s_2^* = \left(\frac{3}{5}, \frac{2}{5}\right). \quad \blacksquare$$

Example 7.8 In the following game, there are two pure-strategy NEs: (R_1, L_2) and (L_1, R_2).

P1\P2	L_2	R_2
L_1	-2, -1	1, 0
R_1	-1, 2	1, 1

To find a mixed-strategy NE, assume that P1 mixes his two pure strategies. Let P2's strategy be $\sigma_2 = (t, 1-t)$ for any $t \in [0, 1]$. Since P1 mixes his two pure strategies, he must be indifferent between his two pure strategies, implying

$$-2t + 1 - t = -t + 1 - t,$$

implying $t^* = 0$. Conversely, let P1's strategy be $\sigma_1 = (r, 1-r)$ for any $r \in [0, 1]$. Since L_2 is taken with zero probability $t^* = 0$ and R_2 is taken with positive probability (actually with probability 1), according to (3), we have

$$-1r + 2(1-r) \leq 0r + 1(1-r),$$

implying $r^* \geq 0.5$. For $r \in [0.5, 1)$, P1 indeed mixes his two strategies, which is consistent with our initial assumption. Hence, we have a mixed-strategy NE:

$$NE3: \quad \sigma_1^* = (r, 1-r), \quad s_2^* = R_2, \quad \text{for any } r \in [0.5, 1).$$

Further, assume that P2 mixes her two pure strategies. Then,

$$-1r + 2(1-r) = 0r + 1(1-r),$$

implying $r = 0.5$. But, if P1 mixes his two pure strategies (with $r = 0.5$), then $t = 0$, which is shown before. This contradicts our initial assumption that P2 mixes her two strategies. Hence, P2 will never take a mixed strategy in equilibrium. ∎

How about the uniqueness and existence of NE? NEs are obviously not unique. The existence of a mixed-strategy NE can be guaranteed for a finite game. The existence of a pure-strategy NE can also be guaranteed under more stringent conditions. The proofs of the following two existence results can be found in Mas-Colell et al. (1995, p. 260).

Proposition 7.4 (Existence). *If $\mathbb{S}_1, \ldots, \mathbb{S}_n$ are finite sets, there exists a Nash equilibrium in $\Gamma_N = [\mathbb{N}, \{\Delta(\mathbb{S}_i)\}, \{u_i\}]$.* ∎

Proposition 7.5 (Existence). *A Nash equilibrium exists in $\Gamma_N = [\mathbb{N}, \{\mathbb{S}_i\}, \{u_i\}]$ If, for all i,*

(1) $\mathbb{S}_i$ *is nonempty, convex, and compact subset of some Euclidean space $\mathbb{R}^m$.*
(2) $u_i(s_1, \ldots, s_n)$ *is continuous in $(s_1, \ldots, s_n)$ and quasiconcave in each s_i.* ∎

NE has two logical problems. First, Nash requires that each player knows other players' strategies when she chooses her own strategy.[6] However, Nash also assumes that all players move simultaneously, which means that the player does not know other players' strategies when she chooses her own strategy. Second, if others react to my deviation from the NE, I may do better by deviating from my NE strategy. If so, an equilibrium outcome may not be the NE. Nash's rationality does not take into account others' possible reactions.

2.2 Dominant-Strategy Equilibrium

We now look for stronger versions of NEs, i.e., those satisfying stronger conditions. We have two stronger versions of NEs: dominant-strategy NE, and trembling-hand perfect NE.

Definition 7.7

- A strategy $\sigma_i \in \Delta(\mathbb{S}_i)$ for player i is strictly <u>dominated</u> in game $\Gamma_N = [\mathbb{N}, \{\Delta(\mathbb{S}_i)\}, \{u_i\}]$ if there exists another strategy $\sigma'_i \in \Delta(\mathbb{S}_i)$ such that

---

[6]Other players' strategies may be functions of some random variables. In this case, these functions are required to be known to the player.

2 Equilibria in Normal-Form Games

$u_i(\sigma'_i, \sigma_{-i}) > u_i(\sigma_i, \sigma_{-i})$ for all $\sigma_{-i} \in \Delta(\mathbb{S}_{-i})$ In this case, we say that σ'_i strictly dominates σ_i.

- A strategy σ_i is a strictly <u>dominant strategy</u> for player i in game $\Gamma_N = [\mathbb{N}, \{\Delta(\mathbb{S}_i)\}, \{u_i\}]$ if it strictly dominates every other strategy in $\Delta(\mathbb{S}_i)$.
- A strategy $\sigma_i \in \Delta(\mathbb{S}_i)$ is <u>weakly dominated</u> if there exists another strategy $\sigma'_i \in \Delta(\mathbb{S}_i)$ such that $u_i(\sigma'_i, \sigma_{-i}) \geq u_i(\sigma_i, \sigma_{-i})$ for all $\sigma_{-i} \in \Delta(\mathbb{S}_{-i})$ and with strict inequality for some σ_{-i}. ∎

In words, a strategy is a strictly dominant strategy for player i if it the best strategy for him no matter what strategies his rivals may play.

In the prisoner's dilemma mentioned in Chap. 6, the Nash equilibrium is that both choose to confess and this is a dominant strategy equilibrium. It is a paradigmatic example of self-interested rational behavior that leads to a Pareto inferior result for the players.

Proposition 7.6 *Player i's strategy $\sigma_i \in \Delta(\mathbb{S}_i)$ is strictly dominated by σ'_i in game $\Gamma_N = [\mathbb{N}, \{\Delta(\mathbb{S}_i)\}, \{u_i\}]$ iff $u_i(\sigma'_i, s_{-i}) > u_i(\sigma_i, s_{-i})$ for all $s_{-i} \in \mathbb{S}_{-i}$.*

Proof By (1), given any $\sigma_i \in \mathbb{S}_i$, we have

$$u_i(\sigma_i, \sigma_{-i}) = \sum_{s_{-i} \in \mathbb{S}_{-i}} \sigma_{-i}(s_{-i}) u_i(\sigma_i, s_{-i}) = \sum_{s_{-i} \in \mathbb{S}_{-i}} \prod_{j \neq i} \sigma_j(s_j) u_i(\sigma_i, s_{-i}).$$

Hence,

$$u_i(\sigma'_i, \sigma_{-i}) - u_i(\sigma_i, \sigma_{-i}) = \sum_{s_{-i} \in \mathbb{S}_{-i}} \sigma_{-i}(s_{-i}) [u_i(\sigma'_i, s_{-i}) - u_i(\sigma_i, s_{-i})].$$

This expression is positive for all σ_{-i} iff $u_i(\sigma'_i, s_{-i}) - u_i(\sigma_i, s_{-i})$ is positive for all s_{-i}. ∎

This proposition indicates that when we show that a strategy σ_i strictly dominates another strategy σ'_i for player i, we can assume that other players take pure strategies only.

Definition 7.8 A mixed strategy profile $\sigma^* = (\sigma^*_1, \ldots, \sigma^*_n)$ is a <u>dominant-strategy equilibrium</u> (DSE) in $\Gamma_N = [\mathbb{N}, \{\Delta(\mathbb{S}_i)\}, \{u_i\}]$ if for every $i \in \mathbb{N}$, $u_i(\sigma^*_i, \sigma_{-i}) \geq u_i(\sigma_i, \sigma_{-i})$ for all $\sigma_i \in \Delta(\mathbb{S}_i)$ and $\sigma_{-i} \in \Delta(\mathbb{S}_{-i})$.

Proposition 7.7 *A DSE is a NE in which each player's equilibrium strategy is a dominant strategy.* ∎

The following proposition suggests that, to reduce the complexity of finding NEs, we can first eliminate some dominated pure strategies since they will never be played in a NE.

Proposition 7.8 *For player i, if a pure strategy $\bar{s}_i$ is strictly dominated by a mixed strategy that never uses $\bar{s}_i$ (zero probability to $\bar{s}_i$), then every mixed strategy that uses $\bar{s}_i$ (positive probability to $\bar{s}_i$) is strictly dominated by a mixed strategy that never uses $\bar{s}_i$.*

Proof Suppose $u_i(\sigma_i, s_{-i}) > u_i(\bar{s}_i, s_{-i})$ for all $s_{-i} \in \mathbb{S}_{-i}$, with $\sigma_i(\bar{s}_i) = 0$. That is, for all $s_{-i} \in \mathbb{S}_{-i}$,

$$\sum_{s_i \neq \bar{s}_i} \sigma_i(s_i) u_i(s_i, s_{i-1}) > u_i(\bar{s}_i, s_{-i}). \tag{5}$$

Suppose that σ'_i is a mixed strategy of player i that assigns a positive probability to $\bar{s}_i$. Then, we can design another mixed strategy σ''_i that assigns $\sigma'_i(s_i) + \sigma_i(s_i)\sigma'_i(\bar{s}_i)$ to any $s_i \neq \bar{s}_i$, but nothing to $\bar{s}_i$, i.e.,

$$\sigma''_i(s_i) = \sigma'_i(s_i) + \sigma_i(s_i)\sigma'_i(\bar{s}_i), \quad \text{for } s_i \neq \bar{s}_i, s_i \in \mathbb{S}_i;$$
$$\sigma''_i(\bar{s}_i) = 0.$$

The condition

$$u_i(\sigma''_i, s_{-i}) > u_i(\sigma'_i, s_{-i}) \tag{6}$$

is equivalent to

$$\sum_{s_i \neq \bar{s}_i} [\sigma'_i(s_i) + \sigma_i(s_i)\sigma'_i(\bar{s}_i)] u_i(s_i, s_{-i}) > \sum_{s_i \in \mathbb{S}_i} \sigma'_i(s_i) u_i(s_i, s_{-i}),$$

or

$$\sum_{s_i \neq \bar{s}_i} \sigma_i(s_i)\sigma'_i(\bar{s}_i) u_i(s_i, s_{-i}) > \sigma'_i(\bar{s}_i) u_i(\bar{s}_i, s_{-i}),$$

for all $s_{-i} \in \mathbb{S}_{-i}$. This inequality is implied by (5). Thus, (6) holds. That is, σ'_i is strictly dominated. ∎

Mas-Colell et al. (1995) offer the following corollary. We provide a general result in Proposition 7.8, which is much more useful. Some pure strategies can be dominated by a mixed strategy even though they cannot be dominated by a pure strategy.

Corollary 7.1 *If, for player i, the pure strategy $\bar{s}_i$ is strictly dominated, then so is every mixed strategy that assigns a positive probability to this strategy.* ∎

With Proposition 7.8, we can iteratively eliminate strictly dominated strategies when we try to find Nash equilibria in a normal-form game. We can eliminate not only strictly dominated strategies and strategies that are strictly dominated after the

2 Equilibria in Normal-Form Games

first deletion of strategies but also strategies that are strictly dominated after the next deletion of strategies, and so on. One feature of this process of iteratively eliminating strictly dominated strategies is that the order of deletion does not affect the set of strategies that remain in the end. That is, if at any given point several strategies are strictly dominated, then we can eliminate them all at once or in any sequence without changing the set of strategies that we ultimately end up with. However, we cannot use the iteratively eliminating process for weakly dominated strategies since the final set of strategies depends on the order of deletion.

Example 7.9 Consider the following game:

P1\P2	L_2	M_2	R_2
L_1	2, 3	-2, 2	5, 2
M_1	-2, 3	4, 5	2, 3
R_1	1, 4	-3, -1	8, 1

We can first eliminate R_2 since it is strictly dominated by $\sigma_2 = (0.5, 0.5, 0)$. Once R_2 is eliminated, we can further eliminate R_1 since it is strictly dominated by L_1. The game is now reduced the one in Example 7.7, from which we know all the NEs of the reduced game. These NEs are all the NEs in the original game. ∎

2.3 Trembling-Hand Perfect Nash Equilibrium

Trembling-hand perfection is a term given to consideration of the robustness of Nash equilibria. In particular, it is concerned with the possibility that players may deviate slightly from their Nash strategies by mistakes. If so, will the Nash equilibrium be destroyed?

Example 7.10 There are situations in which a player is indifferent between two alternative strategies, one of which is the equilibrium strategy. This player has no incentive to deviate if other players do not make any mistakes. However, the situation changes if possible mistakes by other players are taken into account. Consider the following simple game:

P1\P2	L_2	R_2
L_1	1, 2	0, 2
R_1	0, 1	3, 3

There two pure-strategy NEs: (L_1, L_2) and (R_1, R_2). In (L_1, L_2), given L_1, player 2 is indifferent between L_2 and R_2. However, if player 1 may make some mistakes by taking R_1 with probability $\varepsilon > 0$, no matter how small ε is, player 2 will strictly prefer R_2 to L_2. Thus, (L_1, L_2) is not an error-proof equilibrium, while (R_1, R_2) is. ∎

We now formally model an error-proof equilibrium concept. A <u>totally mixed strategy</u> is a mixed strategy in which every pure strategy receives a strictly positive probability.

Definition 7.9 A NE σ^* is <u>trembling-hand perfect (THP)</u> if there is a sequence of totally mixed strategies $\{\sigma^k\}_{k=1}^{\infty}$ such that

- $\lim_{k \to \infty} \sigma^k = \sigma^*$,
- For each i and when k is large enough, σ_i^* is the best response to σ_{-i}^k. ∎

By definition, σ_i^* is the best response to σ_{-i}^*. By THP, σ_i^* is the best response to σ_{-i}^k when σ_{-i}^k is close to σ_{-i}^*, implying that other players are allowed to make small mistakes.

Example 7.11 Reconsider the above example. For (R_1, R_2), we consider totally mixed strategies $\sigma_1 = (\varepsilon_1, 1 - \varepsilon_1)$ and $\sigma_2 = (\varepsilon_2, 1 - \varepsilon_2)$.

P1\P2	$L_2(\varepsilon_2)$	$R_2(1 - \varepsilon_2)$
$L_1(\varepsilon_1)$	1, 2	0, 2
$R_1(1 - \varepsilon_1)$	0, 1	3, 3

We suppose that $\varepsilon_1, \varepsilon_2 > 0$ and $\varepsilon_1 \to 0$ and $\varepsilon_2 \to 0$. Under this circumstance, P1 will still choose $s_1^* = R_1$ iff $0\varepsilon_2 + 3(1 - \varepsilon_2) \geq 1\varepsilon_2 + 0(1 - \varepsilon_2)$, which holds when ε_2 is close to 0. Also, P2 will continue to choose $s_2^* = R_2$ iff $2\varepsilon_1 + 3(1 - \varepsilon_1) \geq 2\varepsilon_1 + 1(1 - \varepsilon_1)$, which always holds. Hence, (R_1, R_2) is trembling-hand perfect.

For (L_1, L_2), consider totally mixed strategies $\sigma_1 = (1 - \varepsilon_1, \varepsilon_1)$ and $\sigma_2 = (1 - \varepsilon_2, \varepsilon_2)$.

P1\P2	$L_2(1 - \varepsilon_2)$	$R_2(\varepsilon_2)$
$L_1 (1 - \varepsilon_1)$	1, 2	0, 2
$R_1(\varepsilon_1)$	0, 1	3, 3

We suppose that $\varepsilon_1, \varepsilon_2 > 0$ and $\varepsilon_1 \to 0$ and $\varepsilon_2 \to 0$. Under this circumstance, P1 will still choose $s_1^* = L_1$ iff $1(1 - \varepsilon_2) + 0\varepsilon_2 \geq 0(1 - \varepsilon_2) + 3\varepsilon_2$, which holds when ε_2 is close to 0. Also, P2 will continue to choose $s_2^* = L_2$ iff $2(1 - \varepsilon_1) + 1\varepsilon_1 \geq 2(1 - \varepsilon_1) + 3\varepsilon_1$, which does not hold. Hence, (L_1, L_2) is not trembling-hand perfect. ∎

The following proposition dramatically simplifies the issue on trembling-hand perfection.

2 Equilibria in Normal-Form Games

Proposition 7.9 *When $n = 2$, a NE is trembling-hand perfect iff none of its equilibrium strategies is weakly dominated.*[7] ∎

This result is consistent with the conclusion in Example 7.11. Using Proposition 7.9, we can easily obtain the following proposition.

Proposition 7.10 (Existence). *A game in which every strategy space is a finite set has a THP NE.* ∎

2.4 Reactive Equilibrium

Some NEs do not make much sense. The following example shows such a case.

Example 7.12 (Meeting in an Airport). Wang and Yang are to meet in an airport. However, they do not know whether they are to meet at door A or door B. It is better for them to meet at door A since it is closer to a parking lot. The payoffs are specified in the following normal form game:

Wang\Yang	A	B
A	20, 20	0, 0
B	0, 0	10, 10

There are two pure-strategy NEs: (A, A) and (B, B). There is also a mixed-strategy NE:

$$NE3: \quad \sigma_1^* = \frac{1}{3}A + \frac{2}{3}B, \quad \sigma_2^* = \frac{1}{3}A + \frac{2}{3}B.$$

These NEs do not give us much idea about what outcome to expect in this game. Further, this mixed-strategy NE does not make much sense. While they both prefer A, they both are more likely to choose B.

One problem with mixed-strategy NEs is that they often do not make sense. In fact, a player takes a mixed strategy in equilibrium only when she does not care what to do. It is precisely this indifference that determines a mixed-strategy NE. In this example, why do the players go to door B with a higher probability? The answer is that, when a player does this, the other player will then be indifferent between the two choices, and only in this situation do we have a NE. ∎

Nash assumes that other players would never react to a player's deviation from a NE. For example, suppose (σ_1^*, σ_2^*) is a NE in a game of two players P1 and P2, and denote $\hat{\sigma}_1(\sigma_2)$ and $\hat{\sigma}_2(\sigma_1)$ respectively as the optimal solutions of the following problems:

---
[7]This result does not hold when $n \geq 3$.

$$\max_{\sigma_1} u_1(\sigma_1, \sigma_2), \quad \max_{\sigma_2} u_2(\sigma_1, \sigma_2).$$

If P2 never reacts to P1's deviation, then P1 should stick to σ_1^*. However, if P2 does react to P1's deviation, then, if P1 takes σ_1', P2 will take $\hat{\sigma}_2(\sigma_1')$. If $u_1(\sigma_1', \hat{\sigma}_2(\sigma_1')) > u_1(\sigma_1^*, \sigma_2^*)$, which is possible, then P1 should deviate from her equilibrium strategy. In fact, P1 should take σ_1^{**}, which is the optimal solution of the following problem:

$$\max_{\sigma_1} u_1(\sigma_1, \hat{\sigma}_2(\sigma_1)).$$

We define this σ_1^{**} as P1's strategy in reactive equilibrium. To choose σ_1^{**}, P1 does not need to know P2's equilibrium strategy. Hence, the reactive equilibrium avoids the two problems in NE as mentioned at the end of Sect. 2.1.

Definition 7.10 A reactive equilibrium (RE) is a situation in a two-person game in which each player plays a Stackelberg strategy, by which the player takes into account possible reactions of the other play one step ahead.

Example 7.13 Reconsider the game in Example 7.7. There are two pure-strategy NEs: (L_1^*, L_2^*) and (R_1^*, R_2^*).

P1\P2	L_2	R_2
L_1	2, 3	-2, 2
R_1	-2, 3	4, 5

To reach a RE, P1 first figures out P2's reactions: the reaction strategies of P2 are

$$\hat{s}_2(L_1) = L_2, \quad \hat{s}_2(R_1) = R_2.$$

Then, from $u_1[L_1, \hat{s}_2(L_1)]$ and $u_1[R_1, \hat{s}_2(R_1)]$, we find $s_1^{**} = R_1$. Symmetrically, P2 figures out P1's reactions: the reaction strategies of P1 are

$$\hat{s}_1(L_2) = L_1, \quad \hat{s}_1(R_2) = R_1.$$

Then, from $u_2[\hat{s}_1(L_2), L_2]$ and $u_2[\hat{s}_1(R_2), R_2]$, we find $s_2^{**} = R_2$. Hence, there is only one pure-strategy RE, which is a pure-strategy NE (R_1, R_2). In fact, this is the unique RE (no mixed-strategy RE) and its outcome is Pareto efficient. ∎

Example 7.14 Reconsider the airport game in Example 7.12. It turns out that there is only one pure-strategy RE, which is the pure-strategy NE (A, A). In fact, this is the unique RE (no mixed-strategy RE) and its outcome is Pareto efficient. ∎

3 Equilibria in Extensive-Form Games

In the extensive form, players generally take turns to choose their strategies. That is, an extensive-form game is a dynamic game. For extensive-form games, we introduce three equilibrium concepts: the Nash equilibrium (NE), the subgame perfect Nash equilibrium (SPNE), and the Bayesian equilibrium (BE).

3.1 Nash Equilibrium

For extensive-form games, we define Nash equilibrium first. Nash treats all games as simultaneous-move games, including dynamic games. In a Nash equilibrium, all players simultaneously choose their own strategies.

Definition 7.11 A behavior strategy profile $\sigma^* = (\sigma_1^*, \ldots, \sigma_n^*)$ is a <u>Nash equilibrium (NE)</u> in behavior strategies of an extensive-form game Γ_E if, for every $i \in \mathbb{N}$, we have $u_i(\sigma_i^*, \sigma_{-i}^*) \geq u_i(\sigma_i, \sigma_{-i}^*)$ for all player i's behavior strategies σ_i. ∎

A game is usually initially defined by the extensive form. When there is a need, it is then converted into the normal form. This conversion is unique. By the outcome-equivalence between behavior strategies for extensive-form games and mixed strategies for normal-form games (Kuhn's Theorem), we have the following result.

Proposition 7.11 *Every Nash equilibrium in behavior strategies of an extensive-form game Γ_E has an output-equivalent Nash equilibrium in mixed strategies of the normal-form game of Γ_E.* ∎

Since it is easier to identify NEs in mixed strategies, from now on, we will refer to NEs of an extensive-form game Γ_E as the NEs in mixed strategies of the normal-form game of Γ_E.

The NE concept ignores the timing and sequence of actions and assumes that all players act simultaneously. Also, NE assumes that when a player is choosing his actions, he supposes that others will not react to whatever actions chosen. In particular, if he deviates from his equilibrium strategy, he still supposes that others will not react accordingly. This is a strong assumption. Although it is optimal for a player to stick to his equilibrium strategy assuming that others will never change their strategies, if others do change their strategies accordingly when he changes his strategy, it may be better for him to change. Let us see an example in which a problematic equilibrium occurs due to the simplicity of the Nash equilibrium concept.

Example 7.15 (Selten). For the game in Fig. 5, the normal form has two pure-strategy NEs: (L, L) and (R, R). However, (L, L) is not a sensible equilibrium. From the game tree, we know that P2 will choose R when its his turn to make a

Fig. 5 Game tree and NEs

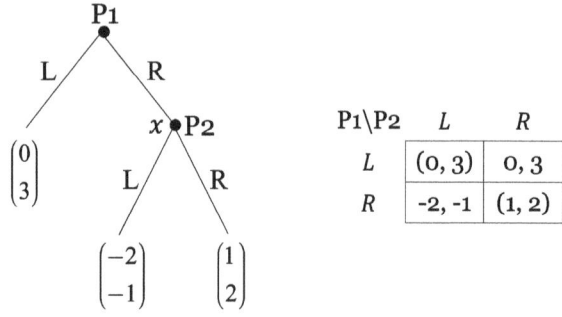

move. P1 understands this. Hence, P1 should choose R. Hence, the NE (L, L) does not make sense.

In this equilibrium, P1 decides to choose L since P2 is to choose L. However, once P1 has chosen R, P2 prefers R. Hence, P2 will deviate from his equilibrium strategy if P1 deviates. Here, the problem is that the NE concept assumes that when a player chooses a move, he assumes that others will stick to their equilibrium strategies no matter what he does. For (L, L), Nash assumes that P2 will take L and stick to it. If so, P1 will choose L. Conversely, if P1 has chosen L, there is no reason for P2 to change. Hence, (L, L) is a NE. This is due to the weakness of the NE concept: a player always takes his opponents' equilibrium strategies as given without asking whether it is rational for the opponents not to change their strategies. ∎

As extensive-form games are dynamic games, we will naturally rely on dynamic principles to rule out unreasonable NEs. In particular, to rule out NEs such as (L, L) in Example 7.15, we impose the principle of sequential rationality.

Sequential rationality under perfect information. *Under perfect information, the players' strategies imply an optimal action at every decision node, assuming so in subsequent decision nodes, i.e., rationality at every decision node.*

This sequential rationality focuses on the fact that the players take their actions in sequence. Hence, it is natural to suppose that each player must take an optimal action when its time for her action, and this optimal action will be conditional on the fact that subsequent players will also take optimal actions at their decision nodes. By assuming optimality at subsequent decision nodes, this rationality at each decision node takes into account possible reactions by subsequent players.

Since the extensive-form game in Example 7.15 belongs to the class of finite games of perfect information, we can use a procedure called backward induction to find a solution that satisfies this principle. Such a solution satisfies the principle of sequential rationality and is called a sequentially rational (SR) solution. We first look at the last decision node x. P2 will choose R at this decision node; see the left side of Fig. 6. Once this is done, we can then determine P1's optimal choice at the top decision node given the anticipation of what will happen after his choice. This

3 Equilibria in Extensive-Form Games   229

Fig. 6 Sequential rationality
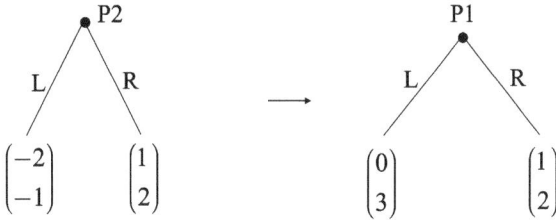

second step is accomplished by considering a reduced extensive form game where P2's decision node is replaced by the payoffs that will result from P2's optimal decision; see the right side of Fig. 6. We can see that P1's optimal decision is to play R from this reduced game. Therefore, we find the SR solution (R, R), which turns out to be one of the NEs.

Proposition 7.12 *Under perfect information, SR solutions are NEs.*

Proof A NE requires that given that others have taken the equilibrium actions, any player's equilibrium action is optimal. A SR solution satisfies this condition and is hence a NE. ∎

The above example shows that the procedure of backward induction for a finite game with perfect information proceeds as follows. We start by determining the optimal actions at the final decision nodes in the tree. Then, given that these will be the actions taken at the final decision nodes, we can proceed to the next-to-last decision nodes and determine the optimal actions to be taken by the players there, and so on backward through the game tree.

Example 7.16 Consider the game in Fig. 7.

This is a finite game with perfect information. By backward induction, on the left side of Fig. 8 is the reduced game formed by replacing the final decision nodes by the payoffs that result from the optimal play once these nodes have been reached. On the right side of Fig. 8 is the reduced game derived in the next stage of the backward induction procedure when the final decision nodes of the reduced game on the left side are replaced by the payoffs arising from optimal play at these nodes.

By this backward induction, we find the Nash equilibrium (s_1^*, s_2^*, s_3^*) that satisfies the principle of sequential rationality:

$$s_1^* = R_1, \quad s_2^* = L_2, \quad s_3^* = \langle \bar{L}_3, R_3, \hat{L}_3 \rangle,$$

with payoff vector $(1, 1, 2)$. ∎

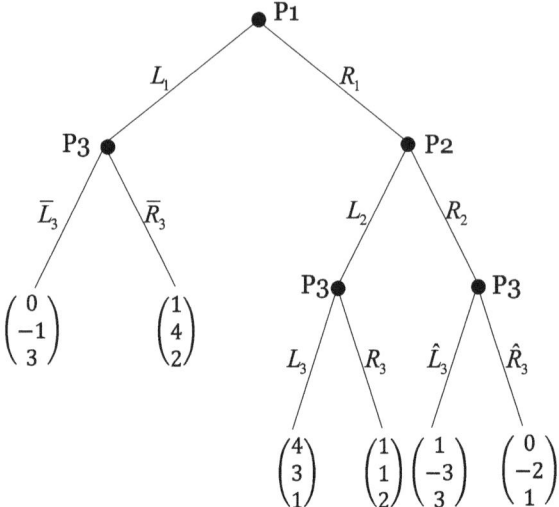

Fig. 7 Backward induction under perfect information

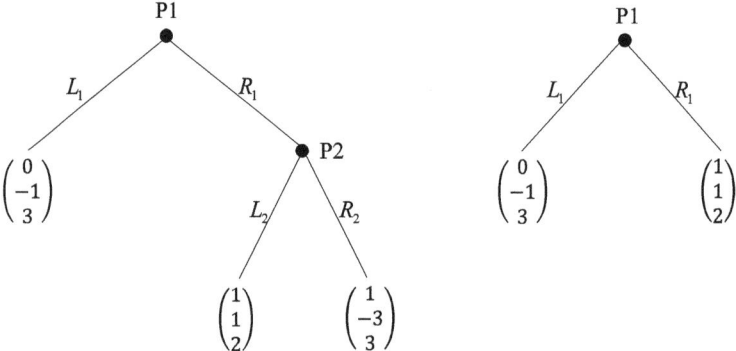

Fig. 8 Backward induction

3.2 Subgame Perfect Nash Equilibrium

To apply backward induction to finite games with imperfect information, we need the concept of the subgame.

Definition 7.12 A <u>subgame</u> of an extensive-form game Γ_E is a subset of a game having the following properties:

(a) It begins with an information set that contains a single decision node, contains all the decision nodes that are successors (both immediate and later) of this node, and contains only these nodes.

3 Equilibria in Extensive-Form Games

(b) If a decision node x of an information set H is in the subgame, then every $x' \in H$ is also. That is, there are no broken information sets. ∎

Note that the game as a whole is a subgame by definition. A subgame that is not the whole game is called a <u>real subgame</u>. Generally, a subgame starts from a single decision node and proceeds to include everything following this node. In finite games with perfect information, every decision node initiates a subgame.

We denote $SG(x)$ as the subgame that starts at decision node x. We say that a strategy profile σ^* in an extensive-form game Γ_E induces a NE in a subgame $SG(x)$ of Γ_E if the strategies in σ^* constitute a NE when this subgame is considered in isolation.

Definition 7.13 A strategy profile $\sigma^* = (\sigma_1^*, \ldots, \sigma_n^*)$ in an extensive-form game Γ_E is a <u>subgame perfect Nash equilibrium (SPNE or PNE)</u> if it induces a NE in every subgame of Γ_E.

Sequential rationality under imperfect information. *Under imperfect information, the players' strategies imply a NE in every subgame, assuming so in subsequent subgames, i.e., rationality in every subgame.*

This sequential rationality focuses on the fact that the players take their actions in sequence. Hence, it is natural to suppose that each player must take an optimal action when its time for her action in a subgame, and this optimal action will be conditional on her opponents' strategies in this subgame and the fact that this will also happen in subsequent subgames. By assuming NEs in subsequent subgames, this rationality in each subgame takes into account possible reactions by subsequent players.

To identify SPNEs in a finite game Γ_E, we use the <u>generalized backward induction</u> procedure, which derives NEs backwards a subgame at a time, instead of step by step. Specifically,

1. Start at the end of the game tree and identify NEs of each of the final subgames.
2. Select one NE in each of these final subgames and derive the reduced extensive-form game where these final subgames are replaced by the payoffs that result in these subgames when players use equilibrium strategies.
3. Repeat steps 1 and 2 for the reduced game. Continue the procedure until every move in Γ_E is determined. This collection of moves at the various information sets of Γ_E constitutes a profile of SPNE strategies.
4. If multiple equilibria are never encountered in any step, this strategy profile is the unique SPNE. If multiple equilibria are encountered, the full set of SPNEs is identified by repeating the procedure for each possible equilibrium that could occur for the subgames in question.

For an infinite game, the definition of subgame perfection remains the same. However, backward induction can no longer be used. Instead, we need a recursive structure on a game.

Example 7.17 Consider the game in Fig. 9. It has two subgames: the whole game and $SG(x)$ as shown in Fig. 10.

$SG(x)$ has a two NEs: $(\hat{L}_1, R_2)$ and $(\hat{R}_1, R_2)$. For each pair of the resulting payoffs, after replacing this subgame with the payoffs, the reduced game implies P1's optimal choice, implying one SPNE. The two NEs in the subgame lead to two SPNEs. The SPNEs are

$$\begin{aligned} SPNE1: \quad & s_1^* = \langle R_1, \hat{L}_1 \rangle, \quad s_2^* = R_2. \\ SPNE2: \quad & s_1^* = \langle R_1, \hat{R}_1 \rangle, \quad s_2^* = R_2. \end{aligned} \tag{7}$$

In fact, in $SG(x)$, since P1 is indifferent between $\hat{L}_1$ and $\hat{R}_1$ when P2 chooses R_2 and P2 will choose R_2 for certain no matter what P1 chooses, for each $r \in [0, 1]$, there is a mixed-strategy NE $(r\hat{L}_1 + (1-r)\hat{R}_1, R_2)$ in $SG(x)$. Each of the NE yields a SPNE. Hence, we have a third SPNE:

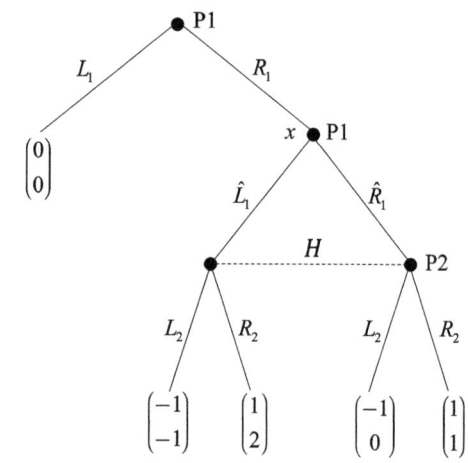

Fig. 9 Subgame perfection under imperfect information

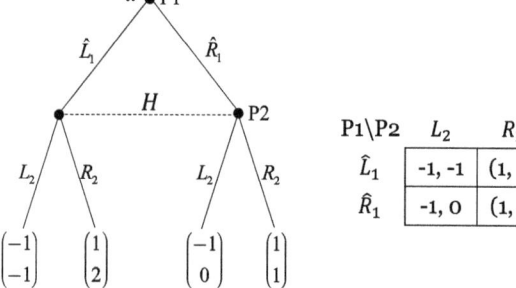

Fig. 10 Subgame $SG(x)$

3 Equilibria in Extensive-Form Games

$$\text{SPNE3}: \quad s_1^* = \langle R_1, r\hat{L}_1 + (1-r)\hat{R}_1\rangle, \quad s_2^* = R_2, \quad r \in [0,1].$$

On the other hand, the normal form of the whole game is

P1\P2	L_2	R_2
$\langle L_1, \hat{L}_1\rangle$	(0, 0)	0, 0
$\langle L_1, \hat{R}_1\rangle$	(0, 0)	0, 0
$\langle R_1, \hat{L}_1\rangle$	-1, -1	(1, 2)
$\langle R_1, \hat{R}_1\rangle$	-1, 0	(1, 1)

There are total four pure-strategy NEs:

NE1: $s_1^* = \langle L_1, \hat{L}_1\rangle$, $s_2^* = L_2$. NE2: $s_1^* = \langle L_1, \hat{R}_1\rangle$, $s_2^* = L_2$.
NE3: $s_1^* = \langle R_1, \hat{L}_1\rangle$, $s_2^* = R_2$. NE4: $s_1^* = \langle R_1, \hat{R}_1\rangle$, $s_2^* = R_2$.

The first two NEs are unreasonable and they are not SPNEs. ∎

However, the SPNE concept has a few problems. We show the problems in the following two examples.

Example 7.18 We first show a problem with the concept of subgame perfection when there are multiple equilibria in a subgame. Consider the game in Fig. 11. $SG(x)$ obviously has a two pure-strategy NEs: $(\hat{L}_1, L_2)$ and $(\hat{R}_1, R_2)$. The resulting SPNEs are

$$\begin{aligned}\text{SPNE1}: \quad s_1^* &= \langle L_1, \hat{L}_1\rangle, \quad s_2^* = L_2.\\ \text{SPNE2}: \quad s_1^* &= \langle R_1, \hat{R}_1\rangle, \quad s_2^* = R_2.\end{aligned} \quad (8)$$

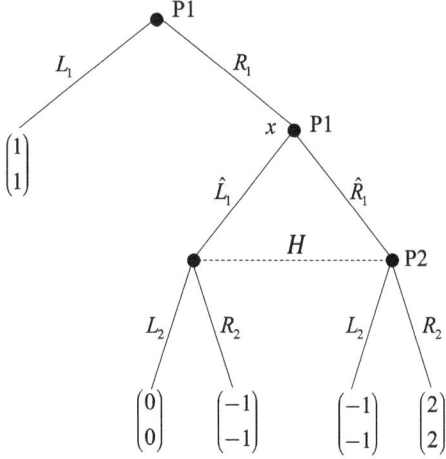

Fig. 11 A problem with subgame perfection

However, this example shows a problem of the definition of SPNE when there are multiple NEs in a subgame. For example, for SPNE1 in (8), when P1 is making a decision at the starting node, P1 must assume that the NE in $SG(x)$ is $(\hat{L}_1, L_2)$, otherwise P1 may make a different choice. If P1 expects $(\hat{R}_1, R_2)$ to have a better chance than $(\hat{L}_1, L_2)$ (which is a sensible expectation), than P1 will choose R_1 at the start. Hence, SPNE1 may not be a sensible solution. In practice, we cannot rule out the possibility that a player makes expectations on which NEs in a subgame are more likely to happen and such expectations affect the player's choice. SPNE does not take into account this possibility. ∎

Example 7.19 We modify the game in Example 7.17 slightly so that there are no real subgames, as shown in Fig. 12.
The normal form is

P1\P2	L_2	R_2
L_1	(0, 0)	0, 0
M_1	-1, -1	(1, 2)
R_1	-1, 0	(1, 1)

There are three pure-strategy NEs: (L_1, L_2), (M_1, R_2), and (R_1, R_2). Since there is no real subgame, all these NEs are SPNEs.

Again, the first NE is not reasonable. Once P1 decides to go to information set H, P2 will definitely choose R_2. Even though P2 does not know whether it is M_1 or R_1, the choice R_2 is always strictly better than L_2. P1 should understand this and, if so, L_1 is an inferior choice. Unfortunately, this unreasonable NE cannot be ruled out by subgame perfection. This failure calls for a different equilibrium concept for extensive-form games. ∎

Fig. 12 A game with a bad spne

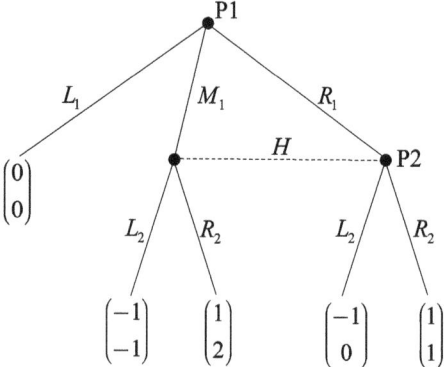

3.3 Bayesian Equilibrium

How can we eliminate unreasonable equilibria? We find that if we allow players to form beliefs about his opponent's strategies, we can then eliminate some unreasonable equilibria.

Example 7.20 Reconsider the game in Fig. 13 from Example 7.12. When Yang is to make a move, she does not know what Wang has chosen. However, Yang can take a guess and assign a probabilistic distribution over the two possible choices by Wang. Suppose that the probabilities are μ_1 and μ_2 for the two possible choices as shown in the diagram, with $\mu_1 + \mu_2 = 1$. Given the beliefs, for Yang,

$$A \succ B \Leftrightarrow 20\mu_1 + 0\mu_2 > 0\mu_1 + 10\mu_2 \quad \text{or} \quad \mu_1 > 1/3.$$

That is, if Yang believes that Wang will go to gate A with a probability higher than $1/3$, she will choose gate A. Here, since Wang prefers to meet at gate A than gate B, such a belief makes sense. Interestingly, if Yang chooses A with probability one, Wang should choose A with probability one as well, which is consistent with the condition $\mu_1 > 1/3$. In other words, there is an equilibrium where $\mu_1 = 1$ and $\mu_2 = 0$ and this belief is consistent with the equilibrium strategies derived from this belief. This solution seems more acceptable than other solutions in Example 7.12. ∎

Definition 7.14 A system of <u>beliefs</u> μ in an extensive-form game Γ_E is a specification of a probability $\mu(x) \in [0, 1]$ for each decision node x in Γ_E such that $\sum_{x \in H} \mu(x) = 1$ for all information sets H. ∎

A system of beliefs specifies, for each information set H, a probabilistic distribution μ_H over H by the player who moves at H on the relative likelihood of being at each of H's various decision nodes, conditional upon play having reached H.

Definition 7.15 Given a belief system μ, denote $E[u_i | H, \mu, \sigma_i, \sigma_{-i}]$ as player i's expected utility at her information set H if she uses strategy σ_i and her rivals use

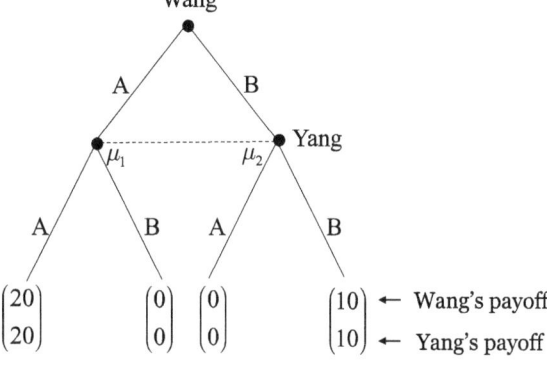

Fig. 13 Bayesian solution to the airport meeting game

strategies σ_{-i}. Denote $i(H)$ as the player who moves at information set H. A strategy profile σ^* in Γ_E is <u>sequentially rational (SR)</u> at information set H given a belief system μ if

$$E\left[u_{i(H)}\Big|H,\mu,\sigma^*_{i(H)},\sigma^*_{-i(H)}\right] \geq E\left[u_{i(H)}\Big|H,\mu,\sigma_{i(H)},\sigma^*_{-i(H)}\right], \tag{9}$$

for all $\sigma_{i(H)} \in \Delta(\mathbb{S}_{i(H)})$. A strategy profile σ^* in Γ_E is sequentially rational (SR) given belief system μ if it satisfies (9) for all the information sets in the game. ■

This definition can be alternatively described by the following principle of sequential rationality.

Sequential rationality in Bayesian games. *In a Bayesian game, the players' strategies imply an optimal action at every information set, assuming so in subsequent information sets, i.e., rationality at every information set.*

This sequential rationality focuses on the fact that the players take their actions in sequence. Hence, it is natural to suppose that each player must take an optimal action when its time for her action at one of her information sets, and this optimal action will be conditional on the fact that subsequent players will also take optimal actions at their information sets. By assuming optimality at subsequent information sets, this rationality at each information set takes into account possible reactions by subsequent players. In other words, a strategy profile σ^* is sequentially rational if no player finds it worthwhile, once one of her information sets has been reached, to revise her strategy given her rivals' strategies and her beliefs about the future (as embodied in μ).

A key problem with NEs is that they may not be SR. The following example shows that SR can rule out some unreasonable NEs.

Example 7.21 For the game in Fig. 14 from Example 7.19, given a belief system μ, we now find a SR strategy profile σ^*. Given arbitrary beliefs μ_1 and μ_2 for the information set H, with $\mu_i \geq 0$ and $\mu_1 + \mu_2 = 1$, P2 strictly prefers L_2 over R_2 iff

Fig. 14 A SR strategy profile

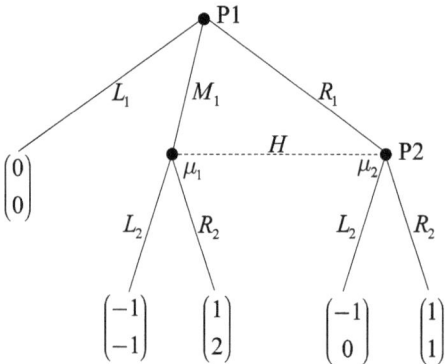

$$-1\mu_1 + 0\mu_2 > 2\mu_1 + 1\mu_2.$$

This is completely impossible. Hence, no matter what belief system P2 has, he will always choose R_2 once information set H has been reached. That is, $\hat{s}_2 = R_2$ is a rational choice. As P2 chooses R_2 for sure, L_1 is an inferior choice for P1. Hence, P1 will choose either M_1 or R_1. In fact, M_1 and R_1 are indifferent to P1. Thus, $\hat{\sigma}_1 = (0, \tau, 1 - \tau)$ is a rational strategy, where $\tau \in [0, 1]$. Therefore, given an arbitrary belief system μ, the SR strategies are

$$\hat{\sigma}_1 = (0, \tau, 1 - \tau), \quad \hat{s}_2 = R_2, \quad \text{for } \tau \in [0, 1]. \tag{10}$$

The unreasonable SPNE is not included. That is, SR rules out the unreasonable SPNE. ∎

Example 7.22 For the game in Fig. 15, given a belief system μ, we now find SR solutions.

First, P2 chooses L_2 over R_2 iff $\mu_1 > -\mu_1 + \mu_2$, i.e., $\mu_1 > 1/3$. Thus, if $\mu_1 > 1/3$, P2 will play L_2 for sure. If so, P1 will choose R_1 for certain.

Symmetrically, if $\mu_1 < 1/3$, P2 will choose R_2 for sure. If so, P1 will choose M_1 for certain.

If $\mu_1 = 1/3$, P2 is indifferent between L_2 and R_2. That is, any strategy $\sigma_2 = (t, 1 - t)$ is rational, where $t \in [0, 1]$. It is obvious that L_1 is an inferior choice for P1. Then, P1 chooses M_1 over R_1 iff $t + 2(1 - t) > 3t + (1 - t)$, which is $t < 1/3$. Symmetrically, P1 chooses R_1 iff $t > 1/3$. Also, when $t = 1/3$, P1 is indifferent between M_1 and R_1.

In summary, the SR solutions are

Fig. 15 A SR strategy profile

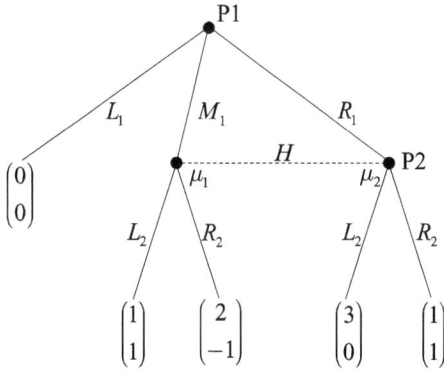

If $\mu_1 > 1/3$, then $\hat{s}_1 = R_1$ and $\hat{s}_2 = L_2$.
If $\mu_1 < 1/3$, then $\hat{s}_1 = M_1$ and $\hat{s}_2 = R_2$.
If $\mu_1 = 1/3$, then $\hat{\sigma}_2 = (t, 1-t)$ for any $t \in [0, 1]$ and
$\quad \hat{s}_1 = M_1$ if $t < 1/3$,
$\quad \hat{s}_1 = R_1$ if $t > 1/3$,
$\quad \hat{\sigma}_1 = (0, \tau, 1-\tau)$ for any $\tau \in [0, 1]$ if $t = 1/3$. ■
(11)

With a belief system, we can now derive optimal strategies backward step by step, instead of subgame by subgame. This process can go on repeatedly from the last step until we reach the starting point of the game. Such a solution from backward induction will be sequentially rational. Here, the belief system is assumed to be public knowledge so that each player can figure out what future moves will imply.

When a player takes a certain mixed strategy to go to a node, this node will be reached with certain probability. This is the objective probability that the node will be reached. At the same time, there is a belief that this node would be reached with certain probability; this is a subjective probability. The objective probability actually depends on this subjective probability (as embodied in the sequentially rational strategies). These two probabilities should be consistent with each other, otherwise the players will find the error over time and then adjust their beliefs. Hence, besides sequential rationality, a Bayesian equilibrium requires certain consistencies in the equilibrium path.

SR solutions and NEs do not imply each other. Under perfect information, SR solutions are NEs. However, under imperfect information, SR solutions may not be NEs; it depends on the belief system. For example, in Example 7.22, although there are many SR solutions, there is only one NE: $\sigma_{11}^* = 1/3$, $\sigma_{12}^* = 1/3$. Conversely, NEs may not be SR. For example, in Example 7.21, imposing SR rules out some unreasonable NEs.

A path of a game is defined as a sequence of decision nodes by which the game goes from the initial node to a terminal node. Here, mixed strategies are allowed, meaning that some paths are taken with a certain probability. Notice that each decision node belongs to one and only one information set and each information set belongs to one and only one player. Given an equilibrium σ^*, which can be any type of equilibrium, we say that an information set H is on the equilibrium path if $\Pr(H) > 0$, and H is off-equilibrium or is on an off-equilibrium path if $\Pr(H) = 0$. Since a path can be taken with a certain probability, an equilibrium may involve several paths.

A strategy is a series of planned actions by a player. Those planned actions in equilibrium strategies (planned to be taken with positive probability) are called equilibrium actions. Those equilibrium actions on the equilibrium path are called realized equilibrium actions, and those equilibrium actions on off-equilibrium paths are called unrealized equilibrium actions.

We say that an equilibrium, or equivalently an equilibrium strategy profile, is rational along a path if every player behaves rationally at his/her information sets along the path. Rationality may be based on mixed strategies or beliefs. Sequential rationality is based on beliefs.

Definition 7.16 A strategy-belief pair (σ^*, μ^*) is a <u>Bayesian equilibrium (BE)</u> [or a weak perfect Bayesian equilibrium (weak PBE)] in Γ_E if it has the following properties:

(a) Sequential Rationality: Strategy profile σ^* is sequentially rational given the belief system μ^*.[8]
(b) Equilibrium-path (EP) consistency: The belief system μ^* is consistent with the strategy profile σ^* and Nature's odds ρ through the Bayes rule along the equilibrium path: for any $H \in \mathbb{H}$ with $\Pr(H) > 0$ for σ^*, we have $\mu^*(x) = \Pr(x|H), \forall x \in H$.[9] ∎

We call a strategy profile σ^* a BE if there exists a belief μ^* such that (σ^*, μ^*) is a BE.

A Bayesian equilibrium requires that at any point in the game, a player's strategy prescribes optimal actions from that point on, given a consistent belief system along the equilibrium path.

The conditional probability $\Pr(x|H)$ is derived based on the Bayes rule, i.e.,

$$\Pr(x|H) = \frac{\Pr(x, H)}{\Pr(H)} = \frac{\Pr(x)}{\Pr(H)} = \frac{\Pr(x)}{\sum_{x' \in H} \Pr(x')}.$$

For example, in Fig. 16, if $\Pr(H) > 0$, where $\Pr(H) = \sigma_{21} + \sigma_{31}$, the Bayes rule implies $\mu_1 = \Pr(x_1|H) = \frac{\sigma_{21}}{\sigma_{21} + \sigma_{31}}$. For another example, in Fig. 17, if $\Pr(I_1) = \alpha_1 \rho_1 + \beta_1 \rho_2 > 0$, we have

$$\Pr(x|I_1) = \frac{\alpha_1 \rho_1}{\alpha_1 \rho_1 + \beta_1 \rho_2}.$$

Consistency means

$$\mu_1 = \frac{\alpha_1 \rho_1}{\alpha_1 \rho_1 + \beta_1 \rho_2}.$$

---

[8] Here, sequential rationality needs to be satisfied for all information sets, including those on off-equilibrium paths. This is important to know since, even with a proper belief system, some Nash equilibria do not satisfy sequential rationality on off-equilibrium paths. By Proposition 7.13, this happens when a NE is not a BE.

[9] The words "whenever applicable" in Gibbons' (1992) definition are vague. A strategy is only a plan. Even for an equilibrium strategy, a planned action may never be taken in equilibrium. For example, given the equilibrium strategy profile σ^* defined by (12), for point y in Fig. 23, we have $\Pr(y|H) = 1$. But, $\mu_2^* = 0$. We can either say that σ^* is a BE since H is an off-equilibrium path or say that σ^* is not a BE since $\Pr(y|H) \neq \mu_2^*$. The weak version of BE in this book takes the former definition, while the strong version of BE in Gibbons (1992) takes the latter definition.

Fig. 16 Consistency in a Bayesian equilibrium

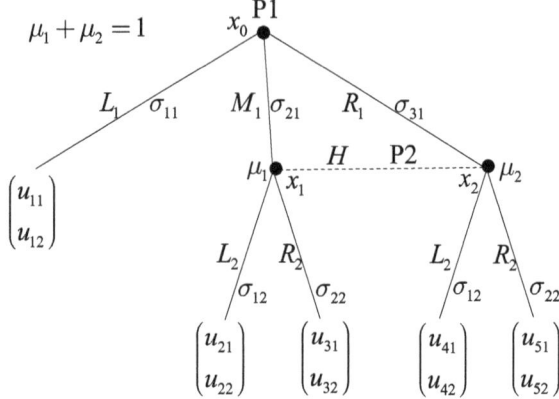

Fig. 17 Consistency in a Bayesian equilibrium

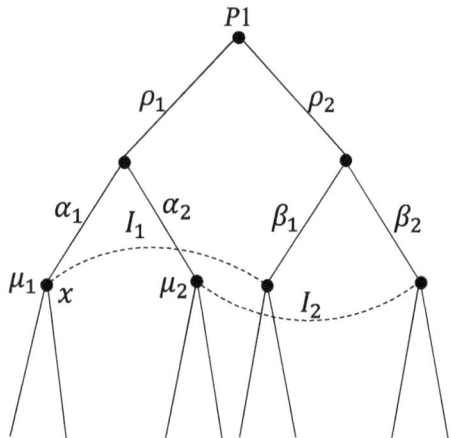

For a BE, the required consistency condition between strategies and beliefs is $\mu^*(x) = \Pr(x|H)$, where $\Pr(x|H)$ is the (objective) probability that node $x \in H$ will be reached given the fact that H has been reached under the equilibrium strategy profile σ^*. This means that the belief $\mu^*(x)$ is indeed correct. However, if H is not on the equilibrium path, then the probability that any node x in H is reached is zero. Given the requirement of $\sum_{x \in H} \mu^*(x) = 1$ for a belief system, there is no way for $\mu^*(x) = \Pr(x|H)$ to hold for this H. Alternatively, by the Bayes rule, $\Pr(x|H) = \Pr(x)/\Pr(H)$; but when $\Pr(H) = 0$, the Bayes rule is not applicable and hence $\Pr(x|H)$ is meaningless. This is intuitive; if H never happens, then we have no way to assess the probability of the occurrence of a node in H. Hence, the consistency condition applies only to those information sets H that are on the equilibrium path. As we will see later, since those beliefs on off-equilibrium paths can be quite arbitrary, they can cause some problems.

3 Equilibria in Extensive-Form Games

Fig. 18 BEs in the airport meeting game

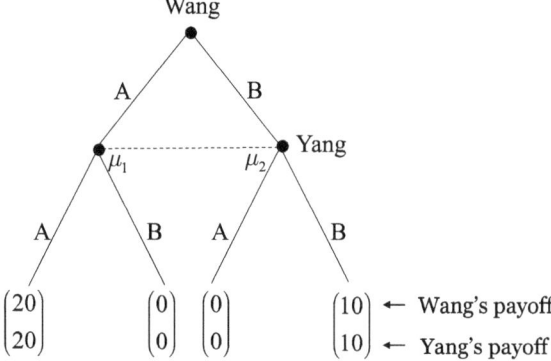

Example 7.23 Let us find all BEs in the airport game (Fig. 18).

Yang will go to gate A iff $\mu_1 > 1/3$. If so, Wang will go to A. Consistency implies $\mu_1^* = 1$,[10] which can be satisfied. Hence, one BE is: $s_w^* = A$, $s_y^* = A$, and $\mu_1^* = 1$.

Yang will go to gate B iff $\mu_1 < 1/3$. If so, Wang will go to B. Consistency implies $\mu_1^* = 0$, which can be satisfied. Hence, another BE is: $s_w^* = B$, $s_y^* = B$, and $\mu_1^* = 0$.

Yang is indifferent between A and B iff $\mu_1 = 1/3$. Let Yang's strategy be $(\sigma_y, 1-\sigma_y)$, where σ_y can be any value in $[0,1]$. Then, for $\mu_1 = 1/3$ to hold, consistency requires Wang to have $\sigma_w^* = 1/3$. If so, Wang must be indifferent between A and B, which implies

$$20\sigma_y + 0(1-\sigma_y) = 0\sigma_y + 10(1-\sigma_y),$$

implying $\sigma_y^* = 1/3$. The choice of $\sigma_w^* = 1/3$ must be optimal for Wang since he is indifferent between the two choices. Hence, we have a third BE: $\sigma_w^* = \sigma_y^* = \mu_1^* = 1/3$. ∎

Example 7.24 Let us find BEs for the following game from Example 7.21 (Fig. 19).

By the SR solution in (10), condition (a) in Definition 7.16 is already satisfied. For condition (b), we only need to impose $\mu_1 = \tau$. Hence, the BE is

$$\sigma_1^* = (0, \tau, 1-\tau), \quad s_2^* = R_2, \quad \mu_1^* = \tau, \quad \text{for any } \tau \in [0,1].$$

Alternatively, we can also solve for BEs directly without finding the SR strategies first. For arbitrary beliefs μ_1 and μ_2 to the information set H, with $\mu_i \geq 0$ and $\mu_1 + \mu_2 = 1$, P2 chooses L_2 over R_2 iff $-1\mu_1 + 0\mu_2 > 2\mu_1 + 1\mu_2$. This is completely impossible. Thus, no matter what belief system P2 has, he will always

---
[10]When we mention "consistency" without specifying which type of consistency, we refer to "equilibrium-path consistency".

Fig. 19 BEs

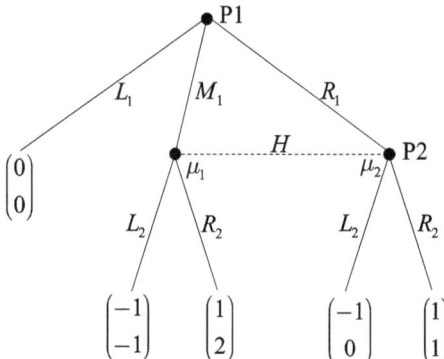

Fig. 20 BEs

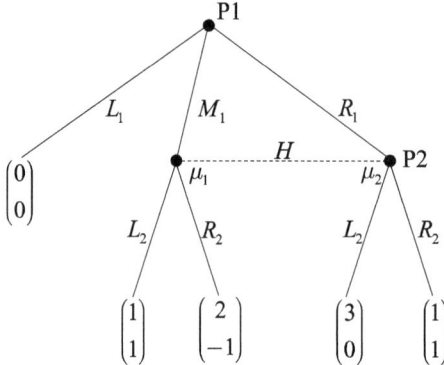

choose R_2 once information set H has been reached. As P2 chooses R_2 for sure, L_1 is an inferior choice for P1. Hence, P1 will choose either M_1 or R_1. In fact, M_1 and R_1 are indifferent to P1. Thus, $\sigma_1^* = (0, \tau, 1 - \tau)$, where $\tau \in [0, 1]$. If so, by the consistency condition, we have $\mu_1 = \tau$. Hence, the BE is:

$$\sigma_1^* = (0, \tau, 1 - \tau), \quad s_2^* = R_2, \quad \mu_1^* = \tau, \text{ for any } \tau \in [0, 1].$$

The unreasonable SPNE is ruled out. ∎

The reason that the BE concept can solve the above problem is that a belief system gives us the opportunity to derive an equilibrium backward one step at a time. A SPNE only allows us to derive backward one subgame at a time, and this backward approach does not apply within a subgame. A subgame itself may contain some unreasonable NEs, as shown in Example 7.19. A belief system allows us to solve the game in two steps, instead of in one step.

Example 7.25 Let us find BEs for the game in Fig. 20 from Example 7.22. By the SR solution in (11), condition (a) in Definition 7.16 is already satisfied. For condition (b), consistency can be satisfied only when $\mu_1 = 1/3$. That is, we find one and only one BE:

$$\sigma_1^* = \left(0, \frac{1}{3}, \frac{2}{3}\right), \quad \sigma_2^* = \left(\frac{1}{3}, \frac{2}{3}\right), \quad \mu_1^* = \frac{1}{3}.$$

Alternatively, we can also solve for BEs directly without finding the SR strategies first. First, P2 chooses L_2 over R_2 iff $\mu_1 > -\mu_1 + \mu_2$, i.e., $\mu_1 > 1/3$. Thus, if $\mu_1 > 1/3$, P2 will play L_2 for sure. If so, P1 will choose R_1 for certain, implying $\mu_1 = 0$ by consistency, which contradicts $\mu_1 > 1/3$. Hence, there is no such a BE.

If $\mu_1 < 1/3$, P2 will choose R_2 for sure. If so, P1 will choose M_1 for certain, implying $\mu_1 = 1$ by consistency, which contradicts $\mu_1 < 1/3$. Hence, there is no such a BE either.

If $\mu_1 = 1/3$, P2 is indifferent between L_2 and R_2. Let $\sigma_2 = (t, 1-t)$ for $t \in [0,1]$. With $\mu_1 = 1/3$, consistency suggests that there are only two possibilities for P1: either L_1 is chosen for certain or M_1 and R_1 are indifferent. In the former case, μ_1 can be any value. However, L_1 is obviously been dominated. Hence, it can only be the second case. If so, consistency must be imposed on H. Consistency implies that P1 must have chosen M_1 with probability $1/3$, implying that P1 must be indifferent between M_1 and R_1, i.e.,

$$1t + 2(1-t) = 3t + 1(1-t),$$

implying $t = 1/3$. Then, we find one and only one BE:

$$\sigma_1^* = \left(0, \frac{1}{3}, \frac{2}{3}\right), \quad \sigma_2^* = \left(\frac{1}{3}, \frac{2}{3}\right), \quad \mu_1^* = \frac{1}{3}.$$

Note that since P1 is indifferent between M_1 and R_1, $\sigma_1^* = (0, 1/3, 2/3)$ is optimal for P1.

Let us find all the NEs in the game. The normal form is:

P1\P2	L_2	R_2
L_1	0, 0	0, 0
M_1	1, 1	2, -1
R_1	3, 0	1, 1

Since P1's strategy L_1 is strictly dominated by his strategy R_1, we only need to consider the following normal form:

P1\P2	L_2	R_2
M_1	1, 1	2, -1
R_1	3, 0	1, 1

This game has no pure-strategy NE and, by (2), it has one mixed-strategy NE, which is: $\sigma_1^* = (1/3, 2/3)$, $\sigma_2^* = (1/3, 2/3)$. This mixed-strategy NE is the same as the BE. ∎

There is an implicit assumption: once a belief system is introduced, the calculation of the expected utility is based on the belief system whenever possible, rather than on the mixed strategy profile. That is, instead of $\sum (\prod_k \sigma_{ki}) u_i(s_1,\ldots,s_n)$ as the expected utility value, we now use $\sum (\prod_k \mu_{ki}) u_i(s_1,\ldots,s_n)$. For example, in Fig. 16, at information set H, P2's expected payoff for taking action L_2 is $\mu_1 u_{22} + \mu_2 u_{42}$, where the mixed strategies σ_{11}, σ_{21} and σ_{31} are not used in the calculation. Given P2's mixed strategy $\sigma_2 = (\sigma_{12}, \sigma_{22})$, P1's expected payoff for taking R_1 is $\sigma_{12} u_{41} + \sigma_{22} u_{51}$. However, with a sequentially rational strategy profile σ^*, σ_{12}^* and σ_{22}^* are to be derived using the beliefs, implying that both players' expected payoffs will be based on the belief system. If there is no belief system, the calculation will be based on the mixed strategies. The belief system consists of subjective probabilities, while the mixed strategy consists of actual probabilities. Hence, consistency between strategies and beliefs is crucial.

Unfortunately, such consistency cannot be easily imposed on off-equilibrium paths, which can result in various problems. That is, although BE can be used to deal with some problems that SPNE cannot deal with, some problems remain. The following three examples show two problems with BE.

Example 7.26 One problem with the BE concept is that some beliefs need not make sense. In the following game, given μ_2, P2 will play L_2 over R_2 iff $5 > 2\mu_2 + 10(1 - \mu_2)$, i.e., $\mu_2 > 5/8$. If so, P1 will choose L_1. In this case, $\Pr(H_2) = 0$; hence consistency of μ_2 is not required. Consistency of μ_1 is satisfied. Thus, we have a BE:

$$BE1: \quad s_1^* = L_1, \quad s_2^* = L_2, \quad \mu_1^* = 0.5, \quad \mu_2^* > 5/8.$$

But $\mu_2 > 5/8$ is not sensible. Since the two R_1's are the same action, the sensible μ_2 should be $\mu_2 = 0.5$. For this sensible μ_2, we have a completely different BE (Fig. 21). ∎

What are the NEs in this game? The normal form is[11]

P1\P2	L_2	R_2
L_1	(2, 10)	2, 10
R_1	0, 5	(5, 6)

We can see that BE1 is nevertheless a NE or a SPNE. ∎

Example 7.27 One more example of a senseless belief system in Fig. 22.

In this case, P2 will choose L_2 over R_2 for certain iff $-\mu_1 + \mu_2 > -2\mu_1 + 3\mu_2$ or $\mu_1 > 2/3$. If so, P1 will choose L_1 for certain. In this case, consistency is not

---

[11] We have expected payoffs in the cells since Nature offers a half-half chance. This NE is actually a BNE.

Fig. 21 Problem with BEs

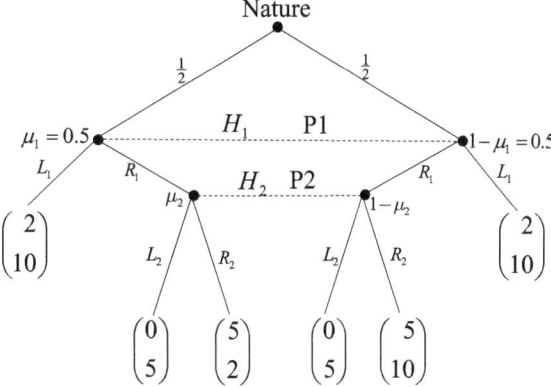

Fig. 22 Problem with off-equilibrium paths

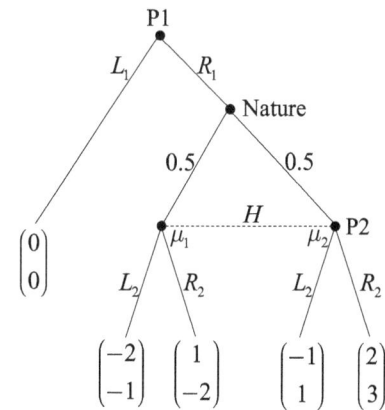

required on H. Hence, a belief system with $\mu_1 > 2/3$ supports a BE with $s_1^* = L_1$ and $s_2^* = L_2$ and payoff pair $(0,0)$. This belief system does not make sense (obviously we should have $\mu_1 = 0.5$); yet, it can support a BE. The reason is that consistency between strategies and Nature's odds are not required on off-equilibrium paths. A sensible belief system should have $\mu_1 = 0.5$, which supports a different BE with $s_1^* = R_1$ and $s_2^* = R_2$. ∎

Example 7.28 Another problem with the BE concept is that a BE need not be subgame perfect. Consider the game in Fig. 23. Given $\mu = (\mu_1, \mu_2)$, P2 chooses L_2 over R_2 iff $-\mu_1 + \mu_2 > -2\mu_1 + 3\mu_2$ or $\mu_1 > 2/3$. If so, P1 chooses $\hat{R}_1$ at node x. Then, since choosing R_1 means a payoff of -1, P1 chooses L_1 at the beginning. Hence, any belief system with $\mu_1 > 2/3$ can support a BE that leads to the payoff pair $(0,0)$, i.e., we have the following BE:

$$BE1: \quad s_1^* = L_1, \hat{R}_1, \quad s_2^* = L_2, \quad \mu_1^* > 2/3. \tag{12}$$

Fig. 23 BE may not be subgame perfect

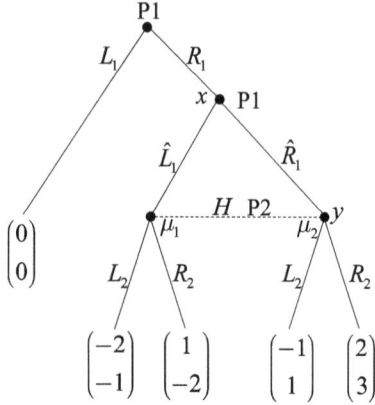

Here, since the information set H is off the equilibrium path, consistency is not required on H.

What are the SPNEs? The normal form for $SG(x)$ is

P1\P2	L_2	R_2
$\hat{L}_1$	-2, -1	1, -2
$\hat{R}_1$	-1, 1	(2, 3)

The only pure-strategy NE is $(\hat{R}_1, R_2)$. There are no mixed-strategy NEs, which can be easily shown by the fact that $\hat{R}_1$ is the dominant strategy for P1. Hence, the only SPNE is: $s_1^* = \langle R_1, \hat{R}_1 \rangle$ and $s_2^* = R_2$ with payoff pair $(2, 3)$. Therefore, the BE above is not subgame perfect.

The problem with the BE comes from the fact that consistency between strategies and beliefs is not ensured off the equilibrium paths. For the BE in (12), P2 chooses L_2 because she assumes that P1 will choose $\hat{L}_1$. P2's choice is a rational response to her own belief rather than to P1's strategies. Since $\hat{R}_1$ dominates $\hat{L}_1$, P2 should firmly believe that P1 will choose $\hat{R}_1$ at node x. Due to the inconsistency between P2's beliefs and P1's strategies at the information set H on an off-equilibrium path, the BE strategies within $SG(x)$ are not NE strategies. Therefore, the BE is not a SPNE. ∎

In fact, BE and SPNE do not imply each other. In Example 7.24, a SPNE is not sequentially rational and hence not a BE; in Example 7.28, a BE is not a SPNE. These problems call for a stronger equilibrium notion that can unite the two concepts. In a BE, players react optimally based on beliefs; in a NE, players react optimally based on strategies. Hence, consistency between strategies and beliefs is crucial. However, in a BE, this consistency is not guaranteed on off-equilibrium paths. One naturally looks for a suitable condition for off-equilibrium paths. The following proposition strengthens the view that what happens on off-equilibrium paths is crucial.

3 Equilibria in Extensive-Form Games

To consider stronger versions of BE, let us first put NE in the context of a Bayesian game. Nash equilibrium is based on rationality. But it is not based on the rationality at every information; instead, each player is rational on the equilibrium path, along which all others take their equilibrium strategies.

The belief system μ^* is consistent with the strategy profile σ^* and Nature's odds ρ through the Bayes rule along the equilibrium path: for any $H \in \mathbb{H}$ with $\Pr(H) > 0$ for σ^*, we have $\mu^*(x) = \Pr(x|H)$, $\forall x \in H$

Proposition 7.13 *A strategy profile σ^* is a NE of Γ_E iff there exists a belief system μ^* such that*[12]

(a) *The strategy profile σ^* is sequentially rational given belief system μ^* at those information sets H with $\Pr(H|\sigma^*) > 0$.*
(b) *The belief system μ^* is consistent with the strategy profile σ^* and Nature's odds ρ through the Bayes rule along the equilibrium path: for any $H \in \mathbb{H}$ with $Pr(H) > 0$ for σ^*, we have $\mu^*(x) = \Pr(x|H)$, $\forall x \in H$.*

Proof Necessity. Given a NE σ^*, we can easily assign beliefs using σ^* such that (b) is satisfied. Specifically, on the equilibrium path, we assign beliefs μ using σ^* based on the Bayes rule; on off-equilibrium paths, we can assign arbitrary beliefs. Since the expected utility on an off-equilibrium path is zero, the players in a NE are guaranteed to be rational based on σ^* only on the equilibrium path. Further, since σ^* and μ are consistent along the equilibrium path, the players must be rational based on μ along the equilibrium path. That is, (a) holds.

Sufficiency. Conversely, by (a), each player acts rationally based on the beliefs along the equilibrium path. By (b), the beliefs are consistent with the strategies along the equilibrium path. Thus, each player will act rationally based on the strategies along the equilibrium path. Since the NE concept requires rationality along the equilibrium path only, the strategies must be a NE.[13] ∎

Proposition 7.13 indicates that differences among the three equilibrium concepts, NE, SPNE and BE, are due to differences in the required conditions for off-equilibrium paths. The difference between NE and BE is in rationality. NE

---
[12]See Mas-Colell et al. (1995, p. 285).
[13]Given that everyone else is staying on the equilibrium path, those players whose information sets are on off-equilibrium paths have no need to deviate from their equilibrium strategies (since what they do does not matter assuming that others will stick to their current strategies). Thus, for a NE, we only need to verify rationality for those information sets on the equilibrium path. In other words, if rationality is guaranteed on the equilibrium path, we have a NE. Since we do not solve for a NE by backward induction, there is no guarantee that a NE is rational on off-equilibrium paths. In Example 7.15, for NE $\sigma^* = (L, L)$, the initial decision node is on the equilibrium path and P1 is indeed rational at that node. But the second decision node is on an off-equilibrium path and P2 is not rational. In fact, given P1 choosing L, P2 is indifferent among his choices. P2 simply reasons that since the equilibrium path will not pass through his information set, whatever he decides to do does not matter and he has no need to deviate from the equilibrium strategy. The situation in Example 7.17 for NE $\sigma^* = (\langle L_1, \hat{L}_1 \rangle, L_2)$ is also the same.

requires rationality along the equilibrium path, while BE requires rationality at any information set. Since BE has a stronger version of rationality, A BE must be a NE. However, the converse is not true. In Example 7.15 and Example 7.19, the NEs with payoffs $(0, 3)$ and $(0, 0)$ respectively violate sequential rationality at an off-equilibrium information set, and therefore they are not BEs.

The difference between NE and SPNE is also due to restrictions on off-equilibrium paths. A SPNE is rational along the equilibrium path within every subgame, including those subgames on off-equilibrium paths of the whole game. However, a NE guarantees rationality only along the equilibrium path of the whole game; in particular, rationality may fail in an off-equilibrium subgame. Only when $\Pr(H) > 0$ for all information sets H, we can be sure that a NE is a SPNE.

The difference between BE and SPNE is again due to restrictions on off-equilibrium paths. In a BE, each player reacts rationally at his information sets via beliefs. In a SPNE, each player reacts rationally at his information sets via strategies along the equilibrium path within each subgame (including those subgames on off-equilibrium paths of the whole game). If the beliefs and strategies are inconsistent, the players in a BE may not act rationally via strategies. This means that for a BE, on an off-equilibrium path of the whole game, even though part of it is on an equilibrium path within a subgame, the players may not act rationally against strategies. Thus, a BE may not be a SPNE. Only if all beliefs at information sets on off-equilibrium paths are also consistent with the strategies, we can be sure that a BE is a SPNE. In Example 7.28, although BE1 is rational based on the beliefs at all information sets (as required by definition), it is not rational based on strategies at information set H. The reason is that the beliefs at H are not consistent with P1's strategies leading to H. Due to this, BE1 is not a SPNE.

4 Refinements of Bayesian Equilibrium

In this section, we offer three kinds of refinements to BEs. NEs have three kinds of refinements: subgame perfection, trembling hand perfection, and dominance. We will also offer these three kinds of refinements to BEs and they are called respectively perfect BEs, sequential equilibria, and equilibrium dominant BEs. The focus of the refinements on BEs is to strengthen the consistency condition.

4.1 Perfect Bayesian Equilibrium

This part mainly follows Gibbons (1992, pp. 183–244). It is traditionally called perfect Bayesian equilibrium (PBE) and sometimes called a strongly perfect Bayesian equilibrium. It adds one more condition to the definition of the weak PBE in Definition 7.16. This definition is first provided by Gibbons (1992, p. 180). Fudenberg and Tirole (1991) provide a formal definition of PBE for a broad class of dynamic games of incomplete information. Our definition here is based on Gibbons

4 Refinements of Bayesian Equilibrium

(1992). However, Gibbons (1992, p. 180) does not give a precise definition; he uses words "where possible." We interpret it as meaning "all subgames," which is what we will use in our definition.[14]

One defining feature of a subgame is that there is only one branch from the game tree that connects to it. Hence, if the subgame is on the equilibrium path, the strategy on this branch or action must be a positive probability. Hence, a BE always induces a BE in any subgame on its equilibrium path. However, this property does not hold on off-equilibrium paths. This leads to an idea that one way to strengthen consistency between strategies and beliefs is to require that a BE induces a BE in any subgame, especially subgames on off-equilibrium paths.

Proposition 7.14 *A BE always induces a BE in any subgame on its equilibrium path.* ∎

Given an equilibrium, if the equilibrium strategy that takes action a_i is a positive probability, we call a_i an equilibrium action. Within a subgame $SG(x)$, although the subgame may or may not be on the equilibrium path, we say that an information set H in the subgame is on the equilibrium path of $SG(x)$ if $\Pr[H|SG(x)] > 0$. Given a BE (σ^*, μ^*), we are interested in whether the induced strategy-belief pair (σ_s^*, μ_s^*) in a subgame is a BE or not. It is obviously sequentially rational within this subgame. The question is whether (σ_s^*, μ_s^*) is consistent along the equilibrium path in this subgame. If it is consistent in this subgame, we say that (σ^*, μ^*) has subgame consistency in this subgame.

Definition 7.17 A perfect Bayesian equilibrium (PBE) (a strong PBE or a subgame perfect BE) is a pair (σ^*, μ^*) of strategies σ^* and beliefs μ^* that satisfy the following conditions:

(a) Sequential Rationality: σ^* is sequentially rational given the belief system μ^*.
(b) Subgame Consistency: μ^* is consistent with σ^* through the Bayes rule in all subgames. ∎

To explain the definition of PBE, consider the game in Fig. 24. This game has only one real subgame $SG(x)$. We find a BE by solving it backward. P3's problem is

$$L' \succ R' \Leftrightarrow \mu + 2(1-\mu) > 3\mu + 1 - \mu \Leftrightarrow \mu < 1/3.$$

Hence, if $\mu < 1/3$, P3 will choose L', P2 will then choose L, and P1 will then choose A. In this case, since information set H is not on the equilibrium path, there is no restriction on μ. Hence, we have a BE:

---

[14]No idea what Gibbons actually means by "where possible", at least our definition is consistent with the examples in Gibbons (1992, 180–183).

Fig. 24 PBE

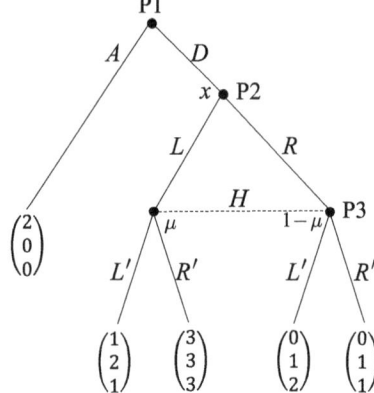

$$BE1: \quad s_1^* = A, \quad s_2^* = L, \quad s_3^* = L', \quad \mu^* < 1/3. \tag{13}$$

To have a PBE, we apply subgame consistency on $SG(x)$. This means that we must have $\mu = 1$, which cannot satisfy the requirement of $\mu^* < 1/3$ in BE1. Hence, BE1 is not a PBE. In $SG(x)$, the normal form is

P2\P3	L'	R'
L	2, 1	(3, 3)
R	1, 2	1, 1

We can see that there is a unique NE is this subgame, which is $s_2 = L$ and $s_3 = R'$. Hence, there is a unique SPNE:

$$s_1^* = D, \quad s_2^* = L, \quad s_3^* = R'.$$

This shows that BE1 is not subgame perfect.

If $\mu > 1/3$, P3 will choose R', P2 will then choose L, and P1 will then choose D. In this case, since information set H is on the equilibrium path, consistency is required, which implies $\mu = 1$. This $\mu = 1$ satisfies $\mu > 1/3$. Hence, we have another BE:

$$BE2: \quad s_1 = D, \quad s_2 = L, \quad s_3 = R', \quad \mu = 1. \tag{14}$$

Since there is no subgame on an off-equilibrium path for this BE, consistency on off-equilibrium paths is automatically satisfied. Hence, this BE is a PBE. Notice that this PBE is subgame perfect. In fact, according to the following proposition, any PBE is a SPNE.

By definition, since a BE is rational at all information sets, a PBE must induce a BE in any subgame. Hence, we have the following result.

Proposition 7.15 *A BE is a PBE iff it induces a BE in every subgame. That is, a PBE is a subgame perfect BE.* ∎

4 Refinements of Bayesian Equilibrium

Given a game, we have two ways to solve for all PBEs. First, using the definition, we can solve for all BEs first and then eliminate those that fail to satisfy subgame consistency. Second, using Proposition 7.15, we can solve for PBEs backwards subgame by subgame and require a solution to be a BE in every subgame.

Proposition 7.16 *For any PBE (σ^*, μ^*), σ^* is a SPNE. Conversely, if (σ^*, μ^*) is a BE and σ^* is a SPNE, then there is a belief system μ such that (σ^*, μ) is a PBE, but (σ^*, μ^*) may not be a PBE.*

Proof Since a PBE ensures it is a BE in every subgame, it ensures that it is a NE in every subgame. Hence, a PBE must be a SPNE. The converse is from Wang (2017). ∎

We now apply subgame consistency to problems in Examples 7.26–7.28.

Example 7.29 Reconsider Example 7.26. Since there is no real subgame in the game, subgame consistency cannot help. Hence, all BEs in Example 7.26 are PBEs. ∎

Example 7.30 Reconsider the game in Fig. 25 from Example 7.27. We have two pure-strategy BEs:

$$BE1: \quad s_1^* = L_1, \quad s_2^* = L_2, \quad \mu_1^* > 2/3;$$
$$BE2: \quad s_1^* = R_1, \quad s_2^* = R_2, \quad \mu_1^* = 0.5.$$

By subgame consistency, BE1 is not a PBE, but BE2 is. In fact, this is the unique PBE. This PBE indeed has sensible beliefs.

We also have a mixed-strategy BE. When $\mu_1 = 2/3$, P2 takes a mixed strategy $\hat{\sigma}_2 = (t, 1-t)$ for any $t \in [0, 1]$. Since equilibrium-path (EP) consistency cannot be satisfied, H cannot be on the equilibrium path, which means that P1 has to take

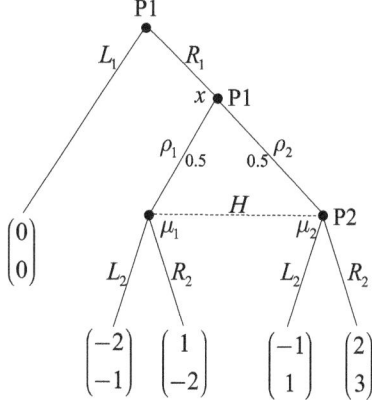

Fig. 25 PBE

$s_1 = L_1$ for sure. For this choice to be rational, we need the choice of L_1 to be better than the choice of R_1, which requires

$$0 \geq 0.5(-2t+1-t) + 0.5(-t+2(1-t)),$$

i.e., $t \geq 0.5$. Hence, there is a mixed-strategy BE:

$$BE3: \quad s_1^* = L_1, \quad s_2^* = (t, 1-t), \quad \mu_1^* = 2/3, \quad \text{for } t \geq 0.5.$$

This BE is obviously not a PBE. ∎

Example 7.31 Reconsider the game in Fig. 26 from Example 7.28. In the unique SPNE, $s_2^* = R_2$. Since a PBE must be a SPNE, a PBE must have $s_2^* = R_2$. Hence, BE1 in (12) is not a PBE.

If $\mu_1 < 2/3$, P2 chooses R_2 for sure. Then, P1 chooses $\hat{R}_1$ at node x and R_1 at the starting node. EP-consistency requires $\mu_2 = 1$, which can be satisfied. Hence, we find a second BE:

$$BE2: \quad s_1^* = \langle R_1, \hat{R}_1 \rangle, \quad s_2^* = R_2, \quad \mu_1^* = 0.$$

Since there is no real subgame on off-equilibrium paths, subgame consistency is automatically satisfied. Hence, BE2 is a PBE. As expected, we can see that this PBE is a SPNE.

If $\mu_1 = 2/3$, P2 will have a mixed strategy $\sigma_2 = (t, 1-t)$ for any $t \in [0, 1]$. Subgame consistency requires P1 to take $\hat{L}_1$ with probability $2/3$. Hence, P1 must be indifferent between $\hat{L}_1$ and $\hat{R}_1$, implying

$$-2t+1-t = -t+2(1-t),$$

i.e., $t = 0.5$. If so, P1 will take L_1 over R_1. Hence, we have a mixed-strategy PBE:

Fig. 26 PBE

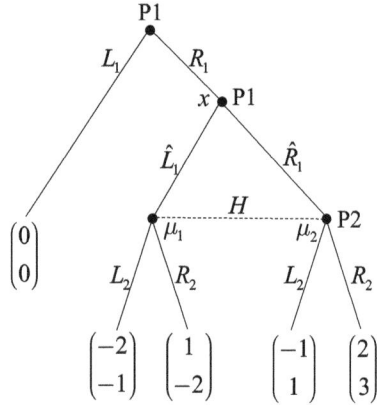

BE3: $\sigma_1^* = \langle L_1, (2/3, 1/3) \rangle$, $\sigma_2^* = (0.5, 0.5)$, $\mu_1^* = 2/3$.

Note that taking $\hat{L}_1$ with probability $2/3$ is rational since P1 is indifferent between $\hat{L}_1$ and $\hat{R}_1$.

For $\mu_1 = 2/3$, we now find all BEs. P1's preferences at node x is

$$\hat{L}_1 \succ \hat{R}_1 \Leftrightarrow -2t + 1 - t > -t + 2(1-t) \Leftrightarrow t > 0.5.$$

If $t > 0.5$, P1 takes $\hat{L}_1$ at node x. Then, P1 will take L_1 at the starting node if

$$0 > -2t + 1 - t \quad \text{or} \quad t > 1/3,$$

which is true. In this case, EP-consistency is not required on H. Hence, we have a fourth BE:

BE4: $s_1^* = \langle L_1, \hat{L}_1 \rangle$, $\sigma_2^* = (t, 1-t)$, $\mu_1^* = 2/3$, for $t > 0.5$.

If $t < 0.5$, P1 takes $\hat{R}_1$ at node x. Then, P1 will take L_1 at the starting node if

$$0 > -t + 2(1-t) \quad \text{or} \quad t > 2/3,$$

which cannot hold. On the other hand, if P1 takes R_1 with a positive probability, EP-consistency will be required, but EP-consistency cannot be satisfied. Hence, there is no BE when $t < 0.5$. If $t = 0.5$, P1 takes a mixed strategy $\hat{\sigma}_1 = (\tau, 1 - \tau)$ for any $\tau \in [0, 1]$ at mode x. Then, P1 will take L_1 at the starting node if

$$0 > -0.5\tau + 0.5(1-\tau) \quad \text{or} \quad \tau > 0.5.$$

When $\tau > 05$ or when $\tau = 0.5$ but P1 will take L_1 at the starting node, since EP-consistency on H is not required in this case, we have a fifth BE:

BE5: $s_1^* = \langle L_1, (\tau, 1-\tau) \rangle$, $\sigma_2^* = (t, 1-t)$, $\mu_1^* = 2/3$, for $\tau \geq 0.5$, $t = 0.5$.

We can see that BE3 is a special case of BE5 when $\tau = 2/3$. On the other hand, when $\tau < 0.5$ or when $\tau = 0.5$ but P1 takes R_1 with a positive probability, EP-consistency will be required. EP-consistency requires $\tau = \mu_1 = 2/3$, which cannot be satisfied. Hence, there is no BE when $\tau < 0.5$ or when $\tau = 0.5$ but P1 takes R_1 with a positive probability. ∎

The following proposition is implied by Propositions 7.18 and 7.19.

Proposition 7.17 *In any finite game, there exists a PBE.* ∎

4.2 Sequential Equilibrium

PBE is stronger than BE, in the sense that all PBEs are BEs. In this section, we introduce another even stronger equilibrium concept, called sequential equilibrium, which is slightly stronger than PBE.

Recall that we used trembling-hand perfection to strength NEs. We now use trembling-hand perfection to strength BEs. This work is done by Kreps and Wilson (1982), who propose the concept of sequential equilibrium. They impose certain consistencies between strategies and beliefs to restrict μ^* to be the limit of a sequence of totally mixed strategy profiles $\{\sigma^k\}$. For each of the strategy profiles σ^k, any information set H will be reached with a positive probability. Thus, given each σ^k, the Bayes rule can be used to derive a consistent belief system μ^k over all the information sets. If $\lim_{k \to \infty}(\sigma^k, \mu^k) = (\sigma^*, \mu^*)$, although $\Pr(H) = 0$ can still happen at some information set H under σ^*, there is some consistency between σ^* and μ^* for all information sets. Such consistency allows us to rule out many unreasonable equilibria. This leads to a stronger equilibrium concept called sequential equilibrium.

Definition 7.18 A strategy-belief pair (σ^*, μ^*) is a <u>sequential equilibrium (SE)</u> or <u>trembling-hand perfect BE</u>[15] of a game Γ_E if

(1) Strategy profile σ^* is sequentially rational given belief system μ^*.
(2) Trembling-hand consistency: there exists a sequence of totally mixed strategy profiles $\{\sigma^k\}_{k=1}^{\infty}$ with beliefs μ^k derived from σ^k and Nature's odds using the Bayes rule such that[16]

$$\lim_{k \to \infty}(\sigma^k, \mu^k) = (\sigma^*, \mu^*). \quad \blacksquare$$

We can think of SE as a kind of stable BE, just like a trembling-hand perfect NE. The SE concept requires beliefs to be justifiable as coming from a sequence of totally mixed strategies that are close to the equilibrium strategy profile σ^*. Using the Bayes rule, condition (2) makes sure that off-equilibrium beliefs are sensible, which prevents unreasonable off-equilibrium beliefs from producing a BE that is not a SPNE.

Since condition (2) in the definition of SE is stronger than condition (b) in the definition of BE, a SE must be a BE. But, the converse is generally not true.

---

[15] A trembling-perfect BE is different from Selton's (1975) trembling-perfect NE for an extensive-form game. NE does not involve beliefs.
[16] We can assume that Nature's odds are always strictly positive. That is, if $\rho = (\rho_1, \ldots, \rho_k)$ is Nature's probability distribution over an information set, we will have $\rho_i > 0$ for all i. There is no point for Nature to have zero probability.

Fig. 27 Sequential equilibrium

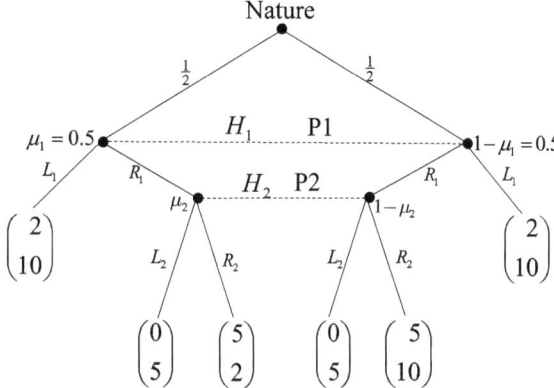

Proposition 7.18 *Any SE is a PBE.*

Proof Within each subgame SG, since σ^k and μ^k are consistent and $\sigma^k \to \sigma^*$ and $\mu^k \to \mu^*$, when $\Pr(H|SG) > 0$ under σ^* for an information set H in the subgame, σ^* and μ^* relating to H must still be consistent. To see this, suppose that H has m nodes and they are reached by conditional probabilities $\sigma^*_{1H}, \ldots, \sigma^*_{mH}$ under σ^* conditional on H. If $\sigma^*_{1H} + \cdots + \sigma^*_{mH} > 0$, i.e., H is on the equilibrium path of the subgame, then $\mu^k_{iH} = \frac{\sigma^k_{iH}}{\sigma^k_{1H} + \cdots + \sigma^k_{mH}}$ by definition and $\mu^*_{iH} = \frac{\sigma^*_{iH}}{\sigma^*_{1H} + \cdots + \sigma^*_{mH}}$ after taking the limit with $k \to \infty$. Thus, we have EP-consistency within the subgame, implying that a SE must be a BE within the subgame. Since this SE is a BE in any subgame, it must be a PBE. ∎

Proposition 7.19 *In any finite game, there exists a SE.*[17] ∎

We now use SE to revisit the problems in Examples 7.26–7.28. In particular, Example 7.34 gives a good illustration of the proof of Proposition 7.18.

Example 7.32 Reconsider the game in Fig. 27 from Example 7.26. Given any totally mixed strategies $(\sigma^k_1, 1 - \sigma^k_1)$ and $(\sigma^k_2, 1 - \sigma^k_2)$ for the two players on their own information sets, respectively, consistency on H_1 implies $\mu^k_1 = 0.5$, and consistency on H_2 implies $\mu^k_2 = 0.5$, where the beliefs are derived from the strategies and Nature's odds using the Bayes rule.

Consider the first BE:

$$BE1: \quad s^*_1 = L_1, \quad s^*_2 = L_2, \quad \mu^*_1 = 0.5, \quad \mu^*_2 > 5/8.$$

BE1 has already satisfied rationality; we now check trembling-hand consistency. For this BE, trembling-hand consistency requires

$$\sigma^k_1 \to 1, \quad \sigma^k_2 \to 1, \quad \mu^k_1 \to 0.5, \quad \mu^k_2 \to \mu^*_2.$$

---

[17] It is from Kreps and Wilson (1982).

Obviously, since $\mu_2^k = 0.5$, it is impossible to have $\mu_2^k \to \mu_2^*$. Hence, this BE is not a SE. Since this BE is a PBE, it shows that a PBE may not be a SE.

Consider the second pure-strategy BE:

$$BE2: \quad s_1^* = R_1, \quad s_2^* = R_2, \quad \mu_1^* = \mu_2^* = 0.5.$$

For this BE, trembling-hand consistency requires

$$\sigma_1^k \to 0, \quad \sigma_2^k \to 0, \quad \mu_1^k \to 0.5, \quad \mu_2^k \to 0.5. \tag{15}$$

For $k \geq 2$, define $\sigma_1^k = 1/k$ and $\sigma_2^k = 1/k$. Then, condition (15) is satisfied. Hence, BE2 is a SE.

Consider the third BE. When $\mu_2 = 5/8$, P2 is indifferent and hence takes a strategy $\sigma_2 = (t, 1-t)$ for any $t \in [0,1]$. If H_2 is on the equilibrium path, EP-consistency will fail. Hence, we must have $s_1 = L_1$. For this, rationality requires

$$2 \geq 0.5(0t + 5(1-t)) + 0.5(0t + 5(1-t)) \quad \text{or} \quad t \geq 3/5,$$

which is feasible. Hence, we have a third BE:

$$BE3: \quad s_1^* = L_1, \quad \sigma_2^* = (t, 1-t), \quad \mu_1^* = 0.5, \quad \mu_2^* = 5/8, \quad \text{for } t \geq 3/5.$$

For this BE, trembling-hand consistency requires

$$\sigma_1^k \to 1, \quad \sigma_2^k \to t, \quad \mu_1^k \to 0.5, \quad \mu_2^k \to 5/8, \quad \text{for } t \geq 3/5.$$

However, since $\mu_2^k = 0.5$, this consistency condition cannot be satisfied. Hence, BE3 is not a SE. ∎

BE1 in Example 7.32 shows that PBE is strictly weaker than SE. This example also shows that PBE defined by continuation games instead of subgames is also strictly weaker than SE.

Example 7.33 Reconsider the game in Fig. 28 from Example 7.27. Given any totally mixed strategies $(\sigma_1^k, 1 - \sigma_1^k)$ and $(\sigma_2^k, 1 - \sigma_2^k)$ for the two players on their own information sets, respectively, consistency on H implies $\mu_1^k = 0.5$, where the beliefs are derived from Nature's odds using the Bayes rule.

The three BEs are solved in Example 7.30. Consider the first BE:

$$BE1: \quad s_1^* = L_1, \quad s_2^* = L_2, \quad \mu_1^* > 2/3.$$

For this BE, trembling-hand consistency requires

$$\sigma_1^k \to 1, \quad \sigma_2^k \to 1, \quad \mu_1^k \to \mu_1^*.$$

Fig. 28 Sequential equilibrium

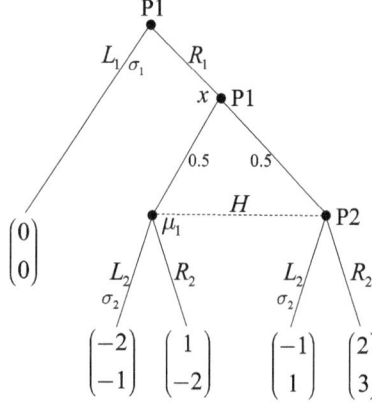

Since $\mu_1^k = 0.5$, it is impossible to have $\mu_1^k \to \mu_1^*$. Hence, this BE is not a SE. In fact, by Example 7.30, since this BE is not a PBE, we already know that it is not a SE.

Consider the second BE:

$$BE2: \quad s_1^* = R_1, \quad s_2^* = R_2, \quad \mu_1^* = 0.5.$$

For this BE, trembling-hand consistency requires

$$\sigma_1^k \to 0, \quad \sigma_2^k \to 0, \quad \mu_1^k \to \mu_1^*.$$

For $k \geq 2$, define $\sigma_1^k = \sigma_2^k = 1/k$. Then, the above consistency condition is satisfied. Hence, this BE us a SE.

Consider the third BE:

$$BE3: \quad s_1^* = L_1, \quad s_2^* = (t, 1-t), \quad \mu_1^* = 2/3, \quad \text{for } t \geq 0.5.$$

For this BE, trembling-hand consistency requires

$$\sigma_1^k \to 1, \quad \sigma_2^k \to t, \quad \mu_1^k \to \mu_1^*, \quad \text{for } t \geq 0.5.$$

Since $\mu_1^k = 0.5$, it is impossible to have $\mu_1^k \to \mu_1^*$. Hence, this BE is not a SE. In fact, since this BE is not a PBE, we already know that it is not a SE. ∎

Example 7.34 Reconsider the game in Fig. 29 from Example 7.28. Given any totally mixed strategies $(\sigma_1^k, 1 - \sigma_1^k)$, $(\hat{\sigma}_1^k, 1 - \hat{\sigma}_1^k)$ and $(\sigma_2^k, 1 - \sigma_2^k)$ at the top node, at node x and at H, respectively, consistency on H implies $\mu_1^k = \hat{\sigma}_1^k$, where the beliefs are derived using the Bayes rule.

Fig. 29 Sequential equilibrium

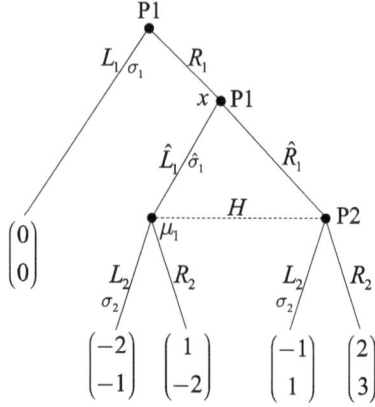

The five BEs are solved in Example 7.31. Consider the first BE:

$$BE1: \quad s_1^* = <L_1, \hat{R}_1>, \quad s_2^* = L_2, \quad \mu_1^* > 2/3.$$

For this BE, trembling-hand consistency requires

$$\sigma_1^k \to 1, \quad \hat{\sigma}_1^k \to 0, \quad \sigma_2^k \to 1, \quad \mu_1^k \to \mu_1^*.$$

Since $\mu_1^k = \hat{\sigma}_1^k \to 0$, it is impossible to have $\mu_1^k \to \mu_1^*$. Hence, this BE is not a SE. In fact, by Example 7.31, since this BE is not a PBE, we already know that it is not a SE.

Consider the second BE:

$$BE2: Rs_1^* = \langle R_1, \hat{R}_1 \rangle, \quad s_2^* = R_2, \quad \mu_1^* = 0.$$

For this BE, trembling-hand consistency requires

$$\sigma_1^k \to 0, \quad \hat{\sigma}_1^k \to 0, \quad \sigma_2^k \to 0, \quad \mu_1^k \to 0.$$

For $k \geq 2$, define $\sigma_1^k = \hat{\sigma}_1^k = \sigma_2^k = 1/k$. Then, the above consistency condition is satisfied. Hence, this BE is a SE.

Consider the third BE:

$$BE3: \quad \sigma_1^* = \langle L_1, (2/3, 1/3) \rangle, \quad \sigma_2^* = (0.5, 0.5), \quad \mu_1^* = 2/3.$$

For this BE, trembling-hand consistency requires

$$\sigma_1^k \to 1, \quad \hat{\sigma}_1^k \to 2/3, \quad \sigma_2^k \to 0.5, \quad \mu_1^k \to 2/3.$$

4 Refinements of Bayesian Equilibrium

For $k \geq 3$, define $\sigma_1^k = 1 - 1/k$, $\hat{\sigma}_1^k = 2/3 - 1/k$, and $\sigma_2^k = 0.5 - 1/k$. Then, the above consistency condition is satisfied. Hence, this BE us a SE.

Consider the fourth BE:

$$BE4: \quad s_1^* = <L_1, \hat{L}_1>, \quad \sigma_2^* = (t, 1-t), \quad \mu_1^* = 2/3, \quad \text{for } t > 0.5.$$

For this BE, trembling-hand consistency requires

$$\sigma_1^k \to 0, \quad \hat{\sigma}_1^k \to 0, \quad \sigma_2^k \to t, \quad \mu_1^k \to 2/3, \quad \text{for } t > 0.5.$$

Since $\mu_1^k = \hat{\sigma}_1^k \to 0$, it is impossible to have $\mu_1^k \to 2/3$. Hence, this BE is not a SE.

Consider the fifth BE:

$$BE5: \quad s_1^* = <L_1, (\tau, 1-\tau)>, \quad \sigma_2^* = (0.5, 0.5), \quad \mu_1^* = 2/3, \quad \text{for } \tau \geq 0.5.$$

For this BE, trembling-hand consistency requires

$$\sigma_1^k \to 0, \quad \hat{\sigma}_1^k \to \tau, \quad \sigma_2^k \to t, \quad \mu_1^k \to 2/3, \quad \text{for } \tau \geq 0.5, \, t = 0.5.$$

When $\tau = 2/3$, BE5 is the same BE3, which is shown to be a SE. When $\tau \neq 2/3$, since $\mu_1^k = \hat{\sigma}_1^k \to \tau$, it is impossible to have $\mu_1^k \to 2/3$. Hence, this BE is not a SE when $\tau \neq 2/3$. ∎

It is actually very easy to judge whether or not a BE is a SE. We can easily identify 'inconsistency' between strategies and beliefs without relying on the Bayes rule. For example, in BE1 with $\mu_1^* > 2/3$ in the above example, P1 is taking $\hat{R}_1$ for certain, but P2 believes $\mu_2^* < 1/3$. Such inconsistency cannot happen in a SE even though the information set is not on the equilibrium path.

Intuitively, the BE and SE concepts require rationality via beliefs at every information set including off-equilibrium paths. The SPNE concept requires rationality via strategies on the equilibrium path within each subgame. Hence, when beliefs are always consistent with strategies at every information set, especially in every subgame, this BE is a SPNE. In a SE, such consistency is guaranteed at every information set and thus it implies SPNE.

A SPNE is a NE within each subgame; hence, rationality is guaranteed along each subgame's equilibrium path, but rationality may not be guaranteed on off-equilibrium paths of a subgame. Conversely, a NE is rational along the whole game's equilibrium path, but it may not be rational along a subgame's equilibrium path; of course, it may not be rational on off-equilibrium paths of a subgame or the whole game. Hence, a SPNE must be a NE, but the converse is not true.

Example 7.17 is a good example. For SPNE1 $(R_1, \hat{L}_1, R_2)$ in (7), the equilibrium path for the whole game is $R_1 \to \hat{L}_1 \to R_2 \to (1,2)$, where $(1, 2)$ means the payoff cell. Information set H is on the equilibrium path. First, the players are rational via strategies along the equilibrium path. Second, the equilibrium path of $SG(x)$ is

$\hat{L}_1 \to R_2 \to (1,2)$ and H is on this equilibrium path. The players are again rational via strategies along this equilibrium path. Let's now look at NE1 $(\langle L_1, \hat{L}_1 \rangle, L_2)$. The equilibrium path for the whole game is $L_1 \to (0,0)$ and the equilibrium path of $SG(x)$ is $\hat{L}_1 \to L_2 \to (-1,-1)$. In this case, H is not on the equilibrium path of the whole game, but it is on the equilibrium path of $SG(x)$. The players are rational via strategies on the equilibrium path of the whole game, but P2's choice of L_2 is not rational within the subgame via strategies in $SG(x)$ given P1's choice of $\hat{L}_1$. Hence, this NE is not a SPNE.

Example 7.19 is also interesting. NE (L_1, L_2) is a SPNE. However, P2 at H is not rational since L_2 is always worse than R_2. The equilibrium path is $L_1 \to (0,0)$ and H is not on the equilibrium path. That is, a SPNE may not be rational on off-equilibrium paths of a subgame, although it is rational on the equilibrium path of a subgame.

In summary, we have introduced five equilibrium concepts for an extensive-form game. Their relations are shown in Fig. 30.

SE strengthens BE by trembling-hand consistency; PBE strengthens BE by subgame consistency; and SPNE strengthens NE by subgame perfection. These equilibrium concepts requires different conditions as described verbally in the following:

1. NE requires rationality via strategies at those information sets along the whole game's equilibrium path.
2. SPNE requires rationality via strategies at those information sets along every subgame's equilibrium path.
3. BE requires rationality via beliefs at every information set and consistency between beliefs and strategies along the whole game's equilibrium path.
4. SE requires rationality via beliefs at every information set and trembling-hand consistency between beliefs and strategies at every information set.

The SE concept also has its problems. In fact, the SE concept seems too strong in some aspects but too weak in other aspects. For example, it implies that any two players with the same information must have exactly the same beliefs regarding the

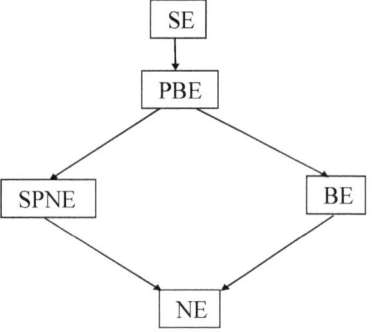

Fig. 30 Five equilibrium concepts

deviations by other players that have caused play to reach a given part of the game tree. Also, a SE may not be trembling-hand perfect, as shown in Question 7.14 in the Problem Set. Due to these problems, there are further refinements of equilibrium concepts in the literature. However, for applications under imperfect information, depending on the issues and the particular model specifications, researchers generally use either the SPNE concept or the BE concept. In fact, from Example 7.27 and Example 7.28, we see that the problems with BE occur only if some beliefs are obviously not sensible. Thus, as long as beliefs look sensible, a BE will generally be a SE.

4.3 BE Under Complete Dominance: CDBE

In many games, especially games of incomplete information, there are no real subgames. For these games, subgame consistency cannot help to rule out senseless equilibria. In this and next subsections, we introduce two dominance concepts for such games. These two dominance concepts help us eliminate some unreasonable BEs in incomplete-information games when subgame consistency is not helpful. Note that, in an incomplete-information game, each type of a player is treated as a separate player when we discuss PBEs. By this, an incomplete-information game becomes an imperfect-information game.

Based on the concept of dominance in actions, we have another two strong versions of BE. These concepts are completely independent of the concepts based on subgames and equilibrium paths.

Some PBEs are unreasonable. The following example is from Gibbons (1992, p. 233).

Example 7.35 Consider the game in Fig. 31. For P2, $L_2 \succ R_2$ iff $\mu > 1 - \mu$. That is, P2 chooses L_2 if $\mu > 0.5$. Hence, we have a pure-strategy PBE:

$$\text{BE1}: \quad s_1 = M_1, \quad s_2 = <R'_2, L_2>, \quad \mu = 1.$$

Fig. 31 Unreasonable PBEs

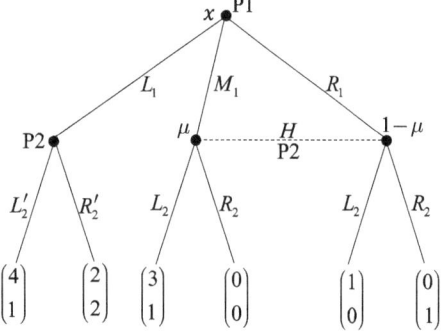

P2 chooses R_2 if $\mu<0.5$. Hence, we have another pure-strategy PBE:

$$\text{BE2}: \quad s_1 = L_1, \quad s_2 = <R'_2, R_2>, \quad \mu<0.5.$$

P2 is indifferent between L_2 and R_2 if $\mu = 0.5$. In this case, P2 takes a mixed strategy $(t, 1-t)$ for any $t \in [0,1]$ at H, and P1 will never take R_1 with a positive probability. Since $\mu = 0.5$, the information H cannot be on the equilibrium path. Hence, P1 must choose L_1. In order for P1 to choose L_1, we need

$$2 \geq 3t \quad \text{and} \quad 2 \geq t.$$

That is, when $t \leq 2/3$, P1 will choose L_1. Hence, we have the third PBE:

$$\text{BE3}: \quad s_1 = L_1, \quad \sigma_2 = \langle R'_2, (t, 1-t) \rangle, \quad \mu = 0.5, \quad \text{where } t \in [0, 2/3].$$

However, for P1, R_1 is completely dominated by L_1 in payoffs. That is, whatever P2 does, P1 is strictly better by choosing L_1 over R_1. Hence, it is not reasonable for P2 to have $1 - \mu > 0$. That is, we should have $\mu = 1$. Thus, BE2 and BE3 are not reasonable. The CD criterion defined below intends to rule out such BEs. ∎

An action a_i at a node x of player i is <u>completely dominated</u> if there is another action a'_i at the same node x such that i's worst possible payoff following a'_i is strictly better than i's best possible payoff following a_i whatever subsequent players do,[18] i.e.,

$$\min_{\sigma_{\Gamma(a'_i,x)}} u_i\left(a'_i, \sigma_{\Gamma(a'_i,x)}\right) > \max_{\sigma_{\Gamma(a_i,x)}} u_i\left(a_i, \sigma_{\Gamma(a_i,x)}\right).$$

where $\Gamma(a,x)$ is the remaining part[19] of the game tree following action a at node x, and $\sigma_{\Gamma(a,x)}$ is the remaining part of the strategy profile σ on $\Gamma(a,x)$.

Complete dominance (CD) criterion: *Zero belief should be placed on a node of an information set H if the action preceding this node is completely dominated. However, in an all-dominance case when all actions preceding this information set H are completely dominated, no additional restriction should be placed on beliefs over this information set.*[20]

---

[18] See the definition in Gibbons (1992, p. 236), called strict dominance. For games of incomplete information, each type of a player is treated as a separate player in the definition of complete dominance.

[19] More clearly, the "remaining part" means the remaining part relevant to the calculation of payoffs implied by the action at the node. Note that the same action can be taken at different nodes in the same information. Hence, it is an action and a node together that determine the remaining part of the game tree. Player i may be one of the subsequent players.

[20] This condition is Requirement 5 in Gibbons (1992, p. 235).

4 Refinements of Bayesian Equilibrium

Example 7.13 illustrates an example. The focus is whether we should impose $\mu = 0$ at the node in information set H_3. The question is whether action R_1 preceding the node is completely dominated. That is, the question is whether action L_1 at node x_1 completely dominates R_1 at x_1; if so, we should impose $\mu = 0$ unless we encounter an all-dominance situation. The all-dominance situation relates to the question of whether action L_2 preceding H_3 is also completely dominated, or whether action R_2 at x_2 completely dominated L_2 at x_2. If indeed both R_1 and L_2 are completely dominated, we cannot impose $\mu = 0$ (Fig. 32).

The CD criterion argues that a player will never take a completely dominated action and hence other players should play zero belief on the node immediately following this action. However, if all nodes in an information set H would have zero belief by this principle, since zero belief on all nodes of an information is impossible (the total sum of beliefs at any information is 1), we cannot apply this principle to this H, and hence no additional restriction should be placed on beliefs over H.

A BE under the CD criterion is called a <u>CDBE</u> or a completely dominant BE.

Example 7.36 For the game in Fig. 31, our concern is whether or not the beliefs on information set H make sense. The preceding actions of H are from the initial node x. Action M_1 from this node is not completely dominated since taking the alternatively actions L_1 or R_1 from the same node is not strictly better for the player P1. For example, taking L_1, P1 may end up with payoff 2, while taking M_1, P1 may end up with payoff 3.

However, action R_1 is completely dominated, since L_1 from the same node completely dominates R_1. Taking L_1, P1 will at least get payoff 2, while taking R_1, P1 can at most get payoff 1. Hence, P1 is not expected to take action R_1 at the node, and consequently the belief $1 - \mu$ should be zero. That is, we should impose $\mu = 1$ on all equilibria. This means that in Example 7.35, BE1 is a CDBE, while BE2 and BE3 are not CDBEs.

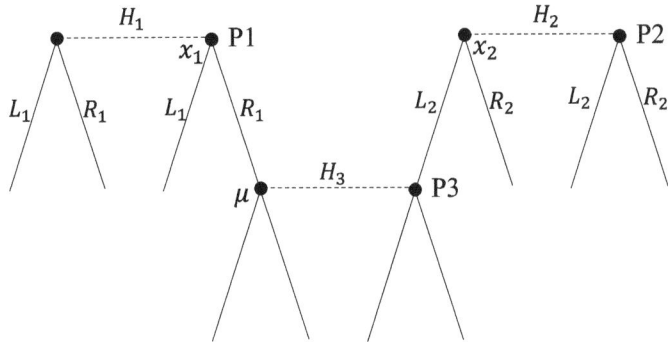

Fig. 32 Complete dominance

If we replace payoff 3 in Fig. 31 by 1.5, then BEs are:

BE1': $s_1 = L_1$, $s_2 = \langle R'_2, L_2 \rangle$, $\mu > 0.5$;
BE2': $s_1 = L_1$, $s_2 = \langle R'_2, R_2 \rangle$, $\mu < 0.5$.
BE3': $s_1 = L_1$, $\sigma_2 = \langle R'_2, (t, 1-t) \rangle$, $\mu = 0.5$, where $t \in [0, 1]$.

Since both M_1 and R_1 are strictly dominated by L_1 in this case, we cannot apply the CD criterion on μ. Hence, all these three BEs are CDBEs. This is the special case mentioned in the CD definition. In the meantime, this example also shows that the CD criterion cannot be used to rule out all unreasonable BEs. This leads to the intuitive criterion by Cho and Kreps (1987) on equilibrium dominance in the following subsection. ∎

Note that SE and CDBE do not imply each other. For example, in Example 7.35, BE2 is a SE, but not a CDBE; in Example 7.34, BE1 is a CDBE, but not a SE.

4.4 BE Under Equilibrium Dominance: EDBE

Given a BE (σ^*, μ^*), an action a_i at a node $x \in H$ of player i is <u>equilibrium-dominated</u> if i's equilibrium payoff from taking the equilibrium behavior strategy σ_{iH}^* at H while subsequent players also take their equilibrium strategies is strictly greater than i's highest possible payoff by taking a_i whatever subsequent players do,[21] i.e.,

$$u_i\left(\sigma_{iH}^*, \sigma_{\Gamma(\sigma_{iH}^*, x)}^*\right) > \max_{\sigma_{\Gamma(a_i, x)}} u_i(a_i, \sigma_{\Gamma(a_i, x)}), \qquad (16)$$

where σ_{iH}^* is player i's equilibrium behavior strategy at H, and $\sigma_{\Gamma(\sigma_{iH}^*, x)}^*$ is the remaining part of the equilibrium profile σ^* on $\Gamma(\sigma_{iH}^*, x)$.

Equilibrium dominance at an information set H_i of player i means that, given the equilibrium strategy, there is no reason for player i to deviate from the equilibrium behavior strategy at H_i even if other players may follow suit by deviating from their equilibrium strategies. If so, zero belief should be placed on a node on off-equilibrium paths.

Note that at each information set, at least one action is used in equilibrium, although the information set may not be on the equilibrium path. In particular, the information set H_i involved in (16) may not be on the equilibrium path. Note also that by rationality at all information set, player i's optimal behavior strategy at H_i is

---

[21]Player i may take a mixed behavior strategy at this information set H in equilibrium. However, this action a_i will never be used by player i in equilibrium. If player i places a positive probability on a_i in equilibrium, as indicated by (2), this a_i will be indifferent to player i's other actions used in equilibrium at H. This means that player i will do equally well by placing 100% probability on this action, given the remaining part of σ^*. This is impossible by condition (16).

her equilibrium strategy given others' equilibrium strategies, even though H_i may not be on the equilibrium path.

For games of incomplete information, each type of a player is treated as a separate information set in the definition of equilibrium dominance. See the definition in Gibbons (1992, p. 239).

Equilibrium dominance (ED) criterion: *Given a BE, zero belief should be placed on a node of an information set H if the action preceding this node is equilibrium-dominated. However, in an all-dominance case when all actions preceding this information set H are equilibrium-dominated, no additional restriction should be placed on beliefs over this information set.*

The ED criterion argues that, given an equilibrium, a player will never take an equilibrium-dominated action and hence other players should play zero belief on the node immediately following this action. However, if all nodes in an information set H would have zero belief by this principle, since zero belief on all nodes of an information is not possible (the total sum of beliefs on any information is 1), we cannot apply this principle to this H, and hence no additional restriction should be placed on the beliefs over H.

A BE under the ED criterion is called an EDBE or an equilibrium dominant BE.

Proposition 7.20 *A completely dominated action is an equilibrium-dominated action. Hence, an EDBE is generally a CDBE, except the special case in which all actions preceding an information are equilibrium dominated.* ∎

We use the following figure to illustrate the proof of Proposition 7.20. Suppose that action M is an equilibrium action and action L completely dominates R. Since L completely dominates R, the worst outcome of L must be better than that of R. Since M is an equilibrium action, M must be better than the worst outcome of L. Hence, M must also equilibrium-dominate R. That is, R is equilibrium-dominated. In other words, if an action is completely dominated, it is generally equilibrium-dominated. Hence, if a node is given zero belief by the CD criterion, it is generally also given zero belief by the ED criterion. Hence, the ED criterion generally imposes zero belief on more nodes than the CD criterion. Therefore, an EDBE is generally a CDBE. For example, BE1 is Example 7.35 is consistent with this explanation (Fig. 33).

However, even if complete dominance imposes zero belief on a node, equilibrium dominance may not. Suppose that L completely dominates R but not M in the above game, implying $\mu = 1$ by complete dominance. However, it is possible that L equilibrium-dominates both R and M, implying no restriction on μ by equilibrium dominance. That is, even if CD dominance implies $\mu = 1$, ED dominance may not. This leads to the possibility that a BE is not a CDBE but is an EDBE. In other words, if a node is given zero belief by the CD criterion, it is generally given zero belief by the ED criterion, except the special case in which all actions preceding the information set are equilibrium-dominated.

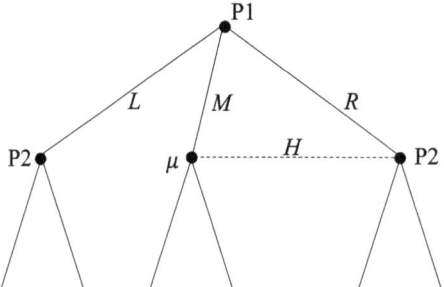

Fig. 33 CD dominance implies ED dominance

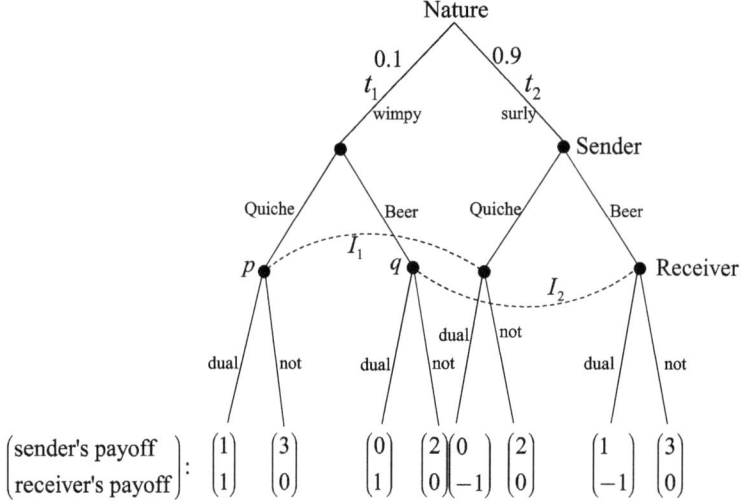

Fig. 34 BE with the ED criterion

Note that SE and EDBE do not imply each other. For example, in Example 7.35, BE2 is a SE, but not an EDBE; in Example 7.34, BE1 is an EDBE, but not a SE.

Notice that for BE1 in Example 7.34 the equilibrium path is to take L_1 and exit. However, when we consider possible beliefs for information set H, we should consider the equilibrium path $\hat{R}_1 \to L_2 \to$ exit. By taking this equilibrium path, P1 gets -1, while taking the alternative action $\hat{L}_1$, P1 gets either -2 or 1. Hence, the equilibrium path does not dominate $\hat{L}_1$, by which we cannot impose $\mu_1 = 0$. Therefore, BE1 is an EDBE.

Example 7.37 Consider the game in Fig. 34. We find all BEs first and then use the ED criterion to eliminate some BEs. At I_1, the Receiver's preferences are

$$\text{dual} \succ \text{not} \Leftrightarrow p - (1-p) > 0 \Leftrightarrow p > 0.5.$$

At I_2, the Receiver's preferences are

$$\text{dual} \succ \text{not} \Leftrightarrow q - (1-q) > 0 \Leftrightarrow q > 0.5.$$

There are four possible cases. First, if $p > 0.5$ and $q > 0.5$, then the Receiver picks dual at both I_1 and I_2, i.e., $s_{rec} = <dual, dual>$. Then, $s_{sen} = <Quiche, Beer>$. This cannot be a BE since EP consistency requires $p = 1$ and $q = 0$.

Second, if $p > 0.5$ and $q < 0.5$, then $s_{rec} = <dual, not>$ and $s_{sen} = <Beer, Beer>$. EP consistency then requires $q = 0.1$, which can be satisfied. Hence, we have a BE:

BE1 : $s_{sen} = <Beer, Beer>$, $s_{rec} = <dual, \text{not}>$, $p > 0.5$, $q = 0.1$.

Notice that the wimpy sender is not truthful in BE1.

Third, if $p < 0.5$ and $q > 0.5$, then $s_{rec} = <not, dual>$ and $s_{sen} = <Quiche, Quiche>$. EP consistency then requires $p = 0.1$, which can be satisfied. Hence, we have another BE:

BE2 : $s_{sen} = <Quiche, Quiche>$, $s_{rec} = <not, dual>$, $p = 0.1$,
$q > 0.5$.

Notice that the surly sender is not truthful in BE2.

Fourth, if $p < 0.5$ and $q < 0.5$, then $s_{rec} = <not, not>$ and $s_{sen} = <Quiche, Beer>$. EP consistency then requires $p = 1$ and $q = 0$, which cannot be satisfied. Hence, we have total two pure-strategy BEs.

Since at each of the Sender's information set, none of the Sender's actions is completely dominated, all the BEs satisfy the CD criterion, implying that both BEs are CDBEs.

Will these two BEs satisfy the ED criterion? For BE1, since the Surly's action Quiche is equilibrium-dominated, the ED criterion is satisfied if $p = 1$. Hence, only a special case of BE1 satisfies the ED criterion, which is

BE1' : $s_{sen} = <Beer, Beer>$, $s_{rec} = <dual, \text{not}>$, $p = 1$, $q = 0.1$.

For BE2, the wimpy's action Beer is equilibrium-dominated. Hence, the ED criterion requires $q = 0$. But, BE2 does not allow this. Hence, BE2 is not an EDBE. ∎

Example 7.38 Let us find all EDBEs of the game in Fig. 31. We already fine all BEs. For BE1, R_1 is equilibrium-dominated, implying $\mu = 1$. Since BE1 satisfies this condition, it is an EDBE.

For BE2, R_1 is equilibrium-dominated, implying $\mu = 1$. Since BE2 cannot satisfy this condition, it is not an EDBE.

For BE3, R_1 is again equilibrium-dominated, implying $\mu = 1$ Since BE3 cannot satisfy this condition, it is not an EDBE. ∎

The dominance concept is completely independent of the concepts based on subgames and equilibrium paths. Consequently, SE is not stronger than CDBE and vice versa.

Our definitions of the CD and ED criteria are based on Gibbons (1992) but are more general. We define these two concepts for any game, while Gibbons defines them only for games of incomplete information with two players. We define these two concepts on BEs, while Gibbons defines them on PBEs.

4.5 Alternative Versions of PBE

At the end of this chapter, we briefly mention alternative versions of PBE in the literature.

Definition 7.19 Given a game Γ_E, starting from an information set I, a <u>continuation game</u> starting from I is the part of the Γ_E that

- includes I and all the subsequent nodes;
- no information set is broken (see the left side of Fig. 35). ∎

If including all subsequent nodes after an information set I causes some information sets to be broken, then there is no continuation game starting from that information set I.

The starting information set can be a singleton set. Hence, a subgame is a special continuation game. The concept of continuation game is more general than the concept of subgame.

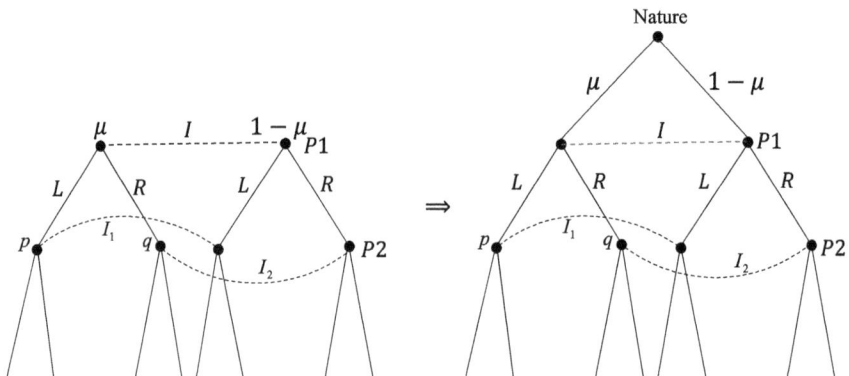

Fig. 35 A continuation game is converted to a typical game

4 Refinements of Bayesian Equilibrium

The definition of a continuation game is almost the same as that of a subgame, except that a subgame starts from a singleton information set. A subgame is a game itself, but a continuation game may not be a game. Even though a continuation game may not be a game, BE can be similarly defined on it. Alternatively, we may convert a continuation game into a game and then define BE on it. The conversion is done by replacing the starting information set by Nature. Figure 35 illustrates this conversion.

There are four versions of PBE in the literature:

(1) A BE (σ^*, μ^*) is a PBE if σ^* induces a NE in every subgame. Call it PBE(1).
(2) A BE is a PBE if it induces a BE in every subgame. Call it PBE(2).
(3) A BE is a PBE if it induces a BE in every continuation game. Call it PBE(3).
(4) A BE is a PBE if it is a PBE(i) and satisfies either the CD criterion or the ED criterion, where i = 1, 2 or 3. Call it PBE(4).

All four versions of PBE are SPNEs. Among these four versions, the earlier versions are weaker than the later versions.

PBE(1) is based on subgame perfection and is weaker than ours, while PBE(2) is equivalent to ours, which is based on subgame consistency. Some versions of PBE are based on continuation games instead of subgames. Note that BE1 in Example 7.32 shows that PBE(3) defined by continuation games instead of subgames is also strictly weaker than SE.

Also, Fudenberg and Tirole (1991) give their version of PBE for a broad class of dynamic games of incomplete information and provide conditions under which their PBE is equivalent to SE.

Notes

Good references for this chapter are Mas-Colell et al. (1995, Chaps. 7–9), Gibbons (1992) and Kreps (1990). Section 2.4 on reactive equilibrium is from Wang (2017).

Incomplete Information Games 8

The focus of this chapter is on games of incomplete information, including games of complete information as a special case. We will present several popular equilibrium concepts.

There are two kinds of asymmetric information: imperfect information and incomplete information, which result in four kinds of games. When each player in a game knows perfectly what actions other players have taken earlier, it is a game of perfect information; otherwise it is a <u>game of imperfect information</u>. If all players' payoffs are all common knowledge in a game, it is a game of complete information; otherwise, it is a <u>game of incomplete information</u>.[1] There are four combinations of these two kinds of asymmetric information, resulting in four kinds of games. The following table shows the popular choices of equilibrium concepts depending on the type of information.

	Normal Form	Extensive Form
Imperfect information	NE	NE, SPNE, BE, PBE, SE
Incomplete information	BNE	BE, CDBE, EDBE

Following Harsanyi (1967, 1968), each player's utility function has a parameter called type. This type is known to the player herself but not to others. Such asymmetric information is called incomplete information. Using this so-called Harsanyi transformation, a game of incomplete information can be represented by an extensive-form game of complete but imperfect information. A frictional player, called Nature, is introduced into the game to randomly assign each player a type at the beginning of the game. This is the so-called a <u>Bayesian game</u>.

More specifically, in a Bayesian game, we treat a player's each possible type as a separate information set of this player. By this, a game of incomplete information

---

[1] There are more general definitions of incomplete information. For the types of games covered in this book, this definition is appropriate and sufficient.

becomes a game of imperfect information and all equilibrium concepts for games of imperfect information are applicable to games of incomplete information. However, since games of incomplete information have many special applications, we often focus on some special issues with them, instead of treating them as typical games of imperfect information.

1 Bayesian Nash Equilibrium

There are situations in which some players have private information. A player's private information is the information that the player knows but others do not. This private information is typically called the player's type. In the Bayesian approach, other players simply treat this unknown information as uncertainty.[2] That is, the player herself knows her information precisely while others only know the distribution function of the information. This distribution function is assumed to be common knowledge.

The presence of incomplete information raises the possibility that we may need to consider a player's beliefs about other players' preferences, his beliefs about their beliefs about his preferences, and so on, much in the spirit of rationalizability. Such a process of rationalizability can be very complicated. Fortunately, there is a widely accepted approach to this problem, originated by Harsanyi (1967, 1968), that makes this process unnecessary. In this approach, one imagines that each player's preferences are determined by the realization of a random variable. A fictional player, named Nature, assigns realizations of this random variable. Through this formulation, the game of incomplete information becomes a game of imperfect information. By this formulation, Nature makes the first move, choosing realizations of the random variable that determine each player's preference type. Each agent then knows her type, but others only know the distribution function of her type. A game of this sort is known as a Bayesian game.

Formally, there are n players $i \in \mathbb{N} = \{1, \ldots, n\}$, each with a type parameter θ_i. The type $\theta_i \in \Theta_i \subset \mathbb{R}$ of player i is random ex ante and is chosen by Nature and its realization is observed only by player i. The joint density function is $\phi(\theta)$ for $\theta = (\theta_1, \ldots, \theta_n) \in \Theta \equiv \Theta_1 \times \cdots \times \Theta_n$, which is common knowledge. Each player i has a strategy space $\mathbb{S}_i$. Each player i has a preference relation $\succsim_{\theta_i}$ over $\mathbb{S} \equiv \mathbb{S}_1 \times \cdots \times \mathbb{S}_n$, or equivalently a payoff function $u_i(x, \theta_i)$ for $x \in \mathbb{S}$.[3] The utility function $u_i(\cdot, \cdot)$ is also common knowledge, but each specific value of $u_i(x, \theta_i)$ is observed only by agent i. A <u>Bayesian game</u> is represented by $[\mathbb{N}, \{\mathbb{S}_i\}, \{u_i\}, \Theta, \phi]$.

---

[2]The fact that unknown information is treated as uncertainty is referred to as awareness, i.e., all players are aware of the existence of the information.
[3]We can actually allow an agent's utility function to take the form $u_i(x, \theta)$ rather than $u_i(x, \theta_i)$. All the concepts on Bayesian implementation are readily extendable to this case.

1 Bayesian Nash Equilibrium

In a Bayesian game, each player maximizes his expected payoff conditional on his rivals' strategies.[4]

A mapping $s_i : \Theta_i \to \mathbb{S}_i$ is called a <u>strategy</u> of agent i. Player i's <u>strategy set</u> $\mathcal{S}_i$ is the set of all the strategy mappings $s_i : \Theta_i \to \mathbb{S}_i$. Player i's ex ante payoff from a profile of strategies $s = (s_1, \ldots, s_n)$ is

$$\tilde{u}_i(s_1, \ldots, s_n) = E_\theta\{u_i[s_1(\theta_1), \ldots, s_n(\theta_n), \theta_i]\},$$

where E_θ is the mathematical expectation over θ. The ex ante time point is before Nature assigns types. At that time point, player i does not know his own type. Although each player knows his own type θ_i when he starts to carry out his strategy, his strategy is decided before Nature assigns his type. The technical reason that we need to derive player i's strategy for all possible values of θ_i, i.e., the strategy function $s_i(\theta_i)$ for all $\theta_i \in \Theta_i$, is because other players need to know $s_i(\theta_i)$ for all $\theta_i \in \Theta_i$. Other players do not know which value θ_i is and hence have to take into account all possible values of θ_i. That is, other players need to know $s_i(\theta_i)$ for all $\theta_i \in \Theta_i$, while player i himself only need to know $s_i(\theta_i)$ for the given value of θ_i. One way to reconcile this difference on the requirement of the knowledge of $s_i(\theta_i)$ is to assume that each player decides his strategy before Nature assigns his type. Proposition 8.1 will show that this approach will not create a problem of time inconsistency. In other words, for each player, the decision made ex post after his type is known and the decision made ex ante before his type is known are the same —they are time consistent.

If we view each type as an information set, then an incomplete information game is an imperfect information game and a strategy in an incomplete information game is defined exactly the same as a strategy in the corresponding imperfect information game. For example, suppose Θ_i has finite values: $\Theta_i = \{\theta_i^1, \ldots, \theta_i^k\}$. Then, player i's strategy is written as a function $s_i(\theta_i)$ under incomplete information, but is written as a vector $s_i = <s_i(\theta_i^1), \ldots, s_i(\theta_i^k)>$ under imperfect information (in Chap. 7).

Definition 8.1 A <u>Bayesian Nash equilibrium (BNE)</u> in a Bayesian game $[\mathbb{N}, \{\mathcal{S}_i\}, \{u_i\}, \Theta, \phi]$ is a strategy profile $(s_1^*, \ldots, s_n^*) \in \mathcal{S}_1 \times \cdots \times \mathcal{S}_n$ that constitutes a Nash equilibrium of the game $\Gamma_N = [\mathbb{N}, \{\mathcal{S}_i\}, \{\tilde{u}_i\}]$. That is, for each i, s_i^* solves

---

[4]Given the density function $\phi(\theta)$ for all players, the conditional density function $\phi_{-i}(\theta_{-i}|\theta_i)$ of θ_{-i} conditional on the knowledge of θ_i is

$$\phi_{-i}(\theta_{-i}|\theta_i) = \frac{\phi(\theta)}{\phi_i(\theta_i)} = \frac{\phi(\theta)}{\int_{\Theta_{-i}} \phi(\theta) d\theta_{-i}}.$$

The expectation operator $E_{\theta_{-i}|\theta_i}$ uses this conditional density function.

$$\max_{s_i \in \mathbb{S}_i} \tilde{u}_i(s_i, s_{-i}^*). \blacksquare \qquad (1)$$

Note that this density function is not from the players' beliefs. In a simultaneous-move game (such as a normal-form game), there are no beliefs; instead, the density function represents true probabilities. Since we do not have beliefs in a BNE, we do not need to discuss whether or not the subjective beliefs are consistent with the objective probabilities.

The following proposition shows that the strategies are time consistent in the sense that the players play according to their plans after they find out their own types.

Proposition 8.1 *A strategy profile* $(s_1^*, \ldots, s_n^*)$ *is a BNE in Bayesian model* $[\mathbb{N}, \{\mathbb{S}_i\}, \{u_i\}, \Theta, \phi]$ *iff, for all i and* $\theta_i \in \Theta_i$ *occurring with a positive probability,* $s_i^*(\theta_i)$ *solves*

$$\max_{s_i \in \mathbb{S}_i} E_{\theta_{-i} | \theta_i} \{ u_i [s_i, s_{-i}^*(\theta_{-i}), \theta_i] \} \qquad (2)$$

for any $i \in \mathbb{N}$ *and* $\theta_i \in \Theta_i$, *where* $E_{\theta_{-i} | \theta_i}$ *is the expectation operator over* θ_{-i} *conditional on* θ_i.

Proof Problem (1) can be written as

$$\max_{s_i \in \mathbb{S}_i} \int_\Theta u_i[s_i(\theta_i), s_{-i}^*(\theta_{-i}), \theta_i] \phi(\theta) d\theta$$

$$= \max_{s_i \in \mathbb{S}_i} \int_{\Theta_i} \int_{\Theta_{-i}} u_i[s_i(\theta_i), s_{-i}^*(\theta_{-i}), \theta_i] \frac{\phi(\theta)}{\phi_i(\theta_i)} d\theta_{-i} \phi_i(\theta_i) d\theta_i$$

$$= \max_{s_i \in \mathbb{S}_i} \int_{\Theta_i} E_{\theta_{-i}} \{ u_i[s_i(\theta_i), s_{-i}^*(\theta_{-i}), \theta_i] | \theta_i \} \phi_i(\theta_i) d\theta_i,$$

where $\phi_{-i}(\theta_{-i}) = \phi(\theta)/\phi_i(\theta_i)$. By the Pontryagin Theorem,[5] the above problem is equivalent to problem (2), where the Hamiltonian function is

$$H(s_i, \theta_i) = E_{\theta_{-i}} \{ u_i[s_i, s_{-i}^*(\theta_{-i}), \theta_i] | \theta_i \}. \blacksquare$$

Example 8.1 (Battle of the Sexes). Consider an example from Gibbons (1992, p. 152), in which the two players, Chris and Pat, do not know each other's payoffs very well. The payoffs are listed in the following normal-form game, where θ_c is Chris' private information and θ_p Pat's private information. θ_c and θ_p are independent and each follows $U[0, x]$ for $x > 0$, where $U[0, x]$ is the uniformly distribution on $[0, x]$.

---

[5]See Wang (2008, 2015, Theorem 4.8).

1 Bayesian Nash Equilibrium

Chris\Pat	Opera	Boxing
Opera	$2+\theta_c$, 1	0, 0
Boxing	0, 0	1, $2+\theta_p$

We try to find a pure-strategy BNE in which Chris chooses Opera if $\theta_c \geq c$ and chooses Boxing otherwise and Pat chooses Boxing if $\theta_p \geq p$ and chooses Opera otherwise, where c and p are some positive constants. Let us first calculate Chris' payoffs in the two choices:

$$u_c(\text{Opera}) = (2+\theta_c)\int_0^p \frac{1}{x}d\theta_p = \frac{p}{x}(2+\theta_c), \quad u_c(\text{Boxing}) = 1 \cdot \int_p^x \frac{1}{x}d\theta_p = 1 - \frac{p}{x}.$$

Then,

$$\text{Opera} \succ_{Chris} \text{Boxing} \Leftrightarrow \theta_c > x/p - 3.$$

Hence,

$$c = x/p - 3. \tag{3}$$

Similarly, Pat's payoffs in the two choices are:

$$u_p(\text{Opera}) = 1 \cdot \int_c^x \frac{1}{x}d\theta_c = 1 - \frac{c}{x},$$

$$u_p(\text{Boxing}) = (2+\theta_p)\int_0^c \frac{1}{x}d\theta_c = \frac{c}{x}(2+\theta_p).$$

Then,

$$\text{Boxing} \succ_{Pat} \text{Opera} \Leftrightarrow \theta_p > x/c - 3.$$

Hence,

$$p = x/c - 3. \tag{4}$$

From (3) and (4), we find $p = c$ and $p^2 + 3p - x = 0$, which implies

$$p^* = \frac{\sqrt{9+4x} - 3}{2}.$$

We find that, as $x \to 0$, the ex ante probability $\frac{x-c^*}{x}$ that Chris chooses Opera and the ex ante probability $\frac{x-p^*}{x}$ that Pat chooses Boxing approach $\frac{2}{3}$. This limit is the mixed-strategy NE of the complete information game in the following table when $x = 0$:

Chris\Pat	Opera	Boxing
Opera	2, 1	0, 0
Boxing	0, 0	1, 2

This example shows that a mixed-strategy NE of a complete-information game can be viewed as the limit of a pure-strategy BNE of an incomplete information game as the incompleteness of information disappears. ∎

Example 8.2 Consider a buyer-seller problem from Gibbons (1992, pp. 158–163) under double-sided private information. A buyer and a seller is making a trade on an item for a price p. Each of them knows his/her own valuation, v_s for the seller and v_b for the buyer, on the item. The seller names an asking price p_s and the buyer simultaneously names an offer price p_b. If $p_b \geq p_s$, then trade occurs at price $p = (p_s + p_b)/2$; if $p_b < p_s$, then no trade. In a BNE, the buyer's strategy is $p_b(v_b)$ and the seller's strategy is $p_s(v_s)$. A pair of $\{p_b(v_b), p_s(v_s)\}$ is a BNE if the buyer solves the following problem:

$$\max_{p_b} \int_{p_b \geq p_s(v_s)} \left[v_b - \frac{p_b + p_s(v_s)}{2} \right] f(v_s) dv_s, \tag{5}$$

and the seller solves the following problem:

$$\max_{p_s} \int_{p_s \leq p_b(v_b)} \left[\frac{p_s + p_b(v_b)}{2} - v_s \right] f(v_b) dv_b, \tag{6}$$

where $f(\cdot)$ is the density function of both v_s and v_b. Assume that f follows $U[0,1]$, which is the uniform distribution on $[0,1]$. Let us find the BNE with linear pricing strategies:

$$p_s(v_s) = \alpha_s + \beta_s v_s \qquad p_b(v_b) = \alpha_b + \beta_b v_b. \tag{7}$$

Then, the FOC of (5) is

$$0 = \frac{\partial}{\partial p_b} \int_{p_b \geq p_s(v_s)} \left[v_b - \frac{p_b + p_s(v_s)}{2} \right] f(v_s) dv_s = \frac{\partial}{\partial p_b} \int_0^{\frac{p_b - \alpha_s}{\beta_s}} \left(v_b - \frac{p_b + \alpha_s + \beta_s v_s}{2} \right) dv_s$$

$$= \frac{1}{\beta_s} (v_b - p_b) - \int_0^{\frac{p_b - \alpha_s}{\beta_s}} \frac{1}{2} dv_s = \frac{v_b - p_b}{\beta_s} - \frac{p_b - \alpha_s}{2\beta_s},$$

which implies

$$p_b = \frac{2v_b + \alpha_s}{3}. \tag{8}$$

The FOC of (6) is

$$0 = \frac{\partial}{\partial p_s} \int_{\frac{p_s - \alpha_b}{\beta_b}}^{1} \left(\frac{p_s + \alpha_b + \beta_b v_b}{2} - v_s \right) dv_b = -\frac{p_s - v_s}{\beta_b} + \frac{1}{2} \left(1 - \frac{p_s - \alpha_b}{\beta_b} \right),$$

which implies

$$p_s = \frac{2v_s + \alpha_b + \beta_b}{3}. \tag{9}$$

Matching (7) with (8) and (9), we find

$$\alpha_b = \frac{\alpha_s}{3}, \quad \beta_b = \beta_s = \frac{2}{3}, \quad \alpha_s = \frac{\alpha_b + \beta_b}{3},$$

which implies a linear BNE:

$$p_s(v_s) = \frac{1}{4} + \frac{2}{3} v_s \quad p_b(v_b) = \frac{1}{12} + \frac{2}{3} v_b.$$

One question is: can any other BNE do better? Myerson and Satterthwaite (1983) show that, under uniform distributions, the linear BNE is the most efficient. ∎

We have considered normal-form games in the above two examples. The following example is an extensive-form game from Example 7.37. For an extensive-form game, to find BNEs, we first convert it to the normal form, using the

278   8 Incomplete Information Games

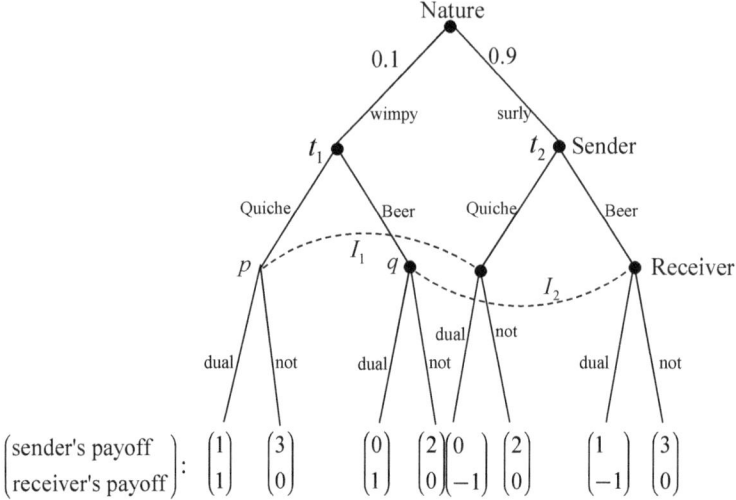

Fig. 1 BNE

given density function ϕ. When we convert an extensive form into a normal form under incomplete information, we treat each type of a player as a separate information set of the player.

Example 8.3 Find BNEs for the game in Fig. 1. Its normal form is

Sender\Receiver	<D,D>	<D,N>	<N,D>	<N,N>
<Q,Q>	0.1(1,1)+0.9(0,-1)	0.1(1,1)+0.9(0,-1)	0.1(3,0)+0.9(2,0)	0.1(3,0)+0.9(2,0)
<Q,B>	0.1(1,1)+0.9(1,-1)	0.1(1,1)+0.9(3,0)	0.1(3,0)+0.9(1,-1)	0.1(3,0)+0.9(3,0)
<B,Q>	0.1(0,1)+0.9(0,-1)	0.1(2,0)+0.9(0,-1)	0.1(0,1)+0.9(2,0)	0.1(2,0)+0.9(2,0)
<B,B>	0.1(0,1)+0.9(1,-1)	0.1(2,0)+0.9(3,0)	0.1(0,1)+0.9(1,-1)	0.1(2,0)+0.9(3,0)

or

Sender\Receiver	<D,D>	<D,N>	<N,D>	<N,N>
<Q,Q>	0.1, -0.8	0.1, -0.8	(2.1, 0)	2.1, 0
<Q,B>	1, -0.8	2.8, 0.1	1.2, -0.9	3, 0
<B,Q>	0, -0.8	0.2, -0.9	1.8, 0.1	2, 0
<B,B>	0.9, -0.8	(2.9, 0)	0.9, -0.8	2.9, 0

We have two pure-strategy BNEs: (<B,B>, <D,N>) and (<Q,Q>, <N,D>). ∎

1 Bayesian Nash Equilibrium

BNEs are actually NEs when an incomplete information game is treated as an imperfect information game where a type is treated as an information set. For example, if we treat the game in Example 8.1 as an imperfect information game, the NEs are exactly the same as the BNEs. That is, BNEs for an incomplete information game are exactly the same as NEs when we treat the game as an imperfect information game. Hence, BEs must be BNEs. Indeed, the strategies of the two BEs in Example 7.37 are the BNEs in Example 8.1.

2 Signalling Games

In extensive-form games of incomplete information, players form beliefs over other players' types. With beliefs, signals can play an important role. A player with private information may use signals to influence other players' beliefs.

We will focus on two-stage signalling games, in which a message sender sends messages in the first stage, and a message receiver receives messages and acts in the second stage. The signalling games are typical games of incomplete information. Instead of finite games in Chap. 7, we will allow infinite games in this chapter. Due to this, we will restate the definition of BEs for our signalling games. Since BEs allow beliefs, BEs are adopted as the equilibrium concept in our signalling games.

2.1 Pure Strategies in Signalling

As in Gibbons (1992), in our signalling game, there are a set $\mathbb{M}$ of possible messages, a set $\mathbb{A}$ of possible actions, a set $\mathbb{T}$ of possible types of the sender, Nature's probability distribution $p(t)$ over types, and payoffs $u_S(m, a, t)$ and $u_R(m, a, t)$ of the sender and receiver, respectively. A <u>signalling game</u> $[\mathbb{M}, \mathbb{A}, \mathbb{T}, p(t), u_S, u_R]$ involves:

- Nature assigns the sender's type $t \in \mathbb{T}$ based on a probability distribution $p(t)$, where $p(t) > 0$ for all $t \in \mathbb{T}$.
- The sender observes his type t and then decides to send a message $m(t) \in \mathbb{M}$.
- The receiver cannot observe the type t, but she can observe the message m She then forms a belief that the probability/density of type t is $\mu(t|m)$, with $\int_{\mathbb{T}} \mu(t|m) dt = 1$ for any $m \in \mathbb{M}$, and then takes an action $a(m) \in \mathbb{A}$.
- The payoffs for the sender and receiver are respectively $u_S(m, a, t)$ and $u_R(m, a, t)$.

Definition 8.2 In a signalling game, a pure-strategy BE is $[m^*(t), a^*(m), \mu^*(t|m)]$ that satisfies the following conditions:

(a) For each $m \in \mathbb{M}$, the receiver takes the following action:

$$a^*(m) \in \underset{a \in \mathbb{A}}{Arg\max} \int_{\mathbb{T}} u_R(m, a, t) \mu^*(t|m) dt.$$

(b) For each $t \in \mathbb{T}$, the sender takes the following action:

$$m^*(t) \in \underset{m \in \mathbb{M}}{Arg\max}\, u_S[m, a^*(m), t].$$

(c) EP consistency: For each $m \in \mathbb{M}$, if there exists $t_0 \in \mathbb{T}$ such that $m^*(t_0) = m$, then for every type t who sends m we require:

$$\mu^*(t|m) = \frac{p(t)}{\int_{m^*(\tau)=m} p(\tau) d\tau}. \quad \blacksquare$$

Condition (c) requires consistency to hold for those who send m, where

$$\frac{p(t)}{\int_{m^*(\tau)=m} p(\tau) d\tau} = \Pr(\text{type is } t \mid \text{message is } m).$$

The condition "there exists $t_0 \in \mathbb{T}$ such that $m^*(t_0) = m$" means that consistency is required only on the equilibrium path. Only when a message m is actually been sent by someone in equilibrium, the information set defined by this message m is on the equilibrium path. In the following game, suppose $p(t_i) > 0$ for all types. If L is sent by one of the types, EP consistency requires (Fig. 2)

$$p = \begin{cases} \frac{p(t_2)}{p(t_2) + p(t_3)} & \text{if both } t_2 \text{ and } t_3 \text{ send } L; \\ 1 & \text{if only } t_2 \text{ sends } L. \end{cases}$$

2.2 Mixed Strategies in Signalling

In the last subsection, we discuss pure-strategy BEs only. Following Crawford and Sobel (1982), we now allow mixed messages.

A signalling game involves:

- Nature assigns the sender's type $t \in \mathbb{T}$ based on a probability distribution $p(t)$, where $p(t) > 0$ for all $t \in \mathbb{T}$.
- The sender observes his type t and then decides to take a strategy $\sigma(m|t)$, which is the probability/density of sending message m by type t with $\int_\mathbb{M} \sigma(m|t) dm = 1$ for any $t \in \mathbb{T}$.

2 Signalling Games

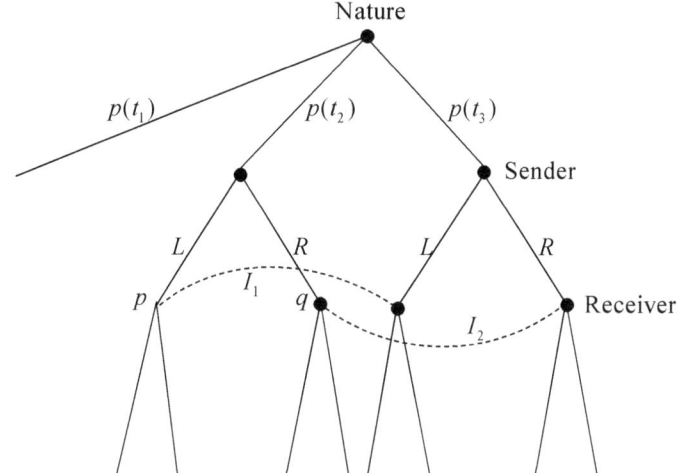

Fig. 2 BEs in a signalling game

- The receiver cannot observe t but she can observe the message m. She then forms a belief $\mu(t|m)$ that the probability/density of type t is $\mu(t|m)$, with $\int_\mathbb{T} \mu(t|m) dt = 1$ for any $m \in \mathbb{M}$, and then takes an action $a(m) \in \mathbb{A}$.
- The payoffs for the sender and receiver are respectively $u_S(m,a,t)$ and $u_R(m,a,t)$.

Definition 8.3 A mixed-strategy BE with mixed messages in a signalling game is $[a^*(m), \sigma^*(m|t), \mu^*(t|m)]$ that satisfies the following conditions:

(a) For each $m \in \mathbb{M}$, the receiver takes the following action:

$$a^*(m) \in \text{Arg} \max_{a \in \mathbb{A}} \int_\mathbb{T} u_R(m,a,t) \mu^*(t|m) dt. \tag{10}$$

(b) For each $t \in \mathbb{T}$, the sender's mixed strategy $\sigma^*(m|t)$ satisfies the condition: if $\sigma^*(m^*|t) > 0$,[6] then

$$m^* \in \text{Arg} \max_{m \in \mathbb{M}} u_S[m, a^*(m), t]. \tag{11}$$

---

[6]Condition $\sigma^*(m^*|t) > 0$ means that those messages that are actually been sent must be optimal for the sender. This requirement follows Crawford and Sobel (1982). An alternative in Matthews (1989) is $\int_\mathbb{T} \sigma^*(m|t) p(t) dt > 0$.

(c) EP consistency: For each $m \in \mathbb{M}$, if there exists $t \in \mathbb{T}$ such that $\sigma^*(m|t) > 0$, then

$$\mu^*(t|m) = \frac{p(t)\sigma^*(m|t)}{\int_\mathbb{T} p(\tau)\sigma^*(m|\tau)d\tau}. \blacksquare \qquad (12)$$

Condition (a) is the requirement of sequential rationality at all information sets. For each message, no matter whether it is on the equilibrium path or not, rationality is required. For condition (b), if there is a unique optimal message for a type, the sender will have a pure strategy; if there are multiple optimal messages for a type, the sender will be happy to mix these optimal messages arbitrarily. The equilibrium mix will be pinned down by the consistency condition (c).

Note that, if we denote $\sigma(m)$ as the density of m, $\sigma(m,t)$ as the density of (m,t) and $\sigma(m|t)$ as the density of m conditional on t etc., then, by the Bayes rule,

$$\sigma(t|m) = \frac{\sigma(t,m)}{\sigma(m)} = \frac{\sigma(m|t)\sigma(t)}{\int_\mathbb{T} \sigma(m|\tau)\sigma(\tau)d\tau}.$$

This implies the consistency condition in (12). Many authors in the literature do not mention the belief $\mu(t|m)$; instead, they use the consistency condition to replace $\mu(t|m)$ by $\frac{p(t)\sigma(m|t)}{\int_\mathbb{T} p(\tau)\sigma(m|\tau)d\tau}$ in the receiver's rationality condition. By this, the consistency condition is not needed.

Note also that, if m^* is taken with a positive probability $(\sigma^*(m^*|t) > 0)$, it must be strictly better than or indifferent from any other alternative message m, otherwise the sender can do better by moving the assigned probability $\sigma^*(m^*|t)$ from m^* to m. This explains condition (11).

In Fig. 3, suppose $p(t_i) > 0$ for all types. If L is sent with a positive probability by one of the types, then EP consistency in (12) requires

$$p = \begin{cases} \dfrac{p(t_2)\sigma(L|t_2)}{p(t_2)\sigma(L|t_2) + p(t_3)\sigma(L|t_3)} & \text{if both } t_2 \text{ and } t_3 \text{ send } L; \\ 1 & \text{if only } t_2 \text{ sends } L. \end{cases}$$

We can also allow both mixed messages and actions.[7]

Definition 8.4 A mixed-strategy BE with both mixed actions and messages in a signalling game is $[\sigma^*(a|m), \sigma^*(m|t), \mu^*(t|m)]$ that satisfies the following conditions:

(a) For each $m \in \mathbb{M}$, the receiver's mixed strategy $\sigma^*(a|m)$ satisfies the condition: if $\sigma^*(a^*|m) > 0$, then

---

[7]We have not seen this extension in the literature.

2 Signalling Games

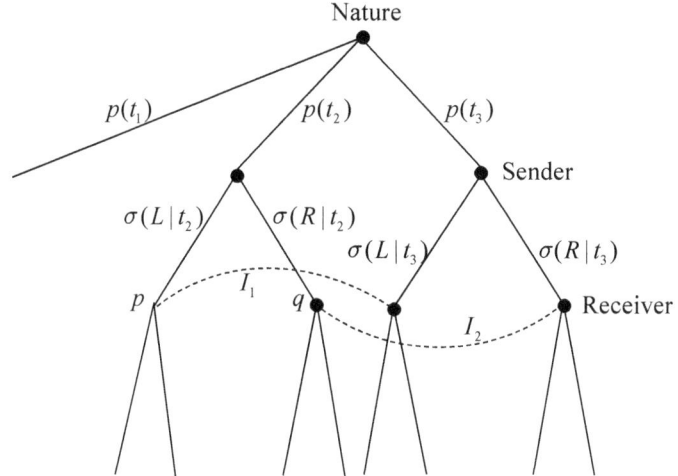

Fig. 3 Consistency

$$a^* \in \text{Arg} \max_{a \in \mathbb{A}} \int_{\mathbb{T}} u_R(m, a, t) \mu^*(t|m) dt.$$

(b) For each $t \in \mathbb{T}$, the sender's mixed strategy $\sigma^*(m|t)$ satisfies the condition: if $\sigma^*(m^*|t) > 0$, then

$$m^* \in \text{Arg} \max_{m \in \mathbb{M}} u_S[m, a^*(m), t].$$

(c) EP consistency: For each $m \in \mathbb{M}$, if $\exists t \in \mathbb{T}$ such that $\sigma^*(m|t) > 0$, then

$$\mu^*(t|m) = \frac{p(t)\sigma^*(m|t)}{\int_{\mathbb{T}} p(\tau)\sigma^*(m|\tau) d\tau}. \quad \blacksquare$$

Example 8.4 (BEs in a Signaling Game). In Gibbons (1992), signaling games are treated as a special type of games, called dynamic games of incomplete and imperfect information. We treat them as typical games of imperfect information. This example illustrates our point. Consider the game in Fig. 4 from Gibbons (1992, p. 189).

We solve it backward. At I_1, the receiver's preferences are

$$u \succ d \quad \Leftrightarrow \quad 3p + 4(1-p) > 1-p,$$

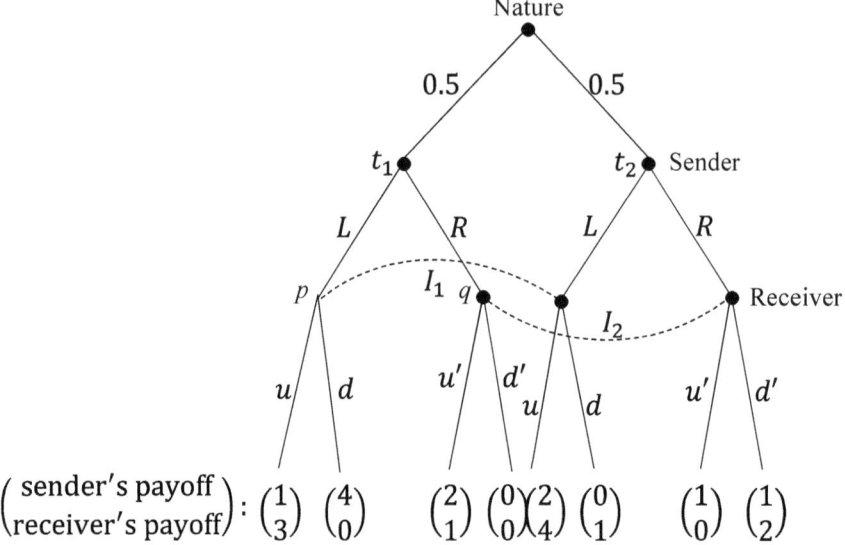

Fig. 4 BEs in a signalling game

which holds for any $p \in [0, 1]$. At I_2, the receiver's preferences are

$$u' \succ d' \Leftrightarrow q > 2(1-q),$$

which holds iff $q > 2/3$.

If $q > 2/3$, then $s_r = \,<u, u'>$, and then the sender t_1 takes R and the sender t_2 takes L, i.e., $s_s = \,<R, L>$. Since both information sets I_1 and I_2 are on the equilibrium path, EP consistency requires $q = 1$ and $p = 0$, which can satisfy $q > 2/3$. Hence, we have a BE:

BE1: $s_s = \,<R, L>$, $s_r = \,<u, u'>$, $p = 0$, $q = 1$.

If $q < 2/3$, then $s_r = \,<u, d'>$, implying $s_s = \,<L, L>$. Since only information set I_1 is on the equilibrium path, EP consistency requires $p = 0.5$, which can satisfy $q < 2/3$. Hence, we have a second BE:

BE2: $s_s = \,<L, L>$, $s_r = \,<u, d'>$, $p = 0.5$, $q < 2/3$.

If $q = 2/3$, then $s_r = \,<u, (t, 1-t)>$ for $t \in [0, 1]$. For type t_1, $L \succ R$ iff $1 > 2t$ or $t < 0.5$. For type t_2, $L \succ R$ iff $2 > 1$. Hence, $s_s = \,<L, L>$ if $t < 0.5$, and $s_s = \,<R, L>$ if $t > 0.5$, and $s_s = \,<(r, 1-r), L>$ if $t = 0.5$ for $r \in [0, 1]$. If $t < 0.5$, consistency requires $p = 0.5$, which can be satisfied. Hence, we have a third BE:

BE3: $s_s = <L,L>$, $s_r = <u,(t,1-t)>$, $p = 0.5$, $q = 2/3$,
for $t \in [0,1]$.

If $t > 0.5$, consistency requires $p = 0$ and $q = 1$, which cannot be satisfied. If $t = 0.5$ and if $r < 1$, consistency requires $p = \frac{0.5r}{0.5r + 0.5} = \frac{r}{1+r}$ and $q = 1$, which cannot be satisfied. If $t = 0.5$ and if $r = 1$, consistency requires $p = 0.5$, which can be satisfied and leads to BE3.

BE1 is a separating BE, since the sender's action reveals her type; BE2 and BE3 are pooling BEs, since the sender's action does not reveal her type.

A PBE also requires consistency on off-equilibrium paths whenever possible. However, since there are no real subgames, there are no additional requirements on the beliefs. Hence, the three BEs are also PBEs.

Since none of the sender's actions is completely dominated, the BEs are CDBEs. However, only BE1 is an EDBE. Since for BE2 and BE3 the equilibrium path $t_2 \to L \to u$ equilibrium-dominates R, the ED criterion requires $q = 1$. This condition cannot be satisfied by BE2 and BE3. Hence, BE2 and BE3 are not EDBEs. ∎

Example 8.5 (Gibbons 1992, Exercise 4.3). Find all BEs in the game of Fig. 5. We solve backwards. Consider the receiver's problem first. Given L, the receiver's preferences are

$$u \succ d \quad \Leftrightarrow \quad p_1 + p_1 + p_3 > 0,$$

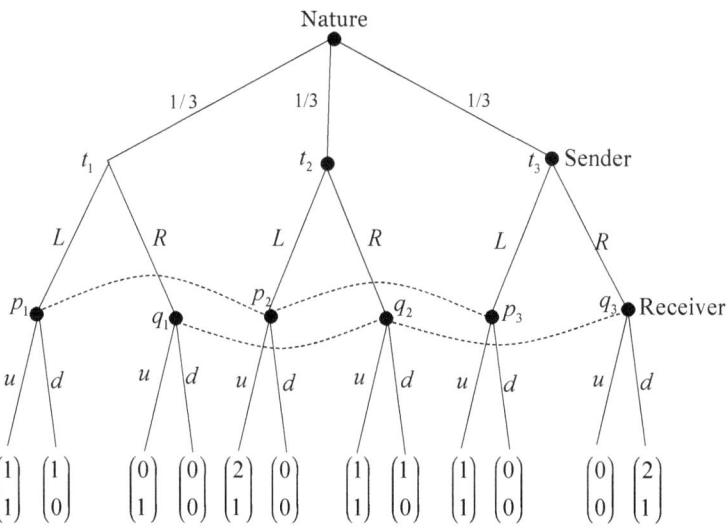

Fig. 5 BEs

which is always true. Given R, the receiver's preferences are

$$u \succ d \quad \Leftrightarrow \quad q_1 + q_1 > q_3 \quad \Leftrightarrow \quad q_1 + q_2 > 0.5.$$

If $q_1 + q_2 > 0.5$, then $s_r = <u,u>$, implying $s_s = <L,L,L>$. Consistency requires $p_1 = p_2 = 1/3$, which can be satisfied. Hence, we have a BE:

BE1: $\quad s_s = <L,L,L>, \quad s_r = <u,u>, \quad p_1 = p_2 = 1/3, \quad q_1 + q_2 > 0.5.$

If $q_1 + q_2 < 0.5$, then $s_r = <u,d>$, implying $s_s = <L,L,R>$. Consistency requires $p_1 = p_2 = 0.5$ and $q_3 = 1$, which can be satisfied. Hence, we have a second BE:

BE2: $\quad s_s = <L,L,R>, \quad s_r = <u,d>, \quad p_1 = p_2 = 1/2, \quad q_1 = q_2 = 0.$

BE1 is a pooling BE, while BE2 is a partially pooling BE. Since there is no a real subgame, these BEs are PBEs.

Also, for type t_1, action R is completely dominated. Hence, we should have $q_1 = 0$ in a CDBE. Imposing $q_1 = 0$ on the two BEs, we find two CDBEs:

CDBE1: $\quad s_s = <L,L,L>, \quad s_r = <u,u>, \quad p_1 = p_2 = 1/3, \quad q_1 = 0, \quad q_2 > 0.5.$
CDBE2: $\quad s_s = <L,L,R>, \quad s_r = <u,d>, \quad p_1 = p_2 = 1/2, \quad q_1 = q_2 = 0.$

No complete dominance for other types.

Further, given BE1, for types t_1 and t_2, action R is equilibrium-dominated, implying $q_1 = q_2 = 0$, which cannot be satisfied. Hence, BE1 cannot be an EDBE. Given BE2, for types t_1 and t_2, action R is equilibrium-dominated, implying $q_1 = q_2 = 0$, which is satisfied; also, for type t_3, action L is equilibrium-dominated, implying $p_3 = 0$, which is also satisfied. Hence, BE2 is an EDBE. ∎

2.3 Cheap Talk

We now consider a special type of signalling games, called cheap-talk games. Crawford and Sobel (1982) are the first to discuss cheap-talk games. In a cheap-talk game, the sender's messages have no cost and no benefit to anyone. That is, when messages have no effect on any player's payoffs, the signalling game is called a cheap-talk game.

Given a set $\mathbb{M}$ of possible messages, a set $\mathbb{A}$ of possible actions, a set $\mathbb{T}$ of possible types, Nature's probability distribution $p(t)$ over types, payoffs functions $u_S(a,t)$ and $u_R(a,t)$, a <u>cheap-talk game</u> $[\mathbb{M}, \mathbb{A}, \mathbb{T}, p(t), u_S, u_R]$ involves:

2 Signalling Games

- Nature assigns the sender's type $t \in \mathbb{T}$ based on a probability distribution $p(t)$, where $p(t) > 0$ for all $t \in \mathbb{T}$.
- The sender observes his t and then decides to send a message $m(t) \in \mathbb{M}$.
- The receiver cannot observe the type t, but she can observe a message m After observing a message m she forms a belief that the probability/density of type t is $\mu(t|m)$, and then takes an action $a(m) \in \mathbb{A}$.
- The payoffs for the sender and receiver are respectively $u_S(a,t)$ and $u_R(a,t)$, where a message has no effect on payoffs.

Definition 8.5 A pure-strategy BE in a cheap-talk game is $[m^*(t), a^*(m), \mu^*(t|m)]$ that satisfies the following conditions:

(a) For each $m \in \mathbb{M}$, the receiver takes the following strategy:

$$a^*(m) \in \text{Arg} \max_{a \in \mathbb{A}} \int_{\mathbb{T}} u_R(a,t) \mu^*(t|m) dt.$$

(b) For each $t \in \mathbb{T}$, the sender takes the following strategy:

$$m^*(t) \in \text{Arg} \max_{m \in \mathbb{M}} u_S[a^*(m), t].$$

(c) EP consistency: For each $m \in \mathbb{M}$, if there exists $t_0 \in \mathbb{T}$ such that $m^*(t_0) = m$, then

$$\mu^*(t|m) = \frac{p(t)}{\int_{m^*(\tau)=m} p(\tau) d\tau}. \blacksquare$$

One unique feature of a cheap-talk game is that the payoffs of all players do not depend on the sender's messages. The message is cheap talk since the message has no effect on any player's payoff.

In cheap-talk games, we often assume that the size of the set $\mathbb{M}$ of messages is the same as the size of the set $\mathbb{T}$ of types. By this, $\mathbb{M}$ is rich enough to signal what needs to be signalled.

Since messages have no direct effect on payoffs, a pooling BE always exists. In fact, for any message $m \in \mathbb{M}$, if action $a^* \in \mathbb{A}$ is a solution of

$$\max_{a \in \mathbb{A}} \int_{\mathbb{T}} u_R(a,t) p(t) dt. \tag{13}$$

then $[m, a^*, p(t)]$ is a pooling equilibrium, in which all types of the sender send m and the receiver always takes action a^*. In a pooling BE, the belief must be $\mu^*(t|m) = p(t)$, since the message reveals nothing.

Proposition 8.2 *For a cheap-talk game, for any message $m \in \mathbb{M}$, there is a pooling BE $[m^*(t), a^*, \mu^*(t|m)]$, where a^* maximizes the receiver's expected payoff, and $m^*(t) = m$ and $\mu^*(t|m) = p(t)$ for all $t \in \mathbb{T}$.*

Proof. For an arbitrary $m \in \mathbb{M}$, define $\mu^*(t|m) = p(t)$ and $m^*(t) = m$ for all $t \in \mathbb{T}$, and

$$a^* \in \operatorname*{Arg\,max}_{a \in \mathbb{A}} \int_\mathbb{T} u_R(a,t) p(t) dt,$$

where a^* is independent of m. Then, we can easily verify that $[m^*(t), a^*, \mu^*(t|m)]$ satisfies the three conditions in Definition 8.5, which confirms it to be a BE. ∎

We can also allow mixed messages in a signalling game.

Definition 8.6 A mixed-strategy BE with mixed messages in signalling games is $[a^*(m), \sigma^*(m|t), \mu^*(t|m)]$ that satisfies the following conditions:

(a) For each $m \in \mathbb{M}$, the receiver takes the following strategy:

$$a^*(m) \in \operatorname*{Arg\,max}_{a \in \mathbb{A}} \int_\mathbb{T} u_R(a,t) \mu^*(t|m) dt.$$

(b) For each $t \in \mathbb{T}$, the sender's mixed strategy $\sigma^*(m|t)$ satisfies the condition: if $\sigma^*(m^*|t) > 0$, then

$$m^* \in \operatorname*{Arg\,max}_{m \in \mathbb{M}} u_S[a^*(m), t].$$

(c) EP consistency[8]: For each $m \in \mathbb{M}$, if there exists $t \in \mathbb{T}$ such that $\sigma^*(m|t) > 0$, then

$$\mu^*(t|m) = \frac{p(t)\sigma^*(m|t)}{\int_\mathbb{T} p(\tau)\sigma^*(m|\tau) d\tau}. \qquad ∎$$

We can also allow both mixed messages and actions in a signalling game.

---

[8]This definition is similar to that in Crawford and Sobel (1982, p. 1434), except that Crawford and Sobel's definition is not completely correct on the consistency condition (c).

Definition 8.7 A mixed-strategy BE with both mixed actions and messages in a cheap-talk game is $[\sigma^*(a|m), \sigma^*(m|t), \mu^*(t|m)]$ that satisfies the following conditions:

(a) For each $m \in \mathbb{M}$, the receiver's mixed strategy $\sigma^*(a|m)$ satisfies the condition: if $\sigma^*(a^*|m) > 0$, then

$$a^* \in \text{Arg} \max_{a \in \mathbb{A}} \int_{\mathbb{T}} u_R(a,t) \mu^*(t|m) dt.$$

(b) For each $t \in \mathbb{T}$, the sender's mixed strategy $\sigma^*(m|t)$ satisfies the condition: if $\sigma^*(m^*|t) > 0$, then

$$m^* \in \text{Arg} \max_{m \in \mathbb{M}} u_S[a^*(m), t].$$

(c) EP consistency: For each $m \in \mathbb{M}$, if there exist $t \in \mathbb{T}$ such that $\sigma^*(m|t) > 0$, then

$$\mu^*(t|m) = \frac{p(t)\sigma^*(m|t)}{\int_{\mathbb{T}} p(\tau)\sigma^*(m|\tau) d\tau}. \blacksquare$$

An interesting question in a cheap-talk game is whether nonpooling equilibria exist. The answer is yes. In the following, we show two examples, in which separating and partially pooling equilibria exist.

Example 8.6 (Gibbons 1992, p. 214). There are two types $\mathbb{T} = \{t_L, t_H\}$, two messages $\mathbb{M} = \{L, R\}$, and two actions $\mathbb{A} = \{a_L, a_H\}$. Nature decides the type with $\Pr(t_L) = \delta \in (0,1)$. Assume $x_H > x_L$ and $y_H > y_L$. The payoffs are defined in the following table:

	t_L	t_H
a_L	x_H, 1	y_L, 0
a_H	x_L, 0	y_H, 1

where in each cell the sender's payoff is on the left and the receiver's payoff is on the right. The game is indicated more clearly by the game tree in Fig. 6.

We now try to find pure-strategy BEs. We solve the game backward. We have

$$\text{given } L, \quad a_L \succ a_H \quad \Leftrightarrow \quad p > 1 - p \quad \Leftrightarrow \quad p > 0.5;$$
$$\text{given } R, \quad a_L \succ a_H \quad \Leftrightarrow \quad q > 1 - q \quad \Leftrightarrow \quad q > 0.5.$$

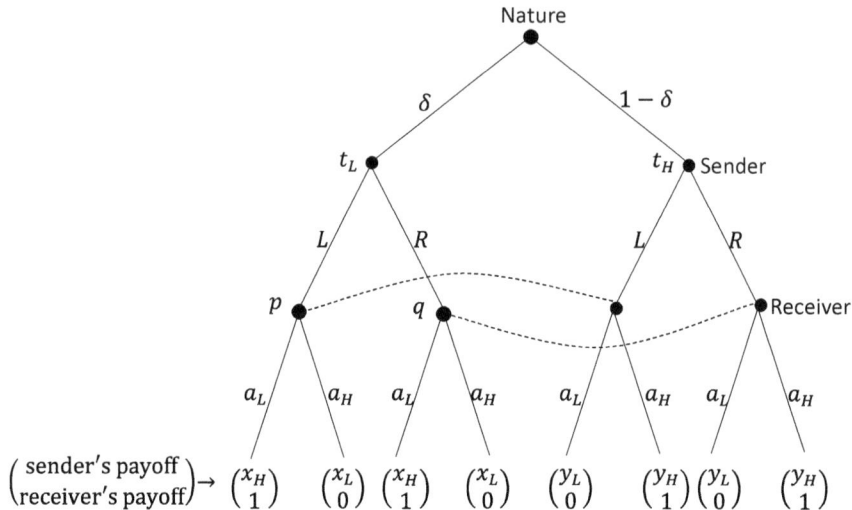

Fig. 6 A cheap-talk game

Then, first,

if $p > 0.5$ and $q > 0.5$, then $s_r = <a_L, a_L>$ and
$s_s = <(t, 1-t), (\tau, 1-\tau)>$, for $t, \tau \in [0, 1]$.

We consider pure-strategy BEs only. If $s_s = <L, L>$, we need $p = \delta$. If $s_s = <R, R>$, we need $q = \delta$. If $s_s = <L, R>$, we need $p = 1$ and $q = 0$. If $s_s = <R, L>$, we need $p = 0$ and $q = 1$. Hence, if $\delta > 0.5$, we have two BEs:

BE1: $s_s = <L, L>$, $s_r = <a_L, a_L>$, $p = \delta$, $q > 0.5$.
BE2: $s_s = <R, R>$, $s_r = <a_L, a_L>$, $p > 0.5$, $q = \delta$.

If $\delta \leq 0.5$, no pure-strategy BE in this case. Second,

if $p < 0.5$ and $q < 0.5$, then
$s_r = <a_H, a_H>$ and $s_s = <(t, 1-t), (\tau, 1-\tau)>$, for $t, \tau \in [0, 1]$.

If $s_s = <L, L>$, we need $p = \delta$. If $s_s = <R, R>$, we need $q = \delta$. If $s_s = <L, R>$, we need $p = 1$ and $q = 0$. If $s_s = <R, L>$, we need $p = 0$ and $q = 1$. Hence, if $\delta < 0.5$, we have two BEs:

BE3: $s_s = <L, L>$, $s_r = <a_H, a_H>$, $p = \delta$, $q < 0.5$.
BE4: $s_s = <R, R>$, $s_r = <a_H, a_H>$, $p < 0.5$, $q = \delta$.

If $\delta \geq 0.5$, no pure-strategy BE in this case. Third,

if $p < 0.5$ and $q > 0.5$, then $s_r = <a_H, a_L>$ and $s_s = <R, L>$.

We need $p = 0$ and $q = 1$, which can be satisfied. Hence, we have a fifth BE:

BE5: $s_s = <R, L>$, $s_r = <a_H, a_L>$, $p = 0$, $q = 1$.

Fourth,

if $p > 0.5$ and $q < 0.5$, then $s_r = <a_L, a_H>$ and $s_s = <L, R>$.

We need $p = 1$ and $q = 0$, which can be satisfied. Hence, we have a sixth BE:

BE6: $s_s = <L, R>$, $s_r = <a_L, a_H>$, $p = 1$, $q = 0$.

We have got all the pure-strategy BEs. Since there is no real subgame, each BE is a PBE. Since there is no complete dominance, all the BEs are CDBEs. Also, since it is not possible to have equilibrium dominance, all the BEs are EDBEs. ∎

In the above example, BE1 to BE4 are trivial solutions since we know the existence of such pooling solutions in cheap-talk games. We do have separating equilibria, in which the receiver reacts differently to different messages.

Proposition 8.3 *BEs in cheap-talk games are always PBEs, CDBEs and EDBEs.* ∎

Example 8.7 This example is from Crawford and Sobel (1982) and Gibbons (1992, pp. 214–218). The sender's type follows $U[0, 1]$, the type is uniformly distributed on $\mathbb{T} = [0, 1]$. The receiver's action space is $\mathbb{A} = [0, 1]$. The sender's payoff is $U_S = -(a - b - t)^2$ and the receiver's payoff is $U_R = -(a - t)^2$, where b is a positive constant. Here, b represents the disparity of the players' preferences—when b is closer to zero, the players' interests are more closely aligned. We do not define a message space $\mathbb{M}$; here we are free to use whatever space necessary.

Under complete information when t is publically known, the receiver's optimal action is $a^{**} = t$, but the sender wishes the receiver to take $\hat{a} = b + t$.

Suppose that there is $t_1 \in \mathbb{T}$ such that all types in $[0, t_1)$ send one common message, say L, while those in $[t_1, 1]$ send another common message, say R. After receiving the messages, the receiver believes that the sender's type follows $U[0, t_1)$ and $U[t_1, 1]$, respectively. After receiving L, the receiver considers

$$\max_a E(U_R) = -\int_0^{t_1} (a - t)^2 \frac{1}{t_1} dt, \quad (14)$$

which implies $a_L^* = t_1/2$. After receiving R the receiver considers

$$\max_{a} E(U_R) = -\int_{t_1}^{1} (a-t)^2 \frac{1}{1-t_1} dt, \tag{15}$$

which implies $a_R^* = \frac{1+t_1}{2}$. In (14) and (15), we have used the true conditional probabilities instead of beliefs. Given these actions, the sender's payoffs are respectively

$$U_S(a_L^*) = -\left(\frac{t_1}{2} - b - t\right)^2, \quad U_S(a_R^*) = -\left(\frac{1+t_1}{2} - b - t\right)^2.$$

To justify that type $t \in [0, t_1)$ will choose to send message L and type $t \in [t_1, 1]$ will choose to send message R, we need

$$\left(\frac{t_1}{2} - b - t\right)^2 \leq \left(\frac{1+t_1}{2} - b - t\right)^2, \quad \text{for all } t \in [0, t_1); \tag{16}$$

$$\left(\frac{t_1}{2} - b - t\right)^2 \geq \left(\frac{1+t_1}{2} - b - t\right)^2, \quad \text{for all } t \in [t_1, 1]. \tag{17}$$

This implies

$$\left(\frac{t_1}{2} - b - t_1\right)^2 = \left(\frac{1+t_1}{2} - b - t_1\right)^2, \tag{18}$$

which implies

$$t_1 = \frac{1}{2} - 2b. \tag{19}$$

Condition (16) is equivalent to

$$t \leq \frac{1}{4} + \frac{t_1}{2} - b.$$

With (19), this condition is equivalent to $t \leq t_1$. Hence, $U_S(a_L^*) \geq U_S(a_R^*)$ iff $t \leq t_1$; and $U_S(a_L^*) \leq U_S(a_R^*)$ iff $t \geq t_1$. Since t_1 has to be positive, we need $b < 0.25$. Hence, when $b \geq 0.25$, the sender and receiver's preferences are too dissimilar to allow a nonpooling solution.

Crawford and Sobel consider mixed strategies. All types in $[0, t_1)$ choose a message randomly based on $U[0, t_1)$; all types in $[t_1, 1]$ choose a message randomly based on $U[t_1, 1]$. If we assume that $\mathbb{M} = \mathbb{T}$, all messages are sent in equilibrium. Hence, consistency by a BE implies that the receiver's belief after observing any

2 Signalling Games

message from $[0, t_1)$ is that t follows $U[0, t_1)$, and the receiver's belief after observing any message from $[t_1, 1]$ is that t follows $U[t_1, 1]$.

Now, suppose there are n steps: $t_0, t_1, \ldots, t_{n-1}, t_n$, implying intervals:

$$[t_0, t_1), \quad [t_1, t_2), \quad \ldots, \quad [t_{n-1}, t_n],$$

where $t_0 = 0$ and $t_n = 1$. Consider any two nearby intervals of type: $[t_{k-1}, t_k)$ and $[t_k, t_{k+1})$. As in (14) and (15), after receiving messages from the types in these two intervals, where all players in an interval send a common message, the receiver's actions are respectively

$$a_k^* = \frac{t_{k-1} + t_k}{2}, \quad a_{k+1}^* = \frac{t_k + t_{k+1}}{2}.$$

Then, as in (18), we have

$$t_k + b - \frac{t_{k-1} + t_k}{2} = \frac{t_k + t_{k+1}}{2} - b - t_k,$$

implying

$$(t_{k+1} - t_k) - (t_k - t_{k-1}) = 4b,$$

implying

$$\sum_{k=1}^{m} [(t_{k+1} - t_k) - (t_k - t_{k-1})] = 4mb,$$

implying

$$(t_{m+1} - t_m) - (t_1 - t_0) = 4mb,$$

implying

$$t_{m+1} - t_m = d + 4mb,$$

where $d \equiv t_1 - t_0$. Then,

$$\sum_{m=0}^{n-1} (t_{m+1} - t_m) = nd + 4b \sum_{m=0}^{n-1} m,$$

implying

$$1 = nd + 4b \frac{n-1}{2} n,$$

implying

$$d = \frac{1 - 2bn(n-1)}{n}.$$

Therefore, if $2bn(n-1) < 1$, there exists an n-step partially pooling equilibrium (pooling within each interval, but separating across intervals). Hence, the largest possible number of steps $n^*(b)$ is the largest value of n such that $2bn(n-1) < 1$, which is

$$n^*(b) = \left\lceil \frac{1 + \sqrt{1 + 2/b}}{2} \right\rceil,$$

where $[x]$ denotes as the largest integer less than or equal to x. We can see that $n^*(b)$ decreases in b. We have $n^*(b) \to \infty$ if $b \to 0$, and $n^*(b) \to 1$ if $b \to \infty$. Perfect communication occurs only when the players' preferences are perfectly aligned. ∎

Cheap-talk games are special. Given each type t, the sender has a wishful action $\hat{a}_S(t)$ from

$$\max_{a \in A} \; u_S(a, t).$$

We call $\hat{a}_S(\cdot)$ the ideal action for the sender. The sender wishes that the receiver would take this action. On the other hand, if the receiver knows the type, she would take $\hat{a}_R(t)$ from

$$\max_{a \in A} \; u_R(a, t).$$

We call $\hat{a}_R(t)$ the complete-information action. If $\hat{a}_S(t) = \hat{a}_R(t)$ for all t, then the sender can induce the receiver to take the ideal action by providing completely separating messages to fully reveal his type. The complete-information action serves the best interest of the sender in this case when the ideal solution is consistent with the complete-information action. Example 8.6 is such an example, in which the sender with type t_L prefers a_L, and type t_H prefers a_H, and the receiver has the same preferences. That is, the ideal action is the same as the complete-information action for all types in this example. BE5 and BE6 imply the ideal action.

If the complete-information action is different from the ideal action, how should the sender send her messages? The following is such an example.

Example 8.8 This example is from Gibbons (1992, Exercise 4.8). There are three types $\mathbb{T} = \{t_1, t_2, t_3\}$, two possible messages $\mathbb{M} = \{L, R\}$, and three possible actions $\mathbb{A} = \{a_1, a_2, a_3\}$. Nature decides the type with equal probabilities. The payoffs are defined in the following table:

2 Signalling Games

	t_1	t_2	t_3
a_1	0, 1	0, 0	0, 0
a_2	1, 0	1, 2	1, 0
a_3	0, 0	0, 0	2, 1

The game is indicated more clearly by the following tree. Here, S means the sender and R means the receiver. The first value in each payoff column is the sender's and the second value is the receiver's (Fig. 7).

When receiving message L, the receiver forms a belief (p_1, p_2, p_3), with $p_i \geq 0$ and $p_1 + p_2 + p_3 = 1$; when receiving message R, the receiver forms a belief (q_1, q_2, q_3), with $q_i \geq 0$ and $q_1 + q_2 + q_3 = 1$. Given message L, the receiver's preferences are

$$a_1 \succ a_2 \succ a_3 \quad \text{iff} \quad p_1 > 2p_2 > p_3. \tag{20}$$

Similarly, given message R, the receiver's preferences are

$$a_1 \succ a_2 \succ a_3 \quad \text{iff} \quad q_1 > 2q_2 > q_3. \tag{21}$$

Regarding ideal actions, the sender t_1 prefers a_2; the sender t_2 also prefers a_2; and the sender t_3 prefers a_3. i.e.,

$$\hat{a}_S(t_1) = a_2, \quad \hat{a}_S(t_2) = a_2, \quad \hat{a}_S(t_3) = a_3.$$

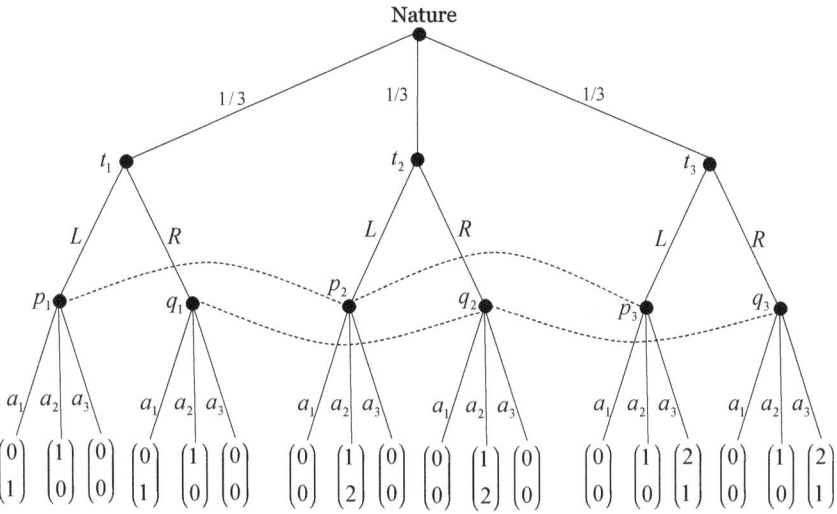

Fig. 7 A cheap-talk game

Given the preferences in (20) and (21), the sender try to choose his messages so that his ideal preferences are satisfied. We guess that the following strategy will induce the receiver to choose what the sender wants:

$$\text{type } t_1 \text{ sends } L; \quad \text{type } t_2 \text{ sends } L; \quad \text{type } t_3 \text{ sends } R.$$

Given this strategy, consistency implies $p_1 = p_2 = 0.5$ and $q_1 = q_2 = 0$. If so, by (20) and (21), the receiver will take a_2 when she sees L and she will take a_3 when she sees R. Hence, the sender's wishes are indeed satisfied. Therefore, we have a BE:

$$BE1: \quad s_s = <L, L, R>, \quad s_r = <a_2, a_3>, \quad p_1 = p_2 = 0.5, \quad q_1 = q_2 = 0,$$

in which the receiver takes the ideal actions.

Given only two possible messages, there are only three possible combinations: (a) the first two types pool; (b) the first and third types pool; and (3) the second and third types pool. BE1 is in case (a). In case (b), assuming the two pooling types send L, $p_1 = p_3 = 0.5$ and $q_2 = 1$. If so, by (20), the receiver will choose either a_1 or a_3 when seeing L and choose a_2 when seeing R. Then, type t_1 will choose to send R instead, which is a deviation. Hence, it cannot be a BE. In case (c), assuming the two pooling types send L, $p_2 = p_3 = 0.5$ and $q_1 = 1$. If so, by (20), the receiver will choose a_2 when seeing L and choose a_1 when seeing R. Then, type t_1 will send L, which is a deviation. Hence, it cannot be a BE.

Although we take a guess to identify the ideal solution in BE1, there is indeed some rules to apply. Since $\hat{a}_S(t_1) = \hat{a}_S(t_2)$, the same message should be sent by these two types; since $\hat{a}_S(t_1) \neq \hat{a}_S(t_3)$, type t_3 should send a different message. One question: should t_1 and t_2 send L or R? Suppose

$$\text{type } t_1 \text{ sends } R; \quad \text{type } t_2 \text{ sends } R; \quad \text{type } t_3 \text{ sends } L.$$

Then, consistency requires $q_2 = q_3 = 0.5$ and $p_3 = 1$. By (20) and (21), we find $s_r = <a_3, a_2>$. This solution is also ideal for the sender. Hence, BEs that imply the ideal actions are not unique. This example suggests that messages themselves are not important; what is important is that different messages should be sent for different ideal actions and the same message should be sent for the same ideal action. ∎

Example 8.9 Reconsider Example 8.6. We look for those BEs that imply the ideal actions. Given message L, the receiver will have the preferences:

$$a_L \succ a_H \quad \text{iff} \quad p > 1 - p. \tag{22}$$

Similarly, given message R, the receiver will have the preferences:

$$a_L \succ a_H \quad \text{iff} \quad q > 1 - q. \tag{23}$$

2 Signalling Games

Regarding ideal actions, the sender t_L prefers a_L; and the sender t_H prefers a_H. i.e.,

$$\hat{a}_S(t_L) = a_L, \quad \hat{a}_S(t_H) = a_H,$$

which turn out to be the same as the complete-information actions.

Given the preferences in (22) and (23), the sender try to choose his message so that his ideal preferences are satisfied. Consider the following strategy of the sender:

$$s_s = <L, R> .$$

Given this strategy, consistency implies $p = 1$ and $q = 0$. If so, by (22) and (23), $s_r = <a_L, a_H>$, which implies the ideal actions. The sender's wishes are satisfied. Therefore, we have a BE that implies the ideal actions, which is

$$BE1: \quad s_s = <L, R>, \quad s_r = <a_L, a_H>, \quad p = 1, \quad q = 0.$$

Similarly, if the sender's strategy is $s_s = <R, L>$, consistency implies $p = 0$ and $q = 1$, implying $s_r = <a_H, a_L>$. The corresponding BE is

$$BE2: \quad s_s = <R, L>, \quad s_r = <a_H, a_L>, \quad p = 0, \quad q = 1.$$

This BE also implies the ideal actions. ∎

Notes

Good references for this chapter are Gibbons (1992) and Kreps (1990). In particular, a good reference for signalling games is Gibbons (1992, 183–218).

Cooperative Games

9

For a group of players, it is possible that they cooperate in some way to improve their individual welfare. They may find a need for cooperation on a common objective, but they may bargain over the sharing of benefits, just like the OPEC. They may form coalitions to cooperate within a coalition but compete between coalitions, just like political parties in practice. They may also seek third-party coordination, such as arbitration and government social programs.

Here, the term 'cooperation' does not mean that the individuals have to set aside their own interests and work for the joint goal such as with a collective utility function. This is not what cooperative games must imply. Instead, in cooperative games, each player is driven by self-interest and a rational decision maker whose behavior is still determined by maximizing his own expected payoff. In other words, players in a cooperative game are driven by self-interest but they realize that certain forms of cooperation can serve their individual interests well.

This chapter discusses several notions of cooperative solutions under complete information.

1 The Nash Bargaining Solution

Consider a situation in which two players bargain for a share of a fixed pie. Since this pie is fixed, it is a zero-sum game: one player's gain is the other player's loss. Information is complete: everything is public knowledge. What is a sensible solution in such a situation? The Nash bargaining solution is a simple and popular solution (popular in theory).

1.1 The Nash Solution

There are n players, $i = 1, \ldots, n$ in $\mathbb{N} \equiv \{1, 2, \ldots, n\}$. Let $\mathbb{X}$ be a set of potential bargaining outcomes, and D be the outcome if the bargaining fails, the disagreement outcome. Assume that, for all $x \in \mathbb{X}$ and i, we have $x_i \succsim_i D$, and there exists $x_0 \in \mathbb{X}$ such that $x_0 \succ_i D$ for all i.

An agreement $x^* \in \mathbb{X}$ is the <u>Nash bargaining solution</u> if

$$\text{If } (x, p; D, 1-p) \succ_i x^* \text{ for some } i \in \mathbb{N}, p \in [0, 1] \text{ and } x \in \mathbb{X}, \\ \text{then } (x^*, p; D, 1-p) \succ_j x \text{ for } j \neq i. \tag{9.1}$$

The Nash bargaining concept can be explained in the following way. Player i makes the argument: "I demand the outcome x rather than x^*; I back up this demand by threatening to quit negotiation with probability $1 - p$, a threat that is credible since it is better for me to do so." However, other players such as player j, $j \neq i$, object to this argument with the counter argument: "Even if you may quit with probability $1 - p$, it is still better for us to insist on x^* rather than to agree to x."

Remark 9.1 Two weak points about the Nash solution. First, the solution is not a limit point of a bargaining process. In other words, the solution is more like a stalemate rather than a final compromise. A bargaining process in reality seems to be a process of revealing and discovering each other's bottom lines. Such a process is difficult to model in a bargaining model with complete information. The alternating-offer bargaining concept in Sect. 9 has some flavor of such a process. Second, the players are treated symmetrically, implying equal bargaining power for all players. Besides the differences in preferences and in the disagreement outcome, there is no heterogeneity in bargaining positions among the players.

Proposition 9.1 (Nash Solution). *For $n = 2$, there exists a unique Nash solution. And, an agreement $x^* \in \mathbb{X}$ is a Nash solution if it is the solution of the following problem:*

$$\max_{x \in \mathbb{X}} [u_1(x) - u_1(D)][u_2(x) - u_2(D)], \tag{9.2}$$

where u_i is the expected utility representation of $\succsim_i$ for $i = 1, 2$.

Proof For simplicity, assume $u_i(D) = 0$. We first prove that (9.2) is sufficient. Let x^* be a solution of (9.2). Since there is $x_0 \in \mathbb{X}$ such that $u_i(x_0) > u_i(D) = 0$, we must have $u_i(x^*) > 0$ for all i. If there is $x \in \mathbb{X}$ such that $pu_2(x) > u_2(x^*)$ for some $p \in [0, 1]$. then

$$pu_1(x^*)u_2(x) > u_1(x^*)u_2(x^*) \geq u_1(x)u_2(x),$$

implying $pu_1(x^*) > u_1(x)$. That is, condition (9.1) is verified and x^* is indeed a Nash solution.

1 The Nash Bargaining Solution

Conversely, given condition (9.1) and x^* defined by it, for any $x \in \mathbb{X}$ such that $u_i(x) > 0$ for all i and $u_i(x) > u_i(x^*)$ for some i, for any $p \in [0,1]$ such that $pu_i(x) > u_i(x^*)$, by (9.1), we have $pu_j(x^*) > u_j(x)$ for $j \neq i$. For any $\varepsilon > 0$, consider p defined by

$$p = \frac{u_i(x^*)}{u_i(x)} + \varepsilon.$$

Since $pu_i(x) > u_i(x^*)$ for this p, we have

$$\left[\frac{u_i(x^*)}{u_i(x)} + \varepsilon\right] u_j(x^*) > u_j(x).$$

Letting $\varepsilon \to 0$ implies

$$u_i(x^*)u_j(x^*) \geq u_i(x)u_j(x).$$

That is, x^* is indeed a solution of (9.2). ∎

Example 9.1 Consider two risk-neutral players with $u_i(D) = r_i$, where r_i is called the reservation value of player i. Assume that the size of the pie is R, called the revenue, with $R > r_1 + r_2$. Then, for an outcome $x = (t_1, t_2)$ with $t_1 + t_2 = R$, the utility values are $u_1(x) = t_1$ and $u_2(x) = t_2$. By Proposition 9.1, the Nash solution is from:

$$\max_{t_i \in [0,R]} (t_1 - r_1)(t_2 - r_2)$$
$$\text{s.t.} \quad t_1 + t_2 = R.$$

The solution is, for $i = 1, 2$,

$$t_i^* = r_i + \frac{1}{2}(R - r_1 - r_2).$$

That is, the Nash solution gives each player his reservation value r_i plus half of the surplus from the trade. ∎

A generalized Nash bargaining solution is, for $i = 1, \ldots, n$,

$$t_i^* = r_i + \theta_i \left(R - \sum_{i=1}^{n} r_i\right),$$

where θ_i represents the bargaining power of player i, $\theta_i \geq 0$ and $\sum_{i=1}^{n} \theta_i = 1$. However, this formula is not obtainable from (9.2), since the Nash bargaining concept treats every player as equal in bargaining power. Instead, this generalized solution is the solution of the following problem:

$$\max_{t_i \in [0,R]} \prod_{i=1}^{n} (t_i - r_i)^{\theta_i}$$

$$\text{s.t.} \sum_{i=1}^{n} t_i = R,$$

where the players have rank-dependent utility functions.

Corollary 9.1 (Risk Aversion). *Suppose $\succsim_1$ is less risk averse than $\succsim'_1$. Let x and x' be the Nash solutions of the bargaining problems $(\mathbb{X}, D, \succsim_1, \succsim_2)$ and $(\mathbb{X}, D, \succsim'_1, \succsim_2)$, respectively. Then, $x \succsim_1 x'$.* ∎

This corollary suggests that risk aversion causes a player to accept a less favorable deal.

An alternative mechanism to this allocation problem under complete information is the general equilibrium approach. In the GE, the price will be 1 and any feasible allocation is a GE allocation. This GE allocation is however uninteresting.

1.2 Implementation of the Nash Solution

The Nash solution means that, if two players somehow end up in such a situation, they cannot possibly deviate from it and hence it must be the solution. However, it does not show how such a solution can be reached. We now present a game tree by which the Nash bargaining solution is a SPNE.

Given a set $\mathbb{X}$ of potential bargaining outcomes and a disagreement outcome D, we want to find a simple procedure to implement the Nash solution x^* for a pair of players with preference relations $(\succsim_1, \succsim_2)$. Given x^*, consider an extensive-form game with complete information and chance moves consisting of the following stages:

- Stage 1: Player 2 chooses an alternative $x \in \mathbb{X}$ and $p \in [0, 1]$ with probability $1 - p$ that game ends with D, and with probability p that it continues.
- Stage 2: Player 1 accepts x or chooses the lottery $(x^*, p;\ D, 1 - p)$; this choice is the outcome.

These two stages are repeated infinitely. Figure 1 shows these two stages, as a two-stage lottery. Figure 1 shows the choices and outcomes, but not a game tree.

Proposition 9.2 (Implementation). *The above game implements the Nash solution as a SPNE.* ∎

1 The Nash Bargaining Solution

Fig. 1 A repeated game

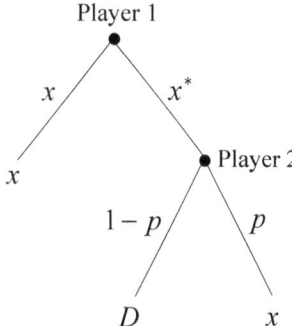

Proof Given the Nash solution x^*, since $(x^*, p;\ D, 1-p) \succ_1 x$, player 1 will insist on x^* in the third stage. Moving backward one step, player 2 is indeed justified to propose x and to threaten to quit with probability $1-p$ since $(x, p;\ D, 1-p) \succ_2 x^*$.

Going back one step further, player 1 will choose x^* since $(x^*, p; D, 1-p) \succ_1 x$. That is, x^* is a sequential outcome of the above game.[1] ∎

The Nash solution can also be implemented by alternating-offer bargaining.

Proposition 9.3 (Implementation). *The solution of the alternating-offer bargaining game with a probability α of breakdown after each rejection converges to the Nash solution as $\alpha \to 0$.* ∎

2 The Alternating-Offer Bargaining Solution

2.1 The Alternating-Offer Solution

Two players P1 and P2 with time preference discount factors δ_1 and δ_2, respectively, bargain alternately over a pie of size 1. The bargaining ends if one player accepts an offer from the other. More specifically, *P1 makes an offer first*. If P2 accepts the offer, the game ends; if not, P2 makes an offer in the next round, and P1 decides to accept or reject, and so on.

What is the SPNE? For each player, the future is always the same at any point of time. Thus, if P1 thinks that x is his best share in period 1, he will still think that x is his best share in period 3 when it is his turn again to make an offer; similarly, if P2 thinks that y is her best share in period 2, she will still think that y is her best share in period 4 when it is her turn again to make an offer. Thus, given the Nash equilibrium (x^*, y^*) in the next subgame, P2 will accept the offer x from P1 in

---
[1]What is actually shown here is that the choice of x^* is a consistent choice for player 1 since he will insist on it. But, player 2 will insist on x too—a stalemate.

period 1 if $\delta_2 y^* \leq 1 - x$. P1 understands this; he will thus make x as large as possible such that the equality holds, i.e.,

$$\delta_2 y^* = 1 - x^*.$$

Symmetrically, P1 will accept the offer y from P2 in period 2 if $\delta_1 x^* \leq 1 - y$. P2 understands this; she will thus make y as large as possible such that the equality holds, i.e.,

$$\delta_1 x^* = 1 - y^*.$$

The solution from the two equations is

$$x^* = \frac{1 - \delta_2}{1 - \delta_1 \delta_2}, \quad y^* = \frac{1 - \delta_1}{1 - \delta_1 \delta_2}. \tag{9.3}$$

This is the <u>alternating-offer bargaining solution</u>. In particular, when $\delta_1 = \delta_2 \equiv \delta$, we have

$$x^* = y^* = \frac{1}{1 + \delta}.$$

The above analysis is based on risk-neutral players. For a general case, let $\mathbb{X}$ be the outcome space and suppose that player i's expected utility for outcome $x \in \mathbb{X}$ is $u_i(x)$ and the discount factor is δ_i. An alternating-offer bargaining solution is a pair $(x, y) \in \mathbb{X}^2$, for which x is proposed by P1 and y is proposed by P2. The following is the general result on alternating-offer bargaining, which can be proven by the same approach as above.

Proposition 9.4 (The Alternating-Offer Bargaining Solution). *An alternating-offer bargaining game satisfying certain axioms (Osborne & Rubinstein, 1994, p. 122) has a SPNE. Let (x^*, y^*) be the unique pair in $\mathbb{X}$ satisfying*

$$u_1(y^*) = \delta_1 u_1(x^*), \quad u_2(x^*) = \delta_2 u_2(y^*). \tag{9.4}$$

Then, (x^*, y^*) is the SPNE, for which P1 will propose x^* and P2 will accept, and P2 will propose y^* and P1 will accept. ∎

Example 9.2 When both u_1 and u_2 are risk neutral, we have

$$u_1(x^*) = x^*, \quad u_2(x^*) = 1 - x^*, \quad u_1(y^*) = 1 - y^*, \quad u_2(y^*) = y^*.$$

Then, Proposition 9.4 implies the solution in (9.3). ∎

There are three problems with this theory. First, there is no bargaining in equilibrium. It cannot explain why people bargain in practice. Second, P1 makes an offer first and hence gets a larger share. However, are you eager to make an offer first when you bargain with a seller? The answer is probably no. Third, since a bargaining process may take only a few seconds to complete, a time discount does not make sense. In such a time span, people are not likely to discount time. If so, the theory predicts a trivial outcome: $x^* = y^* = 0.5$.

One more problem in using the alternating-offer bargaining solution is to justify who should be the first to make an offer. It is often difficult to justify who should be the first mover in an applied bargaining problem, while the outcome clearly favors the first mover. The Nash bargaining solution does not pose such a difficult problem to a modeler. This is one reason that the Nash bargaining solution is used more often in applications than the alternating-offer bargaining solution.

2.2 Extensions

In the Nash bargaining game, there is a disagreement outcome D, which plays an important role in the Nash bargaining solution. However, the alternating-offer bargaining solution has nothing to do with a disagreement option. This is not very ideal since a realistic bargaining process often involves a threat of a disagreement and this threat often plays a crucial role in reaching an agreement.

We now consider an extension of the alternating-offer game by allowing one or both players, at various points in the game, to opt out (without requiring the approval of the other player). We allow a player to opt out only when the player is responding to an offer.[2] We call this option an outside option. This opt-out option is actually an individual rationality (IR) constraint, and in this case (9.4) are incentive compatibility (IC) constraints.

Proposition 9.5 (The Outside Option Principle). *Suppose that P2 can opt out when responding to an offer. Let (x^*, y^*) be the unique SPNE outcome without an outside option, and D be P2's outside option. If $u_2(D) < \delta_2 u_2(y^*)$, the SPNE is (x^*, y^*). If $u_2(D) > \delta_2 u_2(y^*)$, the SPNE is a pair $(\hat{x}, \hat{y})$ satisfying*

$$u_1(\hat{y}) = \delta_1 u_1(\hat{x}), \quad u_2(\hat{x}) = u_2(D),$$

for which P1 will propose $\hat{x}$ and P2 will accept, and P2 will propose $\hat{y}$ and P1 will accept.

Proof Let (x, t) denote an outcome that the game ends t periods later with payoff vector x. Let (D, t) be the outcome by which P2 choose to opt out t periods later.

Let (x^*, y^*) be the unique SPNE. If $(D, 0) \prec_2 (x^*, 0)$, P2's threat of opting out is not credible. In this case, P1 proposes x^* and P2 will accept. If $(D, 0) \succ_2 (x^*, 0)$, P2's threat is credible. In this case, if it is P1's turn to offer, P1 should offer x such that $(x, 0) \succsim_2 (D, 0)$. The efficient offer $\hat{x}$ satisfies $(\hat{x}, 0) \sim_2 (D, 0)$. P2 will accept it since continuing the bargaining cannot lead to a better solution. If it is P2's turn, P2 will offer y such that $(y, 0) \succsim_1 (\hat{x}, 1)$ in order to induce acceptance. The efficient

---

[2] That is, if P1 is making an offer this period, P2 has the option to opt out right away in this period; P2 is not allowed to opt out in the next period when it is P2's turn to make an offer.

offer $\hat{y}$ satisfies $(\hat{y},0) \sim_1 (\hat{x},1)$, which P1 will accept. Thus, the pair $(\hat{x},\hat{y})$ is a SPNE. ∎

Proposition 9.5 indicates that the ability of P2 to exercise an outside option ensures the bargaining outcome to be no worse than the no-trade outcome for P2.

Instead of discount factors δ_i, we now assume that there is a chance of breakdown with probability α at the end of each period. Let the outcome of breakdown be B.

Proposition 9.6 *Suppose $u_i(x) \geq 0$ for all $x \in \mathbb{X}$ and i, and $\delta_1 = \delta_2 = 1$. Let (x^*, y^*) be the unique pair of efficient agreements satisfying*

$$u_1(y^*) = \alpha u_1(B) + (1-\alpha) u_1(x^*), \quad u_2(x^*) = \alpha u_2(B) + (1-\alpha) u_2(y^*).$$

Then, (x^*, y^*) is the SPNE, for which P1 will propose x^* and P2 will accept, and P2 will propose y^* and P1 will accept.

Proof The proof of Proposition 9.6 is the same as the proof of Proposition 9.4, which is in turn just like the derivation for (9.3). In fact, by interpreting δ_1 and δ_2 as probabilities, Proposition 9.4 immediately implies Proposition 9.6. ∎

3 The Core

The general equilibrium uses the price mechanism to allocate resources. We consider an alternative mechanism in which economic agents freely form coalitions. Each coalition is a cooperative unit in which resources are pooled and shared among the members. The question is: without a price system, what kind of equilibrium are we going to get?

Assume that there is no production sector. As in Chap. 4, a pure exchange economy consists of k commodities and n agents in $\mathbb{N} = \{1, 2, \ldots, n\}$. Each agent i has an endowment $w_i \in \mathbb{R}_+^k$ of goods, a consumption space $\mathbb{R}_+^k$, and a utility function $u_i : \mathbb{R}_+^k \to \mathbb{R}$ representing his preferences $\succsim_i$. A consumption profile $x = (x_1, x_2, \ldots, x_n)$, where $x_i \in \mathbb{R}_+^k$ is for agent i, is an allocation. An allocation $x = (x_1, x_2, \ldots, x_n)$ is feasible if $\sum_{i=1}^n x_i \leq \sum_{i=1}^n w_i$.

Definition 9.1 A group of agents $S \subset \mathbb{N}$ is a coalition. We say that coalition S blocks a given allocation x if there is some allocation x' such that

- x' is feasible for S: $\sum_{i \in S} x'_i \leq \sum_{i \in S} w_i$,
- x' dominates x for S: $x'_i \succ_i x_i, \quad \forall i \in S$. ∎

If an allocation x can be blocked, then there is some group of agents that can do better by trading among themselves and hence x will not be implemented.

3 The Core

A feasible allocation x is in the core of the economy, denoted as $x \in$ core, if it cannot be blocked by any coalition. Hence, an allocation in the core is an implementable allocation to the whole economy since no group can block it.

Proposition 9.7 $x \in$ core $\Rightarrow x$ is Pareto optimal and $x_i \succsim_i w_i$, for all i.

Proof If x is not Pareto optimal, then it can be blocked by the coalition consisting of all the agents. Therefore, any allocation in the core must be Pareto optimal. Furthermore, if $x_j \prec_j w_j$ for some j, then agent j himself can block the allocation. Thus, we must have $x_i \succsim_i w_i, \forall i$. ∎

Corollary 9.2 *In a two-agent economy, $x \in$ core $\Leftrightarrow x$ is PO and $x_i \succsim_i w_i, \forall i$.* ∎

For the 2-agent 2-good case, the Edgeworth box in Fig. 2 clearly illustrates the core.

As indicated by Fig. 2, there are many allocations in the core, including the GE allocation. However, if we allow the number of agents in the economy to grow, we will have more possible coalitions and hence more opportunities to rule out some allocations from the core. Hence, we hope that the core can shrink to one point as the number of agents goes to infinity, and we hope that this single allocation is the GE allocation.

To increase the number of agents in a tractable way, we expand the economy by repeatedly replicating it. We say that two agents are of the same type if they have the same preferences and endowments. We say that one economy is an r-replica of another if there are r times as many agents of each type in one economy as in the other. The core of the r-replica of an economy is called the r-core of the economy.

The following proposition shows that the GE allocation is always in an r-core and any allocation that is not a GE allocation must eventually not be in the r-core of the economy when the economy gets larger and larger. Hence, only the GE allocation can stay in the core forever as an economy expands.

Proposition 9.8 *Suppose that utility functions are continuous, strictly monotonic, and strictly concave.*

(1) *If (x^*, p^*) is a GE, then $x^* \in$ r-core, for $r \geq 1$.*
(2) *If a feasible allocation x is not a GE allocation, then there exists an integer m such that $x \notin$ r-core, for $r \geq m$.*

Proof

(1) An agent of type i in the jth replica is denoted as A_i^j, and his consumption bundle will be denoted as x_i^j. If (x^*, p^*) is a GE, suppose $x^* \notin$ r-core for some r.[3] Then, there is some coalition S and some allocation $\hat{x}$ such that

$$\hat{x}_i^j \succ_i x_i^{*j}, \quad \forall A_i^j \in S, \tag{9.5}$$

---

[3] x^* is itself in the 1-replica. Here, we show that x^* cannot be blocked by any coalition in a replica economy.

Fig. 2 The core

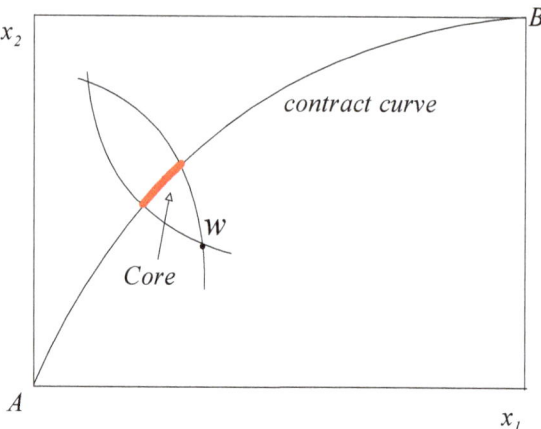

$$\sum_{A_i^j \in S} \hat{x}_i^j \le \sum_{A_i^j \in S} w_i. \tag{9.6}$$

But, by the definition of GE, for any agent $A_i^j \in S$, $\hat{x}_i^j$ must not be affordable. That is, $p \cdot \hat{x}_i^j > p \cdot w_i$. Hence,

$$p \cdot \sum_{A_i^j \in S} \hat{x}_i^j > p \cdot \sum_{A_i^j \in S} w_i,$$

which contradicts (9.6).

(2) For simplicity, we assume that there are only two individuals A and B and two goods y_1 and y_2 so that we can use an Edgeworth box in Fig. 3 to illustrate our proof. If a feasible allocation $\hat{x}$ is not a GE allocation, it must be in a situation like the one in the following Edgeworth box, for which we cannot find a straight line going through the endowment point and separating the two indifference curves going through $\hat{x}$

Consider an m-replica of the original economy. Construct a coalition S with m A-type agents and k B-type agents, $k \le m$, and an allocation (x_A, x_B) that gives $x_B = \hat{x}$ to B-type agents and x_A to A-type agents, where x_A is defined in Fig. 3. Denote ED_i^y as excess demand for good y by agent i, and ES_i^y as excess supply of good y from agent i. Then, as shown in Fig. 3, the excess demand and supply are

$$ES_A^{y_1} = me, \quad ED_A^{y_2} = md, \quad ED_B^{y_1} = kE, \quad ES_B^{y_2} = kD.$$

Fig. 3 A coalition blocks an allocation

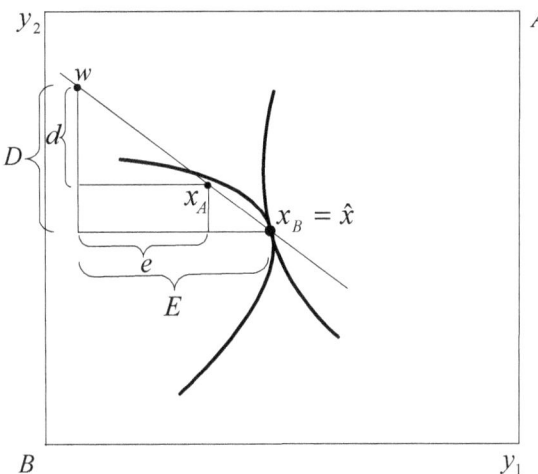

For the allocation (x_A, x_B) to be feasible for coalition S, we must have

$$ES_A^{y_1} = ED_B^{y_1}, \quad ED_A^{y_2} = ES_B^{y_2},$$

that is, $me = kE, md = kD$, implying

$$e = \frac{k}{m}E, \quad d = \frac{k}{m}D.$$

This means that, for any point x_A with $e = \lambda E$, $d = \lambda D$ for some rational number, $\lambda = \frac{k}{m} \leq 1$, no matter how large m is, we can find a coalition to distribute x_A to A-type agents and $\hat{x}$ to B-type agents. When the rational number λ is close to 1, x_A is close to $\hat{x}$. Since the indifference curves are continuous, we can always find a rational number $\lambda \equiv \frac{k}{m}$ such that $x_A \succ_A \hat{x}$. That is, the coalition S consisting of these k and m agents will block the allocation $\hat{x}$. Therefore, $\hat{x} \notin m$-core. Of course, this implies that $\hat{x} \notin r$-core, for any $r \geq m$. ∎

The following result is expected.

Proposition 9.9 (Equal Treatment in the Core). *Suppose that utility functions are continuous, strictly monotonic, and strictly concave. If x is an allocation in the r-core of a given economy, then any two agents of the same type must receive the same consumption bundle.*

Proof Given x defined in the proposition, let

$$\bar{x}_i = \frac{1}{r} \sum_{j=1}^{r} x_i^j, \quad i = 1, \ldots, n.$$

Then, by the concavity of u_i, the utility value of the average is better than the average of the utility values:

$$u_i(\bar{x}_i) = u_i\left(\frac{1}{r} \sum_{j=1}^{r} x_i^j\right) \geq \frac{1}{r} \sum_{j=1}^{r} u_i(x_i^j), \quad i = 1, \ldots, n. \tag{9.7}$$

If, for some type i_0, $x_{i_0}^1 = x_{i_0}^2 = \cdots = x_{i_0}^r$ is not true, then, by strict concavity of u_{i_0}, (9.7) is a strict inequality for i_0. For convenience, suppose $x_i^1 \precsim_1 x_i^j$, $j = 1, \ldots, r$, $i = 1, \ldots, n$. Given this assumption, consider coalition

$$S \equiv \{A_1^1, A_2^1, \ldots, A_n^1\}.$$

Using (9.7), we have

$$u_i(\bar{x}_i) \geq \frac{1}{r} \sum_{j=1}^{r} u_i(x_i^j) \geq \frac{1}{r} \sum_{j=1}^{r} u_i(x_i^1) = u_i(x_i^1), \quad i = 1, \ldots, n$$

and

$$u_{i_0}(\bar{x}_{i_0}) > \frac{1}{r} \sum_{j=1}^{r} u_{i_0}(x_i^j) \geq \frac{1}{r} \sum_{j=1}^{r} u_{i_0}(x_i^1) = u_{i_0}(x_{i_0}^1).$$

The allocation $(\bar{x}_1, \bar{x}_2, \ldots, \bar{x}_n)$ for coalition S is also feasible:

$$\sum_{i=1}^{n} \bar{x}_i = \frac{1}{r}\left(\sum_{i=1}^{n} \sum_{j=1}^{r} x_i^j\right) \leq \frac{1}{r}\left(r \sum_{i=1}^{n} w_i\right) = \sum_{i=1}^{n} w_i.$$

This means that coalition S can block allocation x (note that, by the given conditions in the proposition, PO allocations are strongly PO within S) Therefore, we must have $x_i^1 = x_i^2 = \cdots = x_i^r, \forall i$. ∎

Since any agent of the same type receives the same consumption bundle, we can still use an Edgeworth box to examine the core of a replicated two-agent two-good economy. Instead of an allocation x in the core representing how much A gets and how much B gets, we think of x as telling us how much each agent of type A gets and how much each agent of type B gets.

4 The Shapley Value

4.1 The Balanced Contributions Property

The Shapley value is a value system that efficiently allocates the total obtainable value/welfare of a society to each member of the society. One advantage of the Shapley value is that it has a clearly defined allocation for each cooperative game. This makes it very easy to use in applications.

There are n players $\mathbb{N} = \{1, 2, \ldots, n\}$. A nonempty subset S of $\mathbb{N}$ is called a coalition.

Definition 9.2 A coalitional game $(\mathbb{N}, v)$ with transferable utility[4] consists of

- a finite set $\mathbb{N}$ of players,
- a function v (characteristic function) that assigns every coalition S of $\mathbb{N}$ a real number $v(S)$ (the value of S). ∎

ψ is a value system if (1) it assigns value $\psi(S)$ to each coalition S so that every $i \in S$ gets $\psi_i(S)$, and (2) $\sum_{i \in \mathbb{N}} \psi_i(\mathbb{N}) = v(\mathbb{N})$.

In contrast to the core, the solution concept here restricts the way that an objecting coalition may deviate, by requiring that each possible deviation be balanced by a counter deviation. For the core, any feasible deviation is the end of the story and ignores the fact that a deviation may trigger a reaction that leads to a different final outcome.

A pair of objection by player i and counterobjection by player j to a division of $v(\mathbb{N})$ may take one of the following two forms.

(1) Threat to leave.

    1. Objection by i: give me more otherwise I will leave the coalition, causing you to obtain only $\psi_j(\mathbb{N}\setminus\{i\})$ rather than the larger utility $\psi_j(\mathbb{N})$, so that you lose $\psi_j(\mathbb{N}) - \psi_j(\mathbb{N}\setminus\{i\})$.

    2. Counterobjection by j: it is true that if you leave then I will lose, but if I leave, then you will lose more:
$$\psi_i(\mathbb{N}) - \psi_i(\mathbb{N}\setminus\{j\}) \geq \psi_j(\mathbb{N}) - \psi_j(\mathbb{N}\setminus\{i\}).$$

(2) Threat to exclude.

    1. Objection by i: give me more otherwise I will persuade others to exclude you from the coalition, causing me to obtain $\psi_i(\mathbb{N}\setminus\{j\})$ rather than the smaller utility $\psi_i(\mathbb{N})$, so that I will gain $\psi_i(\mathbb{N}\setminus\{j\}) - \psi_i(\mathbb{N})$.

---

[4]Transferable utility means that all agents' utility values are comparable. We can then represent the total amount of utility available to the members of a coalition $S \subset \mathbb{N}$ by a number $v(S)$. For example, money is a kind of transferable utility that can be shared among individuals.

2. Counterobjection by j: it is true that if you exclude me then you will gain, but if I exclude you, then I will gain more:

$$\psi_j(\mathbb{N}\setminus\{i\}) - \psi_j(\mathbb{N}) \geq \psi_i(\mathbb{N}\setminus\{j\}) - \psi_i(\mathbb{N}).$$

Definition 9.3 A value system ψ satisfies the balanced contributions property if

$$\psi_i(\mathbb{N}) - \psi_i(\mathbb{N}\setminus\{j\}) = \psi_j(\mathbb{N}) - \psi_j(\mathbb{N}\setminus\{i\}), \quad \forall i,j \in \mathbb{N}. \quad \blacksquare \qquad (9.8)$$

The balanced contributions property in some sense suggests that the allocation is fair, a balance of bargaining powers, or a balance of contributions. The Shapley value satisfies this property by which every objection is balanced by a counterobjection: for every objection of any player i against any other player j, there is a counterobjection of player j. And, either of the two forms of objection-counterobjection is balanced, where "balance" means that both the objection and counterobjection are equally effective/convincing.

4.2 The Shapley Value

We now proceed to define the Shapley value system.

Definition 9.4 The marginal contribution of player i to coalition S with $i \notin S$ is

$$\Delta_i(S) \equiv v(S \cup \{i\}) - v(S).$$

Let $|S|$ be the number of players in S. Let $\mathcal{P}$ be the set of all $n!$ permutations π of $\mathbb{N}$, and $S_{\pi,i} \equiv \{j \mid \pi(j) < \pi(i)\}$ be the set of players who precede i in permutation π, where $\pi(i)$ is the position of agent i in permutation π.[5] The Shapley value φ is defined as

$$\varphi_i(\mathbb{N}) \equiv \frac{1}{n!} \sum_{\pi \in \mathcal{P}} \Delta_i(S_{\pi,i}). \quad \blacksquare \qquad (9.9)$$

Suppose that all permutations are equally likely to happen. Then, $\varphi_i(\mathbb{N})$ is the expected marginal contribution of player i to the society:

$$\varphi_i(\mathbb{N}) = \frac{1}{n!} \sum_{S \subset \mathbb{N}\setminus\{i\}} |S|!(n - |S| - 1)!\Delta_i(S). \qquad (9.10)$$

---

[5]For example, for $\pi = (3,1,2)$, we have $\pi(1) = 2, \pi(2) = 3$, and $\pi(3) = 1$.

4 The Shapley Value

Here, given each coalition S, $|S|!(n - |S| - 1)!$ is the number of permutations for which the players in this S are all preceding i and the rest are behind i.[6] Each such permutation represents one possibility of objection-counterobjection, where $|S|$ players form a coalition S without player i and the remaining players form another coalition without player i. Player i can contribute to any of these coalitions S by joining it. Here, $n!$ is the total number of all possible coalitions, and, given S, $|S|!(n - |S| - 1)!$ is the number of coalitions that player i is not involved. Alternatively, by assuming that all the orderings are equally likely, $\varphi_i(\mathbb{N})$ is the expected marginal contribution of player i to the society. We indeed have $\sum_{i \in \mathbb{N}} \varphi_i(\mathbb{N}) = v(\mathbb{N})$.

Proposition 9.10 (Shapley Value). *The Shapley value satisfies the balanced contributions property, and it is the only value system that satisfies the property.*

Proof Step 1. We first show that there is at most one value system that satisfies the balanced contributions property. We will use mathematical induction.

First, since $\sum_{i \in \mathbb{N}} \varphi_i(\mathbb{N}) = v(\mathbb{N})$, when $n = 1$, for any two value systems ψ and ψ', we have $\psi(\mathbb{N}) = \psi'(\mathbb{N}) = v(\mathbb{N})$, implying $\psi = \psi'$ since $\psi(S) = \psi'(S)$ for any coalition S in $\mathbb{N}$ (in this case, $\mathbb{N} = \{1\}$). Hence, it is true when $n = 1$.

Suppose now that it is true when the number of players is $n - 1$ or less. Let $(\mathbb{N}, v)$ be a game with n players. For any two value systems ψ and ψ' satisfying the property, suppose that they are identical for all games with less than n players. That is, $\psi(S) = \psi'(S)$ for any coalition S in $\mathbb{N}$ except for $S = \mathbb{N}$. Since

$$\psi_i(\mathbb{N}\setminus\{j\}) = \psi'_i(\mathbb{N}\setminus\{j\}), \quad \text{for any } i, j \in \mathbb{N},$$

by (9.8), we have

$$\psi_i(\mathbb{N}) - \psi'_i(\mathbb{N}) = \psi_j(\mathbb{N}) - \psi'_j(\mathbb{N}) \quad \text{for any } i, j \in \mathbb{N}.$$

By summing over $j \in \mathbb{N}$, we find

$$\psi_i(\mathbb{N}) - \psi'_i(\mathbb{N}) = 0 \quad \text{for any } i \in \mathbb{N}.$$

Again, we have $\psi = \psi'$ since $\psi(S) = \psi'(S)$ for any coalition S in $\mathbb{N}$.

Step 2. We now verify that the Shapley value φ satisfies (9.8). Given a game $(\mathbb{N}, v)$, we have (9.10). Similarly, for game $(\mathbb{N}\setminus\{j\}, v)$, we have

---

[6] For each such coalition S, agent i has to be at position $|S| + 1$ so that the group S can all fill up the preceding $|S|$ positions. With agent i standing at position $|S| + 1$, the group S pick up positions preceding i [which has $|S|!$ possible combinations] and the rest of the agents $(\mathbb{N}\setminus S)\setminus\{i\}$ then pick up positions behind i [which has $(n - |S| - 1)!$ possible combinations]. Thus, the total number of combinations with group S preceding i and the rest behind i is $|S|!(n - |S| - 1)!$.

$$\varphi_i(\mathbb{N}\setminus\{j\}) = \sum_{S\subset \mathbb{N}\setminus\{i,j\}} \frac{|S|!(n-|S|-2)!}{(n-1)!} \Delta_i(S). \tag{9.11}$$

Since j is either behind or ahead of i in a permutation, we have

$$\varphi_i(\mathbb{N}) = \sum_{S\subset \mathbb{N}\setminus\{i,j\}} \frac{|S|!(n-|S|-1)!}{n!} \Delta_i(S) + \sum_{S\subset \mathbb{N}\setminus\{i,j\}} \frac{(|S|+1)!(n-|S|-2)!}{n!} \Delta_i(S\cup\{j\}),$$

where the first term on the right is for the cases when j is behind i and the second term is for the cases when j is ahead of i. We have a similar formula for $\varphi_j(\mathbb{N})$ as above when we reverse the positions of i and j. Subtracting these two formulas implies

$$\varphi_i(\mathbb{N}) - \varphi_j(\mathbb{N}) = \sum_{S\subset \mathbb{N}\setminus\{i,j\}} \frac{|S|!(n-|S|-1)!}{n!} [\Delta_i(S) - \Delta_j(S)]$$
$$+ \sum_{S\subset \mathbb{N}\setminus\{i,j\}} \frac{(|S|+1)!(n-|S|-2)!}{n!} [\Delta_i(S\cup\{j\}) - \Delta_j(S\cup\{i\})].$$
$$\tag{9.12}$$

Also, by reversing the positions of i and j for the formula in (9.11), we have a similar formula for $\varphi_j(\mathbb{N}\setminus\{i\})$ as in (9.11). Hence,

$$\varphi_i(\mathbb{N}\setminus\{j\}) - \varphi_j(\mathbb{N}\setminus\{i\}) = \sum_{S\subset \mathbb{N}\setminus\{i,j\}} \frac{|S|!(n-|S|-2)!}{(n-1)!} [\Delta_i(S) - \Delta_j(S)]. \tag{9.13}$$

Obviously, by definition, we have

$$\Delta_i(S) - \Delta_j(S) = \Delta_i(S\cup\{j\}) - \Delta_j(S\cup\{i\}),$$

and

$$\frac{|S|!(n-|S|-1)!}{n!} + \frac{(|S|+1)!(n-|S|-2)!}{n!} = \frac{|S|!(n-|S|-2)!}{(n-1)!}.$$

Then, (9.12) and (9.13) imply

$$\varphi_i(\mathbb{N}) - \varphi_j(\mathbb{N}) = \varphi_i(\mathbb{N}\setminus\{j\}) - \varphi_j(\mathbb{N}\setminus\{i\});$$

that is, (9.8) holds for φ. ∎

4 The Shapley Value

Example 9.3 (Cost Sharing). Let $c(\mathbb{N})$ be the cost of providing some service to the community $\mathbb{N}$. How should $c(\mathbb{N})$ be shared among the members? One answer is given by the Shapley value φ, where $\varphi_i(\mathbb{N})$ is the share of cost by member i. ∎

Example 9.4 (Glove Game). Let $\mathbb{N} = \{1,2,3\}$, where players 1 and 2 have right hand gloves and player 3 has a left hand glove. A coalition has value 1 if there is a match of gloves, and 0 otherwise. That is,

$$v(S) = \begin{cases} 1, & \text{if } S = \{1,3\}, \{2,3\}, \text{ or } \{1,2,3\} \\ 0, & \text{otherwise} \end{cases}$$

Then,

π	$S_{\pi,1}$	$\Delta_1(S_{\pi,1})$	$\Delta_3(S_{\pi,3})$
1, 2, 3	∅	0	1
1, 3, 2	∅	0	1
2, 1, 3	{2}	0	1
2, 3, 1	{2,3}	0	1
3, 1, 2	{3}	1	0
3, 2, 1	{2,3}	0	0

Hence, $\varphi_1(\mathbb{N}) = 1/3! \times (0+0+0+0+1+0) = 1/6$. Symmetrically, $\varphi_2(\mathbb{N}) = 1/6$. Since $\sum_{i \in \mathbb{N}} \varphi_i(\mathbb{N}) = v(\mathbb{N}) = 1$, we have $\varphi_3(\mathbb{N}) = 2/3$. ∎

Example 9.5 A firm has n workers $1, \ldots, n$ and an owner $i = 0$ who provides necessary capital. Each worker contributes $p > 0$ to the total profit; without the firm, the workers yield nothing. Then, $\mathbb{N} = \{0, 1, \ldots, n\}$ and

$$v(S) = \begin{cases} (|S| - 1)p, & \text{if } 0 \in S \\ 0, & \text{otherwise} \end{cases}$$

Each permutation π of $\mathbb{N}$ is equivalent to assigning the $n+1$ players to $n+1$ locations. Each π consists of two parts: a permutation π_0 of $\mathbb{N}_0 = \{1, \ldots, n\}$ and a designation of the owner's location in the ordering. Given π_0, the owner can be assigned to one of the $n+1$ locations: in the front, in the back, or between two workers. For example, if $\pi_0 = \{2,3,1\}$, then π can be obtained by assignig 0 to one of the 4 locations, implying $\pi = \{0,2,3,1\}$, $\pi = \{2,0,3,1\}$, $\pi = \{2,3,0,1\}$, or $\pi = \{2,3,1,0\}$. The number of those π's in which 0 is at the first location is obviously $n!$ since the n workers can be arbitrarily assigned to the remaining n locations; the number of those π's in which 0 is at the second location is also $n!$ since the n workers can be arbitrarily assigned to the remaining n locations; the number of those π's in which 0 is at the third location is still $n!$ since the n workers can be arbitrarily assigned to the remaining n locations, and so on. Hence,

$$\Delta_0(S_{\pi,0}) = \begin{cases} 0, & \text{if } 0 \text{ is at the first location} \\ 1p \times n!, & \text{if } 0 \text{ is at the second location} \\ 2p \times n!, & \text{if } 0 \text{ is at the third location} \\ \vdots & \vdots \end{cases}$$

This implies that

$$\varphi_0(\mathbb{N}) \equiv \frac{1}{(n+1)!}\sum_{\pi \in \mathcal{P}}\Delta_0(S_{\pi,0}) = \frac{1}{(n+1)!}\sum_{k=1}^{n}kp \times n! = \frac{np}{2}.$$

Since $v(\mathbb{N}) = np$ and $\sum_{i=0}^{n}\varphi_i(\mathbb{N}) = v(\mathbb{N})$ and the workers' $\varphi_i(\mathbb{N})$ should all be the same, we must have $\varphi_i(\mathbb{N}) = p/2$ for $i = 1, \ldots, n$. That is, the Shapley value is:

$$\varphi_0(\mathbb{N}) = np/2, \quad \varphi_i(\mathbb{N}) = p/2, \quad \text{for } i = 1, \ldots, n.$$

Alternatively, we can also calculate a worker's Shapley value by the following approach. For a worker i, we assign him first to one of the $n+1$ locations, then assign the owner, and then assign the rest. If i is at the first location, then $\Delta_i(S_{\pi,i}) = 0$. If i is at the second location, then $\Delta_i(S_{\pi,i}) = p$ if the owner is at the first location, otherwise $\Delta_i(S_{\pi,i}) = 0$; the rest can be assigned to the remaining $n-1$ locations; there are $(n-1)!$ such possibilities. If i is at the third location, then $\Delta_i(S_{\pi,i}) = p$ only if the owner is at one of the two locations in front of i and the rest can be assigned to the remaining $n-1$ locations; there are $2 \times (n-1)!$ such possibilities. If i is at the fourth location, then $\Delta_i(S_{\pi,i}) = p$ only if the owner is at one of the three locations in front of i and the rest can be assigned to the remaining $n-1$ locations; there are $3 \times (n-1)!$ such possibilities, and so on. This implies that

$$\varphi_i(\mathbb{N}) \equiv \frac{1}{(n+1)!}\sum_{\pi \in \mathcal{P}}\Delta_i(S_{\pi,i}) = \frac{1}{(n+1)!}\sum_{k=1}^{n}kp \times (n-1)! = \frac{p}{2}. \blacksquare$$

Notes Good references for Sect. 1 are Myerson (1991, Chap. 8) and Osborne and Rubinstein (1994, Chaps. 7 and 15). Good references for Sect. 9 are Osborne and Rubinstein (1994, p. 118–130) and Mas-Colell et al. (1995, p. 296–299). A good reference for Sect. 10 is Varian (1992). A good reference for Sect. 11 is Osborne and Rubinstein (1994, p. 289–293). See also Hokari (2000) for extensions of the Shapley value.

Part IV
Information Economics

Market Information 10

Observed market failures and inefficiencies such as those in medical insurance and unemployment insurance are well known. Possible causes include incomplete information, incomplete markets, and incentives. This and the next chapters will focus on incomplete information as a possible cause for market failures. We deal with competitive firms in this chapter and monopolies in the next chapter.

1 The Akerlof Model

Akerlof (1970) illustrates a market failure caused by adverse selection by the following argument:

1. In the used-car market, an individual is more likely to sell her car when she knows that it is not in good condition.
2. Uninformed buyers are wary of this behavior and their willingness to pay is low.
3. This fact further exacerbates the adverse selection problem: if the price of a used car is low, only those sellers with bad cars will offer them for sale.
4. As a result, in the end, there may be no market for used cars.

1.1 The Used Car Market with Incomplete Information

Consider the market for used cars. There are two groups of people, buyers and sellers. In this subsection, assume that the sellers know the quality of their own cars, but the buyers do not know the quality of any car in the market.

The Buyer's Decision
Each buyer decides to buy one or no car, and his utility function is

$$U_b = M + \frac{3}{2}qn,$$

where $n = 0$ or 1 depending on whether or not the buyer buys a car, M is the spending on other goods, and q is the quality of the car. Given income y_b, this individual faces the following budget constraint:

$$M + pn \leq y_b,$$

where p is the price of the car. Here, we have assumed that the price of other goods is 1 in other words, the price of other goods is taken as the numeraire. The buyer does not know the quality q of the car; so he will form expectations $E(q)$ on the quality of a car. Let

$$\bar{q} = E(q | \text{the car is on the market for sale}).$$

Using the budget constraint, the buyer's expected utility is

$$E(U_b) = M + \frac{3}{2}\bar{q}n = y_b + \left(\frac{3}{2}\bar{q} - p\right)n.$$

Thus, the buyer will buy the car, $n = 1$, iff

$$p \leq \frac{3}{2}\bar{q}. \tag{1}$$

Notice that since the quality of the car is unknown, the willingness to pay depends on the average quality of cars on the market.

The Seller's Decision
The seller has a similar utility function and budget constraint. Her utility function is

$$U_s = M + qn,$$

and given income y_s, the budget constraint is

$$M + pn \leq y_s.$$

Again, the seller will try to maximize her own utility. Given the budget constraint, her utility function is

$$U_s = y_s + (q - p)n.$$

1 The Akerlof Model

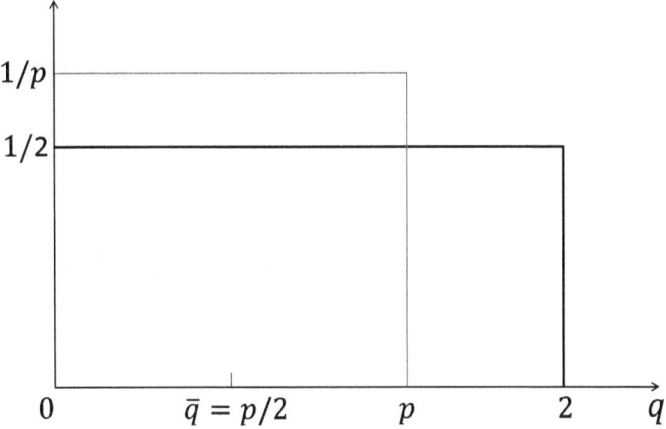

Fig. 1 The distribution of the quality of cars

Thus, since she knows the quality of her own car, she will sell her car, $n = 0$, iff

$$p \geq q. \tag{2}$$

Equilibrium

Assume that the quality of car follows $U[0,2]$, i.e., q is uniformly distributed on $[0, 2]$. Due to the seller's decision, as indicated by (2), only those cars with quality less than p will be on the market, as shown in Fig. 1. Thus, the average quality of cars on the market is $\bar{q} = p/2$.[1] Then, by (1), the buyer will buy the car iff the price of a car satisfies $p \leq 3p/4$, which can only be true iff $p^* = 0$. For this price, only those sellers with the worst cars are willing to sell. In other words, only the worst cars with $q = 0$ will be traded on the market.

Discussion

Under complete information, a buyer's marginal utility (MU_b) of car and a seller's marginal utility (MU_s) of car satisfy

$$MU_b = \frac{3}{2}q > q = MU_s,$$

indicating a potential gain for trade, i.e., a trade can benefit both. Hence, if the quality is known for both groups, there will be a trade for any car. The situation is the same under symmetric information when both sides do not know the quality of each car. However, if there is asymmetric information about the quality of a car, the market fails.

---

[1] It is $E(q) = \int_0^p q \frac{1}{p} dq = \frac{p}{2}$.

The intuition for the market failure is this. Since the quality of a car is not distinguishable to buyers, good and bad (lemons) cars will all trade at the same price. Hence, it becomes difficult to sell a good car at a fair price. Faced with this situation, the owners of the best used cars may decide that it is not worth selling their cars. The average quality of used cars on the market will accordingly be lower as a result, and so will the price. Given this, the owners of second-best used cars may also decide not to sell their cars, so the average quality and price will be lower still, and so on for the third best, fourth best, etc. The fewer good-quality cars there are on the market, the lower the average quality, and so the lower the price will be for used cars. This sort of process might lead to the market unraveling entirely; or, somewhat more plausibly, the system may find an equilibrium in which the only viable market that remains is quite literally the market for lemons—only the poorest quality cars are traded.

The sellers do not put a random sample of cars on the market; they sell only those cars with quality $q \leq p$. Because of the information advantage, the sellers are biased in selecting their cars for sale. This is so-called <u>adverse selection</u>, a selection that results in a bad outcome.

The market price plays a dual role here: it determines the average quality $\bar{q} = p/2$ of the cars on the market, and it serves to equilibrate the demand and supply. This is rather like having to hit two targets with one bullet. One price cannot simultaneously serve two roles well.

In equilibrium, the price is positively correlated with the average quality $\bar{q} = p/2$. Indeed, it is the average quality of used cars on the market that determines the price.

Under asymmetric information, a high-quality car will raise the average quality of cars on the market slightly and thus benefit all sellers (from a higher price), not just the lucky buyer. This is so-called externality, which is caused by asymmetric information in this case.

The situation for medical insurance is the same. An insurance company may not know each individual's health condition, so that it has to offer insurance based on some indicators. Since this company cannot distinguish one individual from another, it has to offer a uniform medical insurance policy for everyone. Since the price of the insurance is based on an average person in the whole population, the price may be too high for healthy individuals. Hence, healthy individuals may choose not to buy the insurance. As a result, the average cost for an insured person will be higher than the original estimate, which forces the company to raise the price. This may cause less-healthy individuals to abandon the insurance and thus cause the price to go up further. This process may go on until the insurance company is forced to abandon any insurance plan.

Under asymmetric information, bad goods can drive out good goods. Supervision of the market becomes necessary. Due to this, many governments impose quality controls on exports. Also, licenses for doctors, lawyers, taxi drivers and accountants are issued to ensure a high level of average quality.

1 The Akerlof Model

1.2 The Used Car Market with Complete Information

What would happen if the quality of a car is known to everyone?

If so, the decision for the seller is still the same; she will sell the car iff $p \geq q$. For the buyer, since his utility function is now

$$U_b = y_b + \left(\frac{3}{2}q - p\right)n,$$

he will buy the car iff $p \leq 3q/2$. Hence, any car can be traded for a price $p \in [q, 3q/2]$ and the settling price of a car is up to the bargaining skills of the two traders. Any car can be sold. The price of a car will generally be higher for high-quality cars. The quality distribution of cars is no longer relevant.

Hence, the market is efficient and the price of a car will be dependent on the quality of the car.

1.3 The Used Car Market with Symmetric Information

Assume now that the sellers are as uninformed as the buyers. Then, the seller's expected utility is

$$E(U_s) = M + \bar{q}n = M + n = y_s + (1-p)n,$$

where $\bar{q} = 1$, which is the average quality for all cars. Since the seller does not have information to do a selection for cars on sale, all cars are on the market. Thus, the seller will sell her car iff $p \geq 1$. The buyer will still buy iff $3\bar{q}/2 \geq p$, i.e., $3/2 \geq p$. Hence, any car can be traded for a price $p \in [1, 3/2]$.

In this case, the price is much higher than that under asymmetric information. From the above three cases, we find that it is indeed asymmetric information that causes adverse selection, which in turn causes the price to fall. The sellers who have the information advantage also suffer. There is thus an incentive for the sellers with high-quality cars to provide information to the buyers. However, the buyers may not trust the sellers since there is also an incentive for the sellers to cheat. It is a dilemma.

1.4 Discussion

How about applying Akerlof's model to the new car market? In this case, the dealers do not have an information advantage on the quality of individual cars. Since no one has driven the car, the dealer knows as little about the car's quality as the buyer knows. Thus, it is a problem under symmetric information. Both the dealers and buyers will trade based on the average quality of cars. Adverse selection does not appear in this case. Besides, there is a warranty on new cars. The manufacturer offers an insurance plan for its cars.

Will the market failure go away if there is a money-back guarantee? A money-back guarantee ensures that a buyer makes a final decision after she know the quality of a car. This is a case under complete information. As we have shown, the outcome will be efficient.

Suppose that the buyers can guarantee a minimum quality of a car by inspection and a test drive. Will adverse selection disappear? We find that it is possible to have a range of cars with different qualities to be traded in this case. See Question 8.1 in the Problem Set.

Will market efficiency improve dramatically if there is a car rental market for new cars? The policy of car rental solves two problems. First, it solves the problem of asymmetric information on quality. Second, it ensures that the buyer will care about the car during the rental period since the car may be his at the end of the rental period. A working paper on this policy shows that the car rental policy is efficient.

Grossman (1981) proposes a warranty in solving Akerlof's lemons problem. He shows that pooling with an optimally designed warranty contract is optimal if the working performance of the good is public information. However, Lutz (1989) argues that such a warranty contract may not be seen in practice, since buyers of the good may undetectably damage the good in order to obtain a large warranty payment.

2 The Rothschild-Stiglitz Model

Insurance is a way for a society to share risks. Insurance companies are particularly oriented toward handling independent individual risks that have no effect on society's aggregate resources but can seriously affect the welfare of the unfortunate individuals.

We discuss Rothschild and Stiglitz (1976) in this section. Consider an exchange economy with n individuals and a single good. An individual has wealth w with probability π of losing an amount L. All the individuals have the same wealth w, the same utility function $u(x)$, and the same potential loss L. Thus, the endowment (wealth before insurance) of any individual is stochastic and is

| w | with probability $1 - \pi$ |
| $w - L$ | with probability π. |

We consider two scenarios. In the first scenario, all individuals have the same probability π. In the second scenario, one group of individuals has a lower risk π_L and the rest have a higher risk π_H. Since π is the only thing that distinguishes the individuals, we call π the type of an individual.

2 The Rothschild-Stiglitz Model

When there is only a single type of individuals, what insurance does is to shift the risk from risk-averse individuals to risk-neutral companies. In this case, the companies can profit from individual risk aversion and the Law of Large Numbers. If there are multiple types of individuals, cross-subsidy between different types becomes possible under incomplete information.

2.1 Insurance with Symmetric Information: A Contingent Market

Assume that there is only one type of individual and that information is symmetric in the sense that all available information is public knowledge. Hence, everyone has the same information and no one has an information advantage over others.

We first consider an economy with contingent markets. Since each of the n individuals has two possible states: good or bad, there are total $S = 2^n$ states of nature in the economy. For complete markets, we need 2^n markets for contingent goods. Let x_s^i be agent i's consumption of the good in state s. Let π_s be the probability of state s occurring:

$$\pi_s = \pi^{n_s}(1-\pi)^{n-n_s},$$

where n_s is the number of agents having an accident in state s. Let p_s be the contingent price for the good at state s.

A general equilibrium is a system of prices $p = (p_1, \ldots, p_S)$ and quantities $x^i = (x_1^i, \ldots, x_S^i)$ for $i = 1, \ldots, n$ such that

1. x^i solves

$$\max_x \sum_{s=1}^{S} \pi_s u(x_s^i)$$

$$\text{s.t.} \sum_{s=1}^{S} p_s x_s^i \leq \sum_{s=1}^{S} p_s w_s^i,$$

where $w_s^i = w$ if i does not have an accident in state s, and $w_s^i = w - L$ otherwise.
2. Equilibrium conditions:

$$\sum_{i=1}^{n} x_s^i \leq \sum_{i=1}^{n} w_s^i, \quad s = 1, \ldots, S.$$

We know that such a GE exists under fairly general conditions and that it is ex-ante Pareto optimal.

By the Law of Large Numbers, when n is large enough, π proportion of agents has accidents with probability 1. Thus, the per capita wealth in state s is

$$\frac{n_s(w-L)+(n-n_s)w}{n} = w - \frac{n_s}{n}L \to w - \pi L, \quad \text{with probability 1.}$$

Since we have identical individuals, all the individuals should have the same consumption in equilibrium, implying $x_s^i = w - \pi L$ for any s with probability 1. That is, the consumption is independent of the state, meaning that the agents will completely insure themselves in a large economy. In other words, based on the Large of Large Numbers, the limiting case of the GE has full insurance.

2.2 Insurance with Symmetric Information: Insurance Market

We can actually obtain the GE allocation in the above contingent market with a much simpler market structure. Consider a perfectly competitive insurance market in which a typical insurance company offers compensation z for a price q. That is, an individual pays qz to the company upfront and the company pays z to the individual when he has an accident. With insurance, the income of the individual becomes

$w - qz$,	if he does not have an accident,
$w - L + z - qz$,	if he has an accident.

The individual's problem is:

$$\max_{z \geq 0} (1-\pi)u(w-qz) + \pi u(w-L+z-qz).$$

The FOC is

$$(1-\pi)qu'(w-qz) = \pi(1-q)u'(w-L+z-qz).$$

The insurance company's expected profit per capita is $qz - \pi z$. Zero profit implies $q = \pi$. Then, the FOC implies $z = L$. In other words, the individual will insure himself completely.

We can illustrate this solution in a diagram. For this purpose, we transform the problem into another equivalent problem. The individual's incomes in the two states are:

$$I_1 = w - qz, \quad I_2 = w - L + z - qz,$$

2 The Rothschild-Stiglitz Model

implying

$$(1-q)I_1 + qI_2 = w - qL.$$

In fact, the conditions for (I_1, I_2) are

$$I_1 \leq w - qz, \quad I_2 \leq w - L + z - qz.$$

Hence, the budget conditions for (I_1, I_2) is

$$(1-q)I_1 + qI_2 \leq w - qL.$$

We can thus present the individual's problem in a different way:

$$\max_{I_1, I_2 \geq 0} (1-\pi)u(I_1) + \pi u(I_2)$$
$$\text{s.t.} \quad (1-q)I_1 + qI_2 = w - qL.$$

This problem is conveniently illustrated in Fig. 2, where the budget line is labeled q in Fig. 2 and is called the q-line.

The FOC of the problem is

$$\frac{(1-\pi)u'(I_1)}{\pi u'(I_2)} = \frac{1-q}{q}, \tag{3}$$

which states that the MRS equals the slope of the budget constraint.

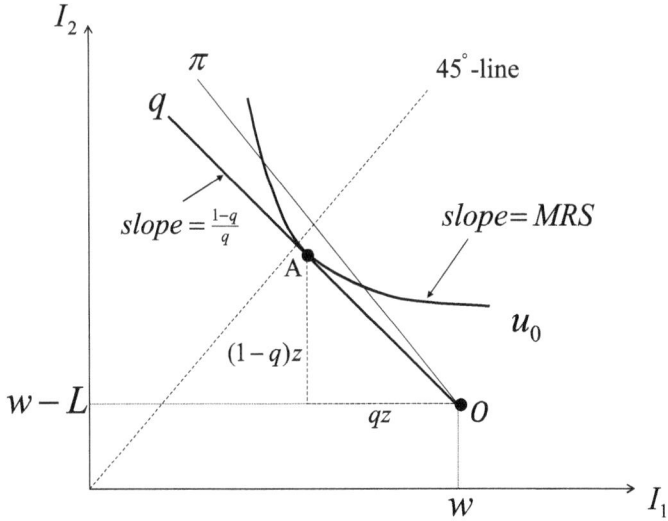

Fig. 2 An insurance problem

When $q = \pi$, the q-line becomes the π-line in Fig. 2. The company makes a profit on the q-line iff $q > \pi$, i.e., iff the q-line is on the left of the π-line. When $q = \pi$, the budget line is the break-even line for the company; in this case, by condition (3), the solution point A must be on 45°-line.

In this insurance trade, each agent gives $(w, w - L)$ to the company in exchange for (I_1, I_2). Hence, the firm's expected profit from each agent is

$$\Pi(I_1, I_2) \equiv (1 - \pi)w + \pi(w - L) - (1 - \pi)I_1 - \pi I_2,$$

which can be shown to be $qz - \pi z$. The following lemma offers much convenience to a graphic analysis of the insurance problem.

Lemma 10.1 *If a firm has a linear profit function $\Pi(x, y) = ax + by + c$, where a, b and c are constants, $a, b, c \in \mathbb{R}$, then $\frac{|\Pi(x_0, y_0)|}{\sqrt{a^2 + b^2}}$ at any point (x_0, y_0) equals the minimum distance of (x_0, y_0) to the zero-profit line defined by $\Pi(x, y) = 0$, as shown in Fig. 3.*

Proof For a given point (x_0, y_0), consider its minimum distance to the line defined by $\Pi(x, y) = 0$:

$$\min_{(x,y)} (x - x_0)^2 + (y - y_0)^2$$
$$\text{s.t.} \quad ax + bx + c = 0.$$

The Lagrangian function is

$$L = (x - x_0)^2 + (y - y_0)^2 + 2\lambda(ax + bx + c).$$

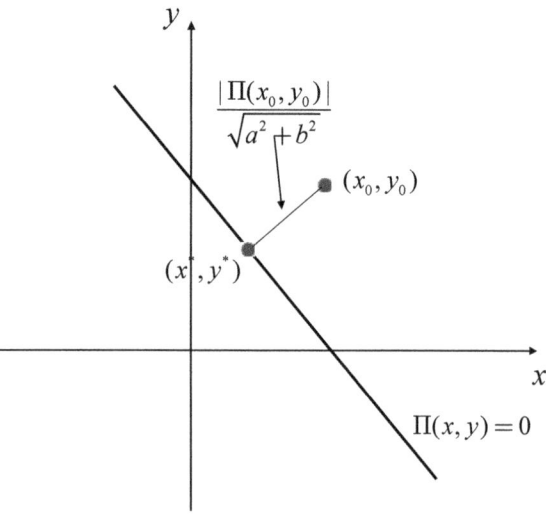

Fig. 3 A linear profit function

The FOCs are

$$(x - x_0) + a\lambda = 0, \quad (y - y_0) + b\lambda = 0.$$

These two equations indicate that the minimum distance is the orthogonal distance of the point to the line. The solution is

$$x^* = x_0 - a\lambda, \quad y^* = y_0 - b\lambda.$$

By the budget constraint,

$$a(x_0 - a\lambda) + b(y_0 - b\lambda) + c = 0,$$

implying

$$\lambda = \frac{ax_0 + bx_0 + c}{a^2 + b^2},$$

implying

$$x^* = x_0 - a\frac{ax_0 + bx_0 + c}{a^2 + b^2}, \quad y^* = y_0 - b\frac{ax_0 + bx_0 + c}{a^2 + b^2}.$$

Thus, the minimum distance is

$$\sqrt{(x^* - x_0)^2 + (y^* - y_0)^2} = \frac{1}{\sqrt{a^2 + b^2}}|ax_0 + bx_0 + c| = \frac{1}{\sqrt{a^2 + b^2}}|\Pi(x_0, y_0)|.$$

∎

Since our profit function $\Pi(I_1, I_2)$ is a linear function, we can easily identify how much $\Pi(I_1, I_2)$ is at point (I_1, I_2) in a diagram. By Lemma 10.1, the minimum distance from point (I_1, I_2) to the zero-profit line $\Pi = 0$ is $\frac{|\Pi(I_1, I_2)|}{\sqrt{(1-\pi)^2 + \pi^2}}$. Hence, $|\Pi(I_1, I_2)|$ is the minimum distance multiplied by the constant $\sqrt{(1-\pi)^2 + \pi^2}$. We can simply ignore the constant and treat the minimum distance as $|\Pi(I_1, I_2)|$.

2.3 Insurance with Asymmetric Information

In the last section, we assumed that all the agents were identical so that the insurance company could easily know the probability of having an accident that is common among all agents. The problem becomes complicated if the individuals have different probabilities of having accidents.

Suppose now that there are two types of agents, with probabilities of having accidents π_L and π_H, $\pi_H > \pi_L$. Assume that the agents know their own types, but that the insurance company only knows the existence of two types and their probabilities but not what type each agent is. Hence, there is asymmetric information instead of symmetric information.

We study subgame perfect Nash equilibria (SPNEs) of the following two-stage game:

Step 1. Firms simultaneously announce contract offers. Each firm may announce any number of contracts.
Step 2. Given the offers, individuals decide whether or not and which contract to accept.

The game can be illustrated by the following game tree. The firms compete in a continuation game (Fig. 4).

The company proposes compensation-price policies or, equivalently, income policies of the form (I_1, I_2). Suppose that a number of policies are offered to the market. Since there are only two types of individuals and individuals of the same type will choose the same policy, at most two policies in the market are chosen by the individuals. We therefore have only two possible (pure-strategy) solutions: either one single policy is accepted by all the individuals in the market, or two policies are accepted separately by the two types of individuals. The first case is called a pooling equilibrium and the second case is called a separating equilibrium.

There is a crucial difference in information revelation between these two types of equilibria. In a separating equilibrium, since the two types of agents accept different

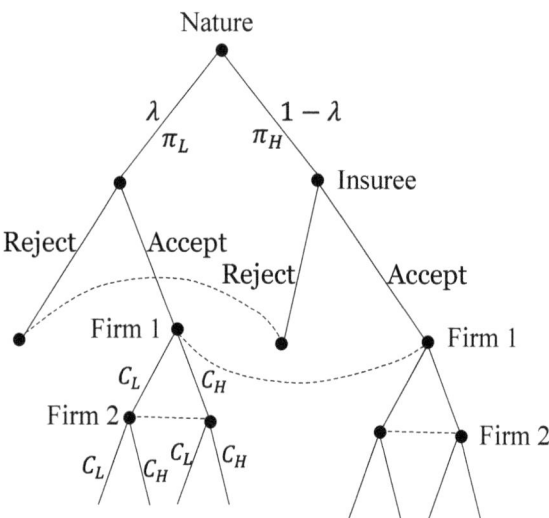

Fig. 4 Insurance with asymmetric information

2 The Rothschild-Stiglitz Model

policies, the company will be able to discovery each agent's type from his choice. There is no such information revelation in a pooling equilibrium.

The Pooling Equilibrium

Let us first consider the possibility of a pooling equilibrium. Only one policy $C_P \equiv (I_1^P, I_2^P)$ is accepted by all agents. In this case, the acceptance of a policy by an agent does not reveal his type. Let A denote "accident". Then, the probability of having an accident for the pooled population is

$$P(A) = P(A|L)P(L) + P(A|H)P(H) = \pi_L \lambda + \pi_H (1 - \lambda),$$

where λ is the population share of the low-risk type. That is, the probability of having accident for the whole population is $\pi_P \equiv \lambda \pi_L + (1 - \lambda) \pi_H$. As explained in Sect. 2.2, zero-profit means $q = \pi_P$ in this case. The corresponding zero-profit line is labeled π_P in Fig. 5. We call this zero-profit line the π_P-line because $q = \pi_P$ on this line. Similarly, the π_L-line and π_H-line are for $q = \pi_L$ and $q = \pi_H$, respectively. We define the u_L-curve and u_H-curve as the indifference curves for type L and type H, respectively.

Zero profit in equilibrium means that a pooling equilibrium has to be on the π_P-line. However, for any point E on the π_P-line, a company can offer another policy that is better for type L and is worse for type H. That policy can be the point C in Fig. 5. Since only type L picks point C and this point is below the π_L-line, the opportunist company makes a positive profit. This means that point E cannot possibly be a NE. There is thus no pooling equilibrium.

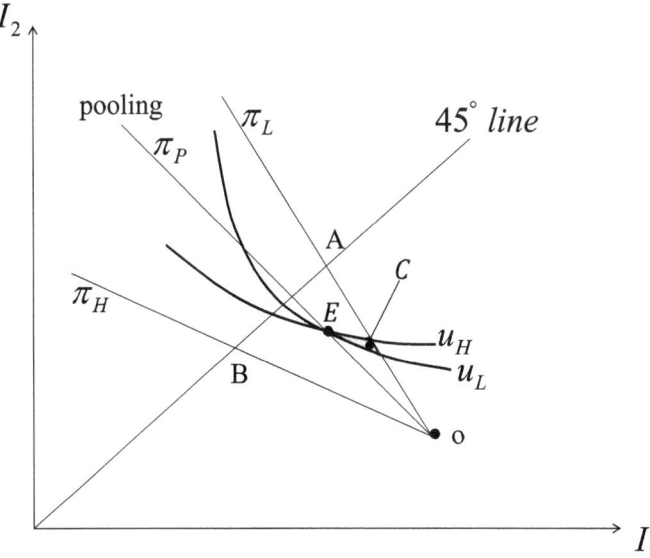

Fig. 5 The pooling equilibrium

The Separating Equilibrium

Consider now a separating equilibrium, where two policies (C_H, C_L) are accepted, one by type L and the other by type H. These two policies can separate the two types since, by accepting a policy, an agent's type is revealed to the company. The pair (C_H, C_L) of policies on the left of Fig. 6 can separate the two types. But they are not an equilibrium since a company can easily offer a policy to attract only the low-risk individuals. A policy in the green area can do just that. Thus, a necessary condition for (C_H, C_L) to be an equilibrium pair of policies is for C_L to be on the π_L-line and at the intersection point of the u_H-curve and the π_L-line, as shown on the right of Fig. 6.

However, the pair (C_H, C_L) is still not in equilibrium since the company will be losing money. Thus, C_H must be on the π_H-line. Furthermore, if policy C_H is not on the 45°-line, the high-risk individual's indifference curve will cut the π_H-line and, if so, another company can offer a policy to beat C_H. Thus, C_H must be on the 45°-line. Hence, the only possible equilibrium is the pair (C_H^*, C_L^*) in Fig. 7.

However, this equilibrium may be beaten by a policy in the shaded area under certain conditions. If so, no equilibrium exists. If λ is small enough such that the pooling zero-profit line does not intersect the u_L-line, then a policy such as D in Fig. 7 cannot exist and thus the separating equilibrium may survive.

More conditions are necessary to rule out other profitable opportunities. In Fig. 8, the pair $\{C_H, C_L\}$ is a separating pair. This pair itself cannot be an equilibrium, but it may be profitable. If this pair is profitable, then the separating pair $\{C_H^*, C_L^*\}$ cannot be an equilibrium, since type H prefers C_H, type L prefers C_L, and firms prefer $\{C_H, C_L\}$ to $\{C_H^*, C_L^*\}$. In fact, we can show that the pair $\{C_H, C_L\}$ is profitable under certain conditions. That is, in order for $\{C_H^*, C_L^*\}$ to survive as an equilibrium, additional conditions are necessary to rule out profitable separating pairs such as $\{C_H, C_L\}$. Finally, according to Rothschild and Stiglitz (1976), if the two situations in Figs. 7 and 8 do not happen, the pair (C_H^*, C_L^*) survives and it is the only separating equilibrium.

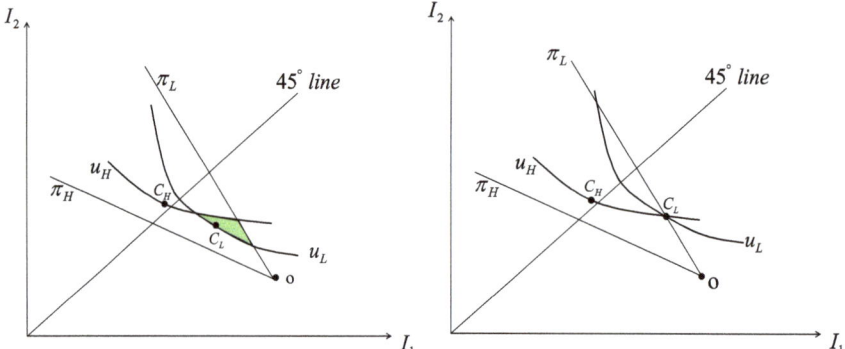

Fig. 6 The separating contracts

Fig. 7 The separating equilibrium

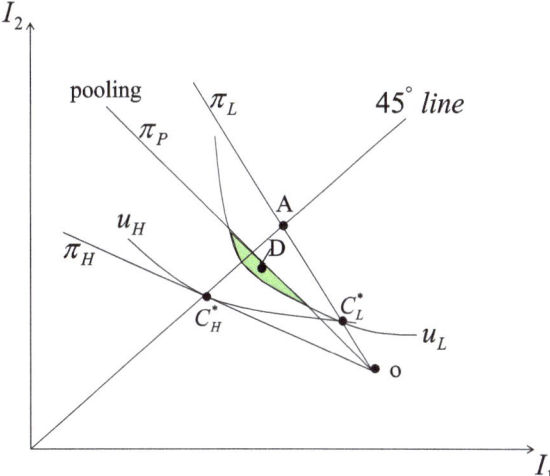

Fig. 8 More profitable opportunities

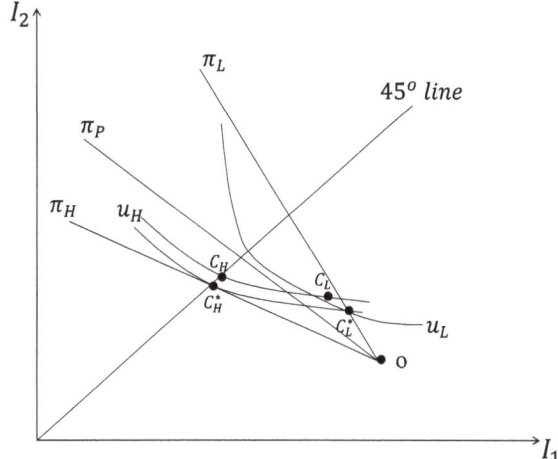

Discussion

The complete information solution is the pair (C_H^*, A) of policies in Fig. 7.

We see that it is type L who suffers from asymmetric information. It is thus in type L's self-interest to reveal themselves to the companies. It is also in the insurance company's self-interest to attract type L and leave type H to other companies. One issue is how type L agents can credibly reveal themselves to the companies.

To deal with the non-existence problem, Riley (1979) proposes a notion of a <u>reactive equilibrium</u>, which takes into account possible reactions from opponents to any deviation from the equilibrium. The idea is to allow only the introduction of those policies that will yield profit after all money-losing policies are removed.

This makes (C_H^*, C_L^*) in Fig. 7 an equilibrium without an additional condition. When a company proposes an alternative policy D against the pair (C_H^*, C_L^*), it is indeed myopic since it should realize that other companies will propose the C policy in Fig. 5 that will render this pooling policy nonviable. Hence, policies such as D will not be proposed, so that (C_H^*, C_L^*) can survive as an equilibrium. By the same argument, any pooling policy on the π_P-line is an equilibrium. When a company proposes an alternative policy C against a pooling policy in Fig. 5 by attracting only type L, it should realize that other companies will have to abandon their offers so that all agents will go to this opportunistic company and in the end this company loses. Hence, this company should not offer the alternative policy C. If so, the pooling policy survives as an equilibrium.

Similar to Riley's reactive equilibrium, Wilson (1997) proposed the concept of an anticipatory equilibrium, in which a policy offer will not be made if a loss is expected after the new offer effectively forces the existing offers to be withdrawn. Each company is assumed to be able to anticipate the implications of a new offer. Offers that become unprofitable as a result of a new offer are simply withdrawn. A company contemplating his new offer must ask whether it will remain profitable in the face of such a withdrawal. If not, the new offer will not be made. By this behavior, Wilson shows the survival of any pooling equilibrium and the separating equilibrium.

Various equilibria have been based on ad hoc nonequilibrium expectations. Hellwig (1987) modeled in a more precise manner the communication of information between insurance companies and the insured and showed the high sensitivity of the results to the extensive form of the game used for modeling.

2.4 Extensions

How about a two-period repeated insurance model? In the separating equilibrium, the insurance company can figure out the types of the agents in the first period; it can thus take advantage of this information in the second period. If so, the high-risk type will not pick C_H^* in the first period. How will the agents behave in a two-period game? If the high-risk agent picks C_L^* in the first period, he could be viewed as a low risk agent and hence is offered A is the second period. To prevent this, the company is to make a better offer to the high-risk agent. One possible solution is a separating solution as shown in Fig. 9. In the first period, the policies are (C_{1H}^*, C_{1L}^*) and in the second period the policies are (C_{2H}^*, C_{2L}^*). In the second period, there is complete information. See Hosios and Peters (1989), who find this solution to the problem.

Is it possible that the complete information outcome (C_H^*, A) in Fig. 7 is achievable in an infinitely repeated environment under adverse selection? The key in a repeated environment is that hidden information can eventually be recovered by

2 The Rothschild-Stiglitz Model

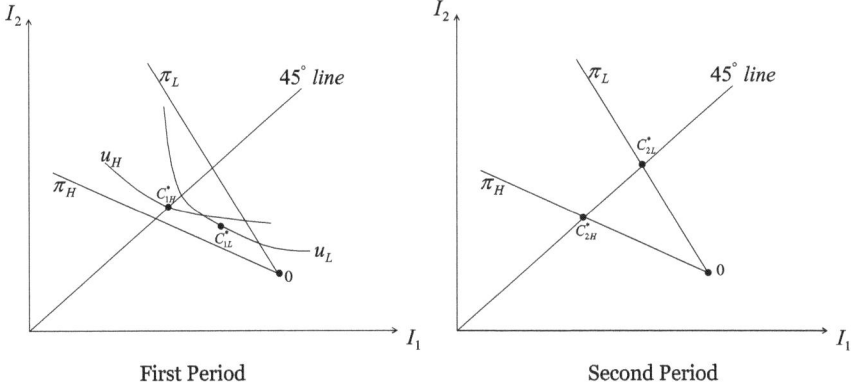

Fig. 9 A two-period insurance model

observing the choices that economic agents make over time. An agent may be able to hide his private information in finite periods, but he may not be able or it may be too costly to do in an infinitely repeated situation. Thus, intuitively, when the discount rate of time preferences is large, the model is approximately a one-period or two-period model and we know that inefficiency will exist; however, when the discount rate is small, it is possible to achieve efficiency. See Fudenberg and Maskin (1986).

If insurance is run by a government that maximizes social welfare as opposed to profit, what should the government do? Since the government is a monopoly, this problem is addressed in the next chapter.

If both the government and commercial companies are allowed to provide insurance, what will be the outcome? The outcome could be that some agents choose a government plan, some choose a commercial plan, and some choose both plans. What are the welfare implications?

What would be the outcome if we extend our two-type model to three types? What is the separating equilibrium? Is it possible that two types may have a pooling policy and the other type chooses a separating policy? Is the separating equilibrium more likely to survive with three types than with two types? This is an exercise in the Problem Set.

3 Job Market Without Signals

This section is from MWG (1995, Chap. 13). A nice feature of this section is that it derives both BEs and SPNEs from the same model, allowing us to compare these two types of solutions. We will also do this to Spence's model in Sects. 4 and 5. In this section, there is no signal, while in Spence's model, there is a signal.

3.1 The Model

By relaxing some of the restrictive assumptions in Akerlof model (e.g., identical consumers), we yields further insights into market processes with asymmetric information.

Consider a labor market in which there are many potential firms that can hire workers. Each firm produces the same output using an identical CRS technology in which labor is the only input. The firms maximize their profits and are perfectly competitive.

There are two types of workers: low productivity θ_l and high productivity θ_h. Assume $\theta_h > \theta_l$. Here θ_i is the number of units of output a worker can produce per hour. The proportion of high-productivity workers is λ_h, and the proportion of low-productivity workers is λ_l. The reservation values of the two workers are respectively r_l and r_h, with $r_h \geq r_l \geq 0$ (they are minimum acceptable wage rates).

Since θ_i is not observable, the wage rate cannot be dependent on it. Hence, there is only one wage rate for all (no signal in this model to separate the workers). There are only three possible situations: (1) everyone is employed when the market wage rate is high; (2) only low-productivity workers are employed when the market wage rate is medium; and (3) no one is employed when the market wage rate is low. Hence, if w is the wage rate, then the expected productivity among employed workers is

$$\bar{\theta}_e = \begin{cases} \lambda_h \theta_h + \lambda_l \theta_l & \text{if } w \geq r_h; \\ \theta_l & \text{if } r_l \leq w < r_h; \\ 0 & \text{if } w < r_l. \end{cases} \quad (4)$$

We denote $\bar{\theta}_e$ as the mean productivity among employed workers, and $\bar{\theta} \equiv \lambda_h \theta_h + \lambda_l \theta_l$ as the unconditional mean of productivity among all workers. Similarly, denote $\bar{r} \equiv \lambda_h r_h + \lambda_l r_l$.

3.2 Bayesian Equilibrium

Firms form expectations on the average productivity of workers who accept their jobs. Let μ be the expected productivity of employed workers. As a standard assumption, assume that all firms have the same expectation. Market competition leads to zero profit, which implies that the firms will offer wage $w = \mu$.[2] In Bayesian equilibrium (BE), we require the firms to have the correct expectation, i.e., $\mu = \bar{\theta}_e$. Combining conditions in (4) and $w = \bar{\theta}_e$, we find three cases of BE[3]:

---

[2]There are no signals to help firms form their expectations.
[3]The unique features of BEs are (1) firms take actions based on their beliefs; (2) they have the same beliefs (a standard assumption); and (3) there are many BEs. With the same beliefs, the firms take the same actions, by which an equilibrium is easy to form. In contrast, without the restriction of common beliefs, SPNEs typically to have fewer equilibria.

3 Job Market Without Signals

$$w^* = \bar{\theta}, \quad \text{if} \quad \bar{\theta} \geq r_h;$$
$$w^* = \theta_l, \quad \text{if} \quad r_l \leq \theta_l < r_h; \qquad (5)$$
$$\text{any } w^* \in [0, r_l], \quad \text{for any} \quad r_l, r_h \geq 0.$$

In the first case, condition $\bar{\theta} \geq r_h$ ensures that all workers will accept the offer $w^* = \bar{\theta}$. Firms make zero profits in this case. Firms expect the productivity to be $\mu = \bar{\theta}$ when they make the offer, which turns out to be correct. It is hence a BE.

In the second case, since $w^* < r_h$, the high-productivity workers will not accept $w^* = \theta_l$. Since $w^* \geq r_l$, the low-productivity workers will accept $w^* = \theta_l$. The firms are making correct expectations by offering $w^* = \theta_l$. Also, the firms have zero profit. Hence, it is a BE.

In the last case, if $w^* < r_l$, no worker will accept the offer from the firms and hence the firms make zero profit. If so, the firms have no way to find out whether its expectation is correct. Hence, it is a BE. If $w^* = r_l$, a low-productivity worker may or may not accept the offer from the firms and the firms still make zero profit. If some low-productivity workers accept the offer, the firms' expectation is correct; if no workers accept the offer, the firms have no way to know if its expectation is correct. Either way, the solution is a BE.

An efficient solution is the one in which a worker with $r_i \leq \theta_i$ should work and a worker with $r_i > \theta_i$ should not work. For example, under complete information, an efficient solution is:

$$w_l = \theta_l, \quad w_h = \theta_h. \qquad (6)$$

where w_l is for a low-productivity worker and w_h is for a high-productivity worker. This solution requires complete information to implement. A BE may not be efficient. For example, the market may fail completely. When $\theta_i > r_i$, one solution from (5) is $w^* = 0$ with no one working. This solution is inefficient since in an efficient solution every worker should work.

For a second example, if $r_l = r_h \equiv r$ and $\bar{\theta} \geq r$, (5) has the solution $w^* = \bar{\theta}$. That is, if $\bar{\theta} \geq r$, there is a BE in which everyone works. But, if $\theta_l < r < \theta_h$, the low-ability workers should not work. Hence, the solution may not be efficient. For this BE, social welfare is $\bar{\theta} - r$. A social welfare-improvement solution is in (6), by which only high-productivity workers will work, even though this solution needs complete information to implement. Its social welfare is $\lambda_h(\theta_h - r)$, which is higher than $\bar{\theta} - r$ when $\theta_l < r$.

A third example is when $r_l \leq \theta_l < r_h < \theta_h$ but $\bar{\theta} < r_h$. In this case, for the BE with $w^* = \theta_l$, only low-ability workers work. Social welfare is $\lambda_l(\theta_l - r_l)$. This solution is inefficient. The solution in (6) under complete information again yields higher social welfare $\bar{\theta} - \bar{r}$. Hence, the BE is inefficient.

3.3 Subgame Perfect Nash Equilibrium

In contrast to BE, in a NE, firms are not constrained by common beliefs and can hence compete by choosing different choices. For a NE, consider a two-stage game:

Stage 1. Firms simultaneously announce their wage offers to maximize profits;
Stage 2. Workers decide whether to work for a firm and, if so, which one (if indifferent, choose each one with probability 0.5).

We consider subgame perfect Nash equilibria (SPNE) from this two-stage game.

In a SPNE, we must have zero profit. Hence, a SPNE must be a solution in (5). For the first case with $w^* = \bar{\theta}$ when $\bar{\theta} \geq r_h$, no firm can attract a worker by offering a lower wage, and no firm can do better by offering a higher wage $w > \bar{\theta}$ either while other firms keep at $w^* = \bar{\theta}$. Hence, $w^* = \bar{\theta}$ is a SPNE when $\bar{\theta} \geq r_h$.

For the second case in (5) with $w^* = \theta_l$ when $r_l \leq \theta_l < r_h$. First, we need $\bar{\theta} \leq r_h$, otherwise a firm can do better by offering $w = \bar{\theta} - \varepsilon$ with $\bar{\theta} - \varepsilon > r_h$ for some $\varepsilon > 0$ while other firms keep at $w^* = \theta_l$. This alternative wage will attract all workers and the firm will make a profit since the average productivity is $\bar{\theta}$ and it is higher than the wage. Hence, we need to impose $\bar{\theta} \leq r_h$. Second, since $\bar{\theta} \leq r_h$, a firm will not try to attract the high-productivity workers by offering a higher wage since such a higher wage must be r_h or higher but the average productivity remains at $\bar{\theta}$. Third, if $\theta_l > r_l$, then some firms can do better by deviating to $w = r_l + \varepsilon$ for some $\varepsilon > 0$ such that $r_l + \varepsilon < \theta_l$. To prevent this, we need $\theta_l = r_l$. That is, if $\theta_l = r_l$ and $\bar{\theta} \leq r_h$, we have a SPNE: $w^* = \theta_l$.

For the last case in (5), if $\theta_l > r_l$, obviously, $w^* = 0$ cannot be a SPNE. If all other firms choose $w^* = 0$, one firm can choose $w^* = \varepsilon$ with $r_l < \varepsilon < \theta_l$ to attract workers and make a profit. Hence, we need to assume $\theta_l = r_l$. Consider any $w^* \in [0, r_l)$. Obviously no worker will accept this wage. Also, since $\theta_l = r_l$ and $\bar{\theta} \leq r_h$, no firm can do better by deviating to a higher wage while other firms keep at w^*. Hence, in this case, $w^* \in [0, r_l)$ is a SPNE. Combining with the above, if $r_l = \theta_l$ and $\bar{\theta} \leq r_h$, any $w^* \in [0, r_l]$ is a SPNE.

In summary, we have the following SPNEs:

$$\begin{aligned} w^* &= \bar{\theta}, & \text{if } \bar{\theta} \geq r_h; \\ \text{any } w^* &\in [0, r_l], & \text{if } r_l = \theta_l \text{ and } \bar{\theta} \leq r_h. \end{aligned} \quad (7)$$

The solutions in (7) indicate that a SPNE wage is always the highest BE wage under the same conditions.

3.4 Constrained Pareto Optimum

Suppose that an equally uninformed government tries to maximize social welfare by central planning. Assume that the government can only observe employment and

3 Job Market Without Signals

unemployment. Due to this, the government's wage package is (w_e, w_u), where w_e is the wage rate for the employed and w_u is the wage rate for the unemployed. With this, the payoff of a worker under employment is w_e and the payoff under unemployment is $w_u + r_i$.

There are only three possible situations: (1) everyone is employed; (2) only low-productivity workers are employed; and (3) no one is employed. We can easily find the conditions for these three cases and their corresponding social welfare. Taking into account budget balance, social welfare for the three cases is

$$SW = \begin{cases} \bar{\theta} - \bar{r} & \text{if } w_e \geq r_h + w_u \text{ and } w_e = \bar{\theta}; \\ \lambda_l(\theta_l - r_l) & \text{if } r_h + w_u > w_e \geq r_l + w_u \text{ and } \lambda_l w_e + \lambda_h w_u = \lambda_l \theta_l; \\ 0 & \text{if } w_e < r_l + w_u \text{ and } w_u = 0. \end{cases}$$

The government is subject to budget balance. In the above specification, the first condition is a worker's incentive condition to accept the offer and the second condition is the budget condition.

What are the social optimal solutions? Obviously, if conditions allow, the first case when all workers are employed is the best. Under conditions, all three cases can be social optimal. We can easily find the social optimal solution:

$$(w_e^*, w_u^*) = \begin{cases} (\bar{\theta}, w_u), & \text{where } w_u + r_h \leq \bar{\theta} & \text{if } \bar{\theta} \geq r_h; \\ \left(\theta_l - \frac{\lambda_h w_u}{\lambda_l}, w_u\right), & \text{where } w_u \leq \lambda_l(\theta_l - r_l) & \text{if } \bar{\theta} < r_h \text{ and } \theta_l \geq r_l; \\ (w_e, 0), & \text{where } w_e < r_l & \text{if } \bar{\theta} < r_h \text{ and } \theta_l < r_l. \end{cases} \quad (8)$$

In the first case, we must have $w_u = \bar{\theta}$ by the budget condition. As long as $\bar{\theta} \geq r_h$, we can find a pair of (w_e, w_u) such that $w_u + r_h \leq w_e$ by which all workers will accept.

In the second case, with $\bar{\theta} < r_h$, a high-productivity worker needs more than $\bar{\theta}$ to accept the job, but the budget condition does not allow this. Hence, the government can at most attract the low-productivity workers. We need two conditions to attract the low-productivity workers and satisfy the budget constraint:

$$w_e \geq r_l + w_u, \quad \lambda_h w_u + \lambda_l w_e = \lambda_l \theta_l.$$

They are equivalent to

$$w_u \leq \lambda_l(\theta_l - r_l), \quad w_e = \theta_l - \frac{\lambda_h w_u}{\lambda_l}.$$

The condition to guarantee such a pair is $\theta_l \geq r_l$.

In the last case, the budget condition does not allow the government to hire any worker. Hence, the government simply provides a wage package that no worker will accept.

The solutions in (8) indicate that a constrained Pareto optimal wage w_e^* is always the highest BE wage.

4 The Spence Model: Job Market Signalling

Given problems with incomplete information, we might expect mechanisms to develop in the marketplace to help firms distinguish between various types of workers. In fact, both firms and some groups of individuals have incentives to do so. The mechanism that we examine in this section is called signalling and it is based on Spence (1973). The basic idea is that high-quality workers may take measures to distinguish themselves from low-quality workers, while low-quality workers try to hide their identities. In this section, we present a model in which an instrument is available to transmit credibly private information from the informed to the uninformed via the use of publicly observable signals.

Suppose that employers are recruiting workers. The employers assess the likely productivity of each applicant and offer each a wage based on their assessment. The problem is: how can employers know the quality of workers? There are some observable characteristics of each worker, say the education level. Able people can get additional education at a relatively low cost, which can serve as a signal. Although low-quality workers can also try to get additional education, the education cost may be relatively high for them and such a signal may be too expensive. Thus, the presence of differences in these signalling costs may make the signal of education credible. In other words, signalling activities may serve to generate information for employers as an endogenous market process.

We use the same model as in Spence (1973), except that we replace some specific values in his paper with arbitrary parameters. The model involves the following three steps:

1. Workers decide on how much education they wish to invest in and they pay the cost.
2. Firms cannot observe a worker's productivity, but they can observe the signal. They form probabilistic beliefs about the relationship between the observed signal and unobserved productivity. Assume that all firms have the same beliefs.
3. With their beliefs, the firms make wage offers, and then workers respond with certain signals.

For a tractable solution, we need some simplifying assumptions:

4 The Spence Model: Job Market Signalling

(1) Firms and workers are all risk neutral.[4] The labor market is perfectly competitive, implying zero profit for firms.[5]
(2) Workers are of two possible types[6]:

| Type L: | productivity $= \theta_L$, | productivity proportion $= 1 - \lambda$, |
| Type H: | productivity $= \theta_H$, | productivity proportion $= \lambda$, |

where $\lambda \in (0, 1)$. Thus, λ is the probability of a worker being of high quality. The firms know this population distribution. Here, the productivity for each worker is assumed to be fixed and the education level has no effect on productivity.[7]

(3) The costs for education level e are respectively $c_L e$ and $c_H e$ for the two types, where $c_L > c_H > 0$. Here, the cost and the marginal cost of education are lower for high-quality workers.
(4) The utility functions are

$$u_L(w, e) = w - c_L e, \quad u_H(w, e) = w - c_H e.$$

We study a two-stage game:

Step 1. Given a belief system, firms simultaneously offer a wage schedule $w(e)$.
Step 2. Given the offers, workers decide to choose an education level.

The model can be illustrated in Fig. 10. Nature decides the worker's type first. The worker knows his type and then decides on a choice of e. The firms then decide on their offers based on the observed choices of e but without the knowledge of each worker's type.

The model consists of a strategy $w(e)$ and a belief system $\mu(H|e) \in [0, 1]$, where $\mu(H|e)$ is the firms' common probability assessment that a worker is of high quality after observing e. A pair $(w(\cdot), \mu(\cdot))$ is a <u>Bayesian equilibrium (BE)</u> if

(i) For each e, the firms' wage offers constitute a Nash equilibrium.
(ii) The workers' strategy $e(\theta)$ is optimal given the firm's strategies.
(iii) EP consistency: for each e^* chosen by some workers in equilibrium, consistency between the belief and the actual probability is required:

---

[4]More precisely, they are risk neutral in income, but their cost functions can be arbitrary convex cost functions. We will use linear cost functions for simplicity, but the same results hold with general cost functions.
[5]We need to have at least two firms to have zero profit in equilibrium.
[6]Here, we can either assume that there is only one worker, with probability λ being of high quality and probability $1 - \lambda$ being of low quality, or assume that there are many workers with population proportions λ and $1 - \lambda$. These two settings are equivalent to our model. We use both settings in the following analysis depending on convenience.
[7]Spence (1974) and Kreps (1990) allow education to affect productivity.

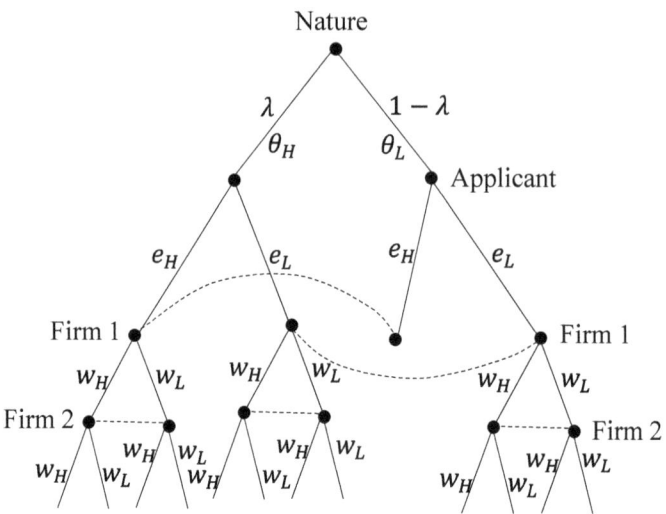

Fig. 10 A signalling model

$$\mu(H|e^*) = \text{actual probability } P(H|e^*) \text{ that a worker with } e^* \text{ is of high quality.} \quad (9)$$

Hence, a BE is formed in four steps:

> Firms form beliefs $\mu(H|e)$ → Firms make wage offers $w(e)$
> → Workers make choices $e(\theta)$ → Consistency condition (9).

We now go through the four steps. First, given beliefs $\mu(H|e)$, the firms' wage offer is determined by a Nash equilibrium. Since the firms are competing in prices (prices of labor), this Nash equilibrium is a Bertrand equilibrium. As shown in Proposition 6.1, if there are at least two firms, this Bertrand equilibrium implies zero profit for all firms. Zero profit means that the firms' wage offers must equal the expected productivity of the worker:

$$w(e) = \mu(H|e)\theta_H + [1 - \mu(H|e)]\theta_L. \quad (10)$$

Conversely, given $w(e) \in [\theta_L, \theta_H]$, we can recover the belief system $\mu(H|e)$ from (10), which is $\mu(H|e) = \frac{w(e) - \theta_L}{\theta_H - \theta_L} \in [0, 1]$. Hence, we can ignore the belief system from now on, since the wage function satisfying $w(e) \in [\theta_L, \theta_H]$ always matches uniquely with an underlying belief system.

Fig. 11 The worker's problem

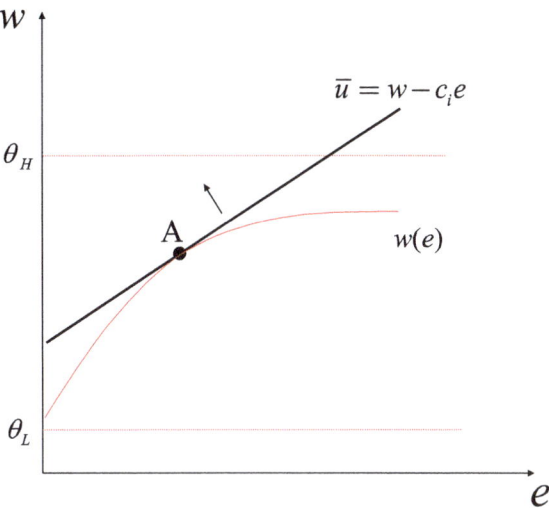

Second, given a wage function $w(e)$, a worker's problem is

$$\max_{w,e} u_i(w, e)$$
$$\text{s.t.} \quad w = w(e).$$

Given type i and reservation utility $\bar{u}$, the indifference equation is a line defined by $\bar{u} = w - c_i e$. We can illustrate the problem in Fig. 11. In the figure, an arbitrary wage curve between the two zero-profit lines is given, and the worker tries to push the indifference curve up and to the left until it touches the wage curve by only one point. That point A determines the optimal choice of e_i and a corresponding wage $w(e_i)$ from firms. Each such wage curve can be justified by a belief system through (10).

Third, consistency condition (9) is required. Since there are only two types of workers, there are only two possible equilibrium outcomes: either every worker chooses the same education level—a <u>pooling equilibrium</u>, or one type chooses one education level and the other type chooses another—a <u>separating equilibrium</u>. In a pooling equilibrium, only one education level e^* is chosen by some workers, and in fact by all workers. Consistency condition (9) requires $\mu(H|e^*) = \lambda$, which in turn by (10) means

$$W(e^*) = \bar{\theta} \tag{11}$$

where $\bar{\theta} \equiv \lambda \theta_H + (1 - \lambda)\theta_L$. In a separating equilibrium, there are two education levels separately chosen by the two types of workers. Suppose that the low type chooses e_L^* and the high type chooses e_H^*. Then, consistency condition (9) requires $\mu(H|e_L^*) = 0$ and $\mu(H|e_H^*) = 1$, which in turn by (10) means

$$w(e_L^*) = \theta_L, \quad w(e_H^*) = \theta_H. \tag{12}$$

Since we have only two possible types of equilibrium, consistency means either (11) or (12).

Education here serves as a signal of the unobservable worker's productivity. The education signal may or may not effectively separate the two groups. In a pooling equilibrium, the education signal does not distinguish the two types and hence everyone is paid the same wage. In a separating equilibrium, firms can correctly use the education signal to distinguish the workers and pay them different wages.

4.1 The Complete Information Solution

When firms can observe each worker's productivity, a Bertrand equilibrium implies

$$w_L^{**} = \theta_L, \quad w_H^{**} = \theta_H.$$

where the double stars indicate choices in this equilibrium. That is, each firm will pay workers for their productivity. Since education does not enhance productivity, the optimal levels of education are

$$e_L^{**} = e_H^{**} = 0.$$

4.2 The No-Signalling Solution

Suppose that education is banned so that

$$\bar{e}_L = \bar{e}_H = 0,$$

where the overbar indicates choices in this equilibrium. The firms cannot distinguish between workers and they have to pay everyone the same wage. Then, the zero-profit condition implies

$$\bar{w} = \bar{\theta}.$$

4.3 Separating Equilibria

We now consider the situation of incomplete information, in which workers know their own types but employers can only infer the types from education signals. There are two types of equilibria: separating equilibria and pooling equilibria. When we have a separating equilibrium, the signal can effectively separate the two

Fig. 12 Separating equilibrium

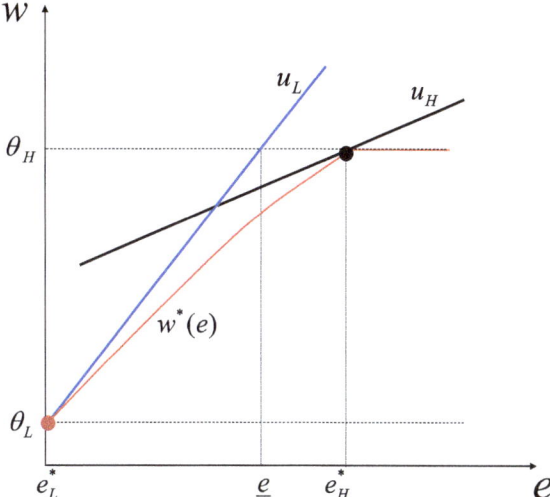

types; when we have a pooling equilibrium, the signal cannot separate the two types.

For each type i, the slope of the indifference curve $\bar{u} = w - c_i e$ is the marginal cost c_i. Since $c_L > c_H$, type L's indifference curve is steeper than is type H's. For convenience, we call these two difference curves the u_L-line and the u_H-line, as shown in Fig. 12.

Let $e^*(\theta)$ be the worker's equilibrium education choice and $w^*(e)$ be the firms' equilibrium wage offer. In a separating equilibrium, the wage offers must satisfy (12). Since having an education will incur costs but will not change the pay $w^* = \theta_L$ for the low type, the low type will choose zero education:

$$e_L^* = 0. \tag{13}$$

Thus, one possible separating equilibrium is where a low-ability worker chooses $e_L^* = 0$ and receives $w^*(e_L^*) = \theta_L$, and a high-ability worker chooses e_H^* and receives $w^*(e_H^*) = \theta_H$, where e_H^* will be determined later.

We now need to find a belief system that supports this equilibrium. By (10), this is equivalent to finding a wage function. We have drawn an arbitrary wage curve $w^*(\cdot)$ in Fig. 12. This wage schedule supports the choices e_H^* and $e_L^* = 0$ as optimal choices of the high type and low type, respectively, and the corresponding wages are θ_H and θ_L, respectively. Since $w^*(e) \in [\theta_L, \theta_H]$, by (10), it can be supported by a belief system. Further, the consistency condition (12) for a separating equilibrium is also satisfied. Therefore, the wage schedule $w^*(\cdot)$ in Fig. 12 and the belief system $\mu^*(\cdot)$ implied by (10) constitute a BE. In this equilibrium, the two types of workers are distinguishable by their different choices of education.

Note that, with wage offers in (12), the firms must have beliefs $\mu^*(H|0) = 0$ and $\mu^*(H|e_H^*) = 1$. These beliefs are indeed consistent with the choices made by the

Fig. 13 A separating equilibrium

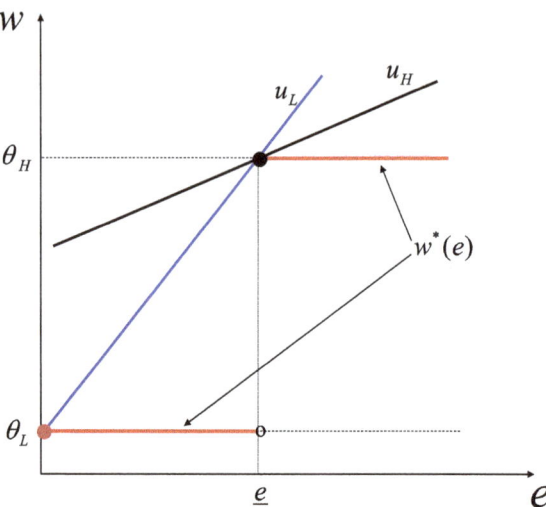

individuals in equilibrium. However, for off-equilibrium education levels, there is no consistency requirement in a BE and the firms' beliefs can be quite arbitrary. Since there is so much freedom to choose beliefs, many wage schedules can arise that support different education choices. For example, Fig. 13 shows a simple wage schedule that supports a separating equilibrium. The corresponding belief system is as follows: the firms believe that a worker with education $e < \underline{e}$ is of low productivity and a worker with education $e \geq \underline{e}$ is of high productivity, where $\underline{e}$ is defined by the intersection of the u_L-line with the θ_H-line.

Note also that, in a BE, each firm's wage offer (strategy) is derived from its beliefs in a Nash equilibrium and all firms are assumed to have the same beliefs. Hence, all firms make the same wage offer; alternative offers from other firms are not allowed. This is different from the Rothschild–Stiglitz model in which different firms can make different wage offers; for example, in Fig. 7, a different firm can make an alternative offer D.

Furthermore, workers may have many equilibrium choices. Type θ_L will always choose $e_L^* = 0$ as shown in (13), but type θ_H has many alternatives. For example, type θ_H may choose $\bar{e}$ as shown in Fig. 14. In fact, any $e_H \in [\underline{e}, \bar{e}]$ can be supported in a separating equilibrium. But, any $e_H \notin [\underline{e}, \bar{e}]$ cannot be an education level for the high type in a separating equilibrium. If e_H is below $\underline{e}$, the low type will choose e_H to pretend to be the high type; if e_H is above $\bar{e}$, the high type will choose $e = 0$, even though she may be viewed as the low type.

We have several observations on the separating equilibria. First, the signal is informative and correct. It separates the two types.

Second, the equilibria are not unique. There are a continuum number of separating equilibria.

Fig. 14 Separating equilibria

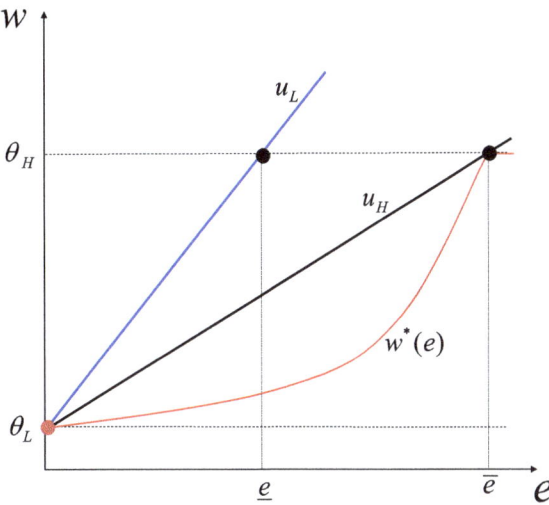

Third, the equilibria can be ranked by the Pareto criterion. In any equilibrium, firms earn zero profit and firms are always indifferent between different equilibria. The low type always chooses $e_L^* = 0$ and is paid θ_L in a separating equilibrium. Hence, the low type is also indifferent between different equilibria. Only the high type has different utility values in different equilibria. Among all the separating equilibria, the equilibrium in which $e_H^* = \underline{e}$ Pareto-dominates all others since it has the lowest education cost for the high type. This is the most efficient separating equilibrium.

Fourth, the solutions do not depend on the population share λ.

Fifth, comparing with the complete information solution, a separating equilibrium is inefficient even though it also has complete information in equilibrium. The complete information equilibrium Pareto-dominates all incomplete information separating equilibria.

Sixth, the no-signalling equilibrium can Pareto-dominate some of the separating equilibria. Without signalling, everyone gets $w^* = \bar{\theta}$ and $e_L^* = e_H^* = 0$. As shown in Fig. 15, type L will be better off without signalling. Type H can also be better off without signalling iff $\bar{\theta} > \hat{\theta}_H$, where $\hat{\theta}_H$ is defined in Fig. 15. The cost of education can make type H worse off with signalling.

Finally, education results in negative externalities, since there is a cost to education but it has no benefit on productivity. Type H may benefit from it by separating themselves from type L. That is, education is unproductive, but people still invest in it.

Example 10.1 Let $\theta_L = 1$ and $\theta_H = 2$ and the cost functions be

$$c_L(e) = e, \quad c_H(e) = e/2.$$

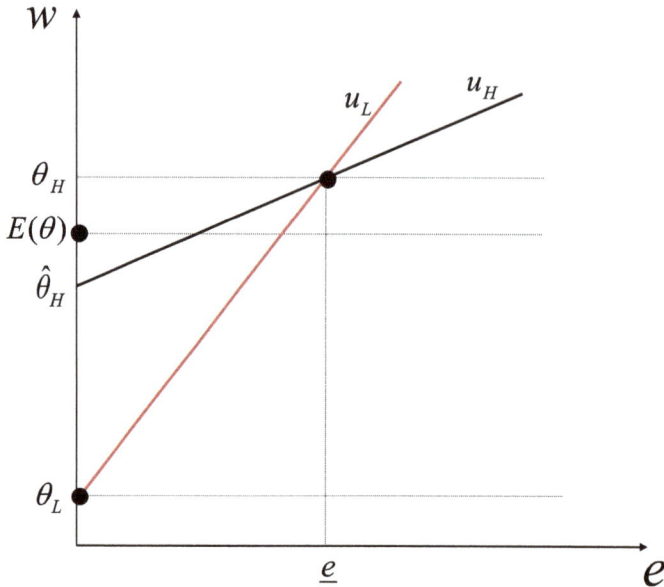

Fig. 15 No-signalling equilibrium

That is, it costs half for type H to obtain the same education level as type L does. Given some constant e_0, suppose that the employers believe that only workers with the education level e_0 or more are of high productivity. With this belief, the wage offer will be

$$w(e) = \begin{cases} 1, & \text{if } e < e_0, \\ 2, & \text{if } e \geq e_0. \end{cases}$$

With this offer, a worker will choose an education level of either 0 or e_0. Let us find the conditions under which type L chooses $e = 0$ and type H chooses $e = e_0$, implying an separating equilibrium. Type L chooses $e = 0$ iff

$$w(0) - c_L(0) \geq w(e_0) - c_L(e_0),$$

which is equivalent to $e_0 \geq 1$. With $e_0 \geq 1$, type L finds that the cost of education is too high so that they voluntarily separate themselves from type H by choosing no education. Type H chooses $e = e_0$ iff

$$w(0) - c_H(0) \leq w(e_0) - c_H(e_0),$$

which is equivalent to $e_0 \leq 2$. The employers' belief is indeed correct at $e = 0$ and at $e = e_0$. Therefore, given the employers' beliefs above, a separating equilibrium holds as long as $1 \leq e_0 \leq 2$, as shown in Fig. 16. ∎

4 The Spence Model: Job Market Signalling   349

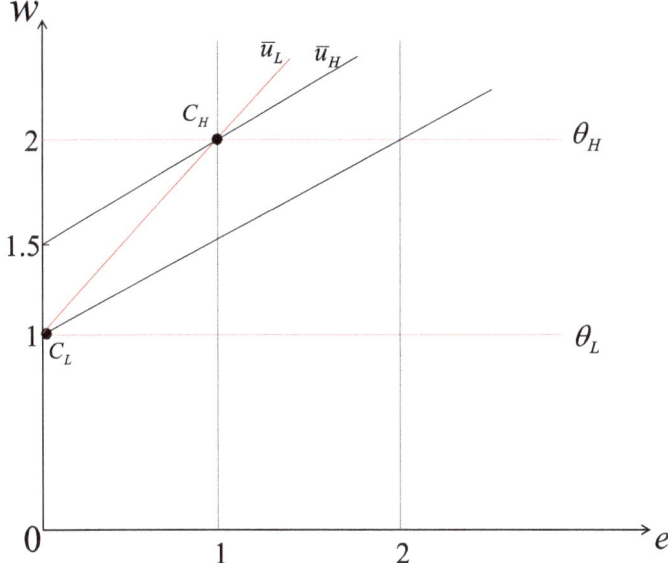

Fig. 16 A separating signalling equilibrium

4.4 Pooling Equilibria

In a pooling equilibrium, both types choose the same education level: $e_H^* = e_L^* = e^*$. By (11), consistency implies $w^* = \bar{\theta}$ for all.

What should be the education level e^* in an pooling equilibrium? It can be easily seen from Fig. 17 that any education level $e \in [0, \hat{e}]$ can be supported by a wage schedule leading to a pooling equilibrium, where $\hat{e}$ is defined in Fig. 17. Any education level $e > \hat{e}$ cannot be supported since the low type would rather choose $e = 0$.

We have several observations on the pooling equilibria. First, all workers send the same signal so that the signal loses its role.

Second, there is a continuum number of pooling equilibria.

Third, the equilibria can be Pareto ranked. Higher values of e^* impose costs on both types. The most Pareto-efficient pooling equilibrium is where $e^* = 0$.

Fourth, the equilibria depend on λ. The wage rate will be higher if λ is higher.

Fifth, the complete information solution does not Pareto-dominate the pooling equilibria, and vice versa. Type L is better off under pooling while type H is worse off.

Sixth, the no-signalling solution Pareto-dominates any pooling equilibrium.

Finally, once again, there is a divergence between the social and private benefits of education. In this case, education does not even convey information.

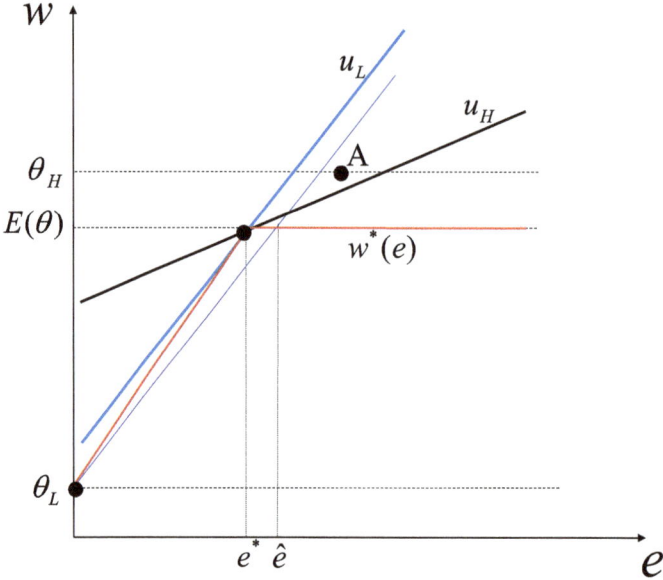

Fig. 17 Pooling equilibrium

Example 10.2 For the model in Example 10.1, suppose now that the firms hold the following belief:

If a job applicant has education $e < e_0$, he is of type L for certain.
If a job applicant has education $e \geq e_0$, he is of type L with probability $1 - \lambda$ and is of type H with probability λ.

This belief implies the following pay scheme:

$$w(e) = \begin{cases} 1, & \text{if } e < e_0 \\ 1 + \lambda, & \text{if } e \geq e_0. \end{cases}$$

The workers will choose either $e = 0$ or $e = e_0$ (no point to choose other levels). For type L, the applicant will choose $e = e_0$ iff

$$w(e_0) - c_L(e_0) \geq w(0) - c_L(0),$$

i.e., $\lambda \geq e_0$. For type H, the applicant will choose $e = e_0$ iff

$$w(e_0) - c_H(e_0) \geq w(0) - c_H(0),$$

i.e., $\lambda \geq e_0/2$. Hence, if $e_0 \leq \lambda$, all workers will choose $e = e_0$. And in this case, the employers' belief is correct. This is thus an equilibrium, a pooling equilibrium, where the signal does not reveal the identity of any worker, as shown in Fig. 18. ∎

4 The Spence Model: Job Market Signalling

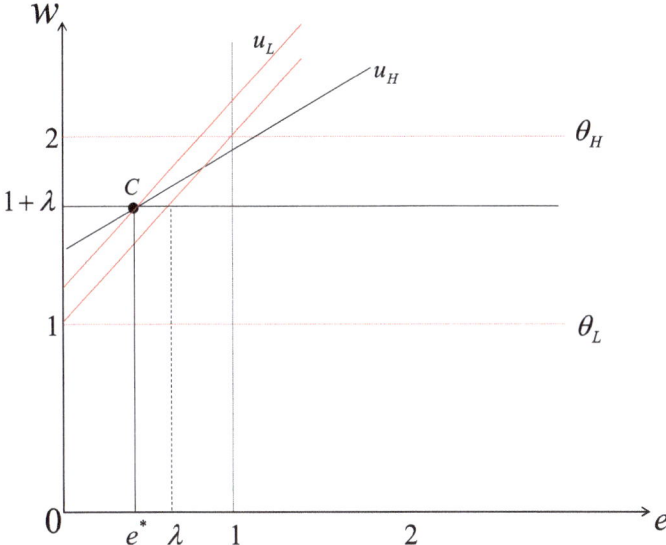

Fig. 18 A pooling equilibrium

4.5 Partial Separating Equilibrium

In a mixed-strategy equilibrium, at least one type chooses a mixed signal. Suppose that the high type chooses e_H for sure, but the low type mixes e_L and e_H. Since the low type mixes the two points (θ_L, e_L) and (w_H, e_H), he must be indifferent between these two choices. Let π be the probability that type L takes (θ_L, e_L). Type L's problem is

$$\max_{\pi \in [0,1]} \pi u_L(\theta_L, e_L) + (1-\pi) u_L(w_H, e_H).$$

If the optimal π is an interior solution, we must have the indifference condition:

$$u_L(\theta_L, e_L) = u_L(w_H, e_H). \tag{14}$$

The consistency condition is

$$\mu(H|e_H) = P(H|e_H) = \frac{P(H, e_H)}{P(e_H)} = \frac{P(e_H|H)P(H)}{P(e_H|H)P(H) + P(e_H|L)P(L)}$$
$$= \frac{\lambda}{\lambda + (1-\pi)(1-\lambda)}.$$

By zero profit,

$$w_H = \mu(H|e_H)\theta_H + [1 - \mu(H|e_H)]\theta_L = \frac{\lambda\theta_H}{\lambda + (1-\pi)(1-\lambda)} + \frac{(1-\pi)(1-\lambda)\theta_L}{\lambda + (1-\pi)(1-\lambda)}. \tag{15}$$

Given an arbitrary e_H, (15) and (14) determine π and e_H. As shown in Fig. 19, given an arbitrary e_H, we can easily find a wage curve to form a BE as described above. In each partial separating equilibrium, there are three variables to determine: π, e_H and w_H. We have only two equations, (14) and (15), which can determine two of the three variables, leaving one of them free. This is expected as we can see from Fig. 19 that there are many such mixed-strategy BEs.

Similarly, given arbitrary $e_L \geq 0$ and e_H, with $e_L < e_H$, we can define hybrid equilibria in which the low type takes $e_L \geq 0$ for sure and the high type mixes e_L with e_H. We have two similar conditions. The indifference condition is

$$u_H(w_L, e_L) = u_H(\theta_H, e_H). \tag{16}$$

The consistency condition is

$$\mu(H|e_L) = P(H|e_L) = \frac{P(H, e_L)}{P(e_L)} = \frac{P(e_L|H)P(H)}{P(e_L|H)P(H) + P(e_L|L)P(L)} = \frac{\pi\lambda}{\pi\lambda + 1 - \lambda}.$$

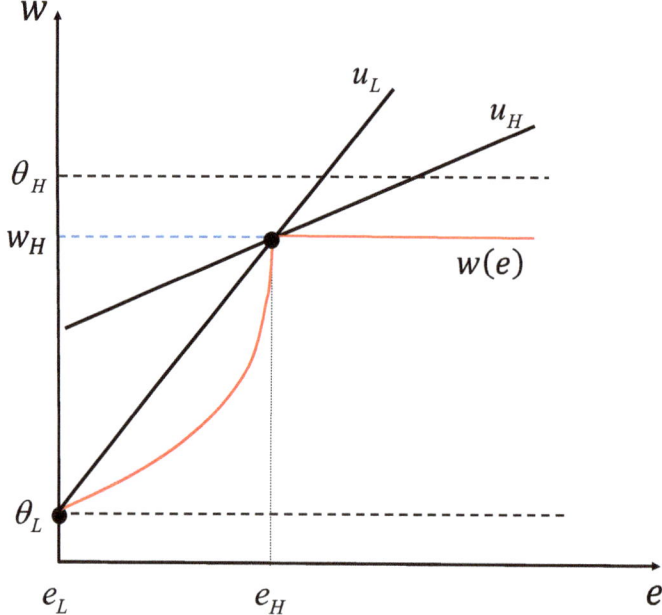

Fig. 19 A hybrid equilibrium

4 The Spence Model: Job Market Signalling

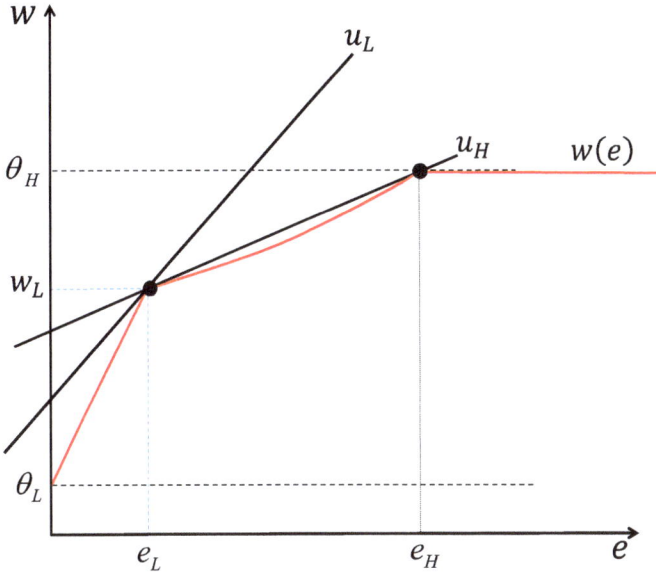

Fig. 20 A hybrid equilibrium

where π is the probability that type H takes (w_L, e_L). The zero-profit condition is

$$w_L = \mu(H|e_L)\theta_H + [1 - \mu(H|e_L)]\theta_L = \frac{\pi\lambda\theta_H}{\pi\lambda + 1 - \lambda} + \frac{(1-\lambda)\theta_L}{\pi\lambda + 1 - \lambda}. \quad (17)$$

Given arbitrary $e_L \geq 0$ and e_H, with $e_L < e_H$, Eqs. (16) and (17) determine π and w_L. This case is shown in Fig. 20.

4.6 Government Intervention

Government intervention may improve efficiency under incomplete information. For example, when $\hat{\theta}_H \leq \bar{\theta}$, the government can impose a ban on signalling activity to achieve a Pareto improvement.

Furthermore, it may be possible for a Pareto improvement even when $\hat{\theta}_H > \bar{\theta}$. In Fig. 21, by offering a low-ability worker the wage $\hat{w}_L$ and a high-ability worker the wage $\hat{w}_H$, both types are better off at $(\hat{w}_L, 0)$ and $(\hat{w}_H, \hat{e}_H)$, respectively.

The central authority can achieve this outcome by mandating that workers with education levels below $\hat{e}_H$ receive wage $\hat{w}_L$ and that workers with education levels of at least $\hat{e}_H$ receive wage $\hat{w}_H$. If so, the low type would choose $e_L^* = 0$ and the high type would choose $e_H^* = \hat{e}_H$. As long as the firms can break even on average under this scheme, this is a sustainable solution and it achieves a Pareto improvement.

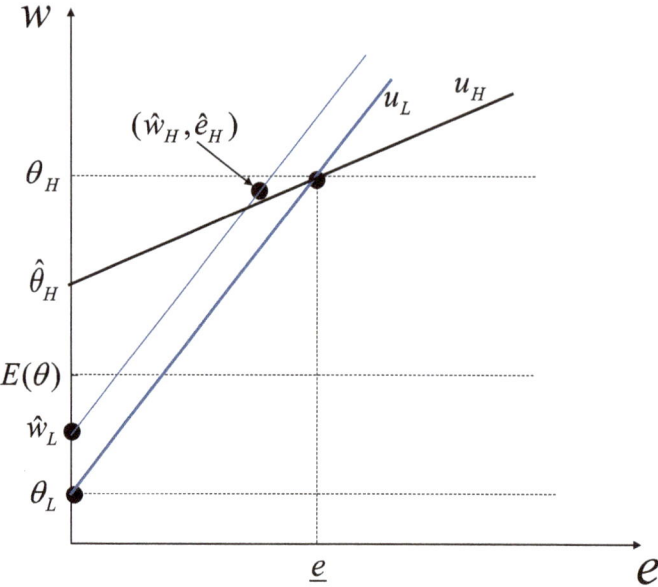

Fig. 21 Government intervention

The key to government intervention is to introduce cross-subsidization, where high-ability workers subsidize low-ability workers. In particular, the no-signalling policy is an extreme measure of cross-subsidization. Of course, these solutions are not feasible in a free and competitive labor market. Opportunistic behaviors will prevent these cross-subsidization cases from happening. For example, in a BE, since the firms' wage offers are derived from their common beliefs, they will not offer $\hat{w}_L$ to $\hat{e}_L = 0$ and offer $\hat{w}_H$ to $\hat{e}_H$.

4.7 Equilibrium Refinement

The multiplicity of equilibria stems from the great freedom that we have to choose beliefs off the equilibrium path, although on the equilibrium path the beliefs are determined by the consistency condition. A great deal of research has focused on finding reasonable further restrictions on beliefs to reduce the number of equilibria.

We can use the dominance criteria to eliminate many BEs. Let us apply the CD criterion first. As shown in Fig. 22, under any circumstance, no matter where is the equilibrium point is, the low type would never take an e with $e > \underline{e}$. By the CD criterion, the firms should believe anyone with e satisfying $e > \underline{e}$ is the high type and pays θ_H (zero belief on the low type). If so, a wage for a worker with an e such

4 The Spence Model: Job Market Signalling

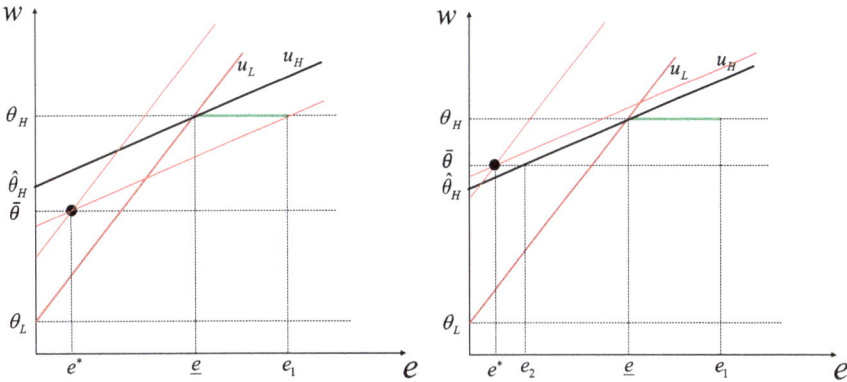

Fig. 22 CDBEs

as $e > \underline{e}$ must be $w = \theta_H$. Hence, the only separating CDBE is the separating pair $(e_L^*, e_H^*) = (0, \underline{e})$.[8]

How about pooling equilibrium under the CD criterion? If $\bar{\theta} < \hat{\theta}_H$, any pooling e^* as shown on the left side of Fig. 22 is impossible to satisfy the CD criterion. Since $\mu(\underline{e}) = 1$ by the CD criterion, the high type will choose $\underline{e}$ instead. Therefore, this e^* cannot be a BE by the CD criterion. Therefore, if $\bar{\theta} < \hat{\theta}_H$, there is no pooling CDBEs. However, if $\bar{\theta} \geq \hat{\theta}_H$, a pooling e^* with $e^* \geq e_2$ is possible. As shown on the right side of Fig. 22, each $e^* \in [0, e_2]$ is a CDBE.

How about BEs under the ED criterion? If $\bar{\theta} < \hat{\theta}_H$, since there is no pooling CDBE, there is no pooling EDBE. If $\bar{\theta} \geq \hat{\theta}_H$, as shown in Fig. 23, any $e > e_1$ is dominated by the equilibrium e^* for the low type; hence, by the ED criterion, we should have $\mu(H|e) = 1$. If so, since this e is better than e^* for the high type, the high type will take this e instead of e^*. That is, by the ED criterion, all pooling BEs are ruled out. Therefore, there is no pooling EDBE. Therefore, there is only one EDBE with or without $\bar{\theta} \geq \hat{\theta}_H$. In sum, the ED criterion, also called the intuitive criterion of Cho and Kreps (1990), predicts the most efficient separating equilibrium as the unique equilibrium.[9]

In summary, the welfare effect of signalling is generally ambiguous. The prediction by Akerlof's model holds if there is no way of credibly transmitting information from the informed to the uninformed. Spence (1973) investigates the possibility of information revelation by signaling. He finds that signalling can have rather surprising effects. Much of the signalling literature involves refinements that

---

[8] Any e, $e > \bar{e}$, is a completely dominated signal for both types (an all-dominance case). Hence, we cannot impose any restriction on the belief. However, since such an education level cannot be an equilibrium education level, we do not need to be concerned about it.

[9] This conclusion is for a case with two types. Stronger conditions are needed to have a unique equilibrium when there are more types; see Cho and Kreps (1987). See also Kreps (1990) and Gibbons (1992) for more details.

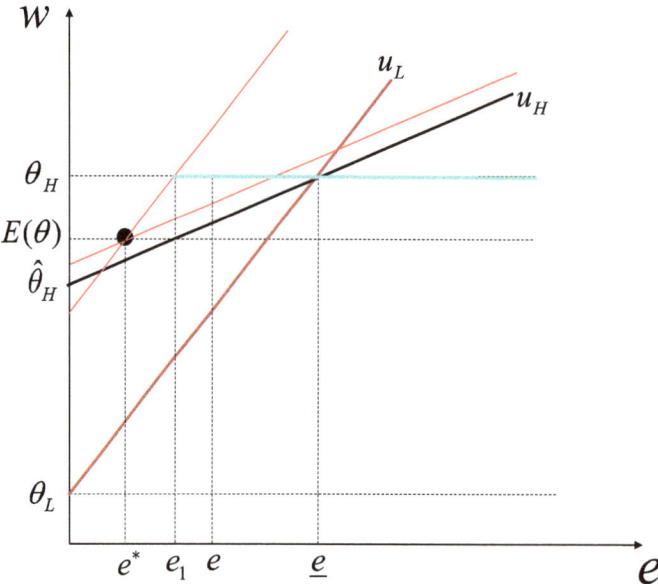

Fig. 23 EDBEs

reject unreasonable beliefs. A strong notion of refinement such as the ED criterion implies a unique equilibrium, which is the Pareto-efficient separating equilibrium.

4.8 Questions

What if education does affect productivity? Since there will be a reward for education itself, can a pooling equilibrium still exist? How about efficiency?

In Spence (1973), all firms have the same belief. What will happen if firms have different beliefs?

In reality, a person's education level at primary school, high school, BA, MS and Ph.D. has discrete values. How will a five-value education system affect our conclusions?

In our analysis, beliefs are given. How are beliefs formed? In a dynamic setting, beliefs may be formed endogenously by a learning process over time.

Finally, Spence uses the BE concept. How about using the SPNE concept? The next section deals precisely with this issue.

5 The Spence Model: Job Market Screening

For Spence's model, instead of solving for BEs, we now solve for SPNEs. Since there are no beliefs in Nash's equilibrium concept, we need to give an education level a different interpretation. We argue that the education level is now used as a screening device.

In Spence (1973), an education level is chosen by workers as a signal, while in Rothschild and Stiglitz (1976), it is specified in a hiring contract by firms as a screening device. That is, instead of offering a wage rate w based on their beliefs, the firms specify a wage rate w together with an education level e in the contract. The education level is used to screen workers in Rothschild and Stiglitz (1976), while it is a signal from workers in Spence (1973). <u>Screening</u> is the ability of the uninformed to use a public signal to screen economic agents.

According to Spence (1973), there are two types of workers of productivities θ_L and θ_H, $\theta_H > \theta_L > 0$, where the fraction of high-ability workers is $\lambda \in (0, 1)$. Now, instead of specifying a wage rate w in a contract, firms offer a pair (w, e) with the wage rate matched to the education level in the contract. Multiple contracts may be offered by a firm. Instead of a wage curve $w(e)$ in the marketplace, a worker faces a number of choices $\{(w_i, e_i)\}$. Of course, we can also describe the wage curve as a set of $\{(w_i, e_i)\}$. Education still has nothing to do with productivity. Utility functions are

$$u_L(w, e) = w - c_L e, \quad u_H(w, e) = w - c_H e,$$

where $c_L > c_H > 0$.

We study subgame perfect Nash equilibria (SPNEs) of the following two-stage game:

Step 1. firms simultaneously announce contract offers. Each firm may announce any number of contracts.
Step 2. Given the offers, workers decide whether or not and which contract to accept.

Again, we have two possible kinds of equilibria: <u>separating equilibria</u>, in which the two types accept different contracts, and <u>pooling equilibria</u>, in which both types accept the same contract.

5.1 Pooling Equilibrium

Does there exist a pooling equilibrium? Let (w^p, e^p) be a pooling equilibrium contract. Zero profit means that it must lie on the pooled break-even line at $w = \bar{\theta}$. Suppose that a firm offers such a contract on this pooled break-even line. As shown in Fig. 24, another firm can easily pick up a contract in the shaded area that attracts

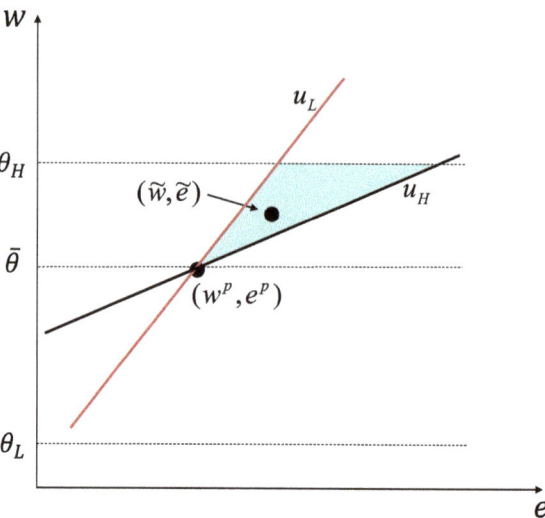

Fig. 24 Pooling equilibrium

only high-ability workers. If so, (w^p, e^p) will attract low-ability workers only so that the firm offering (w^p, e^p) loses money. Therefore, (w^p, e^p) cannot be an equilibrium. Hence, there is no pooling equilibrium.

5.2 Separating Equilibrium

We now try to find a separating equilibrium. In Fig. 25, the pair of contracts (w_L, e_L) and (w_H, e_H) can separate the two types. However, this pair of contracts

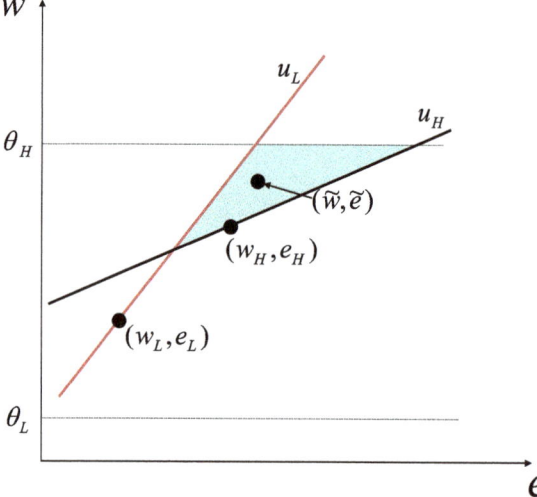

Fig. 25 Separating contracts

Fig. 26 Separating contracts

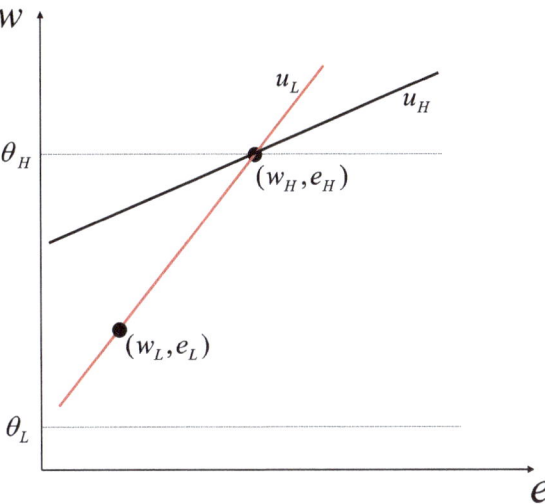

Fig. 27 Separating contracts

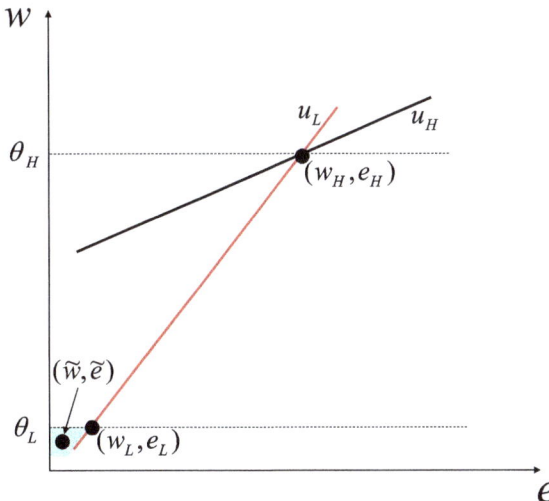

cannot be an equilibrium since another contract $(\tilde{w}, \tilde{e})$ can attract the high type and make a profit.

Thus, given (w_L, e_L), contract (w_H, e_H) has to be in the position shown in Fig. 26 in order to eliminate the shaded area. Again, this pair cannot be an equilibrium since the firm offering the pair of contracts is making a loss. The zero-profit condition implies that (w_L, e_L) must be on the θ_L-line.

However, the pair in Fig. 27 is still not an equilibrium since some firm can offer a contract $(\tilde{w}, \tilde{e})$ in the shaded area to attract the low type and make a profit. Hence, the only possible separating equilibrium is the pair in Fig. 28.

Fig. 28 Separating contracts

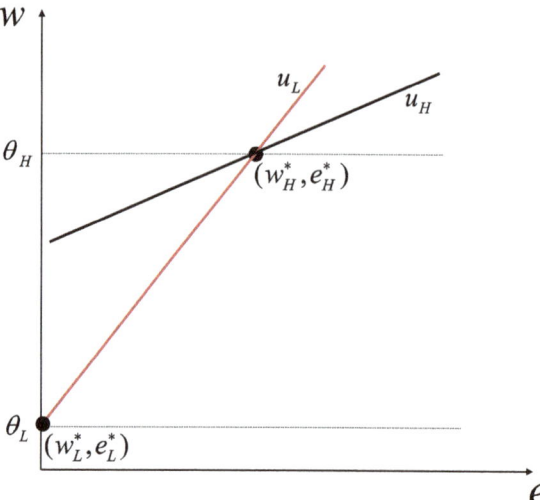

Fig. 29 Nonexistence

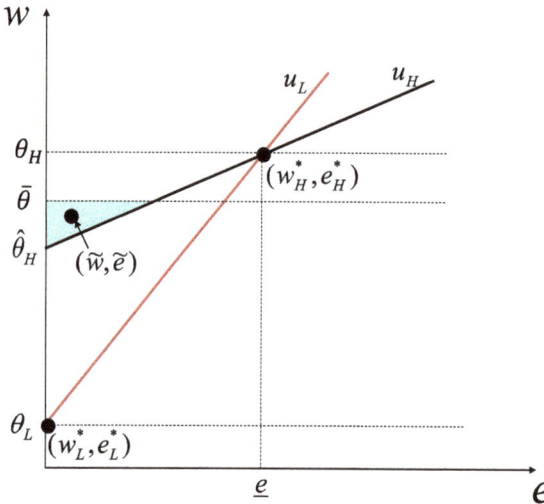

Can this pair survive to become a separating equilibrium? The answer depends on the value of $\bar{\theta}$. If $\bar{\theta} > \hat{\theta}_H$, as shown in Fig. 29, we can easily find a contract $(\tilde{w}, \tilde{e})$ in the shaded area that attracts both types and yet makes a profit. In this case, this separating pair is not sustainable. Hence, there is no separating equilibrium when $\bar{\theta} > \hat{\theta}_H$. Even when $\bar{\theta} \leq \hat{\theta}_H$, we may find an alternative profitable pair of contracts. For example, a pair of contracts $(\tilde{w}_L, \tilde{e}_L)$ and $(\tilde{w}_H, \tilde{e}_H)$ in Fig. 30 may be profitable. If so, the separating pair is not sustainable.

Fig. 30 Cross subsidization

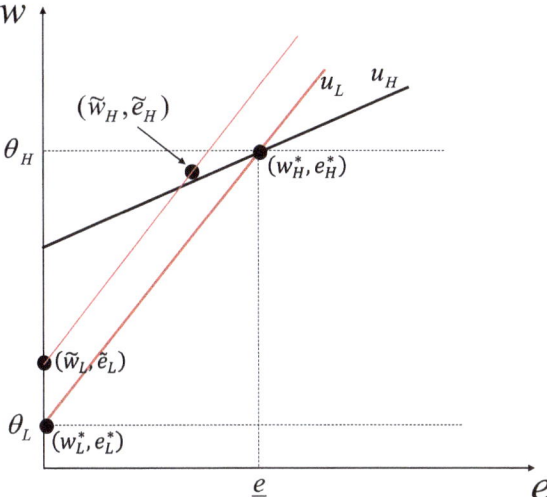

Hence, a separating equilibrium exists if the situations in Figs. 29 and 30 do not happen, i.e., when $\bar{\theta} \leq \hat{\theta}_H$ and there are no such profitable pair $(\tilde{w}_L, \tilde{e}_L)$ and $(\tilde{w}_H, \tilde{e}_H)$ in Fig. 30.

We have several observations on this solution. First, the separating equilibrium is the same as the unique BE in Sect. 4.7 implied by the ED criterion.

Second, by the reactive equilibrium concept of Riley (1979), the separating equilibrium survives even if $\bar{\theta} > \hat{\theta}_H$. Also, any pooling equilibrium survives as a reactive equilibrium. See the discussion in Sect. 2.3.

Third, as in the signalling model, low-ability workers are always worse off with screening, and high-ability workers are better off with screening.[10]

Fourth, the separating equilibrium is Pareto dominated by the complete-information outcome.

Finally, when an equilibrium does exist, the situation in Fig. 30 cannot happen. Thus, the equilibrium is <u>constrained Pareto optimal</u>: a uninformed government can no longer make a Pareto improvement. See Sect. 4.6 on government intervention.

Notes

Besides the cited papers, good references are Mas-Colell et al. (1995, Chap. 13) and Laffont (1995).

---

[10] The conditions that ensure the survival of the separating equilibrium guarantee that high-ability workers will be better off with screening.

Mechanism Design 11

The previous chapter focuses on competitive markets under incomplete information. This chapter focuses on monopoly pricing under incomplete information. The basic modelling approach to monopoly pricing under incomplete information is the revelation principle, by which the uninformed strategically provide incentives for the informed to reveal their types.

1 A Story of Mechanism Design

To motivate the approach, we reconsider the insurance problem of the RS model in the previous chapter. There are two types of individuals who differ in probabilities π_i of having an accident. We call π_i the type of the individual. Each individual knows his own type but the company does not. Instead of a competitive insurance market, we now assume that there is a single monopoly company. Under incomplete information, what kind of contracts should this company offer?

1.1 Market Mechanism Versus Direct Mechanism

Market Mechanism
An insurance company in the real world typically offers contracts of the form (z, q) to the market, where z is the amount of coverage in an accident and q is the price of insurance. Given initial wealth w and a possible loss of an amount L of that wealth, with the insurance, an individual's income becomes

$$w - qz, \quad \text{without an accident,}$$
$$w - L + z - qz, \quad \text{with an accident}$$

Since the company does not know the types of individuals, the contracts cannot be based on individual types. What can the company do?

Take the car insurance market as an example. Since a claim for compensation is observable, a company typically provides a no-claim bonus (NCB) to insurees. Let $\mathbb{S} = \{0, 1\}$ denote the choices of claims: claim or not claim after an accident, $\mathbb{X} = \{0, 1\}$ denote whether or not there is an accident, and Θ denote the set of individual types. The company's yearly contract offers are (z_s, q_s) for $s \in \mathbb{S}$ based on whether or not there is a claim last year. Given the contract offers, each individual $i \in \Theta$ then decides on a claim strategy $s_i \in \mathcal{S}$, where $s_i : \mathbb{X} \to \mathbb{S}$. Then, given last year's outcome $x \in \mathbb{X}$, individual i's claim strategy is $s_i(x) \in \mathbb{S}$, and receives contract $(z_{s_i(x)}, q_{s_i(x)})$. In this case, the company has no direct control over who chooses which contract. In fact, the individuals play strategies against the offers and an individual may not seek compensation from the company even if he has had an accident. This is a market mechanism.

Direct Mechanism

For convenience of theoretical analysis, we consider an artificial mechanism, called a direct mechanism, in which the company simply asks each individual to report his type and is then offered a contract based on his reported type. Since the company does not observe the type, an incentive has to be provided in order for each individual to report his type truthfully.

With two types of individuals in the market, at most two contracts are needed: (z_L, q_L) and (z_H, q_H), with (z_i, q_i) intended for type i, $i = L$ and H. If these two contracts happen to be the same, only one contract is accepted by all individuals. Given the two contracts, each individual decides which contract to choose. Type L will choose (z_L, q_L) only if it is better for him. More precisely, type L will choose (z_L, q_L) iff

$$(1 - \pi_L)u(w - q_L z_L) + \pi_L u(w - L + z_L - q_L z_L) \\ \geq (1 - \pi_L)u(w - q_H z_H) + \pi_L u(w - L + z_H - q_H z_H). \tag{1}$$

Symmetrically, type H will choose (z_H, q_H) iff

$$(1 - \pi_H)u(w - q_H z_H) + \pi_H u(w - L + z_H - q_H z_H) \\ \geq (1 - \pi_H)u(w - q_L z_L) + \pi_H u(w - L + z_L - q_L z_L). \tag{2}$$

We call these two conditions the <u>incentive compatibility (IC) conditions</u>. By these IC conditions, each type will voluntarily choose the contract intended for him in the direct mechanism. If so, when they are asked to report their own types, they will report their types truthfully.

1 A Story of Mechanism Design

1.2 The Optimal Allocation: A Graphic Illustration

To illustrate the optimal solution of the direct mechanism in a diagram, we transform (z, q) to (I_1, I_2), where

$$I_1 = w - qz, \quad I_2 = w - L + z - qz.$$

Here, I_1 and I_2 are the incomes in the two states. Then, given q, the available choices of (I_1, I_2), are on the q-line defined by

$$(1 - q)I_1 + qI_2 = w - qL.$$

In the trade, each individual gives the company his endowment $(w, w - L)$ in exchange for (I_1, I_2). Thus, the company's expected profit from an individual with probability π_i is

$$\Pi_i = (1 - \pi_i)w + \pi_i(w - L) - (1 - \pi_i)I_1 - \pi_i I_2 = (q - \pi_i)z.$$

The company makes a profit from type i iff $q \geq \pi_i$, i.e., iff the contract is on the left of the π_i-line, where the π_i-line is the q-line when $q = \pi_i$ and it is actual the zero-profit line for the company.

Since there are only two types, with each type having one common optimal contract, we can always start with a pair of contracts in our analysis. If the optimal pair of contract turns out to be one contract, we have a pooling solution, otherwise we have a separating solution. In a pooling solution, the acceptance of a contract by an individual does not reveal the individual's type, while in a separating solution, the choice of a contract by an individual will reveal the individual's type.[1]

Consider an arbitrary pair of contracts (C_H, C_L) that separates the types in Fig. 1. Type L will choose C_L and type H will choose C_H. First, using Lemma 10.1, given C_L, the company can increase profits by moving C_H towards the π_H-line until the new u_H-curve passes through C_L. Thus, we can assume equilibrium C_H and C_L to be on the same u_H-curve, as shown in Fig. 2.

Second, given the fact that C_H and C_L are on the same u_H-curve, the company can do better by moving C_L away from the π_L-line (and move C_H towards the π_H-line) until the u_L-curve passes through the initial point w, which is the limit that the u_L-curve can move to the left and thus must be the final position for the u_L-curve, as shown in Fig. 3.

We now know that the position of the u_L-curve, which must go through the initial point w. However, the final position of the u_H-curve is not clear. Its position will affect both C_H and C_L, and we want to have a C_H that is as close to the π_H-line as possible (from the right) and a C_L that is as far from the π_L-line as possible. If we move the u_H-curve closer to the π_H-line, point C_L may move closer to the π_L-line. Thus, the optimal position of the u_H-curve has to balance the distance of C_L to the

---

[1]It generally pays for the company to separate the two types and exploit the differences. Hence, we generally expect a separating solution with a monopoly.

Fig. 1 Separating policies

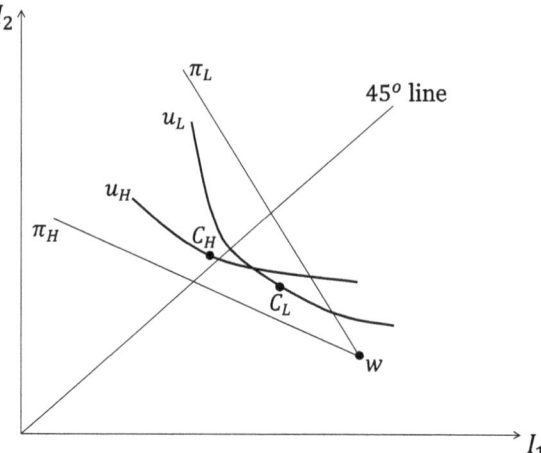

Fig. 2 Separating policies

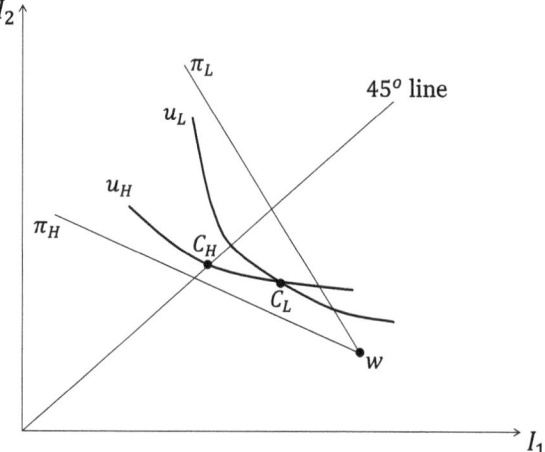

π_L-line and the distance of C_H to the π_H-line. The precise position of C_H for the optimal pair (C_H, C_L) can be known only by actually solving the mechanism design problem. If the optimal pair turns out to be one contract, we have a pooling solution, otherwise we have a separating solution.

1.3 The Optimal Allocation: Mathematical Presentation

We now set up the mechanism design problem. Each consumer gives the company $(w, w - L)$ in exchange for (I_1, I_2). Thus, the company's expected profit from a consumer with π_i is

1 A Story of Mechanism Design

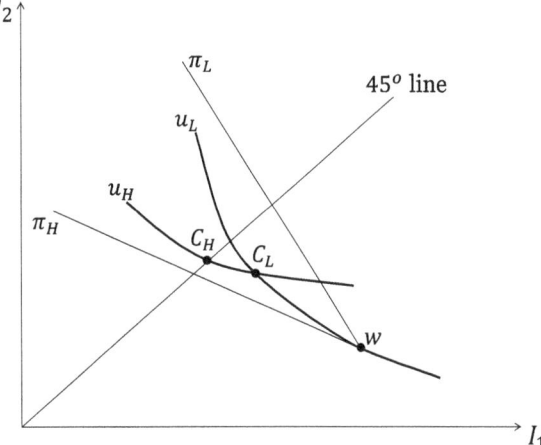

Fig. 3 Separating equilibrium

$$\Pi_i = (1 - \pi_i)w + \pi_i(w - L) - (1 - \pi_i)I_1 - \pi_i I_2.$$

Given two contracts $C_H = (I_{1H}, I_{2H})$ and $C_L = (I_{1L}, I_{2L})$, one intended for type L and the other for type H, the company's per capita profit is

$$\begin{aligned}\Pi &= \lambda \Pi_L + (1-\lambda)\Pi_H \\ &= \pi_p(w-L) + (1-\pi_p)w - \lambda(1-\pi_L)I_{1L} + \pi_L I_{2L} - (1-\lambda)(1-\pi_H)I_{1H} + \pi_H I_{2H},\end{aligned}$$

where λ is the population share of the low-risk individuals. Here, for simplicity, we assume that both individuals have the same initial endowment $(w, w - L)$ and the same utility function $u(\cdot)$. Besides the IC conditions in (1) and (2), the company need to take into account an individual rationality (IR) condition for each type i:

$$(1 - \pi_i)u(I_1) + \pi_i u(I_2) \geq (1 - \pi_i)u(w) + \pi_i u(w - L).$$

This IR condition ensures that type i will choose to participate in the insurance plan. Then, the company's problem is

$$\begin{aligned}\min_{I_{1L}, I_{2L}, I_{1H}, I_{2H} \geq 0} & \quad \lambda[(1-\pi_L)I_{1L} + \pi_L I_{2L}] + (1-\lambda)[(1-\pi_H)I_{1H} + \pi_H I_{2H}] \\ \text{s.t.} & \quad IC_L: (1-\pi_L)u(I_{1L}) + \pi_L u(I_{2L}) \geq (1-\pi_L)u(I_{1H}) + \pi_L u(I_{2H}), \\ & \quad IC_H: (1-\pi_H)u(I_{1H}) + \pi_H u(I_{2H}) \geq (1-\pi_H)u(I_{1L}) + \pi_H u(I_{2L}), \\ & \quad IR_L: (1-\pi_L)u(I_{1L}) + \pi_L u(I_{2L}) \geq (1-\pi_L)u(w) + \pi_L u(w-L), \\ & \quad IR_H: (1-\pi_H)u(I_{1H}) + \pi_H u(I_{2H}) \geq (1-\pi_H)u(w) + \pi_H u(w-L).\end{aligned} \quad (3)$$

That is, the company minimizes the expenditure subject to the two IC conditions and two IR conditions. The two IC conditions ensure that the low-risk individuals prefer C_L to C_H and the high-risk individuals prefer C_H to C_L, and the two IR conditions ensure that they both will participate.[2] In other words, the two IC conditions induce the two types to truthfully report their own types.

By the above graphic analysis, we know that IC_L and IR_H will not be binding and IC_H and IR_L will be binding. According to the Kuhn-Tucker conditions in the Lagrange approach, we can drop the nonbinding conditions. Thus, the problem becomes

$$\min_{I_{1L},I_{2L},I_{1H},I_{2H} \geq 0} \lambda[(1-\pi_L)I_{1L} + \pi_L I_{2L}] + (1-\lambda)[(1-\pi_H)I_{1H} + \pi_H I_{2H}]$$

s.t. IC_H: $(1-\pi_H)u(I_{1H}) + \pi_H u(I_{2H}) = (1-\pi_H)u(I_{1L}) + \pi_H u(I_{2L})$,
      IR_L: $(1-\pi_L)u(I_{1L}) + \pi_L u(I_{2L}) = (1-\pi_L)u(w) + \pi_L u(w-L)$.

The Lagrangian function is

$$L = \lambda[(1-\pi_L)I_{1L} + \pi_L I_{2L}] + (1-\lambda)[(1-\pi_H)I_{1H} + \pi_H I_{2H}]$$
$$- \mu_1(1-\pi_H)u(I_{1H}) + \pi_H u(I_{2H}) - (1-\pi_H)u(I_{1L}) - \pi_H u(I_{2L})$$
$$- \mu_2(1-\pi_L)u(I_{1L}) + \pi_L u(I_{2L}) - (1-\pi_L)u(w) - \pi_L u(w-L).$$

The FOCs are

$$0 = \frac{\partial L}{\partial I_{1L}} = \lambda(1-\pi_L) + \mu_1(1-\pi_H)u'(I_{1L}) - \mu_2(1-\pi_L)u'(I_{1L}), \quad (4)$$

$$0 = \frac{\partial L}{\partial I_{2L}} = \lambda\pi_L + \mu_1\pi_H u'(I_{2L}) - \mu_2\pi_L u'(I_{2L}), \quad (5)$$

$$0 = \frac{\partial L}{\partial I_{1H}} = (1-\lambda)(1-\pi_H) - \mu_1(1-\pi_H)u'(I_{1H}), \quad (6)$$

$$0 = \frac{\partial L}{\partial I_{2H}} = (1-\lambda)\pi_H - \mu_1\pi_H u'(I_{2H}). \quad (7)$$

In particular, by (6)–(7), we find that $I_{1H} = I_{2H}$, indicating that the high-risk type has full insurance.

1.4 The Optimal Allocation: A General Case

More generally, consider an insurance company dealing with a population of individuals each of whom has probability θ of having an accident that costs him L, where $\theta \in \Theta$ and Θ is a set of parameters. Instead of two possible types, if Θ is an

---
[2]The presence of IR conditions must be consistent with the calculation of the objective function. A different set of IR conditions implies a different objective function. For example, if one IR condition ensured that type H would not participate, then our objective function would not include the profits from type H.

interval, the individuals have a continuum of possible types $\theta \in \Theta$ For simplicity, assume that the loss L is the same for all individuals and that all individuals have the same initial wealth w. In exchange for $(w, w - L)$, the company proposes a mechanism $[I_1(\theta), I_2(\theta)]$ that specifies a net income $I_1\left(\hat{\theta}\right)$ in the good state and a net income $I_2\left(\hat{\theta}\right)$ in the bad state for an agent who reported type $\hat{\theta}$.

Given his type θ, an agent reports the $\hat{\theta}$ from

$$\max_{\hat{\theta}} U\left(\hat{\theta}\right) = (1-\theta) u\left[I_1\left(\hat{\theta}\right)\right] + \theta u[I_2\left(\hat{\theta}\right)]. \tag{8}$$

The FOC for the optimal $\hat{\theta}$ is

$$(1-\theta) u'\left[I_1\left(\hat{\theta}\right)\right] I_1'\left(\hat{\theta}\right) + \theta u'\left[I_2\left(\hat{\theta}\right)\right] I_2'\left(\hat{\theta}\right) = 0.$$

In order for the individual to report truthfully, the company must design $I_1(\cdot)$ and $I_2(\cdot)$ in such a way that $\hat{\theta} = \theta$. In other words, we need to have

$$(1-\theta) u'[I_1 \theta] I_1'(\theta) + \theta u'[I_2 \theta] I_2'(\theta) = 0, \quad \forall \theta. \tag{9}$$

We also need the SOC for (8):

$$(1-\theta) u''[I_1(\theta)][I_1'(\theta)]^2 + \theta u''[I_2(\theta)][I_2'(\theta)]^2$$
$$+ (1-\theta) u'[I_1(\theta)] I_1''(\theta) + \theta u'[I_2(\theta)] I_2''(\theta) \le 0, \quad \forall \theta.$$

Taking a derivative of Eq. (9) with respect to (w.r.t.) θ, this SOC becomes

$$u'(I_2) I_2' - u'(I_1) I_1' \ge 0.$$

Using (9) again, we find

$$-\frac{1}{\theta} u'(I_1) I_1' = u'(I_2) I_2' - u'(I_1) I_1' \ge 0.$$

Therefore, the SOC for truthful reporting is satisfied if

$$I_1'(\theta) \le 0, \quad \forall \theta. \tag{10}$$

We have thus derived two conditions (9) and (10) for $I_1(\cdot)$ and $I_2(\cdot)$, by which the insurance policy $[I_1(\theta), I_2(\theta)]$ for $\theta \in \Theta$ will induce truthful reporting. In other words, with such a policy, an individual of type θ will truthfully report his type as θ and thus will be offered the $[I_1(\theta), I_2(\theta)]$ policy.

What is then the optimal allocation scheme $[I_1^*(\cdot), I_2^*(\cdot)]$? Let $F(\theta)$ be the population distribution. Since the revenue is fixed, the insurance company's profit maximizing problem is equivalent to the following cost-minimizing problem:

$$\min_{I_1(\cdot), I_2(\cdot)} \int (1 - \theta I_1(\theta) + \theta I_2(\theta)) dF(\theta)$$

$$\text{s.t.} \quad (1-\theta)u'[I_1(\theta)]I_1'(\theta) + \theta u'[I_2(\theta)]I_2'(\theta) = 0, \tag{11}$$

$$I_1'(\theta) \leq 0,$$

$$(1-\theta)u[I_1(\theta)] + \theta u[I_2(\theta)] \geq (1-\theta)u(w) + \theta u(w-L).$$

1.5 Market Mechanisms Versus the Direct Mechanism

In a typical mechanism $[R, \mathbb{S}, g]$, where R is a set of rules, each agent follows a strategy $s : \Theta \to \mathbb{S}$ and the resulting allocation scheme is $g[s(\theta)]$. The direct mechanism $[R_0, \Theta, f]$ is a special mechanism, in which R_0 is a special set of rules, $\mathbb{S} = \Theta$ and $f(\theta) = (I_1(\theta), I_2(\theta))$. Given a reported $\hat{\theta}$, the contract offer is $f\left(\hat{\theta}\right)$ in the direct mechanism. One natural question is: Can a market mechanism do better than the direct mechanism for the insurance company? Can any other mechanism do better? By the revelation principle, as shown later, if the direct mechanism provides IC conditions for truthful reporting, no other mechanism can possibly do better than the direct mechanism.

2 The Revelation Principle

Given the motivating example in Sect. 1, we now present the revelation principle. The presence of incomplete information raises the possibility that we may need to consider a player's beliefs about other players' preferences, his beliefs about their beliefs about his preferences, and so on, much in the spirit of rationalizability. Fortunately, there is a widely used approach to this problem, originated by Harsanyi (1967–68), that makes this unnecessary. In this approach, one imagines that each player's preferences are determined by the realization of a random variable. Through this formulation, the game of incomplete information becomes a game of imperfect information: Nature makes the first move, choosing realizations of the random variable that determine each player's preference type. A game of this sort is known as a Bayesian game.

Formally, there are n players $i \in \mathbb{N} = \{1, \ldots, n\}$, each with a type parameter θ_i. The type $\theta_i \in \Theta_i$ is random ex ante before Nature assigns its value. After Nature assigns its value, it is observed only by player i. The joint density function is $\phi(\theta)$ for $\theta = (\theta_1, \ldots, \theta_n) \in \Theta \equiv \Theta_1 \times \cdots \times \Theta_n$, which is common knowledge. There is

2 The Revelation Principle

a feasible collection $\mathbb{X}$ of choices, called the <u>allocation set</u>. Each player i has a preference relation $\succsim_{\theta_i}$ over $\mathbb{X}$ or equivalently a payoff function $u_i(x, \theta_i)$ for $x \in \mathbb{X}$.[3] The utility function $u_i(\cdot, \cdot)$ is also common knowledge, but each specific value of $u_i(\cdot, \cdot)$ is observed only by agent i. A <u>Bayesian game</u> is represented by $[\mathbb{N}, \mathbb{X}, \{u_i\}, \Theta, \phi]$.

Although individual i knows his own type, others do not know. Hence, others need to figure out player i's strategy $s_i(\theta_i)$ for all possible $\theta_i \in \Theta_i$; in other words, others need to know player i's strategy function $s_i(\cdot)$. Therefore, for convenience, we can simply assume that each player i has a strategy function $s_i(\cdot)$; and after his type θ_i is assigned by Nature, he takes strategy $s_i(\theta_i)$.

In our allocation problem, a mapping $f : \Theta \to \mathbb{X}$ is called an <u>allocation scheme</u>. Assuming the knowledge of the agents' types θ, an allocation scheme assigns a feasible allocation $f(\theta)$ in $\mathbb{X}$ to the agents. Our problem is to find an optimal allocation scheme that maximizes a given objective. An objective can be the revenue from a piece of art, social welfare from building a bridge, or surplus from trade, etc. One difficulty involved in this problem is that none of the players, especially the player who chooses an optimal allocation scheme, has full knowledge of all agents' types.

What sort of allocation schemes should we be interested? Naturally, one desirable feature of an allocation scheme is ex-post efficiency.

Definition 11.1 An allocation scheme $f : \Theta \to \mathbb{X}$ is <u>ex-post efficient</u> if $f(\theta)$ is Pareto optimal for any $\theta \in \Theta$. That is, there is no $\theta \in \Theta$ and $x \in \mathbb{X}$ such that $u_i(x_i, \theta_i) \geq u_i[f(\theta), \theta_i]$ for all i, and $u_i(x_i, \theta_i) > u_i[f(\theta), \theta_i]$ for some i. ∎

There are three time points in an incomplete information model:

- <u>Ex ante</u>: the time point before Nature moves, when no one knows his own and other players' types.
- <u>Interim</u>: the time point just after Nature has moved, when each player knows his own type, but not others.
- <u>Ex post</u>: the time point at which everyone knows everyone's type (Fig. 4).

Ex-post efficiency requires complete knowledge of each individual's type. In other words, ex-post efficiency means Pareto efficiency ex post when information is complete. In contrast, when a planner can only form expectations of the types, she can only aim at ex-ante efficiency.

To implement an allocation scheme, we need a mechanism. A mechanism is simply a set of rules, including strategy variables, the allocation in equilibrium and other rules. Specifically, a mechanism defines a Bayesian game of incomplete information that consists of a set of rules, a strategy space for each agent and an allocation in equilibrium. With a mechanism, agents interact through an

---
[3] We can actually allow an agent's utility function to take the form $u_i(x, \theta)$ rather than $u_i(x, \theta_i)$. All the concepts on Bayesian implementation are readily extendable to this case.

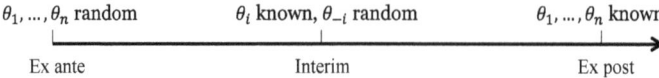

Fig. 4 The time line

arrangement or institution defined by a set of rules governing actions and payoffs. Their interactions eventually reach an equilibrium and then the mechanism allocates payoffs in equilibrium. We may have many mechanisms to implement an allocation scheme. In a special mechanism, called the direct mechanism, each agent i is asked to make an announcement of his own type $\hat{\theta}_i$ and then $f(\hat{\theta})$ is allocated.

Definition 11.2 A <u>mechanism</u> $\Gamma = [R, \{\mathbb{S}_i\}, g(\cdot)]$ is a set of rules, a collection of strategy sets $(\mathbb{S}_1, \ldots, \mathbb{S}_n)$ and an outcome function $g : \mathbb{S}_1 \times \cdots \times \mathbb{S}_n \to \mathbb{X}$. ∎

A mapping $s_i : \Theta_i \to \mathbb{S}_i$ is called a <u>strategy</u> of agent i. The allowed actions are summarized by $\mathbb{S}_i$ and the allocation is given by $g(\cdot)$ in equilibrium. Players determine their strategies before Nature moves; after Nature moves, the players play according to their strategies.

Given a mechanism $\Gamma = [R, \{\mathbb{S}_i\}, g]$, player i's <u>strategy set</u> $\mathcal{S}_i$ is the set of all the strategy mappings $s_i : \Theta_i \to \mathbb{S}_i$. Player i's payoff for a profile of strategies $(s_1, \ldots, s_n)$ is

$$\tilde{u}_i(s_1, \ldots, s_n) = E_\theta u_i\{g[s_1(\theta_1), \ldots, s_n(\theta_n)], \theta_i\},$$

where E_θ is the mathematical expectation operator over randomness defined by θ.

Definition 11.3 A <u>Bayesian Nash equilibrium (BNE)</u> in the Bayesian game $[\mathbb{N}, \mathbb{X}, \{u_i\}, \Theta, \phi]$ under mechanism $\Gamma = [R, \{\mathbb{S}_i\}, g]$ is a strategy profile $(s_1^*, \ldots, s_n^*) \in \mathcal{S}_1 \times \cdots \times \mathcal{S}_n$ that constitutes a Nash equilibrium of the game $\Gamma_N = [\mathbb{N}, \{\mathcal{S}_i\}, \{\tilde{u}_i\}]$. That is, for each i, s_i^* solves

$$\max_{s_i \in \mathcal{S}_i} u_i(s_i, s_{-i}^*). \blacksquare \qquad (12)$$

The following proposition shows that the strategies/plans are time consistent in the sense that the players play according to their plans after they find out their own types.

Proposition 11.1 *A strategy profile* $(s_1^*, \ldots, s_n^*)$ *is a BNE in Bayesian game* $[\mathbb{N}, \mathbb{X}, \{u_i\}, \Theta, \phi]$ *under mechanism* $\Gamma = [R, \{\mathbb{S}_i\}, g]$ *iff, for all i and $\theta_i \in \Theta_i$ occurring with a positive probability, $s_i^*(\theta_i)$ solves*

$$\max_{s_i \in \mathbb{S}_i} E_{\theta_{-i}|\theta_i} u_i\{g[s_i, s_{-i}^*(\theta_{-i})], \theta_i\} \qquad (13)$$

where $E_{\theta_{-i}|\theta_i}$ is the expectation operator over θ_{-i} conditional on θ_i.

Proof Problem (12) can be written as

$$\max_{s_i \in S_i} \int_\Theta u_i\{g[s_i(\theta_i), s^*_{-i}(\theta_{-i})], \theta_i\} \phi(\theta) d\theta$$

$$= \max_{s_i \in S_i} \int_{\Theta_i} \left\{ \int_{\Theta_{-i}} u_i\{g[s_i(\theta_i), s^*_{-i}(\theta_{-i})], \theta_i\} \frac{\phi(\theta)}{\phi_i(\theta_i)} d\theta_{-i} \right\} \phi_i(\theta_i) d\theta_i$$

$$= \max_{s_i \in S_i} \int_{\Theta_i} E_{\theta_{-i}} \{u_i(g[s_i(\theta_i), s^*_{-i}(\theta_{-i})], \theta_i) | \theta_i\} \phi_i(\theta_i) d\theta_i,$$

where $\phi_{-i}(\theta_{-i}|\theta_i) = \frac{\phi(\theta)}{\phi_i(\theta_i)}$. Then, by the Pontryagin Theorem, problem (12) is equivalent to problem (13), since the Hamiltonian function is

$$H(s_i, \theta_i) = E_{\theta_{-i}} \{u_i(g[s_i, s^*_{-i}(\theta_{-i})], \theta_i) | \theta_i\}. \quad \blacksquare$$

Definition 11.4 A mechanism $\Gamma = [R, \{S_i\}, g]$ can implement an allocation scheme f in BNE if there is a BNE $s^*(\cdot) = [s^*_1(\cdot), \ldots, s^*_n(\cdot)]$ of Γ such that $g[s^*(\theta)] = f(\theta)$ for all $\theta \in \Theta$. If such a mechanism exists, we say that f is implementable. $\blacksquare$

That is, we say that a mechanism Γ implements an allocation scheme $f(\theta)$ if there is an equilibrium induced by the mechanism that yields the outcome $f(\theta)$ for all θ.

Definition 11.5 The special mechanism $\Gamma_0 = [R_0, \{\Theta_i\}, f]$ is called the direct mechanism, in which each player i's strategy is to report his type $\hat{\theta}_i$ in his own best interest (which may or may not be the true type), and then $f\left(\hat{\theta}_1, \ldots, \hat{\theta}_n\right)$ is allocated. $\blacksquare$

The reporting strategy in the direct mechanism is a function $\hat{\theta}_i : \Theta_i \to \Theta_i$. Given expectations on others' types, each player derives his report $\hat{\theta}_i$ of his type from his utility maximization problem:

$$\max_{\hat{\theta}_i \in \Theta_i} E_{\theta_{-i}|\theta_i} \left\{ u_i \left[f\left(\hat{\theta}_i, \theta_{-i}\right), \theta_i \right] \right\}. \tag{14}$$

Definition 11.6 An allocation scheme $f(\theta)$ is truthfully implementable in BNE or incentive compatible (IC) if each player's strategy is to report his true type in his own best interest in the direct mechanism Γ_0. That is, for each $i \in \mathbb{N}$ and $\theta_i \in \Theta_i$, the player will always report his true type $\hat{\theta}_i = \theta_i$. ∎

To find the optimal allocation scheme, we first need to identify all implementable allocation schemes. The identification of all implementable allocation schemes may seem like a daunting task since there are so many possible mechanisms. Fortunately, the revelation principle tells us that we can simply restrict our attention to the simplest mechanism, the direct mechanism, in which each agent is asked to report his type $\hat{\theta}_i$ and, given the announcements $\left(\hat{\theta}_1, \ldots, \hat{\theta}_n\right)$, $f\left(\hat{\theta}\right)$ is allocated. Here, the strategy of the player with type θ_i is his report $\hat{\theta}_i(\theta_i)$, which is dependent on his type θ_i. The revelation principle indicates that an allocation scheme is implementable if and only if it is truthfully implementable.

Proposition 11.2 (The Revelation Principle). *An allocation scheme $f(\theta)$ is implementable iff it is truthfully implementable.*

Proof If an allocation is truthfully implementable, it is obviously implementable since it is implementable by the direct mechanism.

Conversely, suppose that a mechanism $\Gamma = [R, \{\mathbb{S}_i\}, g(\cdot)]$ implements an allocation scheme $f(\theta)$. Then, there exists a profile of strategies $s^*(\cdot) = [s_1^*(\cdot), \ldots, s_n^*(\cdot)]$ such that $g[s^*(\theta)] = f(\theta)$ for all θ, where $s_i^*(\theta_i)$ solves the following problem:

$$\max_{s_i \in \mathbb{S}_i} E_{\theta_{-i} | \theta_i} \left\{ u_i \left(g\left[s_i, s_{-i}^*(\theta_{-i}) \right], \theta_i \right) \right\}. \tag{15}$$

We denote s_i as $s_i^*\left(\hat{\theta}_i\right)$ for some $\hat{\theta}_i$.[4] Then, θ_i is the solution of the following problem:

$$\max_{\hat{\theta}_i \in \Theta_i} E_{\theta_{-i} | \theta_i} \left\{ u_i \left(g\left[s_i^*\left(\hat{\theta}_i\right), s_{-i}^*(\theta_{-i}) \right], \theta_i \right) \right\}. \tag{16}$$

Since $g[s^*(\theta)] = f(\theta)$ for all θ, problem (16) can be rewritten as

$$\max_{\hat{\theta}_i \in \Theta_i} E_{\theta_{-i} | \theta_i} \left\{ u_i \left[f\left(\hat{\theta}_i, \theta_{-i}\right), \theta_i \right] \right\}. \tag{17}$$

Hence, θ_i is the solution of (17). That is, $f(\theta)$ is truthfully implementable. ∎

The explanation of the revelation principle is simple: if $s_i^*(\cdot)$ offers a one-to-one relationship between type θ_i and strategy s_i, given others' strategies, choosing the optimal strategy $s_i^*(\theta_i)$ in Γ is equivalent to reporting θ_i in the direct mechanism. More specifically, "$f(\theta)$ is implementable" means that, given θ_{-i}, player i wants

---
[4]Here we need $s_i^*(\cdot)$ to be an onto mapping.

Fig. 5 The direct mechanism

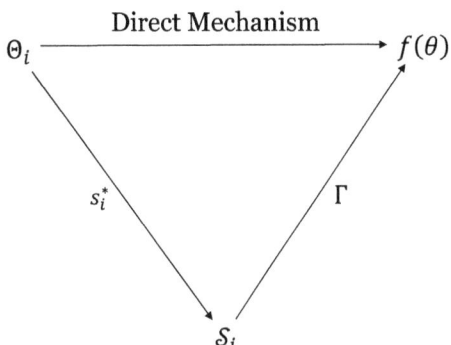

$f(\theta)$. In Γ, given θ_{-i}, player i can choose $s_i^*(\theta_i)$ to get $f(\theta)$; in the direct mechanism, given θ_{-i}, player i can choose $\hat{\theta}_i = \theta_i$ to get $f(\theta)$.[5] The proof follows this explanation.

The revelation principle can be illustrated by the above diagram: in a typical mechanism, you map your type to a strategy and then the mechanism maps a strategy profile to an allocation. However, if the strategy profile and the mechanism are combined into one operation, which is the direct mechanism, you map your type directly to an allocation (Fig. 5).

The significance of this revelation principle is that, for an allocation scheme $f(\theta)$, we have no need to know many mechanisms $\Gamma = [R, \{\mathbb{S}_i\}, g(\cdot)]$, particularly the functions $s_i^*(\theta_i)$ and $g(s_1, \ldots, s_n)$, that can implement $f(\theta)$. If our interest is to identify those implementable allocation schemes $f(\theta)$, we only need to find those allocation schemes $f(\theta)$ that are truthfully implementable. Hence, to identify the optimal allocation scheme under incomplete information, we only need to search among those allocation schemes that are truthfully implementable under the direct mechanisms.

To identify truthfully implementable allocation schemes $f : \Theta \to \mathbb{R}^m$, we only need to identify those $f(\theta) = (f_1(\theta), \ldots, f_m(\theta))$ that imply truthful reporting $\hat{\theta}_i = \theta_i$ in problem (14). We replace problem (14) by its FOC and SOC conditions. The FOC for (14) is

$$E_{\theta_{-i}|\theta_i} \sum_{j=1}^{m} \frac{\partial u_i[f(\theta), \theta_i]}{\partial x_j} \frac{\partial f_j(\theta)}{\partial \theta_i} = 0, \tag{18}$$

for all $i \in \mathbb{N}$ and $\theta_i \in \Theta_i$. The SOC is

---

[5]This depends crucially on the nature of Nash equilibrium. In a Nash equilibrium, you assume that others will play the right strategies whatever you do; hence, if you cheat, you cheat on yourself only.

$$E_{\theta_{-i}|\theta_i} \sum_{j=1}^{m} \left\{ \sum_{k=1}^{m} \left[\frac{\partial^2 u_i[f(\theta), \theta_i]}{\partial x_j \partial x_k} \frac{\partial f_k(\theta)}{\partial \theta_i} \frac{\partial f_j(\theta)}{\partial \theta_i} \right] + \frac{\partial u_i[f(\theta), \theta_i]}{\partial x_j} \frac{\partial^2 f_j(\theta)}{\partial \theta_i^2} \right\} \leq 0.$$

Since (18) holds for all θ_i, by taking a derivative of (18) w.r.t. θ_i, we find

$$0 = E_{\theta_{-i}|\theta_i} \sum_{j=1}^{m} \left\{ \sum_{k=1}^{m} \left[\frac{\partial^2 u_i[f(\theta),\theta_i]}{\partial x_j \partial x_k} \frac{\partial f_k(\theta)}{\partial \theta_i} \frac{\partial f_j(\theta)}{\partial \theta_i} \right] + \frac{\partial u_i[f(\theta),\theta_i]}{\partial x_j} \frac{\partial^2 f_j(\theta)}{\partial \theta_i^2} \right. \\ \left. + \frac{\partial^2 u_i[f(\theta),\theta_i]}{\partial x_j \partial \theta_i} \frac{\partial f_j(\theta)}{\partial \theta_i} \right\}.$$

Hence, the SOC is

$$E_{\theta_{-i}|\theta_i} \left\{ \sum_{j=1}^{m} \frac{\partial^2 u_i[f(\theta), \theta_i]}{\partial x_j \partial \theta_i} \frac{\partial f_j(\theta)}{\partial \theta_i} \right\} \geq 0, \tag{19}$$

for all $i \in \mathbb{N}$ and $\theta_i \in \Theta_i$. Conditions (18) and (19) are necessary for an arbitrary allocation scheme to be truthfully implementable. These FOC and DOC offer a set of equations for identifying truthfully implementable $f(\theta)$. In other words, conditions (18) and (19) define the set of all implementable allocation schemes. This is the set of allocation schemes from which we can conduct a search for the optimal allocation scheme in a mechanism-design problem. Therefore, in a mechanism-design problem, conditions (18) and (19) are the two IC conditions for truthfully implementable allocation schemes, just as the two IC conditions in the mechanism-design problem (11) of the insurance problem.

Besides IC conditions, certain participation conditions or individual rationality (IR) conditions are also necessary. We will discuss IR conditions later.

3 Examples of Allocation Schemes

We now give a few examples of allocation schemes and some real-world mechanisms.

Example 11.1. A Buyer-Seller Problem Consider a trade of a good between a buyer with utility function $U = u(x, \theta) - p$ and a seller with utility function $V = p - c(x, \theta)$. We impose typical assumptions on the functions:

$$u_{xx} < 0, \quad c_{xx} > 0, \quad u_{x\theta} \geq 0, \quad c_{x\theta} \geq 0, \quad u_{x\theta} \leq c_{x\theta},$$

where condition $u_{x\theta} \geq 0$ means that the marginal utility u_x increases as quality improves, and condition $c_{x\theta} \geq 0$ means that the marginal cost c_x increases as quality improves. However, condition $u_{x\theta} \leq c_{x\theta}$ is imposed for convenience only. The seller knows the quality θ of his product, but the buyer does not know this. Let x be the

3 Examples of Allocation Schemes

quantity traded and p be the total payment of the trade. Let $x(\theta)$ be the solution of the ex-post social welfare maximization, which is from

$$\max_x U + V = u(x, \theta) - c(x, \theta),$$

or

$$u_x[x(\theta), \theta] = c_x[x(\theta), \theta]. \tag{20}$$

Define an allocation scheme by $f(\theta) \equiv (x(\theta), p(\theta))$, where $p(\theta)$ is some function. This allocation scheme is ex-post efficient by the definition of $x(\theta)$.

Is this $f(\theta)$ be implementable? To answer this, we look at the seller's problem when the seller is asked to report her type (i.e., product quality):

$$\max_{\hat{\theta}} p\left(\hat{\theta}\right) - c\left[x\left(\hat{\theta}\right), \theta\right].$$

The FOC is

$$p'\left(\hat{\theta}\right) = c_x\left[x\left(\hat{\theta}\right), \theta\right] x'\left(\hat{\theta}\right).$$

Hence, for truth reporting, we need

$$p'(\theta) = c_x[x(\theta), \theta] x'(\theta), \quad \text{for all } \theta. \tag{21}$$

The SOC is

$$p''\left(\hat{\theta}\right) - c_{xx}\left[x\left(\hat{\theta}\right), \theta\right]\left[x'\left(\hat{\theta}\right)\right]^2 - c_x\left[x\left(\hat{\theta}\right), \theta\right] x''\left(\hat{\theta}\right) \leq 0.$$

Hence, for truth reporting, we need

$$p''(\theta) - c_{xx}[x(\theta), \theta][x'(\theta)]^2 - c_x[x(\theta), \theta] x''(\theta) \leq 0, \quad \text{for all } \theta.$$

By taking a derivative of (21) w.r.t. θ, this SOC becomes

$$c_{x\theta}[x(\theta), \theta] x'(\theta) \leq 0,$$

which is satisfied if $x'(\theta) \leq 0$. By (20), we find

$$x'(\theta) = -\frac{u_{x\theta} - c_{x\theta}}{u_{xx} - c_{xx}},$$

which is always negative given our assumptions on the functions. Hence, we have $x'(\theta) \leq 0$ for all θ; that is, the SOC is satisfied. Therefore, the allocation scheme is truthfully implementable if and only if (21) holds, from which we find

$$p(\theta) = p_0 + \int_0^\theta c_x[x(t), t]dx(t), \tag{22}$$

where p_0 can take any value. That is, if the payment function $p(\cdot)$ is defined by (22), the allocation scheme $f(\theta)$ is truthfully implementable. In sum, if the allocation scheme $f(\theta)$ is ex-post efficient and truthfully implementable if $x(\theta)$ is defined by (20) and $p(\theta)$ is defined by (22).

Note that to have an optimal allocation scheme $f(\theta)$, which is ex-post efficient and truthfully implementable, we need to further choose a p_0.

Example 11.2. The Price Mechanism How about the so-called price mechanism used in the general equilibrium theory? Consider a pure exchange economy with k goods and n consumers, in which consumer i has utility function $u_i(x_i, \theta_i)$, consumption set $\mathbb{R}_+^k$ and endowment $w_i \in \mathbb{R}_+^k$. The set of feasible allocations is

$$\mathbb{X} = \left\{ x \in \mathbb{R}_+^{nk} \,\middle|\, \sum_{i=1}^n x_i \leq \sum_{i=1}^n w_i \right\}.$$

Allocation scheme $f(\theta)$ is defined to be the GE allocation when θ is common knowledge. We know that $f(\theta)$ is ex-post efficient under usual conditions. Can this allocation scheme be implemented by some mechanism when θ_i is private information? The answer is no. To see this, Fig. 6 shows three offer curves in an Edgeworth box. The intersection points of the offer curves are the GE allocations. We can see from the figure that, if consumer 1 tells his true type, consumer 2 will generally not tell his true type. By claiming $\hat{\theta}_2$, consumer 2 can obtain $f\left(\theta_1, \hat{\theta}_2\right)$,

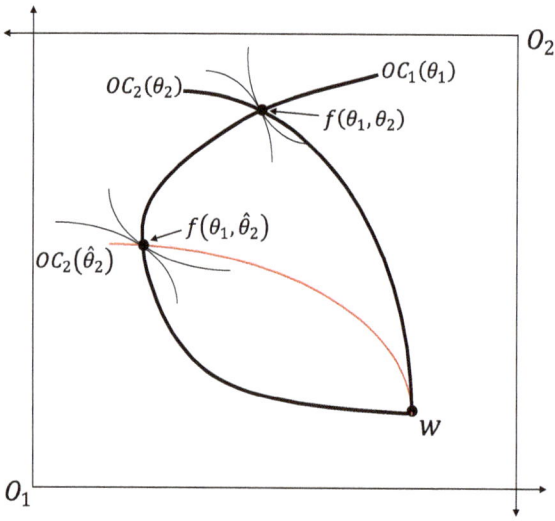

Fig. 6 General equilibria

3 Examples of Allocation Schemes

which is better than $f(\theta_1, \theta_2)$ to consumer 2. This means that $f(\theta)$ is not implementable by the direct mechanism. By the revelation principle, this means that $f(\theta)$ cannot possibly be implementable by any mechanism, including the price mechanism. Akerlof (1970) uses the same mechanism and yields a similar conclusion.

Example 11.3. A Public Project Consider a situation in which n agents decide whether to undertake a public project, whose cost must be funded by the group themselves. For example, a village considers to build a bridge for themselves. An outcome is a vector $x = (\delta, c_1, \ldots, c_n)$, where $\delta = 0$ or 1 indicates whether or not to build the bridge and $c_i \in \mathbb{R}$ is a monetary payment from agent i. The cost of the project is $c \geq 0$. Hence, the set of feasible allocations is

$$\mathbb{X} = \left\{ (\delta, c_1, \ldots, c_n) \,\bigg|\, \sum_i c_i = c\delta, \quad \delta \in \{0, 1\}, \quad c_i \in \mathbb{R} \right\}.$$

Agent i's utility function is

$$u_i(x, \theta_i) = \theta_i \delta - c_i,$$

where θ_i is the agent's benefit from the bridge.

We now define an allocation scheme $f(\theta) = [\delta(\theta), c_1(\theta), \ldots, c_n(\theta)]$. We first define $\delta(\theta)$ by

$$\delta(\theta) = \begin{cases} 1, & \text{if } \sum_i \theta_i \geq c, \\ 0, & \text{if not.} \end{cases} \quad (23)$$

Then, $f(\theta)$ is ex-post efficient.[6]

Next, given that the decision to build is determined by (23), how should the participants allocate the cost? That is, how should $c_1(\theta), \ldots, c_n(\theta)$ be defined? Consider first an equal-sharing rule: $c_i(\theta) = c\delta(\theta)/n$. However, with this cost sharing rule, the allocation scheme is not truthfully implementable. For example, for a θ satisfying $\sum \theta_i \geq c$ and $\theta_1 < c/n$, agent 1 does not want the project even though the project is socially desirable. The best way for agent 1 to discourage taking the project is to report $\hat{\theta}_1 = 0$. That is, agent 1 will not report truthfully. The problem is due to equal cost sharing but unequal benefits. Agent 1's acceptance of the project would have a positive externality on others, but he fails to internalize this effect.

One better cost sharing rule is a proportional sharing rule: $c_i(\theta) = \frac{\theta_i}{\sum_k \theta_k} c\delta(\theta)$. Under this rule, an agent who benefits more pays a proportionally larger cost. By this, when $\sum \theta_k \geq c$, under complete information, every agent is willing to pay his share of the cost to have the project. However, under incomplete information, when cost sharing relies on agents' reported types, an agent may not tell his true type. Let

---
[6]We can define a social welfare function as the sum of the individual's utility functions. This $\delta(\theta)$ maximizes this social welfare function ex post, implying Pareto efficiency for each θ.

us see if agent 1 will report his true type when others report their true types. If agents' reports are observable, when the project is socially worthwhile, agent 1's problem is[7]

$$\max_{\hat{\theta}_1} \theta_1 - \frac{\hat{\theta}_1}{\hat{\theta}_1 + \theta_2 + \cdots + \theta_n} c = \theta_1 - c + \frac{\theta_2 + \cdots + \theta_n}{\hat{\theta}_1 + \theta_2 + \cdots + \theta_n} c$$
$$\text{s.t.} \quad \hat{\theta}_1 + \theta_2 + \cdots + \theta_n \geq c,$$

implying $\hat{\theta}_1 = c - \theta_2 - \cdots - \theta_n$. Hence, when $\sum \theta_i > c$, agent 1 will under-report his type. Hence, the allocation scheme is again ex-post efficient but not implementable.

Example 11.4. The Auction Model There is a single indivisible item to be sold by a seller to one of n agents. The outcome is a vector $x = (\delta_1, \ldots, \delta_n, c_1, \ldots, c_n)$, where $\delta_i = 1$ if agent i gets the item, $\delta_i = 0$ otherwise, and c_i is the monetary payment from agent i.[8] Then, the feasible allocation set is

$$\mathbb{X} = \left\{ (\delta_1, \ldots, \delta_n, c_1, \ldots, c_n) \middle| \delta_i \in \{0, 1\}, \sum_i \delta_i = 1, \quad c_i \in \mathbb{R} \right\}.$$

Again, $u_i(x, \theta_i) = \theta_i \delta_i - c_i$, where θ_i is agent i's benefit from the item. Let $\Theta_i = [0, 1]$. The seller's revenue is $R = \sum_i c_i$. An allocation scheme $f(\theta) = [\delta_1(\theta), \ldots, \delta_n(\theta), c_1(\theta), \ldots, c_n(\theta)]$ is ex-post efficient if the highest type θ_i gets the item.[9]

Assume $n = 2$ and suppose that both buyers' valuations θ_i are drawn independently from $U[0, 1]$ and this fact is common knowledge among the buyers. An allocation scheme $f(\theta) = [\delta_1(\theta), \delta_2(\theta), c_1(\theta), c_2(\theta)]$ is defined by

$$\delta_1(\theta) = \begin{cases} 1 & \text{if } \theta_1 \geq \theta_2; \\ 0 & \text{if not,} \end{cases} \quad \delta_2(\theta) = \begin{cases} 1 & \text{if } \theta_1 < \theta_2; \\ 0 & \text{if not,} \end{cases} \quad (24)$$

$$c_1(\theta) = \theta_1 \delta_1(\theta), \quad c_2(\theta) = \theta_2 \delta_2(\theta). \quad (25)$$

It is ex-post efficient. Is this allocation scheme truthfully implementable? Let $\phi(\theta_i)$ be the density function of $U[0, 1]$. Given buyer 2's report $\hat{\theta}_2 = \theta_2$, buyer 1's report $\hat{\theta}_1$ is from

---

[7]If agents' reports are not publicly observable, agent 1 needs to take expectations over θ_{-1}.
[8]This is the auction model, in which there are many competitive buyers, a single seller and a single indivisible item for sale. This model is popular in the literature of asymmetric information since it is the simplest case of monopoly pricing under asymmetric information.
[9]For each $\theta \in \Theta$, it is Pareto optimal since it maximizes social welfare $SW \equiv R + \sum_i u_i(x, \theta_i)$.

3 Examples of Allocation Schemes

$$\max_{\hat{\theta}_1} \int_{\hat{\theta}_1 \geq \theta_2} \left(\theta_1 - \hat{\theta}_1\right)\phi(\theta_2)d\theta_2 + \int_{\hat{\theta}_1 < \theta_2} 0\phi(\theta_2)d\theta_2$$

$$= \int_0^{\hat{\theta}_1} \left(\theta_1 - \hat{\theta}_1\right)d\theta_2 = \left(\theta_1 - \hat{\theta}_1\right)\hat{\theta}_1,$$

implying

$$\hat{\theta}_1 = \theta_1/2.$$

In other words, if buyer 2 tells the truth, buyer 1 will report only half of his true valuation. Here, a buyer understates his valuation so as to lower his payment for the item, even though this reduces his chance of getting the item. Hence, this $f(\theta)$ is not truthfully implementable.

Let us consider an alternative allocation scheme $\tilde{f}(\theta)$ that satisfies (24) but the payments are changed to

$$c_1(\theta) = \theta_2 \delta_1(\theta), \quad c_2(\theta) = \theta_1 \delta_2(\theta),$$

which is still ex-post efficient. In this case, the winner pays the amount equal to his opponent's valuation (the second highest valuation). Given $\hat{\theta}_2 = \theta_2$, buyer 1 considers his problem:

$$\max_{\hat{\theta}_1} \int_{\hat{\theta}_1 \geq \theta_2} (\theta_1 - \theta_2)\phi(\theta_2)d\theta_2 = \int_0^{\hat{\theta}_1} (\theta_1 - \theta_2)d\theta_2.$$

The FOC is $\theta_1 - \hat{\theta}_1 = 0$, implying $\hat{\theta}_1 = \theta_1$. Hence, truth telling is a BNE. Therefore, the allocation scheme $\tilde{f}(\theta)$ is truthfully implementable.

In the following two examples, we consider two popular real-world mechanisms for the auction model in this example.

Example 11.5. First-Price Sealed-Bid Auction For the auction model in Example 11.4, in a first-price sealed-bid auction, each potential buyer i is allowed to submit a sealed bid $b_i \geq 0$. The bids are then opened and the buyer with the highest bid gets the item and pays an amount equal to his bid to the seller. In the case of multiple highest bids, the item goes to the buyer whose index i is the lowest. Assume again $n = 2$ and both buyers' valuations $\{\theta_i\}$ are drawn independently from $U[0, 1]$. We look for a BNE in which each buyer's strategy $b_i(\theta_i)$ takes the linear form $b_i(\theta_i) = \alpha_i \theta_i$, where $\alpha_i > 0$ is a constant. Given $b_2(\theta_2) = \alpha_2 \theta_2$, buyer 1 considers his problem:

$$\max_{b_1 \geq 0} \int_{b_1 \geq b_2(\theta_2)} (\theta_1 - b_1)\phi(\theta_2)d\theta_2 + \int_{b_1 < b_2(\theta_2)} 0\phi(\theta_2)d\theta_2$$

$$= \int_0^{b_1/\alpha_2} (\theta_1 - b_1)d\theta_2 = (\theta_1 - b_1)\frac{b_1}{\alpha_2}$$

implying $b_1^* = \theta_1/2$. Thus, we have a BNE: $b_1^* = \theta_1/2, b_2^* = \theta_2/2$.[10] Each buyer balances between paying too much and the chance of getting the item. In equilibrium, everyone's bit is half of his valuation. This BNE implies an allocation scheme $f(\theta) = [\delta_1(\theta), \delta_2(\theta), c_1(\theta), c_2(\theta)]$ with

$$\delta_1(\theta) = \begin{cases} 1 & \text{if } \theta_1 \geq \theta_2; \\ 0 & \text{if } \text{not}. \end{cases}, \quad \delta_2(\theta) = \begin{cases} 1 & \text{if } \theta_1 < \theta_2; \\ 0 & \text{if } \text{not}. \end{cases}, \quad (26)$$

$$c_1(\theta) = \tfrac{1}{2}\theta_1\delta_1(\theta), \quad c_2(\theta) = \tfrac{1}{2}\theta_2\delta_2(\theta). \quad (27)$$

Example 11.6. Second-Price Sealed-Bid Auction For the auction model in Example 11.4 again, in a second-price sealed-bid auction, the arrangement is similar to a first-price sealed-bid auction, except that the winner pays the second-highest bid. Again, we look for a BNE in which each buyer's strategy $b_i(\theta_i)$ takes the linear form $b_i(\theta_i) = \alpha_i \theta_i$, where $\alpha_i > 0$ is a constant. Given $b_2(\theta_2) = \alpha_2 \theta_2$, buyer 1 considers his problem:

$$\max_{b_1} \int_{b_1 \geq b_2(\theta_2)} [\theta_1 - b_2(\theta_2)]\phi(\theta_2)d\theta_2 = \max_{b_1} \int_0^{b_1/\alpha_2} (\theta_1 - \alpha_2\theta_2)\phi(\theta_2)d\theta_2.$$

The FOC is $(\theta_1 - b_1)\phi(b_1/\alpha_2)/\alpha_2 = 0$, implying $b_1 = \theta_1$. Thus, we have a BNE: $b_i(\theta) = \theta_i$ for all i. In equilibrium, everyone's bit equals his valuation. This BNE implies an allocation scheme $f(\theta) = [\delta_1(\theta), \delta_2(\theta), c_1(\theta), c_2(\theta)]$ with

$$\delta_1(\theta) = \begin{cases} 1 & \text{if } \theta_1 \geq \theta_2; \\ 0 & \text{if } \text{not}. \end{cases}, \quad \delta_2(\theta) = \begin{cases} 1 & \text{if } \theta_1 < \theta_2; \\ 0 & \text{if } \text{not}. \end{cases},$$
$$c_1(\theta) = \theta_2\delta_1(\theta), \qquad c_2(\theta) = \theta_1\delta_2(\theta).$$

Example 11.7. Mechanism The first-price and second-price sealed-bid auctions are mechanisms defined in Definition 11.2. For the first-price sealed-bid auction, $\mathbb{S}_i = \mathbb{R}_+$, and strategies are $b_i \colon \Theta_i \to \mathbb{S}_i$, for all i, and

---

[10]This BNE turns out to be the unique symmetric BNE when valuations $\{\theta_i\}$ are drawn independently from uniform distributions (Gibbons 1992, p.157).

3 Examples of Allocation Schemes

$$g(b_1, \ldots, b_n) = \{\delta_i(b_1, \ldots, b_n), c_i(b_1, \ldots, b_n)\}_{i=1}^n,$$

with

$$\delta_i(b_1, \ldots, b_n) = \begin{cases} 1 & \text{if } b_i = \max\{b_1, \ldots, b_n\}, \\ 0 & \text{otherwise}, \end{cases}$$

$$c_i(b_1, \ldots, b_n) = \begin{cases} b_i & \text{if } b_i = \max\{b_1, \ldots, b_n\}, \\ 0 & \text{otherwise}. \end{cases}$$

For the second-price sealed-bid auction, everything is the same except that

$$c_i(b_1, \ldots, b_n) = \begin{cases} \max_{j \neq i}\{b_j\} & \text{if } b_i = \max\{b_1, \ldots, b_n\}, \\ 0 & \text{otherwise}. \end{cases}$$

Example 11.8. The Revelation Principle We know that the allocation scheme $f(\theta)$ in (26)–(27) comes from a mechanism, i.e., the allocation is implementable. To verify the revelation principle, we want to show that this allocation scheme is truthfully implementable.

In the direct mechanism, given $\hat{\theta}_2 = \theta_2$, buyer 1's report $\hat{\theta}_1$ is determined by

$$\max_{\hat{\theta}_1} E_{\theta_2|\theta_1}\left\{u_1\left[f\left(\hat{\theta}_1, \theta_2\right), \theta_1\right]\right\} = \int_{\hat{\theta}_1 \geq \theta_2} \left(\theta_1 - \frac{1}{2}\hat{\theta}_1\right) \phi(\theta_2) d\theta_2 = \left(\theta_1 - \frac{1}{2}\hat{\theta}_1\right)\hat{\theta}_1,$$

implying $\hat{\theta}_1 = \theta_1$. Symmetrically, $\hat{\theta}_2 = \theta_2$. Thus, the allocation scheme implied by the first-price sealed-bid auction is truthfully implementable. This confirms the revelation principle.

In mechanism design, we ask three sets of questions:

1. Given an allocation scheme $f(\theta)$, is it ex-post efficient and truthfully implementable?
2. Given a mechanism, what is the BNE and the allocation scheme in equilibrium?
3. Given an objective function, what is the optimal implementable allocation scheme?

We have discussed the first two sets of questions in the examples. In the rest of this chapter, we will focus on the third question.

4 IC Conditions in Linear Environments

When all agents in a model have the following simple form of utility functions:

$$u_i(x, \theta_i) = \theta_i v_i(x) - c_i,$$

where $v_i(x)$ is a real-valued function and c_i is a constant, we say that we have a linear environment. The key to a linear environment is that the parameter θ_i representing the agent's private information has a linear relationship with the agent's utility value.

In a linear environment, we can easily identify the IC conditions that define truthfully implementable allocation schemes. For this purpose, given an allocation scheme $f(\theta) = [x(\theta), c_1(\theta), \ldots, c_n(\theta)]$, denote

$$\bar{c}_i(\theta_i) = E_{\theta_{-i}|\theta_i}[c_i(\theta)], \quad \bar{v}_i(\theta_i) = E_{\theta_{-i}|\theta_i}\{v_i[x(\theta)]\}, \quad U_i(\theta_i) \equiv \theta_i \bar{v}_i(\theta_i) - \bar{c}_i(\theta_i). \tag{28}$$

Proposition 11.3 (IC Conditions). *An allocation scheme $f(\theta) = [x(\theta), c_1(\theta), \ldots, c_n(\theta)]$ is incentive compatible iff, for all i,*

$$(1)\ \bar{v}'_i(\cdot) \geq 0. \tag{29}$$

$$(2)\ U'_i(\cdot) = \bar{v}_i(\cdot). \tag{30}$$

Proof

(1) *Necessity*. Agent i's report $\hat{\theta}_i$ is from

$$\max_{\hat{\theta}_i} \theta_i \bar{v}_i(\hat{\theta}_i) - \bar{c}_i(\hat{\theta}_i).$$

It implies

$$FOC: \quad \theta_i \bar{v}'_i(\theta_i) - \bar{c}'_i(\theta_i) = 0, \quad \text{for all } \theta_i, \tag{31}$$

$$SOC: \quad \theta_i \bar{v}''_i(\theta_i) - \bar{c}''_i(\theta_i) \leq 0, \quad \text{for all } \theta_i. \tag{32}$$

By (31), we immediately have $U'_i(\theta_i) = \bar{v}_i(\theta_i)$. Again, by (31), we have

$$\bar{v}'_i(\theta_i) + \theta_i \bar{v}''_i(\theta_i) - \bar{c}''_i(\theta_i) = 0.$$

Then, by (32), $\bar{v}'_i(\theta_i) \geq 0$.

(2) *Sufficiency*. By condition (30), we have

$$U_i(\theta_i) = U_i(\hat{\theta}_i) + \int_{\hat{\theta}_i}^{\theta_i} \bar{v}_i(s) ds.$$

4 IC Conditions in Linear Environments

No matter whether $\theta_i \geq \hat{\theta}_i$ or $\theta_i \leq \hat{\theta}_i$, by condition (29), we always have

$$\int_{\hat{\theta}_i}^{\theta_i} \bar{v}_i(s) ds \geq \bar{v}_i\left(\hat{\theta}_i\right)\left(\theta_i - \hat{\theta}_i\right).$$

Hence, we have

$$U_i(\theta_i) = U_i\left(\hat{\theta}_i\right) + \int_{\hat{\theta}_i}^{\theta_i} \bar{v}_i(s) ds \geq U_i\left(\hat{\theta}_i\right) + \bar{v}_i\left(\hat{\theta}_i\right)\left(\theta_i - \hat{\theta}_i\right) = \theta_i \bar{v}_i\left(\hat{\theta}_i\right) - \bar{c}_i\left(\hat{\theta}_i\right).$$

Therefore, truth telling is a BNE, i.e., $f(\theta)$ is incentive compatible. ∎

Conditions (29) and (30) impose two conditions on allocation schemes, ensuring that the allocation schemes are incentive compatible. Conditional on these two conditions, we can look for the optimal implementable allocation scheme. These two conditions are used as IC conditions in a mechanism-design problem to ensure implementability of the optimal allocation scheme.

Proposition 11.4 (The Revenue Equivalence Theorem). *In an auction with n buyers and $u_i(x, \theta) = \theta_i \delta_i - c_i$ with $\Theta_i = [\underline{\theta}_i, \overline{\theta}_i]$ and buyers' types being statistically independent with density function $\phi_i(\theta_i) > 0$ for all $\theta_i \in \Theta_i$, if two auction procedures generate BNEs satisfying*

(1) $[\delta_1(\theta), \ldots, \delta_n(\theta)]$ *are the same in the two procedures for all θ,*
(2) $[U_1(\underline{\theta}_1), \ldots, U_n(\underline{\theta}_n)]$ *are the same in the two procedures,*

then these two procedures yield the same expected revenue for the seller.[11] ∎

As an example of Proposition 11.4, consider the first-price and second-price sealed-bid auctions in Examples 11.5 and 11.6. The two conditions in Proposition 11.4 are satisfied: in both auctions, the buyer with the highest valuation always gets the good and a buyer with a zero valuation has an expected utility of zero. Thus, Proposition 11.4 tells us that the seller receives exactly the same expected revenue in these two auctions. We will later show in Sect. 6.4 that the resulting allocation schemes are optimal in maximizing the seller's expected revenue.

---

[11] The proof of this proposition is in the problem set.

5 IR Conditions and Efficiency Criteria

5.1 Individual Rationality Conditions

In a mechanism-design problem, individuals can decide whether or not to participate. The conditions that ensure individuals' participation are called individual rationality (IR) conditions. Hence, the designer of a mechanism-design problem must not only offer IC conditions to ensure implementability of an allocation scheme but also IR conditions to ensure participation of targeted individuals.

There are three types of IR conditions conditional on the three time points. Let $\bar{u}_i(\theta_i)$ be the alternative utility value that agent i with type θ_i can obtain elsewhere after he declines to participate in the mechanism in question. The ex-ante IR condition is to induce agent i to take part in a mechanism ex ante:

$$E_\theta\{u_i[f(\theta), \theta_i]\} \geq E_{\theta_i}[\bar{u}_i(\theta_i)]. \tag{33}$$

The interim IR condition is to induce him to take part in a mechanism after knowing his own type:

$$E_{\theta_{-i}|\theta_i}\{u_i[f(\theta), \theta_i]\} \geq \bar{u}_i(\theta_i). \tag{34}$$

The ex-post IR condition is to induce him to take part in a mechanism ex post:

$$u_i[f(\theta), \theta_i] \geq \bar{u}_i(\theta_i). \tag{35}$$

Proposition 11.5 (Myerson-Satterthwaite). *Consider a bilateral trade in which $u_i(x, \theta) = \theta_i \delta_i - c_i$, the valuations $\{\theta_i\}$ are independently drawn from $[\underline{\theta}_i, \bar{\theta}_i]$ with $\phi_i(\theta_i) > 0$ for all θ_i, and $(\underline{\theta}_1, \bar{\theta}_1) \cap (\underline{\theta}_2, \bar{\theta}_2) \neq \emptyset$. There is no ex-post efficient allocation scheme that is incentive compatible and individually rational in the interim.* ∎

This proposition indicates that, when gains from trade are quite uncertain,[12] it is difficult to achieve ex-post efficiency. This result is not surprising. By Akerlof (1970), we already know that a market outcome under incomplete information can be inefficient. A typical insurance problem also indicates inefficiency under incomplete information.

5.2 Efficiency Criteria

Similar to the three types of IR conditions, we also have three types of efficiency criteria conditional on the three time points.

---

[12] For example, the case with $\bar{\theta}_1 < \underline{\theta}_2$, in which it is certain that a trade can benefit both, is ruled out.

5 IR Conditions and Efficiency Criteria

Definition 11.7 Given a set of feasible allocation schemes $\mathcal{F}$, $f \in \mathcal{F}$ is <u>ex-ante</u> <u>efficient</u> in $\mathcal{F}$ if there is no $\hat{f} \in \mathcal{F}$ such that $E_\theta\{u_i[\hat{f}(\theta), \theta_i]\} \geq E_\theta\{u_i[f(\theta), \theta_i]\}$ for all i and $E_\theta\{u_i[\hat{f}(\theta), \theta_i]\} > E_\theta\{u_i[f(\theta), \theta_i]\}$ for some i. ∎

Definition 11.8 Given a set of feasible allocation schemes $\mathcal{F}$, $f \in \mathcal{F}$ is <u>interim</u> <u>efficient</u> in $\mathcal{F}$ if there is no $\hat{f} \in \mathcal{F}$ such that $E_{\theta_{-i}|\theta_i}\{u_i[\hat{f}(\theta), \theta_i]\} \geq E_{\theta_{-i}|\theta_i}\{u_i[f(\theta), \theta_i]\}$ for all i and $\theta_i \in \Theta_i$, and $E_{\theta_{-i}|\theta_i}\{u_i[\hat{f}(\theta), \theta_i]\} > E_{\theta_{-i}|\theta_i}\{u_i[f(\theta), \theta_i]\}$ for some i and θ_i. ∎

Definition 11.9 Given a set of feasible allocation schemes $\mathcal{F}$, $f \in \mathcal{F}$ is <u>ex-post</u> <u>efficient</u> in $\mathcal{F}$ if there is no $\hat{f} \in \mathcal{F}$ such that $u_i[\hat{f}(\theta), \theta_i] \geq u_i[f(\theta), \theta_i]$ for all i and $\theta \in \Theta$, and $u_i[\hat{f}(\theta), \theta_i] > u_i[f(\theta), \theta_i]$ for some i and θ. ∎

The ex-post efficiency in Definition 11.9 is the same as the ex-post efficiency in Definition 11.1 when $\mathcal{F}$ contains all possible mappings from Θ to $\mathbb{X}$, i.e., $\mathcal{F} = \{f : \Theta \to X\}$. In other words, Definition 11.9 is a general definition, while Definition 11.1 is a special case when $\mathcal{F}$ contains all possible allocation schemes. For example, if f is not ex-post efficient according to Definition 11.1, then there exists x^0 and θ_0 such that $u_i(x^0, \theta_i^0) \geq u_i[f(\theta^0), \theta_i^0]$ for all i, and $u_i(x^0, \theta_i^0) > u_i[f(\theta^0), \theta_i^0]$ for some i. Then, if we define

$$\hat{f}(\theta) = \begin{cases} f(\theta) & \text{if } \theta \neq \theta^0, \\ x^0 & \text{if } \theta = \theta^0, \end{cases}$$

we have $u_i[\hat{f}(\theta), \theta_i] \geq u_i[f(\theta), \theta_i]$ for all i and $\theta \in \Theta$, and $u_i[\hat{f}(\theta), \theta_i] > u_i[f(\theta), \theta_i]$ for θ^0 and some i. That is, f is not ex-post efficient according to Definition 11.9.

Proposition 11.6 *Given a set of feasible allocation schemes $\mathcal{F}$, with continuous ϕ and $\phi(\theta) > 0$ for $\theta \in \Theta$, an ex-ante efficient allocation scheme must be interim efficient, and an interim efficient allocation scheme must be ex-post efficient.*

Proof If $f \in \mathcal{F}$ is not interim efficient, then there exists $\hat{f} \in \mathcal{F}$ such that $E_{\theta_{-i}|\theta_i}\{u_i[\hat{f}(\theta), \theta_i]\} \geq E_{\theta_{-i}|\theta_i}\{u_i[f(\theta), \theta_i]\}$ for all i and θ_i, and $E_{\theta_{-i}|\theta_i}\{u_i[\hat{f}(\theta), \theta_i]\} > E_{\theta_{-i}|\theta_i}\{u_i[f(\theta), \theta_i]\}$ for some i and θ_i. Taking expectation E_{θ_i} on the inequalities gives us $E_\theta\{u_i[\hat{f}(\theta), \theta_i]\} \geq E_\theta\{u_i[f(\theta), \theta_i]\}$ for all i and $E_\theta\{u_i[\hat{f}(\theta), \theta_i]\} > E_\theta\{u_i[f(\theta), \theta_i]\}$ for some i, where the strict inequality is due to continuity of ϕ_i and $\phi_i(\theta_i) > 0$ for all θ_i. Thus, f is not ex-ante efficient.

If $f \in \mathcal{F}$ is not ex-post efficient. Then, there exists $\hat{f} \in \mathcal{F}$ such that $u_i[\hat{f}(\theta), \theta_i] \geq u_i[f(\theta), \theta_i]$ for all i and $\theta \in \Theta$, and $u_i[\hat{f}(\theta), \theta_i] > u_i[f(\theta), \theta_i]$ for some i and θ. Taking expectation $E_{\theta_{-i}|\theta_i}$ on the inequalities gives us $E_{\theta_{-i}|\theta_i}\{u_i[\hat{f}(\theta), \theta_i]\} \geq E_{\theta_{-i}|\theta_i}\{u_i[f(\theta), \theta_i]\}$ for all i and θ_i, and $E_{\theta_{-i}|\theta_i}\{u_i[\hat{f}(\theta),$

$\theta_i]\} > E_{\theta_{-i}|\theta_i}\{u_i[f(\theta), \theta_i]\}$ for some i and θ_i, where the strict inequality is due to continuity of ϕ_{-i} and $\phi_{-i}(\theta_{-i}) > 0$ for all θ_{-i}. Thus, f is not interim efficient. ∎

Comparing Definition 11.7 with Definition 11.9, we can see that condition

$$E_\theta\{u_i[\hat{f}(\theta), \theta_i]\} \geq E_\theta\{u_i[f(\theta), \theta_i]\}, \quad \text{for all } i$$

is easier to satisfy than condition

$$u_i[\hat{f}(\theta), \theta_i] \geq u_i[f(\theta), \theta_i], \quad \text{for all } i \text{ and } \theta.$$

Hence, it is easier to find a Pareto improving $\hat{f}$ ex ante than ex post, implying that it is harder for an allocation scheme f to be ex-ante efficient than ex post efficient. This explains why an ex-ante efficient allocation must be ex-post efficient.

Although we have three time points for IR conditions, there is only one time point for IC conditions. IC conditions are based on Definition 11.6, which is at the interim stage. In other words, when it is time for an agent to report his type, he is always assumed to know his own type but not others' types.

6 Optimal Allocation Schemes

This section discusses mechanism-design problems, in which the designer selects the optimal allocation scheme among incentive compatible and individually rational allocation schemes. We consider four models under different scenarios.

6.1 Monopoly Pricing

We now revisit the problem of monopoly pricing under incomplete information in section 2.3 of Chapter 6. There are two consumers with the demand curves indicated in Fig. 7, where consumer 2's demand curve is above consumer 1's demand curve, with maximum demand $\bar{x}_1$ and $\bar{x}_2$, respectively, at zero price. Assume zero cost.

Suppose that the two demand functions are

$$p = \phi_1(x), \quad p = \phi_2(x), \quad \text{where } \phi_1(x) < \phi_2(x) \text{ for all } x.$$

The monopolist's problem is to solve for (x_1^*, x_2^*, B^*) from

6 Optimal Allocation Schemes

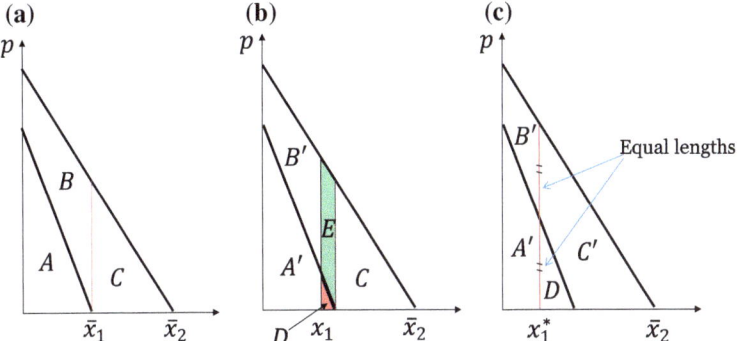

Fig. 7 Monopoly price under Incomplete Information

$$\max_{x_1, x_2, B} \int_0^{x_1} \phi_1(t)dt + \int_0^{x_2} \phi_2(t)dt - B(x_1, x_2)$$

s.t. IC condition,

and make offers[13]:

$$\left(x_1, \int_0^{x_1} \phi_1(t)dt\right), \quad \left(x_2, \int_0^{x_2} \phi_2(t)dt - B(x_1, x_2)\right).$$

To satisfy the IC condition, the monopolist chooses $B = \int_0^{x_1} [\phi_2(t) - \phi_1(t)]dt$. The IC condition ensures that each consumer chooses the package intended for him. Hence, the problem becomes

$$\max_{x_1, x_2} 2 \int_0^{x_1} \phi_1(t)dt + \int_{x_1}^{x_2} \phi_2(t)dt.$$

The FOCs are

$$\frac{\partial \pi}{\partial x_1} = 2\phi_1(x_1) - \phi_2(x_1) = 0,$$

$$\frac{\partial \pi}{\partial x_2} = \phi_2(x_2) = 0.$$

They imply $x_2^* = \bar{x}_2$ and $\phi_1(x_1^*) = \phi_2(x_1^*) - \phi_1(x_1^*)$. This is the solution in Fig. 7c.

---
[13] For simplicity, I have narrowed down the monopolist's admissible package offers.

Under complete information, social welfare is

$$SW^{**} = \int_0^{\bar{x}_1} \phi_1(t)dt + \int_0^{\bar{x}_2} \phi_2(t)dt.$$

With the optimal solution, social welfare is

$$SW^* = 2\int_0^{x_1^*} \phi_1(t)dt + \int_{x_1^*}^{\bar{x}_2} \phi_2(t)dt + \int_0^{x_1^*} [\phi_2(t) - \phi_1(t)]dt$$

$$= \int_0^{x_1^*} \phi_1(t)dt + \int_0^{\bar{x}_2} \phi_2(t)dt.$$

Hence, the optimal solution is inefficient and the dead weight loss is area D:

$$SW^{**} - SW^* = \int_{x_1^*}^{\bar{x}_1} \phi_1(t)dt.$$

6.2 A Buyer-Seller Model with Linear Utility Functions

Suppose that a seller is selling a single product to a buyer. The seller has type θ, which the buyer does not know. The buyer only knows that the type follows $U[0, 1]$. The buyer knows that the relationship between investment I and quality q of the product for the seller of type θ is

$$q = \theta I.$$

The cost of investment for the seller is

$$c(I) = \theta I.$$

For $\beta \in (0, 1)$, the buyer's value from the product of quality q is

$$h(q) = \frac{1}{\beta} q^\beta.$$

The buyer can observe and verify investment I. Hence, the buyer can offer a deal (p, I) to the seller. The direct mechanism is: the seller of type θ reports his type $\hat{\theta}$, and then the buyer pays the price $p(\hat{\theta})$ for the product and demands investment

6 Optimal Allocation Schemes

$I(\hat{\theta})$ for the production of the product. Given a contract (p, I), the seller's surplus is $u = p - c(I)$ and the buyer's surplus is $v = h(q) - p$. We will ignore an IR condition. Instead, we impose the following boundary conditions:

$$I(1) = \bar{I}, \quad p(1) = \bar{p},$$

where we take $\bar{I} \equiv 2^{\frac{1}{p-1}}$ and $\bar{p} = 1$ for convenience.

A contract between the buyer and the seller is an allocation scheme. By the revelation principle, the buyer can confine her search for an optimal contract to the set of incentive-compatible allocation schemes $f(\theta) = [p(\theta), I(\theta)]$. Hence, the buyer's problem is

$$\max_{f(\cdot) = [p(\cdot), I(\cdot)]} E_\theta \{h[\theta I(\theta)] - p(\theta)\}$$
$$\text{s.t.} \quad f(\cdot) \text{ is incentive compatible,}$$
$$\text{other conditions.}$$

As a comparison, we will use two approaches to solve this problem for the optimal contract $(p^*(\theta), I^*(\theta))$.

Approach 1

Given an offer $(p(\theta), I(\theta))$, the value function for the seller is $U(\hat{\theta}, \theta) = p(\hat{\theta}) - \theta I(\hat{\theta})$. Then, the FOC for reporting $\hat{\theta}$ is

$$p'(\hat{\theta}) - \theta I'(\hat{\theta}) = 0.$$

Hence, a truth reporting condition is

$$p'(\theta) - \theta I'(\theta) = 0, \quad \forall \theta. \tag{36}$$

The SOC is $p''(\hat{\theta}) - \theta I''(\hat{\theta}) \leq 0$. Taking a derivative of (36) w.r.t. θ yields $p''(\theta) - \theta I''(\theta) - I'(\theta) = 0$. Then, the SOC becomes:

$$I'(\theta) \leq 0. \tag{37}$$

Hence, the buyer's problem is

$$V = \max_{p(\theta), I(\theta)} \int_0^1 \{h[\theta I(\theta)] - p(\theta)\} d\theta$$
$$\text{s.t.} \quad p'(\theta) - \theta I'(\theta) = 0, \tag{38}$$
$$I'(\theta) \leq 0$$
$$I(1) = \bar{I}, \quad p(1) = 1.$$

We ignore the SOC for the time being and consider the following problem:

$$V = \max_{p(\theta), I(\theta)} \int_0^1 \left\{ \tfrac{1}{\beta}[\theta I(\theta)]^\beta - p(\theta) \right\} d\theta$$
$$\text{s.t.} \quad p'(\theta) - \theta I'(\theta) = 0,$$
$$I(1) = \bar{I}, \quad p(1) = 1.$$

Define the Hamiltonian function:

$$H(p, p', I, I', \theta) = \frac{1}{\beta}(\theta I)^\beta - p + \lambda(\theta)(p' - \theta I'),$$

where λ is the Lagrangian multiplier, dependent on θ. The Euler equations are

$$\frac{d}{d\theta} H_{p'} = H_p : \lambda' = -1, \tag{39}$$

$$\frac{d}{d\theta} H_{I'} = H_I : (-\lambda\theta)' = \theta^\beta I^{\beta-1}. \tag{40}$$

Equation (39) implies

$$\lambda(\theta) = \lambda_0 - \theta, \tag{41}$$

where λ_0 is a constant. Equation (40) implies

$$\lambda'\theta + \lambda + \theta^\beta I^{\beta-1} = 0.$$

By (39) and (41), this equation becomes

$$\lambda_0 - 2\theta + \theta^\beta I^{\beta-1} = 0, \tag{42}$$

implying

$$I(\theta) = \left(\frac{2\theta - \lambda_0}{\theta^\beta}\right)^{\frac{1}{\beta-1}}.$$

Since $I(1) = \bar{I}$, we have $\lambda_0 = 0$, implying

$$I^*(\theta) = \bar{I}/\theta.$$

This solution satisfies the ignored SOC. Hence, this is indeed the solution of problem (38).

We now solve for $p^*(\theta)$. By (36), we have

$$p(1) - p(\theta) = \int_\theta^1 tI'(t)dt = -\bar{I}\int_\theta^1 \frac{1}{t}dt = \bar{I}\ln(\theta),$$

implying

$$p^*(\theta) = 1 - \bar{I}\ln(\theta).$$

The optimal contract $(p^*(\theta), I^*(\theta))$ is solved.

Approach 2

Since we have a linear environment, we can use Proposition 11.3 to set up the problem. The value function for the seller is

$$U(\theta) = p(\theta) - c[I(\theta)] = p(\theta) - \theta I(\theta).$$

Hence, $\bar{v}_i(\theta) = -I(\theta)$ for the $\bar{v}_i$ in Proposition 11.3. To use Proposition 11.3, we also need to replace $p(\theta)$ by $U(\theta)$ in our search for an optimal solution. The value function for the buyer is

$$V(\theta) = v[q(\theta)] - p(\theta) = v[\theta I(\theta)] - p(\theta) = v[\theta I(\theta)] - \theta I(\theta) - U(\theta).$$

By Proposition 11.3, the buyer's problem is

$$\max_{I(\cdot), U(\cdot)} \int_0^1 \left\{\frac{1}{\beta}[\theta I(\theta)]^\beta - \theta I(\theta) - U(\theta)\right\}d\theta$$
$$\text{s.t.} \quad IC_1: I'(\theta) \leq 0, \quad (43)$$
$$IC_2: U'(\theta) = -I(\theta),$$
$$BC: I(1) = \bar{I}, U(1) = \bar{p} - \bar{I}.$$

Instead of finding an optimal pair of $(I(\cdot), p(\cdot))$, we try to find an optimal pair of $(I(\cdot), U(\cdot))$.

To solve the problem, we first eliminate $U(\cdot)$ from the objective function so that only $I(\cdot)$ is left to determine. We have

$$\int_0^1 U(\theta)d\theta = \theta U(\theta)|_0^1 - \int_0^1 \theta U'(\theta)d\theta$$

$$= U(1) + \int_0^1 \theta I(\theta)d\theta = \bar{p} - \bar{I} + \int_0^1 \theta I(\theta)d\theta.$$

We can now eliminate $U(\theta)$ from the problem, which becomes

$$\max_{I(\cdot)} \int_0^1 \left\{ \tfrac{1}{\beta}[\theta I(\theta)]^\beta - 2\theta I(\theta) \right\} d\theta - \bar{p} + \bar{I}$$
$$\text{s.t.} \quad I'(\theta) \leq 0,$$
$$I(1) = \bar{I}.$$
(44)

Here, since the objective has nothing to do with $U(\theta)$, we have dropped the conditions relating to it. Without the IC condition in (44), the optimal solution I^* must satisfy Euler equation:

$$\theta^\beta I^{\beta-1} - 2\theta = 0, \quad \text{for all } \theta,$$

implying

$$I^*(\theta) = 2^{\frac{1}{\beta-1}}/\theta = \bar{I}/\theta.$$

This solution indeed satisfies the IC condition in (44). Hence, it is the solution of problem (44).

We now go back to solve for $p^*(\theta)$. By the IC_2 condition in (43), we have

$$U(1) - U(\theta) = -\int_\theta^1 I^*(t)dt = -\bar{I}\int_\theta^1 \frac{1}{t}dt = \bar{I}\ln\theta,$$

implying

$$p^*(\theta) = U(\theta) + \theta I^*(\theta) = U(1) - \bar{I}\ln\theta + \bar{I} = \bar{p} - \bar{I}\ln\theta.$$

6.3 Labor Market

Consider the Spence model in the previous chapter. There is a single firm, which is now a monopsony in the labor market. Instead of two possible types, we assume that the workers have a continuum of possible types $\theta \in \Theta = [\underline{\theta}, \bar{\theta}]$, where $0 < \underline{\theta} < \bar{\theta}$. The distribution function is Φ and the density function is ϕ, with $\phi(\theta) > 0$ for all $\theta \in \Theta$. Each worker obtains an education level $e \in \mathbb{R}_+$ with utility function $u(e, w, \theta) = w - \theta c(e)$, where w is the wage and $c(e)$ represents the agent's cost for education. Assume that $c(\cdot)$ is differentiable and

$$c(0) = 0, \quad c'(0) = 0, \quad c(e) > 0, \quad c'(e) > 0, \quad c''(e) > 0, \quad \text{for } e > 0.$$

Let the revenue function be $\varphi(\theta)v(e)$ for type θ. The firm's payoff is $\pi = \varphi(\theta)v(e) - w$, where $v(\cdot)$ is differentiable with $v' > 0, v'' \leq 0, \varphi(\theta) > 0$ and

6 Optimal Allocation Schemes

$\varphi'(\theta) \leq 0$.[14] A decreasing $\varphi(\theta)$ means that a high-ability worker is more productive, and an increasing $v(e)$ means that education has a positive effect on productivity. The firm does not observe a worker's type. It uses an allocation scheme to $[e(\theta), w(\theta)]$ to induce each worker to report his type truthfully.

Incomplete Information

Assume that θ is known only to the worker himself. Since each worker's utility function is linear in θ, we can use Proposition 11.3. Let $U(\theta) \equiv w(\theta) - \theta c[e(\theta)]$, which is the utility value if the worker's type is θ and he tells the truth. Suppose that the reservation utility value is $\bar{u}$. By Proposition 11.3, the firm's problem is

$$\max_{e(\cdot), U(\cdot)} E_\theta\{\varphi(\theta)v[e(\theta)] - \theta c[e(\theta)] - U(\theta)\}$$
$$\text{s.t.} \quad IC_1: e'(\theta) \leq 0, \quad (45)$$
$$IC_2: U'(\theta) = -c[e(\theta)],$$
$$IC: U(\theta) \geq \bar{u} \quad \text{for all } \theta.$$

Here, $-c[e(\theta)]$ is the $\bar{v}_i$ in Proposition 11.3. To simplify the problem, we want to eliminate $U(\cdot)$ from the objective function so that we only need to find one function $e(\cdot)$. By the IC_2 condition, $U(\theta)$ is decreasing. Hence, the IR condition is equivalent to $U(\bar{\theta}) \geq \bar{u}$. Also, we have

$$\int_\theta^{\bar{\theta}} U(\theta)\phi(\theta)d\theta = U(\bar{\theta}) - \int_\theta^{\bar{\theta}} \Phi(\theta)U'(\theta)d\theta = U(\bar{\theta}) + \int_\theta^{\bar{\theta}} \frac{\Phi(\theta)}{\phi(\theta)} c[e(\theta)]\phi(\theta)d\theta.$$

We can now eliminate $U(\theta)$ from problem (45). Let $J(\theta) \equiv \theta + \Phi(\theta)/\phi(\theta)$. Then, the problem becomes

$$\max_{e(\cdot), U(\bar{\theta})} \int_\theta^{\bar{\theta}} \{\varphi(\theta)v[e(\theta)] - J(\theta)c[e(\theta)]\}\phi(\theta)d\theta - U(\bar{\theta})$$
$$\text{s.t.} \quad e'(\theta) \leq 0, \quad (46)$$
$$U(\bar{\theta}) \geq \bar{u}.$$

It is obvious that we must have $U(\bar{\theta}) = \bar{u}$ for any solution. Thus, problem (46) can be further written as

$$\max_{e(\cdot)} \int_\theta^{\bar{\theta}} \{\varphi(\theta)v[e(\theta)] - J(\theta)c[e(\theta)]\}\phi(\theta)d\theta - \bar{u}$$
$$\text{s.t.} \quad e'(\theta) \leq 0. \quad (47)$$

---

[14] Here, θ corresponds to c_i in Spence's model and $\theta_i = \varphi(c_i)$. A bad type has a big θ and small productivity $\varphi(\theta)$.

Without condition $e'(\theta) \leq 0$, the optimal solution e^* must satisfy the Euler equation:

$$\varphi(\theta)v'[e^*(\theta)] - c'[e^*(\theta)]J(\theta) = 0, \quad \text{for all } \theta. \tag{48}$$

This implies[15]

$$e^{*\prime}(\theta) = \frac{c'[e^*(\theta)]J'(\theta) - \varphi'(\theta)v'[e^*(\theta)]}{\varphi(\theta)v''[e^*(\theta)] - c''[e^*(\theta)]J(\theta)} \leq 0,$$

which holds if $J(\theta)$ is increasing. By this condition, $e^*(\theta)$ is decreasing in θ, implying that condition $e'(\theta) \leq 0$ is satisfied and $e^*(\cdot)$ is indeed the solution of problem (47).

To find the optimal allocation scheme $f(\theta) = [e^*(\theta), w^*(\theta)]$, we further find $U(\theta)$ from the IC_2 condition:

$$\bar{u} - U(\theta) = -\int_{\theta}^{\bar{\theta}} c[e^*(t)]dt,$$

which then gives us

$$w^*(\theta) = U(\theta) + \theta c[e^*(\theta)] = \bar{u} + \int_{\theta}^{\bar{\theta}} c[e^*(t)]dt + \theta c[e^*(\theta)].$$

Complete Information

When a worker's type is publicly observable, the IC conditions are no longer necessary, since there is no need for workers to tell the truth. Then, the firm's problem (45) becomes[16]

$$\max_{e(\cdot), U(\cdot)} E\{\varphi(\theta)v[e(\theta)] - \theta c[e(\theta)] - U(\theta)\}$$

$$\text{s.t.} \quad U(\theta) \geq \bar{u} \quad \text{for all } \theta.$$

Obviously, we must have $U(\theta) = \bar{u}$ for all θ. Then, the Euler equation for the optimal solution e^{**} is

---

[15]Here, $e^*(\cdot)$ can be a strictly decreasing function, by which the company can recover the applicants' types completely. However, this does not mean that the solution is the complete-information solution. This only means that the solution is a separating equilibrium. The complete-information is $e^{**}(\cdot)$, which is generally more efficient.

[16]We may interpret the model as having many applicants who distribute alone $[\underline{\theta}_i, \bar{\theta}_i]$ with population density function $\phi(\theta)$. Hence, the objective function is the per capita profit.

$$\varphi(\theta)v'[e^{**}(\theta)] - \theta c'[e^{**}(\theta)] = 0, \quad \text{for all } \theta.$$

Since $e^{**}(\theta)$ maximizes the total social surplus $\pi + u - \bar{u} = \varphi(\theta)v[e(\theta)] - \theta c[e(\theta)] - \bar{u}$, it is ex-post efficient. This is not surprising since a solution under complete information is known to be ex-ante efficient and, by Proposition 11.6, ex-ante efficiency implies ex-post efficiency.

Since $\varphi(\theta)v''(e) - \theta c''(e) < 0$ and (48) implies

$$\varphi(\theta)v'[e^{*}(\theta)] - \theta c'[e^{*}(\theta)] = c'[e^{*}(\theta)]\frac{\Phi(\theta)}{\phi(\theta)} \geq 0,$$

we have

$$\begin{aligned} e^{*}(\theta) &< e^{**}(\theta), \quad \text{for all } \theta > \underline{\theta}, \\ e^{*}(\theta) &= e^{**}(\theta) \quad \text{at } \theta = \underline{\theta}. \end{aligned}$$

In other words, under incomplete information, only the best type has an ex-post efficient education level, while all other types have their education level distorted downward.

6.4 Optimal Auction

Consider the auction in Example 11.4 again. There is a seller selling an indivisible object to n buyers $i \in \mathbb{N} = \{1, \ldots, n\}$. Each buyer has a linear utility function of the form $u_i(\delta_i, \theta_i) = \theta_i \delta_i - c_i$, where $\delta_i = 0$ or 1. A feasible allocation scheme is

$$f(\theta) = [\delta_1(\theta), \ldots, \delta_n(\theta), c_1(\theta), \ldots, c_n(\theta)],$$

where

$$\delta_i(\theta) \in \{0, 1\}, \quad \sum_{i \in \mathbb{N}} \delta_i(\theta) = 1, \quad \text{for all } \theta \text{ and } i.$$

Denote $U_i(\theta_i) = \theta_i \bar{\delta}_i(\theta_i) - \bar{c}_i(\theta_i)$, where $\bar{\delta}_i(\theta_i) \equiv E_{\theta_{-i}}[\delta_i(\theta)]$ and $\bar{c}_i(\theta_i) \equiv E_{\theta_{-i}}[c_i(\theta)]$. Each buyer i's type is independently drawn from $\Theta_i = [\underline{\theta}_i, \bar{\theta}_i]$ with density ϕ_i and distribution Φ_i, and $\phi_i(\theta_i) > 0$ for all $\theta_i \in \Theta_i$. Assume that $J_i(\theta_i) \equiv \theta_i - \frac{1-\Phi_i(\theta_i)}{\phi_i(\theta_i)}$ is increasing in θ_i for all i and $\min_{i=1,\ldots,n}\{J_i(\underline{\theta}_i)\} \geq 0$.

The expected revenue is

$$E_\theta[R(\theta)] = \sum_{i \in \mathbb{N}} E_\theta[c_i(\theta)] = \sum_{i \in \mathbb{N}} E_{\theta_i}[E_{\theta_{-i}} c_i(\theta)] = \sum_{i \in \mathbb{N}} E_{\theta_i}[\bar{c}_i(\theta)].$$

Facing the interim IR condition $U_i(\theta_i) \geq 0$ for all $\theta_i \in \Theta_i$ and $i \in \mathbb{N}$, the seller wishes to choose an incentive compatible allocation scheme that maximizes his expected revenue. Thus, by Proposition 11.3, the seller's problem is

$$\max_{\{\delta_i(\cdot), U_i(\cdot)\}_{i \in \mathbb{N}}} \sum_{i \in \mathbb{N}} \int_{\underline{\theta}_i}^{\bar{\theta}_i} [\bar{\delta}_i(\theta_i)\theta_i - U_i(\theta_i)] \phi_i(\theta_i) d\theta_i$$

s.t.
IC_1: $\bar{\delta}_i(\theta_i)$ increasing in θ_i for all $i \in \mathbb{N}$,
IC_2: $U_i'(\theta_i) = \bar{\delta}_i(\theta_i)$, for all $i \in \mathbb{N}$ and θ_i,
IR: $U_i(\theta_i) \geq 0$, for all $i \in \mathbb{N}$ and θ_i,
WN: $\delta_i(\theta) \in \{0, 1\}$, $\sum_{i \in \mathbb{N}} \delta_i(\theta) = 1$, for all $i \in \mathbb{N}$ and θ, (49)

By IC_2, the IR condition is equivalent to $U_i(\underline{\theta}_i) \geq 0$. We also have

$$\int_{\underline{\theta}_i}^{\bar{\theta}_i} U_i(\theta_i) \phi_i(\theta_i) d\theta_i = U_i(\bar{\theta}_i) - \int_{\underline{\theta}_i}^{\bar{\theta}_i} U_i'(\theta_i) \Phi_i(\theta_i) d\theta_i$$

$$= U_i(\underline{\theta}_i) + \int_{\underline{\theta}_i}^{\bar{\theta}_i} U_i'(\theta_i)[1 - \Phi_i(\theta_i)] d\theta_i$$

$$= U_i(\underline{\theta}_i) + \int_{\underline{\theta}_i}^{\bar{\theta}_i} \bar{\delta}_i(\theta_i)[1 - \Phi_i(\theta_i)] d\theta_i.$$

Then, the objective function of (49) becomes

$$E(R) = \sum_{i \in \mathbb{N}} \left[\int_{\underline{\theta}_i}^{\bar{\theta}_i} \bar{\delta}_i(\theta_i) J_i(\theta_i) \phi_i(\theta_i) d\theta_i - U_i(\underline{\theta}_i) \right].$$

Since

$$\bar{\delta}_i(\theta_i) = E_{\theta_{-i}}[\delta_i(\theta)] = \int_{\underline{\theta}_1}^{\bar{\theta}_1} \cdots \int_{\underline{\theta}_{i-1}}^{\bar{\theta}_{i-1}} \int_{\underline{\theta}_{i+1}}^{\bar{\theta}_{i+1}} \cdots \int_{\underline{\theta}_n}^{\bar{\theta}_n} \delta_i(\theta) \prod_{j \neq i} (\phi_j(\theta_j) d\theta_j),$$

we have

$$E(R) = \sum_{i \in \mathbb{N}} \left[\int_{\underline{\theta}_1}^{\bar{\theta}_1} \cdots \int_{\underline{\theta}_n}^{\bar{\theta}_n} \delta_i(\theta) J_i(\theta_i) \phi(\theta) d\theta_1 \cdots d\theta_n - U_i(\underline{\theta}_i) \right]$$

$$= \int_{\underline{\theta}_1}^{\bar{\theta}_1} \cdots \int_{\underline{\theta}_n}^{\bar{\theta}_n} \left[\sum_{i \in \mathbb{N}} \delta_i(\theta) J_i(\theta_i) \right] \phi(\theta) d\theta_1 \cdots d\theta_n - \sum_{i \in \mathbb{N}} U_i(\underline{\theta}_i).$$

Hence, the seller's problem is to choose functions $\delta_i(\theta)$ and values $U_i(\underline{\theta}_i)$ to maximize this expected revenue subject to IC_1, WN and $U_i(\underline{\theta}_i) \geq 0$ conditions for all $i \in \mathbb{N}$. It is evident that the solution must have $U_i(\underline{\theta}_i) = 0$ for all $i \in \mathbb{N}$. Hence, the seller's problem is reduced to choosing functions $\delta_i(\theta)$ for all i to maximize

$$E(R) = \int_{\underline{\theta}_1}^{\bar{\theta}_1} \cdots \int_{\underline{\theta}_n}^{\bar{\theta}_n} \phi(\theta) \sum_{i \in \mathbb{N}} \delta_i(\theta) J_i(\theta_i) d\theta_1 \cdots d\theta_n, \tag{50}$$

subject to IC_1 and WN conditions. We ignore IC_1 for the time being. Then, by the Hamilton approach, an inspection of (50) indicates that the solution $\{\delta_i^*(\cdot)\}$ of the problem without the IC_1 condition is, for all $i \in \mathbb{N}$ and $\theta \in \Theta$,

$$\delta_i^*(\theta) = \begin{cases} 1 & \text{if } J_i(\theta_i) = \max_j \{J_j(\theta_j)\}, \\ 0 & \text{if otherwise} \end{cases} \tag{51}$$

This solution indeed satisfies the IC_1 condition. Thus, $\{\delta_i^*(\cdot)\}$ is the solution of the seller's problem (49).[17]

By the definition of $U_i(\theta_i)$, the optimal transfer scheme is

$$\bar{c}_i^*(\theta_i) = \theta_i \bar{\delta}_i^*(\theta_i) - U_i(\theta_i) = \theta_i \bar{\delta}_i^*(\theta_i) - \int_{\underline{\theta}_i}^{\theta_i} \bar{\delta}_i^*(s) ds. \tag{52}$$

Here, the term $\theta_i \bar{\delta}_i^*(\theta_i)$ indicates the winner's benefit θ_i, but the term $\int_{\underline{\theta}_i}^{\theta_i} \bar{\delta}_i^*(s) ds$ indicates that the payment is lower than the benefit.

We have two observations of the solution. First, the agent who has the largest value of $J_i(\theta_i)$ does not necessarily have the largest valuation θ_i. Thus, the optimal solution need not be ex-post efficient.

Second, in the case of symmetric bidders in which $\underline{\theta}_i = \underline{\theta}$ and $J_i(\cdot) = J(\cdot)$ for all $i \in \mathbb{N}$, where $\underline{\theta} > 0$ is large enough so that $J(\underline{\theta}) > 0$, the optimal auction always

---

[17]When two individuals have the same largest $J_i(\theta_i)$, give the item arbitrarily to one of them. Furthermore, since $J_i(\cdot)$ is increasing, $\delta_i^*(\theta)$ is increasing in θ_i.

gives the object to the bidder with the highest valuation. This means that the solution is ex-post efficient in this case. Further, the payment scheme is based on the second price. To see this, if θ_i is the highest among $\{\theta_j\}$ and θ_{sec} is the second highest, (52) implies

$$c_i^*(\theta) = \theta_i - \int_{\theta_{\text{sec}}}^{\theta_i} 1 ds = \theta_{\text{sec}}.$$

That is, the winner i pays θ_{sec}, which is the second highest bid in the direct mechanism. Hence, the second-price sealed-bid auction is optimal. Using the revenue equivalence theorem, the first-price sealed-bid auction is also optimal.

6.5 A Buyer-Seller Model with Quasi-Linear Utility

Consider a model[18] with one buyer and one seller who have utility functions $U(x, y, \theta)$ and $V(x, y, \theta)$, respectively, where θ is private information of the seller, and x and y are commonly observable variables based on which a contract can be specified. The contract (x, y) is proposed by the buyer. Let $F(\theta)$ be the distribution function for $\theta \in \Theta$, where $\Theta = [\underline{\theta}, \bar{\theta}]$. Using (18) and (19), the buyer's ex-ante optimization problem is

$$\begin{aligned} \max_{x(\cdot), y(\cdot)} \quad & \int U[x(\theta), y(\theta), \theta] dF(\theta) \\ \text{s.t.} \quad & \frac{\partial V}{\partial x} x' + \frac{\partial V}{\partial y} y' = 0, \\ & \frac{\partial^2 V}{\partial \theta \partial x} x' + \frac{\partial^2 V}{\partial \theta \partial y} y' \geq 0, \\ & \int V[x(\theta), y(\theta), \theta] dF(\theta) \geq \bar{v}. \end{aligned} \quad (53)$$

Depending on the interpretation of the problem, we may alternatively consider an ex-post or interim IR condition:

$$V[x(\theta), y(\theta), \theta] \geq \bar{v}, \quad \forall \theta \in \Theta.$$

We may interpret the ex-ante IR condition as a situation in which the seller first agrees to participate before he knows his own type and the trade $\left[x(\hat{\theta}), y(\hat{\theta})\right]$ with a report type $\hat{\theta}$ occurs later when the seller knows his own type. In contrast, the ex-post or interim IR condition ensures that every agent is willing to participate after the agent knows his own type.

Consider quasi-linear utility functions of the form:

$$U(x, y, \theta) = u(x, \theta) - y, \quad V(x, y, \theta) = v(x, \theta) + y.$$

---

[18]This example is from Guesnerie and Laffont (1984). See also Laffont (1989, p.159).

We can interpret y as the price of the good and the buyer purchases x for a price y. Then, the buyer's problem becomes

$$\max_{x(\cdot), y(\cdot)} \int [u(x(\theta), \theta) - y(\theta)] dF(\theta)$$
$$\text{s.t.} \quad IC_1: v_x x' + y' = 0,$$
$$IC_2: v_{x\theta} x' \geq 0, \quad (54)$$
$$IR: \int v dF(\theta) + \int y dF \geq \bar{v}.$$

The IR condition must be binding, by which the objective function becomes

$$\int_\Theta U[x(\theta), y(\theta), \theta] dF(\theta) = \int_\Theta \{u[x(\theta), \theta] + v[x(\theta), \theta]\} dF(\theta) - \bar{v}.$$

By assuming $v_{x\theta} \geq 0,$[19] the IC_2 condition is satisfied if $x'(\theta) \geq 0$. Since y is no longer involved in the objective function, the IC_1 and IR are no longer needed. Therefore, the optimization problem becomes

$$\max_{x(\cdot)} \int_\Theta \{u[x(\theta), \theta] + v[x(\theta), \theta]\} dF(\theta) \quad (55)$$
$$\text{s.t.} \quad x'(\theta) \geq 0.$$

The Lagrangian function of (55) is

$$L = \int_\Theta \{u[x(\theta), \theta] + v[x(\theta), \theta] + \lambda(\theta)\dot{x}(\theta)\} dF(\theta),$$

where $\dot{x}(\theta) \equiv x'(\theta)$ and $\lambda(\theta) \geq 0$. Then, the Hamiltonian function is

$$H(x, \lambda, \theta) = u(x, \theta) + v(x, \theta) + \lambda \dot{x},$$

By the Lagrange theorem, the conditions for an optimal solution $x^*(\cdot)$ are

$$\begin{aligned}
\text{Euler equation}: & \quad \dot{\lambda} = u_x(x^*, \theta) + v_x(x^*, \theta), \\
\text{Transition}: & \quad \dot{x}^*(\theta) \geq 0, \\
\text{Transversality}: & \quad \lambda(\underline{\theta}) = \lambda(\bar{\theta}) = 0, \\
\text{Kuhn-Tucker condition}: & \quad \lambda(\theta)\dot{x}^*(\theta) = 0.
\end{aligned} \quad (56)$$

On the other hand, under complete information, we drop the IC conditions in (53), and the problem becomes

---

[19]Laffont [1989, p.155, expression (7)] has this assumption. It means that the marginal utility v_x increases as θ increases.

$$\max_{x(\cdot),y(\cdot)} \int U[x(\theta),y(\theta),\theta]dF(\theta)$$

$$\text{s.t.} \quad \int V[x(\theta),y(\theta),\theta]dF(\theta) \geq \bar{v}.$$

Since the IR condition must be binding, the problem becomes

$$\max_{x(\cdot)} \int_\Theta \{u[x(\theta),\theta] + v[x(\theta),\theta]\}dF(\theta). \tag{57}$$

Let $x^{**}(\theta)$ be the solution of problem (57). By the Hamiltonian method, $x^{**}(\theta)$ is a solution of the following equation:

$$u_x[x^{**}(\theta),\theta] + v_x[x^{**}(\theta),\theta] = 0, \quad \forall \theta. \tag{58}$$

By the concavity of u and v in x, the solution is unique. Thus, any solution satisfying (58) must be the optimal solution of problem (57).

We now try to solve (56). Due to the condition $\dot{x}(\theta) \geq 0$, $x^*(\theta)$ can either be increasing or constant. Thus, to find a solution, we follow two steps:

First, if $\dot{x}^*(\theta) > 0$ on an interval (A, B), then by the Kuhn-Tucker condition we have $\lambda(\theta) = 0$, and thus $\dot{\lambda}(\theta) = 0$ on (A, B). By the Euler equation, $x^*(\theta) = x^{**}(\theta)$ for $\theta \in (A, B)$.

Second, if $\dot{x}^*(\theta) = 0$, or $x^*(\theta) = c$ for some constant c, on an interval (a, b) and $\dot{x}^*(\theta) > 0$ in nearby regions, by continuity of $\lambda(\theta)$, we must have $\lambda(a) = \lambda(b) = 0$. By the Euler equation, we know that $\lambda(\theta)$ is continuous. Thus, by the Euler equation,

$$\int_a^b [u_x(\theta,c) + v_x(\theta,c)]dF(\theta) = 0. \tag{59}$$

By continuity of x^{**}, we also have

$$x^{**}(a) = c \quad \text{and} \quad x^{**}(b) = c. \tag{60}$$

The conditions (59) and (60) determine a, b and $= c$.

These two steps will give us a solution that can be verified to satisfy the conditions in (56). An illustration of x^* and x^{**} is in the following figure (Fig. 8).

At the end of this chapter, we address a question: if the direct mechanism is so wonderful, why not we see it often in practice? The reason is that the direct mechanism is convenient to use for theorists, but difficult to use for practitioners. For example, suppose that a seller is to sell a piece of painting and there are n buyers with independently uniformly distributed types. By the direct mechanism, the seller needs to know all utility functions $u_i(x, \theta_i)$, $i = 1, \ldots, n$, set up the

6 Optimal Allocation Schemes

Fig. 8 The optimal contract

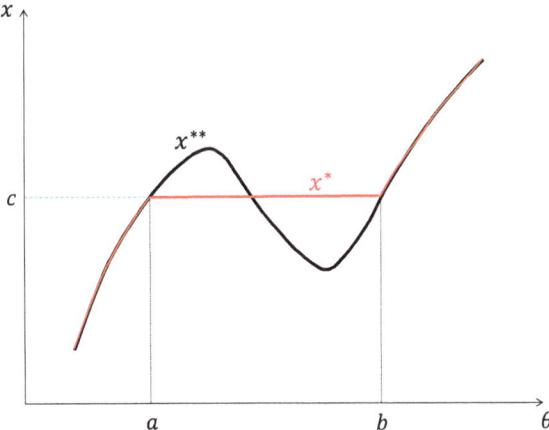

mechanism-design problem with IC and IR conditions, and solve the problem. In practice, the seller may not know the utility functions; and even if she knows all utility functions, solving the mechanism-design problem is difficult. Instead, the seller can use the mechanism called the second-price sealed-bit auction. With this auction, the seller only needs to announce this mechanism, wait for the buyers to play out the game, and then receive the payment. As shown in Sect. 6.4 on optimal auction, this payment has the highest expected value that the seller can expect to receive. That is, the seller's expected incomes from the direct mechanism and the second-price sealed-bit auction are identical. Therefore, the seller will simply use the second-price sealed-bit auction instead of the direct mechanism.

Notes

Good references for this chapter are Mas-Colell et al. (1995, Chapter 23) and Laffont (1995).

Incentive Contracts 12

Besides asymmetric information, there is another type of information problem, called the incentive problem. For example, in an employer-employee relationship, the employee's applied effort may be observable by his employer but not verifiable to a court. If so, the effort cannot be bounded by a contract. In this type of problem, information is symmetric: both the employer and the employee have the same set of information. But a third party, the court, cannot observe the information. So the issue here is: how does the employer provide sufficient incentives in a contract to motivate the employee?

1 The Standard Agency Model

In this section, we present the standard agency model developed by Mirrlees (1974) and Holmström (1979). This section is mainly based on Holmström (1979).[1] The standard contract theory is about an output-sharing rule between a principal and an agent. The principal tries to hire the agent to work on a project. A problem is that the agent's effort may be observable to the principal but is not verifiable to a third party such as a court and is thus non-contractable.[2] Hence, the principal must provide incentives in a contract for the agent to put in a sufficient effort voluntarily. The question is how to provide sufficient incentives in a contract.

Let $\tilde{x}$ be a random output from the firm. Given the agent's effort a, the distribution function of $\tilde{x}$ is $F(x; a)$ and the density function is $f(x; a)$. The effort a is not

---

[1] For more details of the standard principal-agent model, see Ross (1973), Stiglitz (1974), Harris and Raviv (1979), Shavell (1979), and Laffont (1995, pp.180–198).
[2] The word "non-contractable" or "uncontractable" does not mean that a term on the agent's effort cannot be put into a contract; it actually means that such a term is not enforceable by a court. That is, the word "non-contractable" actually means "non-enforceable". See Wang (2016) on the definition of incomplete contracts.

verifiable or contractible; more precisely, it is uncontractable ex ante since it is non-verifiable ex post. In fact, only the output x is contractible (observable and verifiable ex post) in this model. The output is random ex ante, but it becomes known and verifiable ex post. Thus a payment scheme $s(\cdot)$ can be based on the output x; the scheme pays $s = s(x)$ to the agent when output is x. The set of admissible output sharing rules is

$$\mathcal{S} = \{s(x) | s(x) \text{ is Lebesgue integrable and } s(x) \geq 0\}. \tag{1}$$

Here, condition $s(x) \geq 0$ means limited liability to the agent. Let $\mathbb{A}$ be the effort space, which is often taken as $\mathbb{R}_+$. The agent's utility function is

$$u(s) - c(a), \quad \text{with } u' > 0, u'' \leq 0, c' > 0, c'' > 0, \tag{2}$$

where $c(a)$ is the cost of effort which is private. The principal's utility function is

$$v(x - s), \quad \text{with } v' > 0, v'' \leq 0. \tag{3}$$

In addition, assume that at least one of the parties is risk averse.[3] This assumption is necessary for the standard agency theory. We will discuss the special case when both the principal and the agent are risk neutral in Sect. 3.

1.1 Verifiable Effort: The First Best

As a benchmark, we first consider the problem when effort a is verifiable. The principal specifies both an effort and a payment scheme in a contract. That is, the principal offers a contract of the form $[a, s(\cdot)]$ to the agent. Let $\bar{u}$ be the agent's reservation utility, below which the agent will turn down the contract. The principal is constrained by the participation constraint or <u>individual rationality (IR) condition</u>:

$$\int_{-\infty}^{\infty} u[s(x)] f(x; a) dx - c(a) \geq \bar{u}.$$

The principal's problem is then

$$\max_{a \geq 0, s \in \mathcal{S}} \int_{-\infty}^{\infty} v[x - s(x)] f(x; a) dx$$
$$\text{s.t.} \quad IR: \int_{-\infty}^{\infty} u[s(x)] f(x; a) dx \geq c(a) + \bar{u}. \tag{4}$$

---

[3]Many people have the misunderstanding that the agent must be risk averse. Although this is the original assumption, it is actually unnecessary.

1 The Standard Agency Model

Its solution (a^{**}, s^{**}) is said to be the first best and hence the above problem is called the first-best problem. Since this solution maximizes social welfare, it is Pareto optimal. Wang (2012b) shows that this solution is also obtainable as a general equilibrium solution.

Given each a, the *support* of the distribution function is the interval of x on which $f(x; a) > 0$. This interval will generally be dependent on a. The following assumption is imposed in the standard agency model.

Assumption 12.1 The support of the output distribution function is independent of a.

Assumption 12.1 dramatically reduces the complexity of derivation for a solution. However, it has a profound effect on the solution. Section 4 offers a completely different solution when this assumption is dropped.

The Lagrangian for (4) is

$$L \equiv \int_{-\infty}^{\infty} v[x - s(x)] f(x; a) dx + \lambda \left[\int_{-\infty}^{\infty} u[s(x)] f(x; a) dx - c(a) - \bar{u} \right],$$

where $\lambda \geq 0$ is a Lagrange multiplier. The Hamiltonian for L can be defined as

$$H \equiv v(x - s) + \lambda u(s).$$

Given Assumption 12.1, the FOC for a from the Lagrangian and the Euler equation for $s(\cdot)$ from the Hamiltonian are respectively[4]

$$\int_{-\infty}^{\infty} v[x - s(x)] f_a(x; a) dx + \lambda \int_{-\infty}^{\infty} u[s(x)] f_a(x; a) dx = \lambda c'(a), \quad (5)$$

$$\frac{v'[x - s(x)]}{u'[s(x)]} = \lambda. \quad (6)$$

Equation (6) indicates that the marginal rate of substitution $MRS = v'(x-s)/u'(s)$ is constant across all possible states x, implying proportional sharing

---

[4]Since there are no boundary conditions such as $s(-\infty) = \underline{s}$ and $s(\infty) = \bar{s}$ for $s(x)$ for some given constants $\underline{s}$ and $\bar{s}$, we need to satisfy some transversality conditions for the optimal contract. If we impose $s(-\infty) \geq 0$ and $s(\infty) \geq 0$, then the transversality conditions are

$$s(t) H_{\dot{s}}|_{t=-\infty} = 0, \quad s(t) H_{\dot{s}}|_{t=\infty} = 0, \quad H_{\dot{s}}|_{t=-\infty} \leq 0, \quad H_{\dot{s}}|_{t=\infty} \leq 0.$$

Since H does not contain $\dot{s}$, these conditions are all satisfied. Note also that by assumption the optimal solution must be a continuous function if the admissible set is assumed to contain continuous contracts only.

of risk in some sense. We call it the first-best risk sharing. The MRS is constant because the Lagrangian is a linear social welfare function, where the weight λ is endogenous. This result is the same as that in a general equilibrium setting (Wang 2012b), in which the MRS is equal to the price ratio and thus constant in equilibrium.

Equation (5) means that the marginal social utility of a equals the marginal social cost of a. By (5), we have $\lambda > 0$, implying a binding IR condition at the optimum.

By differentiating (6) with respect to x, we have $v'' \cdot [1 - s'(x)] = \lambda u'' \cdot s'(x)$. Hence, if the principal is not risk neutral, we have $0 < s'(x) < 1$, implying that both the agent's income and the principal's income increase strictly with x.

If the principal is risk neutral, the contract will be a constant, i.e., the principal absorbs all the risk. Similarly, if the agent is risk neutral, the agent takes all the risk. This risk sharing arrangement of a risk-neutral party taking all the risk makes sense. When a risk-averse party faces risk, it demands a risk premium (an extra payment to compensate for bearing the risk), while a risk-neutral party cares about expected income only. Hence, the risk-neutral principal can do better by taking all the risk.

1.2 Nonverifiable Effort: The Second Best

Suppose now that a is not verifiable. The principal again offers a contract of the form $[a, s(\cdot)]$ to the agent. Since the principal cannot impose an effort level, she has to offer incentives in the contract to induce the agent to work hard (Fig. 1).The agent is free to choose an effort after signing the contract. Given an income-sharing rule $s(\cdot)$, his problem is

$$\max_{a \geq 0} \int_{-\infty}^{\infty} u[s(x)]f(x;a)dx - c(a). \tag{7}$$

In order for the agent to voluntarily supply the effort a^* indicated in the contract, this a^* must be a solution of (7). Hence, (7) is the incentive compatibility (IC) condition. The principal still decides the income-sharing rule $s(\cdot)$ and effort a. But whereas before, the principal has only to ensure that the agent accepts the contract (the IR condition) and the effort can be imposed upon the agent, now the principal has to provide not only an incentive for the agent to accept the contract but also an incentive to accept the effort (the IC condition). Hence the principal's problem is

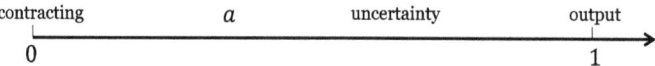

Fig. 1 The timing of events

1 The Standard Agency Model

$$\max_{a \geq 0, s \in S} \int_{-\infty}^{\infty} v[x - s(x)] f(x; a) dx$$

s.t $IR: \int_{-\infty}^{\infty} u[s(x)] f(x; a) dx \geq c(a) + \bar{u},$

$IC: a \in \arg\max_{\hat{a} \geq 0} \int_{-\infty}^{\infty} u[s(x)] f(x; \hat{a}) dx - c(\hat{a}).$

The FOC of the IC condition (7) is

$$\int_{-\infty}^{\infty} u[s(x)] f_a(x; a) dx = c'(a). \tag{8}$$

The SOC of the IC condition (7) is

$$\int_{-\infty}^{\infty} u[s(x)] f_{aa}(x; a) dx < c''(a).$$

The IC condition (7) is too difficult to handle in an optimization problem. As an alternative, the so-called <u>first-order approach (FOA)</u> is to substitute the original IC condition (7) by its FOC (8) in the principal's problem. Under the FOA, we call (8) the IC condition. Thus, the principal's problem becomes

$$\max_{a \geq 0, s \in S} \int_{-\infty}^{\infty} v[x - s(x)] f(x; a) dx$$

s.t. $IR: \int_{-\infty}^{\infty} u[s(x)] f(x; a) dx \geq c(a) + \bar{u},$ \quad (9)

$IC: \int_{-\infty}^{\infty} u[s(x)] f(x; a) dx - c'(a),$

Rogerson (1985) and Jewitt (1988) find conditions under which the FOA is valid, by which the solution of the agency problem (9) is indeed the right one. The solution (a^*, s^*) of (9) is said to be the <u>second best</u> and hence the above problem is called the second-best problem. Due to the additional constraint, this solution is inferior to the first best.

The IC and IR conditions provide different functions. An IR condition induces the agent to participate, but the agent may not work hard or invest an effort up to the principal's desired amount. The principal needs to provide incentives in the contract for the agent to invest the desired amount a^* voluntarily. The IC condition is to provide the incentives. Here a^* can be written into a contract, but there is no need for enforcement.

The Lagrangian for (9) is

$$L \equiv \int_{-\infty}^{\infty} v[x - s(x)]f(x;a)dx + \lambda \left[\int_{-\infty}^{\infty} u[s(x)]f(x;a)dx - c(a) - \bar{u}\right]$$
$$+ \mu \left[\int_{-\infty}^{\infty} u[s(x)]f_a(x;a)dx - c'(a)\right],$$

where λ and μ are Lagrange multipliers, and $\lambda \geq 0$. The Hamiltonian for L is

$$H \equiv v(x - s)f(x;a) + \lambda u(s)f(x;a) + \mu u(s)f_a(x;a).$$

The FOC for a from the Lagrangian and the Euler equation for $s(\cdot)$ from the Hamiltonian are respectively

$$\int_{-\infty}^{\infty} v[x - s(x)]f_a(x;a)dx + \mu\left[\int_{-\infty}^{\infty} u[s(x)]f_{aa}(x;a)dx - c''(a)\right] = 0, \quad (10)$$

$$\frac{v'[x - s(x)]}{u'[s(x)]} = \lambda + \mu \frac{f_a(x;a)}{f(x;a)}. \quad (11)$$

Here, we can see that if both parties are risk neutral, (11) cannot hold. This is the reason that one of the parties must be risk averse.

To understand the Euler Eq. (11), notice that the likelihood function is $\ln f(y,a)$ and

$$\frac{\partial \ln f(x;a)}{\partial a} = \frac{f_a(x;a)}{f(x;a)}, \quad \frac{\partial^2 \ln f(x;a)}{\partial x \partial a} = \frac{\partial}{\partial x}\frac{f_a(x;a)}{f(x;a)}.$$

i.e., $\frac{f_a(x;a)}{f(x;a)}$ is the increase in $\ln f(x;a)$ for a unit increase in a. And, a and x are complementary if $\frac{f_a(x;a)}{f(x;a)}$ is increasing in x. The latter condition is called the monotone likelihood ratio property (MLRP). $\frac{\partial^2 \ln f(x;a)}{\partial x \partial a}$ measures a kind of correlation between a and x. A higher value of $\ln f(x,a)$ means that a higher value of x is more likely to come from a higher value of a rather than by chance. Thus, $\frac{\partial \ln f(x;a)}{\partial a}$ is the marginal increase in the principal's belief that the effort is a from the observation of x.

We further assume that an increase in a generates an improvement in x in the sense of first-order stochastic dominance (FOSD), as stated in the following assumption.

Assumption 12.2 (FOSD). $F_a(x;a) \leq 0$, for all (a,x), with a strict inequality on a set of (a,x) that has a non-zero probability measure.

1 The Standard Agency Model

Theorem 12.1 (Holmström 1979; Shavell 1979). *Under Assumptions 12.1 and 12.2, FOA and $u'' < 0$, we have $\mu > 0$ for the optimal solution of (9).*

Proof Suppose, on the contrary, that $\mu \leq 0$. If $\lambda = 0$, then $f_a(x;a) < 0$ for all x, which is impossible. Hence, $\lambda > 0$. Let $r(x) \equiv x - s(x)$ and define $\bar{r}(x)$ by

$$\frac{v'[\bar{r}(x)]}{u'[x - \bar{r}(x)]} = \lambda. \tag{12}$$

The advantage of $\bar{r}(x)$ over $r(x)$ is that we know $\int v[\bar{r}(x)] f_a(x;a^*) dx > 0$. Then by (11), for x satisfying $f_a(x;a^*) \geq 0$, we have

$$\frac{v'[r(x)]}{u'[x - r(x)]} \leq \lambda = \frac{v'[\bar{r}(x)]}{u'[x - \bar{r}(x)]}.$$

Since $v'' \leq 0$ and $u'' < 0$, we must have $\bar{r}(x) \leq r(x)$. Similarly, for x satisfying $f_a(x;a^*) \leq 0$, we have $\bar{r}(x) \geq r(x)$. Thus,

$$\int_{-\infty}^{\infty} v[r(x)] f_a(x;a^*) dx \geq \int_{-\infty}^{\infty} v[\bar{r}(x)] f_a(x;a^*) dx.$$

Let $[\alpha, \beta]$ be the support of $\tilde{x}$, where α and β are independent of a^*. Then

$$\int_{-\infty}^{\infty} v[\bar{r}(x)] f_a(x;a^*) dx = v[\bar{r}(x)] F_a(x;a^*)\Big|_\alpha^\beta - \int_{-\infty}^{\infty} v'[\bar{r}(x)] F_a(x;a^*) \bar{r}'(x) dx$$

$$= - \int_{-\infty}^{\infty} v'[\bar{r}(x)] F_a(x;a^*) \bar{r}'(x) dx.$$

Since $F(\alpha;a) = 0$ and $F(\beta;a) = 1$ for all a, we must have $F_a(\alpha;a^*) = F_a(\beta;a^*) = 0$. (10) implies that $\bar{r}'(x) > 0$. Also, by Assumption 12.2, $F_a(x;a) < 0$ on a non-zero probability measure set. Therefore,

$$\int_{-\infty}^{\infty} v[\bar{r}(x)] f_a(x;a^*) dx > 0.$$

However, if FOA is valid, the SOC for (7) must hold, which means

$$\int_{-\infty}^{\infty} u[s(y)] f_{aa}(y;a^*) dy - c''(a^*) \leq 0. \tag{13}$$

Then (10) cannot hold. This is a contradiction. We thus must have $\mu > 0$. ∎

Risk sharing in the first best is represented by the expression in (6), while risk sharing in the first best is represented by the expression in (11). By $\mu > 0$, Theorem 12.1 indicates that the risk sharing is no longer the first best. By deviating from the first-best risk sharing, the principal can provide incentives in the contract. To understand the result intuitively, assume that the principal is risk neutral. Then, the first-best risk sharing, as indicated by (6), requires full insurance (no risk) for the agent, i.e., $s(x)$ is constant. However, if the agent's action is not verifiable, under a fixed contract, the agent will choose zero effort. Thus, for the agent to have the incentive to work, the contract with unverifiable effort must relate a payment to output. This means that the agent must share some risk, implying inefficient risk sharing in the case of unverifiable effort.

Let us compare the first-best and second-best contracts. As shown in Fig. 2, as implied by FOSD, an increase in a causes the density curve to shift to the right. This implies that

$$f_a(x;a) > 0 \quad \text{for large } x; \quad f_a(x;a) < 0 \text{ for small } x.$$

Let s^{**} be the first-best contract determined by (6) and s^* be the second-best contract determined by (11). Since the left-hand side of (11) is increasing in s, we have

$$s^*(x) > s^{**}(x) \quad \text{for large } x; \quad s^*(x) < s^{**}(x) \text{ for small } x.$$

Thus, the two contracts should have the relationship shown Fig. 3, which means that, in order to motivate the agent to work hard, the principal must use a steeper contract curve than the first-best contract to provide incentives. The second-best contract rewards high output and punishes low output; at the same time, the second-best contract must also provide a proper risk-sharing scheme.

Example 12.1[5] The principal's problem is

$$\pi \equiv \max_{s \in S, a \geq 0} \int_{-\infty}^{\infty} [x - s(x)] f(x;a) dx$$

$$\text{s.t.} \quad IC: \int_{-\infty}^{\infty} u[s(x)] f_a(x;a) dx = c'(a), \qquad (14)$$

$$IR: \int_{-\infty}^{\infty} u[s(x)] f(x;a) dx \geq c(a) + \bar{u},$$

---

[5]This example is from Holmström (1979, p. 79), but his solution seems to have an error.

1 The Standard Agency Model

Fig. 2 A shift in density function

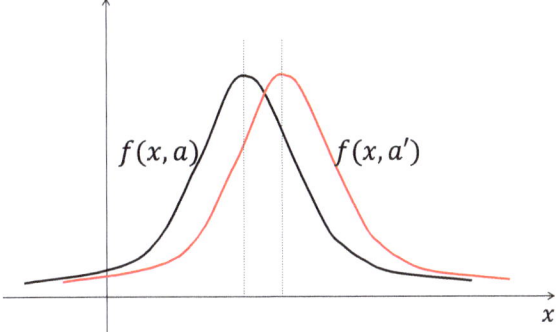

Fig. 3 First- versus second-best contracts

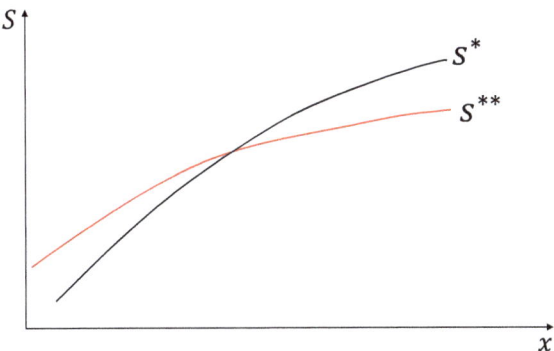

where S is the space of sharing rules that consists of continuous output sharing rules satisfying the limited liability condition: $s(x) \geq 0$. Let

$$u(x) = 2\sqrt{x}; \quad v(x) = x; \quad \bar{u} = \frac{3}{4}; \quad c(a) = a^2;$$
$$f(x;a) = \frac{1}{a} e^{-\frac{x}{a}} \text{ for } x \in [0, +\infty).$$

The density function $f(x;a)$ states that the output follows the exponential distribution with mean $E(x) = a$ and variance $Var(x) = a^2$. We find the second-best solution:

$$a^* = \frac{1}{2}, \quad s^*(x) = \left(x + \frac{1}{4}\right)^2,$$

and the first-best solution:

$$a^{**} = 0.76, \quad s^{**}(x) = 0.44. \blacksquare$$

Example 12.2 Consider
$$u(x) = 2\sqrt{x}; \quad v(x) = x; \quad \bar{u} = \frac{3}{4}; \quad c(a) = a^2;$$
$$f(x,a) = \frac{1}{\sqrt{2\pi}\sigma} e^{-\frac{x-a^2}{2\sigma^2}} \text{ for } x \in \mathbb{R}.$$

We have
$$f_a(x,a) = \frac{1}{\sqrt{2\pi}\sigma} e^{-\frac{x-a^2}{2\sigma^2}} \frac{x-a}{\sigma^2}.$$

Hence,
$$\frac{1}{u'[s(x)]} = \lambda + \mu \frac{x-a}{\sigma^2}. \quad \blacksquare$$

The standard agency theory has two major problems. First, it is based on the FOA. Although Rogerson (1985) and Jewitt (1988) have provided conditions under which the FOA is valid, their conditions are fairly complicated so that people typically give up on verifying those conditions. Second, the standard agency model basically does not offer a closed-form solution. Except the case in Example 12.1, we have not found a second case in which a closed-form solution is obtainable. In most applications, an explicit contractual solution is very much the minimum requirement.

In the rest of this chapter, we modify the standard agency model from several angles in an attempt to address the two major problems in the standard agency model and find simple optimal contracts. These modified versions of the agency model are popular in applications. In other words, these simple agency models are what people actually use in applications.

2 Two-State Agency Models

Two-state models are popular. For example, insurance models typically have only two states. With only two possible states, the agency problem becomes quite tractable.

Again, there is a risk-neutral principal who tries to hire an agent. The output can either be a high output x_H or a low output x_L. The probability of achieving x_H is $p(a)$. Naturally we assume $p'(a) > 0$. The agent's utility function is $u(\cdot)$ and his reservation utility is $\bar{u}$ The agent's private cost of effort is $c(a)$.

2.1 Verifiable Effort

Suppose that the principal pays w_H for x_H and w_L for x_L. The IR condition is

$$p(a)u(w_H) + [1 - p(a)]u(w_L) \geq \bar{u} + c(a). \tag{15}$$

Hence, if effort is verifiable, the principal's problem is

$$\begin{array}{c} \max\limits_{a,w_H,w_L} \quad p(a)x_H - w_H + [1 - p(a)]x_L - w_L \\ \text{s.t.} \quad p(a)u(w_H) + [1 - p(a)]u(w_L) \geq \bar{u} + c(a). \end{array} \tag{16}$$

The Lagrange function for (16) is

$$L = p(a)(x_H - w_H) + [1 - p(a)](x_L - w_L) + \lambda\{p(a)u(w_H) + [1 - p(a)]u(w_L) - \bar{u} - c(a)\},$$

where λ is a Lagrange multiplier and $\lambda \geq 0$. The FOCs for (a, w_H, w_L) are respectively

$$\begin{aligned} & p'(a)(x_H - w_H - x_L + w_L) + \lambda\{p'(a)[u(w_H) - u(w_L)] - c'(a)\} = 0, \\ & p(a) = \lambda p(a) u'(w_H), \\ & 1 - p(a) = \lambda[1 - p(a)] u'(w_L), \end{aligned}$$

which imply

$$\begin{aligned} & \lambda u'(w_H) = \lambda u'(w_L) = 1, \\ & x_H - w_H - x_L + w_L + \lambda\left(u(w_H) - u(w_L) - \frac{c'(a)}{p'(a)}\right) = 0. \end{aligned} \tag{17}$$

We immediately have $\lambda \neq 0$ and

$$w_H^{**} = w_L^{**} = (u')^{-1}(1/\lambda).$$

That is, the optimal salary is independent of output. This is implied by the first-best risk sharing discussed in Sect. 1. Since $\lambda \neq 0$, the IR condition (15) must be binding, implying

$$u(w_i^{**}) = \bar{u} + c(a^{**}). \tag{18}$$

Equation (17) then becomes

$$x_H - x_L = \lambda \frac{c'(a^{**})}{p'(a^{**})},$$

implying

$$u'(w_i^{**})(x_H - x_L) = \frac{c'(a^{**})}{p'(a^{**})}. \tag{19}$$

Equations (18) and (19) determine the first-best solution (a^{**}, w_i^{**}).

2.2 Unverifiable Effort

With unverifiable effort, again suppose that the principal pays w_H for x_H and w_L for x_L. Given the wage contract, the agent considers his own problem:

$$\max_a p(a)u(w_H) + [1 - p(a)]u(w_L) - c(a),$$

which implies the FOC:

$$p'(a)[u(w_H) - u(w_L)] = c'(a).$$

After taking into the agent's IC and IR conditions to induce participation and sufficient effort, the principal's problem is

$$\begin{aligned}\max_{a, w_H, w_L} \quad & p(a)(x_H - w_H) + [1 - p(a)](x_L - w_L) \\ \text{s.t.} \quad & IC : p'(a)[u(w_H) - u(w_L)] = c'(a), \\ & IR : p(a)u(w_H) + [1 - p(a)]u(w_L) \geq \bar{u} + c(a).\end{aligned} \tag{20}$$

The Lagrange function for (20) is

$$L = p(a)(x_H - w_H) + [1 - p(a)](x_L - w_L) + \mu\{p'(a)[u(w_H) - u(w_L)] - c'(a)\} \\ + \lambda\{p(a)u(w_H) + [1 - p(a)]u(w_L) - \bar{u} - c(a)\},$$

where λ is a Lagrange multiplier and $\lambda \geq 0$. The FOCs for (a, w_H, w_L) are respectively

$$p'(a)(x_H - w_H - x_L + w_L) + \lambda\{p'(a)[u(w_H) - u(w_L)] - c'(a)\} = 0, \\ p(a) = \lambda p(a)u'(w_H), \\ 1 - p(a) = \lambda[1 - p(a)]u'(w_L),$$

implying

2 Two-State Agency Models

$$\mu = \frac{p'(a)(x_H - w_H - x_L + w_L)}{c''(a) - p''(a)[u(w_H) - u(w_L)]},$$

$$\frac{1}{u'(w_H)} = \lambda + \mu \frac{p'(a)}{p(a)},$$

$$\frac{1}{u'(w_L)} = \lambda + \mu \frac{-p'(a)}{1 - p(a)}.$$

The third equation implies $\lambda \neq 0$, resulting in a binding IR condition. By Theorem 12.1, $\mu > 0$. These three equations together with the IC and IR conditions determine $(a^*, w_H^*, w_L^*, \lambda, \mu)$.

By the IC condition, we have $w_H^* \neq w_L^*$, i.e., the agent shares some risks.

No matter whether effort is verifiable or not, the agent's welfare is $\bar{u}$ in both solutions. But the principal yields a higher profit from problem (16) than problem (20) since the former has fewer conditions. In other words, since the agent is indifferent but the principal is better off with verifiable effort, the solution with verifiable effort Pareto-dominates the solution with unverifiable effort.

2.3 Example: Insurance

Consider a monopolistic insurance company with a single type of consumers. A consumer can influence his own probability $\pi(a)$ of having an accident by putting in some preventive effort a, with private cost $c(a)$, which is increasing and convex. Assume that $\pi(a)$ is a decreasing function, indicating that a higher effort reduces the chance of an accident occurring. The company offers a price p for compensation z. Laffont (1995, 125–128) and Salanie (1999, 134–135) provide good references for this example.

Insurance under Complete Information: The First Best

If a is verifiable, the company can offer a contract of the form (a, p, z) and its problem is

$$\max_{p,z,a} \quad p - \pi(a)z$$
$$\text{s.t.} \quad \pi(a)u(w - L + z - p) + [1 - \pi(a)]u(w - p) \geq \bar{u} + c(a).$$

The Lagrangian is

$$L = p - \pi(a)z + \lambda\{\pi(a)u(w - L + z - p) + [1 - \pi(a)]u(w - p) - \bar{u} - c(a)\},$$

where λ is a Lagrange multiplier and $\lambda \geq 0$. The FOCs are

$$0 = \tfrac{\partial L}{\partial p} = 1 + \lambda\{-\pi(a)u'(w - L + z - p) - [1 - \pi(a)]u'(w - p)\},$$
$$0 = \tfrac{\partial L}{\partial z} = -\pi(a) + \lambda\pi(a)u'(w - L + z - p),$$
$$0 = \tfrac{\partial L}{\partial a} = -\pi'(a)z + \lambda\{\pi'(a)u(w - L + z - p) - \pi'(a)u(w - p) - c'(a)\},$$

implying

$$u'(w - L + z - p) + \pi(a)[u'(w - p) - u'(w - L + z - p)] = u'(w - p),$$
$$zu'(w - L + z - p) = u(w - L + z - p) - u(w - p) - \tfrac{c'(a)}{\pi'(a)}.$$

Then, if $\pi(a) \neq 1$, we have

$$u'(w - L + z - p) = u'(w - p),$$

implying full insurance: $z = L$. That is, the company will offer full insurance under complete information. This is implied by the standard agency theory, in particular, by (6). The intuition is clear. Since the company is risk neutral and the consumer is risk averse, to avoid having to compensate the consumer for taking risk, the company takes all the risk. There is no incentive problem here since the company can force the consumer to take action a^{**}. Then,

$$u'(w - p) + \frac{1}{L}\frac{c'(a)}{\pi'(a)} = 0.$$

The IR condition must be binding, which implies

$$p = w - u^{-1}[\bar{u} + c(a)],$$

and

$$u'\{u^{-1}[\bar{u} + c(a)]\} + \frac{1}{L}\frac{c'(a)}{\pi'(a)} = 0.$$

This equation determines the optimal a^{**} under complete information.

Insurance under Incomplete Information: The Second Best

If effort a is not verifiable so that the company cannot impose it in accident prevention, it can only offer incentives in a contract (a, p, z) to induce a desirable a. After accepting the contract, the consumer's preferred effort a is determined by:

$$\max_a U \equiv \pi(a)u(w - L + z - p) + [1 - \pi(a)]u(w - p) - c(a). \qquad (21)$$

2 Two-State Agency Models

The consumer's marginal utility in accident prevention is:

$$U_a = \pi'(a)[u(w - L + z - p) - u(w - p)] - c'(a).$$

If $z = L$, we will have $U_a < 0$ for any a, implying that the consumer's preferred effort is $a = 0$. That is, with full insurance, the consumer would have no incentive to make any effort to prevent accidents from occurring. Hence, the company cannot offer full insurance in this case.

To provide incentives for accident prevention, the company ensures that the a in a contract must be the consumer's preferred a. To do this, the company includes the FOC of (21) in its optimization problem, where the SOC of (21) is automatically satisfied. Hence, the company's problem is

$$\max_{p,z,a} \quad p - \pi(a)z$$
$$\text{s.t.} \quad \pi'(a)[u(w - L + z - p) - u(w - p)] = c'(a),$$
$$\pi(a)u(w - L + z - p) + [1 - \pi(a)]u(w - p) \geq \bar{u} + c(a).$$

The IR condition must be binding. Thus the two constraints become

$$u(w - L + z - p) - u(w - p) = \frac{c'(a)}{\pi'(a)}, \quad \pi(a)\frac{c'(a)}{\pi'(a)} + u(w - p) = \bar{u} + c(a).$$

Let $x \equiv p - z$. Then

$$p - \pi(a)z = p - \pi(a)(p - x) = [1 - \pi(a)]p + \pi(a)x,$$
$$p = w - u^{-1}\left[\bar{u} + c(a) - \frac{\pi(a)}{\pi'(a)}c'(a)\right],$$
$$x = w - L - u^{-1}\left[\bar{u} + c(a) + \frac{1-\pi(a)}{\pi'(a)}c'(a)\right].$$

Then the company's problem becomes

$$\max_a \quad w - \pi(a)L - [1 - \pi(a)]u^{-1}\left[\bar{u} + c(a) - \frac{\pi(a)}{\pi'(a)}c'(a)\right]$$
$$- \pi(a)u^{-1}\left[\bar{u} + c(a) + \frac{1-\pi(a)}{\pi'(a)}c'(a)\right].$$

This problem determines the optimal a^* under incomplete information.

3 Linear Contracts Under Risk Neutrality

Linear contracts are the simplest form of contracts and they are very popular in applications. Examples of linear contracts are many: contractual joint ventures, equity joint ventures, crop-sharing contracts, and fixed-price contracts, etc.

A key theoretical question is: under what conditions is a linear contract optimal? Here, an optimal linear contract is better than other admissible contracts, not just better than other linear contracts. When both parties in a contractual relationship are risk neutral, we say there is double risk neutrality; similarly, when both parties in a contractual relationship have moral hazard, we say there is double moral hazard. This section shows that a linear contract can be optimal only under double risk neutrality, and it can be the first best only under single moral hazard.

From (11), we know that the optimal contract is generally nonlinear if one of the parties is risk averse. For example, if the principal is risk neutral, then (11) implies

$$s^*(x) = (u')^{-1}\left[\left(\lambda + \mu \frac{f_a(x;a)}{f(x;a)}\right)^{-1}\right].$$

For some popular output distributions, $f_a(x;a)/f(x;a)$ is linear in x. For example, if output follows the normal distribution $N(a, \sigma^2)$, then $f_a(x;a)/f(x;a) = (x-a)/\sigma^2$. If that's the case, the optimal contract is nonlinear for a risk-averse agent. Hence, to find optimal linear contracts, we assume double risk neutrality in this section and specify utility functions $u(z) = z$ and $v(z) = z$, for all $z \in \mathbb{R}$, for the agent and principal, respectively.

3.1 Single Moral Hazard

Let a be the agent's effort and $f(x;a)$ be the density function of output x. Let $R(a)$ be the expected revenue $R(a) = \int_{-\infty}^{\infty} x f(x;a) dx$.

The First Best

With verifiable effort, the principal offers a contract of the form $(a, s(\cdot))$. The principal's problem is

$$\pi^{**} = \max_{s \in S, a \geq 0} \int_{-\infty}^{\infty} [x - s(x)] f(x;a) dx$$

$$\text{s.t.} \int_{-\infty}^{\infty} s(x) f(x;a) dx \geq c(a) + \bar{u}.$$

If the IR condition is not binding in equilibrium, given an optimal contract $s^*(x)$, the principal can offer contract $s^*(x) - \varepsilon$ instead for some $\varepsilon > 0$ to satisfy the IR condition. This contract raises the profit's profit, which contradicts the fact that contract $s^*(x)$ is optimal. Hence, the IR condition must be binding at the optimum. With a binding IR condition, the problem becomes a problem of social welfare maximization:

3 Linear Contracts Under Risk Neutrality

$$\pi^{**} = \max_{s \in S, a \geq 0} R(a) - c(a) - \bar{u}$$

$$\text{s.t.} \int_{-\infty}^{\infty} s(x) f(x; a) dx = c(a) + \bar{u}.$$

Since the principal's objective function is independent of the contract $s(x)$, this problem can be solved in two steps. First, by maximizing $R(a) - c(a) - \bar{u}$, the first-best effort a^{**} is determined by

$$R'(a^{**}) = c'(a^{**}).$$

Second, given the optimal effort, an optimal contract needs to satisfy the IR condition only. To find an optimal contract, consider a fixed contract $s(x) = \gamma$, where γ is a constant. By the binding IR condition, $s^{**}(x) = c(a^{**}) + \bar{u}$. This is a first-best contract since it supports the first-best effort.

The Second Best

With unverifiable effort, the principal will still offer a contract of the form $(a, s(\cdot))$. But the principal has to provide incentives for the agent to accept this a. For this purpose, an IC is introduced. Hence, the principal's problem is

$$\pi^* = \max_{s \in S, a \geq 0} \int_{-\infty}^{\infty} [x - s(x)] f(x; a) dx$$

$$\text{s.t.} \quad IC: \int_{-\infty}^{\infty} s(x) f_a(x; a) dx = c'(a)$$

$$IR: \int_{-\infty}^{\infty} s(x) f(x; a) dx \geq c(a) + \bar{u}.$$

If the IR condition is not binding in equilibrium, given an optimal contract $s^*(x)$, the principal can offer contract $s^*(x) - \varepsilon$ instead for some $\varepsilon > 0$ to satisfy the IR condition. This contract also satisfies the IC condition. It raises the profit's profit, which contradicts the fact that contract $s^*(x)$ is optimal. Hence, the IR condition must be binding at the optimum. With a binding IR condition, the problem becomes a problem of social welfare maximization:

$$\pi^* = \max_{s \in S, a \geq 0} R(a) - c(a) - \bar{u}$$

$$\text{s.t.} \quad IC: \int_{-\infty}^{\infty} s(x) f_a(x; a) dx = c'(a) \qquad (22)$$

$$IR: \int_{-\infty}^{\infty} s(x) f(x; a) dx = c(a) + \bar{u}.$$

Since the principal's objective function is independent of $s(x)$, this problem can be solved in two steps. First, by maximizing $R(a) - c(a) - \bar{u}$, the second-best effort a^* is determined by

$$R'(a^*) = c'(a^*),$$

implying $a^* = a^{**}$. Second, given the optimal effort, an optimal contract needs to satisfy the IC and IR conditions only. We can find many contracts that can satisfy these conditions for a given a^*. We look for a linear contract. For a linear contract $s(x) = \alpha + \beta x$, the IC and IR conditions become

$$\beta \int_{-\infty}^{\infty} x f_a(x; a^*) dx = c'(a^*), \qquad \alpha + \beta \int_{-\infty}^{\infty} x f(x; a^*) dx = c(a^*) + \bar{u},$$

which imply

$$\beta = \frac{c'(a^*)}{R'(a^*)} = 1, \qquad \alpha = \bar{u} + c(a^*) - \beta R(a^*) = -\pi^*.$$

This $s(x) = \alpha + \beta x$ is a second-best contract since it supports the second-best effort.

Proposition 12.1 *With single moral hazard and double risk neutrality, the linear contract $s^*(x) = x - \pi^*$ is optimal, implying the first-best effort a^{**}.*

Interestingly, this solution can be implemented by a change of ownership: the principal can simply sell the firm to the agent for payment π^*. When the agent is the owner, the incentive problem disappears. This solution leads to an important idea: incentive problems may be solved through an organizational approach. Coase (1960) contributes precisely by proposing this idea. When we use incomplete contracts, ownership transfers are allowed. However, in this chapter, all contracts are complete contracts. Refer to Wang (2016) on incomplete contracts.

3.2 Double Moral Hazard

So far the agent is the only person who invests in a project. In this subsection, we allow both parties in a joint venture to invest. In addition, we will use a partnership setting instead of a principal-agent setting. In a principal-agent setting, the principal is the designer of a contract, by which the principal has 100% bargaining power ex ante and the agent can only decide to take or leave. In a partnership setting, the two parties are partners and they bargain to settle a contract ex ante.

Consider two agents, M_1 and M_2, engaged in a joint project. Efforts (investments) e_1 and e_2 respectively from M_1 and M_2 are private information. Let $\mathbb{E}$ be the

3 Linear Contracts Under Risk Neutrality

effort space. Let $c_1(e_1)$ and $c_2(e_2)$ be the private costs of effort. Let $h = h(e_1, e_2)$ be the joint effort, and $\tilde{x} = X(\omega, h)$ be the ex post revenue depending on the state ω and joint effort h. Let $\tilde{x}$ follow a density function $f(x, h)$ ex ante. Thus, the expected revenue is

$$R(e_1, e_2) \equiv \int_{-\infty}^{\infty} xf[x; h(e_1, e_2)]dx.$$

Assumption 12.3 $h(e_1, e_2)$ is strictly increasing in e_1 and e_2.

Assumption 12.4 $c_1(e_1)$ and $c_2(e_2)$ are convex and strictly increasing.

Assumption 12.5 $R(e_1, e_2)$ is concave and strictly increasing in e_1 and e_2.

Given a sharing rule $[s_1(x), s_2(x)]$, where $s_1(x) + s_2(x) = x$ for all x, that specifies payments $s_1(x)$ and $s_2(x)$ to agents 1 and 2, respectively, the two agents play a Nash game to determine their efforts. In other words, given e_2, agent 1 chooses his effort e_1 by maximizing his own expected utility:

$$\max_{e_1 \in \mathbb{E}} \int_{-\infty}^{\infty} s_1(x) f[x; h(e_1, e_2)] dx - c_1(e_1),$$

which implies the FOC and SOC:

$$h_1' \int_{-\infty}^{\infty} s_1(x) f_h[x; h(e_1, e_2)] dx = c_1'(e_1),$$

$$h_1'' \int_{-\infty}^{\infty} s_1(x) f_{hh}[x; h(e_1, e_2)] dx + (h_1')^2 \int_{-\infty}^{\infty} s_1(x) f_h[x; h(e_1, e_2)] dx < c_1''(e_1).$$

Similarly, given e_1, agent 2 chooses his effort e_2 by maximizing his own utility.

Assume that contracting negotiation leads to social welfare maximum. This assumption is standard in the literature and is imposed on any negotiation outcome. Then the problem of maximizing social welfare can be written as

$$V = \max_{s_i \in \mathcal{S}, e_1, e_2 \in \mathbb{E}} R(e_1, e_2) - c_1(e_1) - c_2(e_2)$$

$$\text{s.t.} \quad IC_1 : h_1' \int_{-\infty}^{\infty} s_1(x) f_h[x; h(e_1, e_2)] dx = c_1'(e_1),$$

$$IC_2 : h_2' \int_{-\infty}^{\infty} s_2(x) f_h[x; h(e_1, e_2)] dx = c_2'(e_2), \qquad (23)$$

$$SOC_1 : \text{for } e_1,$$
$$SOC_2 : \text{for } e_2,$$
$$RC : s_1(x) + s_2(x) = x, \quad \text{for all } x \in \mathbb{R}_+,$$

where the last constraint is the resource constraint (RC). Since we allow a fixed transfer in the contract, IR conditions are unnecessary.

The first-best problem is

$$\max_{e_1,e_2 \in \mathbb{E}} R(e_1,e_2) - c_1(e_1) - c_2(e_2)$$
$$\text{s.t.} \quad RC : s_1(x) + s_2(x) = x, \text{ for all } x \in \mathbb{R}_+,$$

where the contract needs to satisfy RC only.

Proposition 12.2 *Under Assumptions 12.3–12.5, with double moral hazard and double risk neutrality, there exists a linear output-sharing rule*

$$s_i^*(x) = \alpha_i^* x,$$

where $\alpha_i^* = \frac{c_i'(e_i^*)}{R_i'(e_1^*,e_2^*)}$, *that induces the second-best efforts* $e_i^* > 0$ *determined by*

$$\max_{e_1,e_2 \in \mathbb{E}} R(e_1,e_2) - c_1(e_1) - c_2(e_2)$$
$$\text{s.t.} \quad R_1'(e_1,e_2) = c_1'(e_1) + \frac{h_1'(e_1,e_2)}{h_2'(e_1,e_2)} c_2'(e_2).$$

In addition, $0 < \alpha_i^* < 1$, *and the first-best outcome is not obtainable.*

Proof Conditions IC_1 and IC_2 imply[6]

$$R_1'(e_1,e_2) = c_1'(e_1) \mathrm{r} \frac{h_1'(e_1,e_2)}{h_2'(e_1,e_2)} c_2'(e_2). \tag{24}$$

So the problem is equivalent to

$$\max_{s_i \in S, e_1, e_2 \in \mathbb{E}} R(e_1,e_2) - c_1(e_1) - c_2(e_2)$$
$$\text{s.t.} \quad R_1'(e_1,e_2) = c_1'(e_1) + \frac{h_1'(e_1,e_2)}{h_2'(e_1,e_2)} c_2'(e_2), \tag{25}$$
$$IC_2, SOC_1, SOC_2, RC.$$

Since the objective function is not related to the contract, this problem can be solved in two steps. First, we find a solution (e_1^*, e_2^*) from the following problem:

$$\max_{e_1,e_2 \in \mathbb{E}} R(e_1,e_2) - c_1(e_1) - c_2(e_2)$$
$$\text{s.t.} \quad R_1'(e_1,e_2) = c_1'(e_1) + \frac{h_1'(e_1,e_2)}{h_2'(e_1,e_2)} c_2'(e_2). \tag{26}$$

This problem is not related to the contract. Second, given (e_1^*, e_2^*), we look for a contract $s_i(x)$ that satisfies IC_2, SOC_1, SOC_2 and RC. There are many such

---

[6] We also have $R_2'(e_1,e_2) = c_2'(e_2) + \frac{h_2'(e_1,e_2)}{h_1'(e_1,e_2)} c_1'(e_1)$.

3 Linear Contracts Under Risk Neutrality

contracts. We look for a simple sharing contract of the form $s_i(x) = \alpha_i x$ for $i = 1, 2$. We find one with

$$\alpha_i = \frac{c'_i(e^*_i)}{R'_i(e^*_1, e^*_2)}.$$

It is easy to verify that this contract satisfies IC_2. For RC, by (24), we have

$$\frac{c'_1}{h'_1} + \frac{c'_2}{h'_2} = \frac{R'_1}{h'_1}.$$

Thus,

$$\alpha_1 + \alpha_2 = \frac{c'_1}{R'_1} + \frac{c'_2}{R'_2} = \frac{c'_1}{h'_1 \int x f_h dx} + \frac{c'_2}{h'_2 \int x f_h dx} = \frac{R'_1}{h'_1 \int x f_h dx} = \frac{R'_1}{R'_1} = 1.$$

Since $\alpha_i > 0$ for $i = 1, 2$, this implies $0 < \alpha_i < 1$. Further, by Assumption 12.5, we have

$$\frac{\partial^2}{\partial e_1^2} \int_{-\infty}^{\infty} s_1(x) f[x; h(e_1, e_2)] dx \bigg|_{(e^*_1, e^*_2)} = \alpha_1 \frac{\partial^2}{\partial e_1^2} \int_{-\infty}^{\infty} x f[x; h(e_1, e_2)] dx \bigg|_{(e^*_1, e^*_2)}$$
$$= \alpha_1 R''_1(e^*_1, e^*_2) \le 0.$$

Hence, by Assumptions 12.4 and 12.5, condition SOC_1 is satisfied. Condition SOC_2 can also be verified similarly. Finally, since the first-best solution (e^{**}_1, e^{**}_2) satisfies

$$R'_1(e^{**}_1, e^{**}_2) = c'_1(e^{**}_1) = c'_2(e^{**}_2),$$

by condition (24), the solution (e^*_1, e^*_2) of problem (26) cannot be the first best. The proposition is thus proven. ∎

The model in (23) is a model of two partners. An alternative setup is a principal-agent model. If one of the agents, say M_1, is the principal and the other is the agent, we need to add an IR condition of the form

$$\int_{-\infty}^{\infty} s_2(x) f[x; h(e_1, e_2)] dx - c_2(e_2) \ge \bar{u}_2$$

for the agent into problem (23). The principal's problem becomes

$$V = \max_{s_i \in \mathcal{S}, e_1, e_2 \in \mathbb{E}} \int_{-\infty}^{\infty} [x - s_2(x)] f[x; h(e_1, e_2)] dx - c_1(e_1)$$

s.t. $IC_1 : h_1' \int_{-\infty}^{\infty} s_1(x) f_h[x; h(e_1, e_2)] dx = c_1'(e_1),$

$IC_2 : h_2' \int_{-\infty}^{\infty} s_2(x) f_h[x; h(e_1, e_2)] dx = c_2'(e_2),$ \hfill (27)

$IC_3 : SOC_1 :$ for $e_1,$

$IC_4 : SOC_2 :$ for $e_2,$

$RC : s_1(x) + s_2(x) = x,$ for all $x \in \mathbb{R}_+,$

$IR : \int_{-\infty}^{\infty} s_2(x) f[x; h(e_1, e_2)] dx \geq c_2(e_2) + \bar{u}_2.$

The *IR* condition in (27) must be binding for the optimal solution. Hence, the problem becomes

$$V = \max_{s_i \in \mathcal{S}, e_1, e_2 \in \mathbb{E}} R(e_1, e_2) - c_1(e_1) - c_2(e_2)$$

s.t. $IC_1 : h_1' \int_{-\infty}^{\infty} s_1(x) f_h[x; h(e_1, e_2)] dx = c_1'(e_1),$

$IC_2 : h_2' \int_{-\infty}^{\infty} s_2(x) f_h[x; h(e_1, e_2)] dx = c_2'(e_2),$ \hfill (28)

$IC_3 : SOC_1 :$ for $e_1,$

$IC_4 : SOC_2 :$ for $e_2,$

$RC : s_1(x) + s_2(x) = x,$ for all $x \in \mathbb{R}_+,$

$IR : \int_{-\infty}^{\infty} s_2(x) f[x; h(e_1, e_2)] dx \geq c_2(e_2) + \bar{u}_2.$

Comparing (28) with (23), problem (28) has an extra condition. In this case, Proposition 12.2 still holds except that the optimal contract is a linear contract with the form $s_2^*(x) = \alpha_2^* x + \beta_2^*$, where β_2^* is determined by the IR condition, where $s_1^*(x)$ is implied by the RC condition.

A principal-agent setting assumes that the principal has 100% bargaining power ex ante. This is what the word "principal-agent" means. In this setting, an IR condition for the agent is necessary in the model in order to protect the agent. In a partnership setting, all partners have bargaining powers. Since the partners can take care of themselves, IR conditions are unnecessary in the model. The solution will be the same if IR conditions are added.

In all contract models, an ex-ante monetary transfer (more precisely, a predetermined fixed monetary transfer) is allowed. For example, in Proposition 12.2, for any $\beta \in \mathbb{R}$, the following set of contracts is still the optimal solution:

$$s_1(x) = \alpha_1^* x + \beta, \quad s_2(x) = \alpha_2^* x - \beta.$$

Here, β can be regarded as an ex-ante monetary transfer between the two parties. It can also be regarded as a predetermined fixed monetary transfer if the transfer is not

3 Linear Contracts Under Risk Neutrality

made upfront. This β can be determined by the bargaining powers of the two partners. If there are IR conditions for the two parties, as long as the project is socially viable (with positive social welfare in equilibrium), then this β can ensure that both partners' IR conditions are satisfied. This is the reason that we do not explicitly put the two IR conditions into our model (23). Since there is no interest in this β, we do not bother to mention this β in our model. In fact, in a principal-agent model, we are supposed to an IR condition for the principal as well. However, as long as the project is socially viable, this IR condition will be automatically satisfied in equilibrium. Hence, there is no need to explicitly put the principal's IR condition into a principal-agent model.

In summary, an optimal linear contract exists under double risk neutrality; it is the first best under single moral hazard and it is the second best under double moral hazard.

Example 12.3 Consider the following parametric case:

$$h(e_1, e_2) = \mu_1 e_1 + \mu_2 e_2, \quad \tilde{X}(h) = \tilde{A}h, \quad c_i(e_i) = e_i^2/2,$$

where $\mu_1, \mu_2 > 0$, $\tilde{A}$ is a random variable with $\tilde{A} > 0$ and $E(\tilde{A}) = 1$. The second-best solution is

$$e_1^* = \frac{\mu_1^3}{\mu_1^2 + \mu_2^2}, \quad e_2^* = \frac{\mu_2^3}{\mu_1^2 + \mu_2^2}, \quad \alpha_i^* = \frac{\mu_i^2}{\mu_1^2 + \mu_2^2}.$$

The first-best solution is

$$e_1^{**} = \mu_1, \quad e_2^{**} = \mu_2. \quad \blacksquare$$

There is no problem of risk sharing in a model under double risk neutrality since both parties care about expected incomes only. Hence, under single moral hazard, when a single mechanism (a contract) is sufficient to handle the incentive problem, the optimal solution achieves the first best. However, under double moral hazard, when a single mechanism is not sufficient to handle the two incentive problems, the optimal solution cannot achieve the first best.

For the optimal linear solution in Proposition 12.2, the value of α_i^* reflects the relative importance of player i in the project. Example 12.3 indeed shows this, where individual i's marginal contribution to the project is represented by the parameter μ_i and the output share α_i^* indeed reflects his importance.

Bhattacharyya and Lafontaine (1995) are the first to provide such a result in Proposition 2 in a special case. Their result is for a special output process of the form $\tilde{x} = h + \tilde{\varepsilon}$ with a distribution function of the form $F(x; h) = F_{\tilde{\varepsilon}}(x - h)$, where $\tilde{\varepsilon}$ is a random shock with distribution function $F_{\tilde{\varepsilon}}$. Kim and Wang (1998) are the first to provide the general theory in Proposition 12.2, where the proof is in Wang and Zhu (2005).

The optimality of linear contracts in this section is based on risk neutrality. Without double risk neutrality, linear contracts are generally not optimal. However, linear contracts are very popular in reality even though most parties involved are likely to be risk averse. This is puzzling within the framework of complete contracts. All contracts in this chapter are complete contracts. Hence, I cannot not explain this puzzle. The theory on incomplete contracts in Wang (2016) provides a resolution to this puzzle. The explanation is that linear contracts in practice must be optimal incomplete contracts.[7]

Contract theory is based on either a principal-agent model or a partnership model. The partnership model is more general than the principal-agent model. In a partnership model, both parties have bargaining powers and they bargain to reach a contractual agreement. In contrast, in a principal-agent model, the principal has 100% bargaining power ex ante and she offers a take-it-or-leave-it contract, which often causes the agent to have zero surplus (with a binding IR condition). However, the principal-agent model is generally equivalent to the partnership model: they lead to the same solution if the IR condition is binding. For example, we have shown the equivalence of problems (23) and (27).

In the principal-agent model, the objective of the problem is to maximize the principal's profits; in the partnership model, the objective is to maximize social welfare. See, for example, the theory of the Nash bargaining solution (Myerson 1991, Chap. 8; Osborne and Rubinstein 1994, Chaps. 7 and 15), in which the players maximize social welfare. In the partnership model, the principal's profit maximizing behavior is reflected in the principal's IC condition. For example, in the partnership model (23), the objective is to maximize social welfare; in the principal-agent model (27), the objective is to maximize the principal's expected profit.

A contractual solution implicitly allows a upfront transfer (in money, payoff or benefit). For example, if $s^*(x)$ is an optimal contract, the actual contract can be $\alpha + s^*(x)$, where $\alpha \in \mathbb{R}$ is a upfront transfer. This transfer will depend on the bargaining powers of the two parties. In the partnership model, we may or may not have explicit IR conditions for both the principal and the agent; in the principal-agent model, we will have an explicit IR condition for the agent. In a principal-agent model, since the agent has no bargaining power, an IR condition is needed to protect the agent's interest; in a partnership model, since both parties have bargaining powers, no IR condition is necessary. In sum, in the principal-agent model, a upfront transfer and the principal's IR condition are implicit; but in the partnership, these two items may or may not be explicit. In model (23), we do not explicitly mention IR conditions. This means that the actual income scheme should be $s_i^*(x) = \alpha_i^* x + \beta_i^*$, where α_i^* is defined in Proposition 12.2 and β_1^* and β_2^* satisfy the IR conditions of the two parties and $\beta_1^* + \beta_2^* = 0$. There are an infinite number

---

[7]Wang (2016) offers a definition on all contracts, including complete and incomplete contracts. By this definition, all contracts in this chapter are complete contracts.

of such pairs (β_1^*, β_2^*); which pair will be in the actual solution depends on the bargaining powers of the two parties.

4 A Conditional Fixed Contract

We now modify the standard agency model from yet another angle in an attempt to find a simple optimal contract. We have so far found a simple optimal contract, *a linear contract*, in a model under uncertainty only when there is no risk aversion. However, risk aversion is an important parameter in many applications. In this section, we find two conditions under which the agency model with risk aversion has a simple optimal contract known as *a conditional fixed contract*. This contract is very convenient for applied agency problems.

Real-world contracts between principals and agents are typically very simple and often have a bonus structure that specifies wage increases at certain minimum levels of performance. How can contracts be so simple in reality? We provide an answer by establishing the optimality of a conditional fixed contract for the standard agency model under risk aversion.

4.1 The Model

We now abandon Assumption 12.1, and instead assume that the support of output is dependent on effort. Specifically, the support of output $\tilde{x}$ is an interval $\mathbb{X}(a) \equiv [A(a), B(a)]$, where the boundaries of the support $A(a)$ and $B(a)$ are generally dependent on effort a. We allow the special cases with $A(a) = -\infty$ and/or $B(a) = +\infty$ for any a and also those special cases with $A(a)$ and/or $B(a)$ being independent of a. We will also abandon the FOA.

Given a utility function $u : \mathbb{R}_+ \to \mathbb{R}_+$, the agent's expected utility is

$$U(a, s) \equiv \int_{A(a)}^{B(a)} u[s(x)] f(x; a) dx - c(a).$$

Hence, the principal's problem is

$$\pi \equiv \max_{s_i \in \mathcal{S}, a \geq 0} \int_{A(a)}^{B(a)} [x - s(x)] f(x; a) dx$$

$$\text{s.t.} \quad IR : \int_{A(a)}^{B(a)} u[s(x)] f(x : a) dx \geq c(a), \qquad (29)$$

$$IC : a \in \arg\max_{\hat{a}} U(\hat{a}, s).$$

Here, we assume that $\bar{u}=0$. If $\bar{u}\neq 0$, we can replace $c(a)$ by $c(a)+\bar{u}$ and the model remains the same.

This model is the same as the standard agency model except that we allow the boundaries of the output domain to be dependent on effort. It turns out that this dependence is important for our alternative solution to the standard agency problem.

Assumption 12.6 Utility function $u(x)$ is concave, onto and strictly increasing.

Assumption 12.7 The expected revenue $R(a) \equiv \int_{A(a)}^{B(a)} xf(x;a)dx$ is increasing and concave.

Assumption 12.8 (FOSD). $F_a(x;a) \leq 0$ for any $x \in \mathbb{R}$ and $a \in \mathbb{A}$.

These three assumptions are natural requirements. In particular, the utility function u being onto and strictly increasing ensures that $u^{-1}[c(a)]$ is well defined, and Assumption 12.8 is required for any sensible contract theory.

4.2 The Optimal Contract

The optimal conditional fixed contract is stated in the following proposition.

Proposition 12.3 Let a^* be the solution of the following equation:

$$\frac{\partial u^{-1}[c(a^*)]}{\partial a} = R'(a^*), \tag{30}$$

and suppose that the following two conditions are satisfied:

$$\frac{\partial \text{In} u^{-1}[c(a^*)]}{\partial a} \geq \frac{\partial \text{In} c(a^*)}{\partial a}, \tag{31}$$

$$-F_a[A(a^*);a^*] > \frac{\partial \text{In} c(a^*)}{\partial a}. \tag{32}$$

Then under Assumptions 12.6–12.8, the optimal effort is a^* and the optimal contract is

$$s^*(x) = \begin{cases} 0 & \text{if } x < A(a^*), \\ u^{-1}[c(a^*)] & \text{if } x \geq A(a^*). \end{cases}$$

Furthermore, (a^*, s^*) is the first-best solution. ∎

4 A Conditional Fixed Contract

This optimal contract is called a <u>conditional fixed contract</u> since it makes a fixed payment conditional on the performance to be above a cutoff point. The result is from Wang (2012c).

As shown in Fig. 4, this solution looks distinctly different from the solution of the standard agency problem. It, in fact, looks puzzling at first glance because of its simplicity. However, the intuition for the solution turns out to be quite simple. To induce the optimal effort a^*, the FOA suggests that the principal should associate each output level with a payment. Our alternative approach suggests that the principal can induce a^* by simply offering a bonus at a minimum level of performance. For the latter strategy to work, the dependence of the distribution function on effort at the left boundary of the performance domain is crucial. Condition (32) precisely serves to establish this dependence.

We can immediately see some interesting features of the solution. First, the solution achieves the first best. This is impressive considering the simplicity of the contract.

Second, it is a closed-form solution with a clear promotion component. In fact, the pay at the minimum level of performance is a jump, i.e., it could be in the form of a promotion, bonus, or a change of nature of the contract. This solution looks very much like a typical employment contract, in which a fixed wage rate is given based on a certain level of education and working experience plus a potential bonus or promotion based on a minimum level of performance.

Third, the solution has an interesting form with simple and intuitive expressions determining the optimal contract and effort. Based on the fact that $s^*(x) = u^{-1}[c(a^*)]$ for $x \geq A(a^*)$, (31) means that, for a percentage increase in a^*, the percentage increase in pay s^* is greater than the percentage increase in cost. Condition (32) means that, for an increase in a^*, the increase in the probability of an output being larger than $A(a^*)$ is greater than the percentage increase in cost.

Fourth, since both $c(\cdot)$ and $u^{-1}(\cdot)$ are convex, $u^{-1}[c(a)]$ is convex in a, implying that $\frac{\partial u^{-1}[c(a)]}{\partial a}$ is increasing in a. Also, since $R'(a)$ is decreasing in a, a^* from (30) must be unique.

Finally, condition (31) means that the bonus elasticity of effort is larger than the cost elasticity of effort. Since u^{-1} is convex, $u^{-1}[c(a)]$ is more convex than $c(a)$ is. This indicates that condition (31) can be easily satisfied for any $a \in \mathbb{A}$. For example, for $u(x) = x^{1-\alpha}$ and $c(a) = a^\beta$, where $\alpha \in [0,1]$ is the relative risk aversion and $\beta \geq 1$, (31) is satisfied for any $a \in \mathbb{R}_+$.

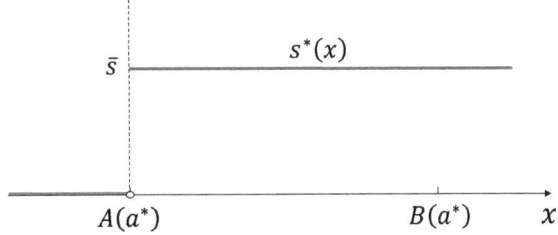

Fig. 4 The optimal conditional fixed contract

We consider two examples: one uses the exponential distribution and the other uses the uniform distribution.

Example 12.4 Suppose

$$u(x) = x^{1-\alpha}, \quad c(a) = a^\beta, \quad f(x;a) = \frac{1}{a} e^{-\frac{x-A(a)}{a}} \quad \text{for } x \geq A(a),$$

where $A(a)$ is an arbitrary increasing function, $\alpha \in [0, 1]$ is the relative risk aversion, and $\beta \geq 1$. These functions include those in Example 12.1 as a special case. The key difference between the two examples is that we allow the boundary $A(a)$ to be dependent on effort. We have

$$u^{-1}[c(a)] = a^{\frac{\beta}{1-\alpha}}, \quad R(a) = A(a) + a, \quad F_a[A(a);a] = -\frac{A'(a)}{a}.$$

Equation (30) becomes

$$\frac{\beta}{1-\alpha}(a^*)^{\frac{\beta}{1-\alpha}-1} = 1 + A'(a^*). \tag{33}$$

Condition (31) is satisfied for any $a \geq 0$. Condition (32) becomes

$$\beta < A'(a^*). \tag{34}$$

Let $A(a) = \gamma a$ for some $\gamma > 1$. Then (34) is satisfied for any $a \in \mathbb{A}$ and for $\beta \leq \gamma$, and (33) implies

$$a^* = \left[\frac{(1+\gamma)(1-\alpha)}{\beta}\right]^{\frac{1-\alpha}{\alpha+\beta-1}}, \quad \bar{s} = \left[\frac{(1+\gamma)(1-\alpha)}{\beta}\right]^{\frac{\beta}{\alpha+\beta-1}}.$$

Example 12.5 Suppose

$$u(x) = x^{1-\alpha}, \quad c(a) = a^\beta, \quad f(x;a) = \frac{1}{\sigma} \quad \text{for } A(a) \leq x \leq A(a) + \sigma,$$

where $A(a)$ is an arbitrary increasing function, $\alpha \in [0, 1]$ is the relative risk aversion, $\sigma > 0$, and $\beta \geq 1$. We have

$$u^{-1}[c(a)] = a^{\frac{\beta}{1-\alpha}}, \quad R(a) = A(a) + \frac{\sigma}{2}, \quad F_a[A(a);a] = -\frac{1}{\sigma}A'(a).$$

4 A Conditional Fixed Contract

Equation (30) becomes

$$\frac{\beta}{1-\alpha}(a^*)^{\frac{\beta}{1-\alpha}-1} = A'(a^*). \tag{35}$$

Condition (31) is satisfied for any $a \geq 0$. Condition (32) becomes

$$\sigma < \frac{1}{\beta}a^*A'(a^*),$$

which is satisfied when σ is sufficiently small. Again, if we let $A(a) = \gamma a$ for some $\gamma > 1$, then (35) implies

$$a^* = \left[\frac{\gamma(1-\alpha)}{\beta}\right]^{\frac{1-\alpha}{\alpha+\beta-1}}, \quad \bar{s} = \left[\frac{\gamma(1-\alpha)}{\beta}\right]^{\frac{\beta}{\alpha+\beta-1}}.$$

5 A Suboptimal Linear Contract

In many published papers, researchers often use a limited set of admissible contracts for the standard agency model and find an optimal contract from this admissible set. In particular, researchers often limit admissible contracts to linear contracts and find an optimal linear contract from this set of linear contracts. This optimal contract is obviously inferior to the optimal contract in the standard agency model and hence we call it a <u>suboptimal linear contract</u> or a <u>third-best linear contract</u>.

Specifically, for the standard agency model in Sect. 1, let the set of admissible contracts be

$$S_0 \equiv \{s(x) | s(x) \text{ is a linear function of the form } s(x) = \alpha + \beta x, \text{ where } \alpha, \beta \in \mathbb{R}\}.$$

With this admissible set, the principal's problem can be written as

$$\max_{a,\alpha,\beta \in \mathbb{R}} \int_{-\infty}^{\infty} v[(1-\beta)x - \alpha]f(x;a)dx$$

$$\text{s.t.} \quad IR: \int_{-\infty}^{\infty} u(\alpha + \beta x)f(x;a)dx \geq c(a) + \bar{u}, \tag{36}$$

$$IC: \int_{-\infty}^{\infty} u(\alpha + \beta x)f_a(x;a)dx = c'(a).$$

The Lagrangian for (36) is

$$L \equiv \int_{-\infty}^{\infty} v[(1-\beta)x - \alpha]f(x;a)dx + \lambda \left[\int_{-\infty}^{\infty} u(\alpha + \beta x)f(x;a)dx - c(a) - \bar{u}\right]$$

$$+ \mu \left[\int_{-\infty}^{\infty} u(\alpha + \beta x)f_a(x;a)dx - c'(a)\right],$$

where λ and μ are Lagrange multipliers and $\lambda \geq 0$. Under Assumption 12.1, the FOCs for problem (36) are

$$0 = \frac{\partial L}{\partial a} = \int_{-\infty}^{\infty} v[(1-\beta)x - \alpha]f_a(x;a)dx + \mu \left[\int_{-\infty}^{\infty} u(\alpha + \beta x)f_{aa}(x;a)dx - c''(a)\right],$$

$$0 = \frac{\partial L}{\partial \alpha} = -\int_{-\infty}^{\infty} v'[(1-\beta)x - \alpha]f(x;a)dx + \lambda \int_{-\infty}^{\infty} u'(\alpha + \beta x)f(x:a)dx$$

$$+ \mu \int_{-\infty}^{\infty} u'(\alpha + \beta x)f_a(x;a)dx,$$

$$0 = \frac{\partial L}{\partial \beta} = -\int_{-\infty}^{\infty} v'[(1-\beta)x - \alpha]xf(x;a)dx + \lambda \int_{-\infty}^{\infty} u'(\alpha + \beta x)xf(x;a)dx$$

$$+ \mu \int_{-\infty}^{\infty} u'(\alpha + \beta x)xf_a(x;a)dx.$$

From these three equations, we can try to solve for (a^*, α^*, β^*). This solution is suboptimal and the suboptimal linear contract is $s^*(x) = \alpha^* + \beta^* x$. This linear contract is the "optimal contract" in many papers.

Notes

For Sect. 1, good references are Ross (1973), Stiglitz (1974), Harris and Raviv (1979), Shavell (1979), and Laffont (1995, pp. 180–198). For Sect. 2, a good reference to is Mas-Colell et al. (1995, Chap. 14). For Sect. 3, good references are Kim and Wang (1998) and Wang and Zhu (2005). Section 4 is based on Wang (2012c).

Appendix
Optimization Methods

A.1 General Optimization

In this appendix, we list some useful mathematical results without proofs and without much explanation. All the results and proofs can be found in Wang (2015).

A.1.1 Directional Differentiation

A popular definition of differentiation for a multi-variable vector-valued function $f: \mathbb{R}^n \to \mathbb{R}^m$ is the so-called directional differentiation.[1] Given $x_0 \in \mathbb{R}^n$ and $t \in \mathbb{R}$, define

$$\psi(h) \equiv \frac{d}{dt} f(x_0 + th) \bigg|_{t=0}, \quad \text{for} \quad h \in \mathbb{R}^n.$$

If this $\psi(h)$ exists, then we say that f has a directional derivative at x_0 in the direction of h, i.e., the directional derivative exists. If all partial derivatives of f exist and are continuous, it turns out that ψ is linearly homogenous in h. That is, there is a matrix $A \in \mathbb{R}^{m \times n}$ such that $\psi(h) = Ah$ for any $h \in \mathbb{R}^n$. We call this A the directional derivative (or Gateaux derivative) of f w.r.t. x at x_0, denoted by $f'(x_0)$ or $\frac{\partial f(x_0)}{\partial x}$. That is,

$$f'(x_0)h \equiv \frac{d}{dt} f(x_0 + th) \bigg|_{t=0}, \quad \text{for all } h \in \mathbb{R}^n.$$

For function $f': \mathbb{R}^n \to \mathbb{R}^{m \times n}$, we can again define its directional derivative as above, called it the second-order directional derivative, denoted as $f''(x_0)$ or $\frac{\partial^2 f(x_0)}{\partial x^2}$.

---

[1] A stronger notion of differentiation is the Fréchet differentiation, which requires more stringent conditions. When the directional derivative exists and is continuous on a open set, then it is also the Fréchet derivative. That is, Fréchet differentiability further requires the directional derivative to be continuous.

© Springer Nature Singapore Pte Ltd. 2018
S. Wang, *Microeconomic Theory*, Springer Texts in Business and Economics,
https://doi.org/10.1007/978-981-13-0041-7

Directional derivatives can be expressed in partial derivatives. For $f : \mathbb{R}^n \to \mathbb{R}^m$, if all partial derivatives of f exist and are continuous, then $f'(x)$ is the so-called Jacobian matrix:

$$f'(x) = \begin{pmatrix} \frac{\partial f_1(x)}{\partial x_1} & \frac{\partial f_1(x)}{\partial x_2} & \cdots & \frac{\partial f_1(x)}{\partial x_n} \\ \frac{\partial f_2(x)}{\partial x_1} & \frac{\partial f_2(x)}{\partial x_2} & \cdots & \frac{\partial f_2(x)}{\partial x_n} \\ \vdots & \vdots & & \vdots \\ \frac{\partial f_m(x)}{\partial x_1} & \frac{\partial f_m(x)}{\partial x_2} & \cdots & \frac{\partial f_m(x)}{\partial x_n} \end{pmatrix}_{m \times n}.$$

For $f : \mathbb{R}^n \to \mathbb{R}$, if the second-order partial derivatives of f exist and are continuous, we have

$$f'(x) = \left(\frac{\partial f(x)}{\partial x_1} \quad \frac{\partial f(x)}{\partial x_2} \quad \cdots \quad \frac{\partial f(x)}{\partial x_n} \right),$$

$$f''(x) = \left(\frac{\partial^2 f(x)}{\partial x_i \partial x_j} \right) \equiv \begin{pmatrix} \frac{\partial^2 f(x)}{\partial x_1 \partial x_1} & \frac{\partial^2 f(x)}{\partial x_1 \partial x_2} & \cdots & \frac{\partial^2 f(x)}{\partial x_1 \partial x_n} \\ \frac{\partial^2 f(x)}{\partial x_2 \partial x_1} & \frac{\partial^2 f(x)}{\partial x_2 \partial x_2} & \cdots & \frac{\partial^2 f(x)}{\partial x_2 \partial x_m} \\ \vdots & \vdots & & \vdots \\ \frac{\partial^2 f(x)}{\partial x_n \partial x_1} & \frac{\partial^2 f(x)}{\partial x_n \partial x_2} & \cdots & \frac{\partial^2 f(x)}{\partial x_n \partial x_n} \end{pmatrix}_{n \times n}.$$

In this case, $f'(x)$ is called the gradient, and $f''(x)$ is called the Hessian matrix.

Theorem A.2 (Composite Mapping Theorem). *Given two open sets $X \subset \mathbb{R}^n$ and $Y \subset \mathbb{R}^m$ and two mappings $f : X \to Y$ and $g : Y \to \mathbb{R}^k$ differentiable at $x_0 \in X$ and $y_0 \equiv f(x_0)$, respectively, then the mapping $h \equiv g \circ f : X \to \mathbb{R}^k$ is also differentiable at x_0, and*

$$(g \circ f)'(x_0) = g'(y_0) f'(x_0),$$

where $g'(y_0) f'(x_0)$ is a matrix product. ■

Proposition A.4

- For any vector α, $\frac{\partial (\alpha \cdot x)}{\partial x} = \alpha$.
- For any matrix A, $\frac{\partial (Ax)}{\partial x} = A$.
- For any matrix A, $\frac{\partial (x'Ax)}{\partial x} = x'(A + A')$. ■

A.1.2 Positive Definite Matrix

Convexity of functions is directly related to minimization, and concavity of functions is directly related to maximization. Convexity and concavity of functions mean respectively positive and negative definiteness of the second-order directional derivatives.

We deal with symmetric matrices only when we talk about definiteness. The notation $A \in \mathbb{R}^{m \times n}$ or $A = (a_{ij})_{m \times n}$ indicates that A is an $m \times n$ matrix. Our matrices contain real entries only. For a symmetric matrix $A \in \mathbb{R}^{n \times n}$,

A is positive semi-definite $(A \geq 0)$ if $x'Ax \geq 0$, $\forall x$;
A is positive definite $(A > 0)$ if $x'Ax > 0$, $\forall x \neq 0$;
A is negative semi-definite $(A \leq 0)$ if $x'Ax \leq 0$, $\forall x$;
A is negative definite $(A < 0)$ if $x'Ax < 0$, $\forall x \neq 0$.

This definition has to do with the Taylor expansion for a function $f : \mathbb{R}^n \to \mathbb{R}$:

$$f(x) = f(x_0) + f'(x_0)(x - x_0) + \frac{1}{2}(x - x_0)'f''(\xi)(x - x_0),$$

where ξ is some point on the connecting line between x and x_0, $f'(\xi)$ is the gradient, and $f''(\xi)$ is the Hessian matrix. At optimum, $f'(x_0) = 0$. Hence, the sign of $(x - x_0)'f''(\xi)(x - x_0)$, or the definiteness of $f''(\xi)$, determines whether or not x_0 is optimal.

Given a matrix $A = (a_{ij})_{n \times n}$, for $i_1, \ldots, i_k \in \{1, 2, \ldots, n\}$ with $i_1 < i_2 < \cdots < i_k$, define a $k \times k$ <u>minor</u> as

$$d_{\{i_1,\ldots,i_k\}} \equiv \begin{vmatrix} a_{i_1 i_1} & a_{i_1 i_2} & \cdots & a_{i_1 i_k} \\ a_{i_2 i_1} & a_{i_2 i_2} & & a_{i_2 i_k} \\ \vdots & \vdots & & \vdots \\ a_{i_k i_1} & a_{i_k i_2} & \cdots & a_{i_k i_k} \end{vmatrix}.$$

In particular, define the <u>principal minors</u> as $d_1 \equiv d_{\{1\}}$, $d_2 \equiv d_{\{1,2\}}$, $\ldots$, $d_n \equiv d_{\{1,2,\ldots,n\}}$. That is,

$$d_1 \equiv a_{11}, \quad d_2 \equiv \begin{vmatrix} a_{11} & a_{12} \\ a_{21} & a_{22} \end{vmatrix}, \quad d_3 \equiv \begin{vmatrix} a_{11} & a_{12} & a_{13} \\ a_{21} & a_{22} & a_{23} \\ a_{31} & a_{32} & a_{33} \end{vmatrix}, \quad \ldots, \quad d_n \equiv |A|.$$

Theorem A.3 *For a symmetric matrix $A \in \mathbb{R}^{n \times n}$,*

(a) $A > 0 \Leftrightarrow d_k > 0$, *for all k.*
(b) $A < 0 \Leftrightarrow (-1)^k d_k > 0$, *for all k.*
(c) $A \geq 0 \Leftrightarrow d_{\{i_1,\ldots,i_k\}} \geq 0$, *for all permutations $\{i_1,\ldots,i_k\}$ and all k.*
(d) $A \leq 0 \Leftrightarrow (-1)^k d_{\{i_1,\ldots,i_k\}} \geq 0$, *for all permutations $\{i_1,\ldots,i_k\}$ and all k.* ■

Proposition A.5

- If $A > 0$, then $A^{-1} > 0$.
- $A \geq 0 \Leftrightarrow -A \leq 0$.
- $A > 0 \Leftrightarrow -A < 0$.
- If A is a tall full-rank matrix, then $A'A > 0$ and $AA' \geq 0$.
- If $A > 0$ and $|B| \neq 0$, then $B'AB > 0$.
- $A \geq 0$ iff all its eigenvalues are positive.
- $A > 0$ iff all its eigenvalues are strictly positive. ∎

A.1.3 Concavity

Given vectors $x^k \in \mathbb{R}^n, k = 1, \ldots, n$, a <u>convex combination</u> of the vectors is

$$y \equiv \lambda_1 x^1 + \cdots + \lambda_m x^m, \quad \lambda_k \geq 0, \quad \sum_{k=1}^{m} \lambda_k = 1.$$

Given any two points $x, y \in \mathbb{R}^n$, the set of all convex combinations of x and y is called the <u>interval</u> connecting x and y and is denoted as $[x, y]$, i.e.,

$$[x, y] \equiv \{z | z = \lambda x + (1 - \lambda) y, \quad \lambda \in [0, 1]\}.$$

We can similarly define $(x, y], [x, y)$ and (x, y). A set $S \subset \mathbb{R}^n$ is called a <u>convex set</u> if the interval of any two points in S is contained in S:

$$\forall x, y \in S, \quad [x, y] \subset S.$$

Proposition A.6 (Properties of Convex Sets)

1. *Any intersection of convex sets is also convex.*
2. *The Cartesian product of convex sets is also convex.* ∎

A function $f : X \to \mathbb{R}$ is <u>concave</u> if its domain $X \subset \mathbb{R}^n$ is convex and

$$f[\lambda x + (1 - \lambda) y] \geq \lambda f(x) + (1 - \lambda) f(y), \quad \forall \lambda \in (0, 1), x, y \in X.$$

If the inequality holds strictly, we say that f is <u>strictly concave</u>. A function f is <u>(strictly) convex</u> if $-f$ is (strictly) concave. See Fig. A.1.

Theorem A.4 (Properties of Concave Functions). *Let the domain $X \subset \mathbb{R}^n$ be a convex set.*

Fig. A.1 A concave curve

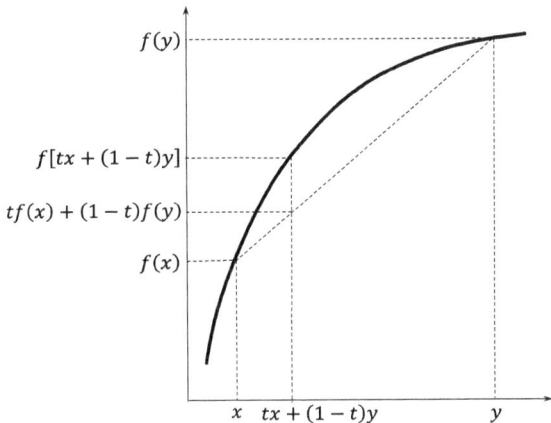

1. A function $f : X \to \mathbb{R}$ is concave iff the lower image $I(f) \equiv \{(x,t) \in (X, \mathbb{R}) | f(x) \geq t\}$ is convex.
2. A concave functions is continuous in the interior of its domain.
3. A function $f : X \to \mathbb{R}$ is concave iff

$$f(\lambda_1 x^1 + \cdots + \lambda_k x^k) \geq \lambda_1 f(x^1) + \cdots + \lambda_k f(x^k),$$

for all $k \geq 1$ and all convex combinations $\lambda_1 x^1 + \cdots + \lambda_k x^k$. ∎

Theorem A.5 *Given convex $X \subset \mathbb{R}^n$, twice differentiable $f : X \to \mathbb{R}$,*

1. *f is convex $\Leftrightarrow f''(x) \geq 0, \quad \forall x \in X$.*
2. *$f''(x) > 0, \quad \forall x \in X \Rightarrow f$ is strictly convex.*
3. *f is concave $\Leftrightarrow f''(x) \geq 0, \quad \forall x \in X$.*
4. *$f''(x) < 0, \quad \forall x \in X \Rightarrow f$ is strictly concave.* ∎

Corollary A.3 *Given convex $X \subset \mathbb{R}^n$, twice differentiable $f : X \to \mathbb{R}$, and let $d_{\{i_1,\ldots,i_k\}}(x)$ be a $k \times k$ principal minors of $f''(x)$,*

1. *f is convex $\Leftrightarrow d_{\{i_1,\ldots,i_k\}}(x) \geq 0$, for any $x \in X, k \leq n$ and $\{i_1, \ldots, i_k\}$.*
2. *f is concave $\Leftrightarrow (-1)^k d_{\{i_1,\ldots,i_k\}}(x) \geq 0$, for any $x \in X$, $k \leq n$ and $\{i_1, \ldots, i_k\}$.*
3. *$d_k(x) > 0$, for any $x \in X$ and $k \leq n \Rightarrow f$ is strictly convex.*
4. *$(-1)^k d_k(x) > 0$, for any $x \in X$ and $k \leq n \Rightarrow f$ is strictly concave.* ∎

Example A.1 For function $f : \mathbb{R}^2_{++} \to \mathbb{R}$, defined by

$$f(x,y) = x^\alpha + y^\beta, \quad \alpha, \beta \in \mathbb{R},$$

we have

$$f \text{ is} \begin{cases} \text{concave}, & \text{if } 0 \leq \alpha, \beta \leq 1; \\ \text{strictly concave}, & \text{if } 0 < \alpha, \beta < 1. \end{cases}$$

■

Example A.2 For Cobb–Douglas function $f : \mathbb{R}^2_{++} \to \mathbb{R}$, defined by

$$f(x,y) = x^\alpha y^\beta, \quad \alpha, \beta \geq 0,$$

we have

$$f \text{ is} \begin{cases} \text{concave}, & \text{if } \alpha, \beta \geq 0, \quad \alpha + \beta \leq 1; \\ \text{strictly concave}, & \text{if } \alpha, \beta > 0, \quad \alpha + \beta < 1. \end{cases} \blacksquare$$

Proposition A.7 *Let $f : X \to \mathbb{R}$ be differentiable. Then,*

(1) *f is concave $\Leftrightarrow f'(x) \cdot (y-x) \geq f(y) - f(x)$, for all $x, y \in X$.*
(2) *f is strictly concave $\Leftrightarrow f'(x) \cdot (y-x) > f(y) - f(x)$, for all $x, y \in X$, $x \neq y$.* ■

This proposition is illustrated in Fig. A.2.

A.1.4 Quasi-concavity

Given a convex set $X \subset \mathbb{R}^n$, we say that $f : X \to \mathbb{R}$ is <u>quasi-concave</u> if

$$f(y) \geq f(x) \Rightarrow f(z) \geq f(x), \quad \forall x, y \in X, \quad z \in (x, y).$$

f is <u>strictly quasi-concave</u> if

$$f(y) \geq f(x) \Rightarrow f(z) > f(x), \quad \forall x, y \in X, \quad z \in (x, y).$$

f is <u>(strictly) quasi-convex</u> if $-f$ is (strictly) quasi-concave. See Fig. A.3.

Theorem A.6 *Given convex set $X \subset \mathbb{R}^n$,*

- *$f : X \to \mathbb{R}$ is quasi-concave iff the <u>upper contour set</u> $\{x \in X | f(x) \geq t\}$ is convex for all $t \in \mathbb{R}$.*
- *$f : X \to \mathbb{R}$ is quasi-convex iff the <u>lower contour set</u> $\{x \in X | f(x) \leq t\}$ is convex for all $t \in \mathbb{R}$.* ■

Appendix: Optimization Methods

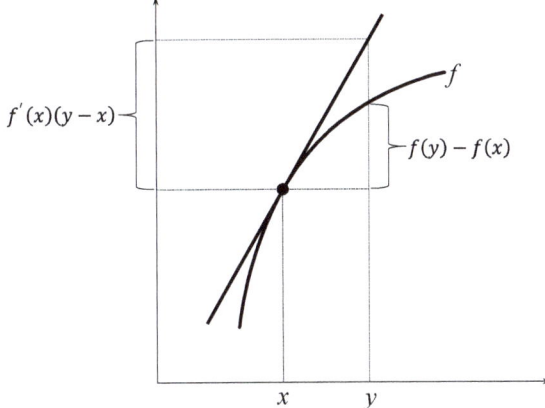

Fig. A.2 Characterization of concave functions

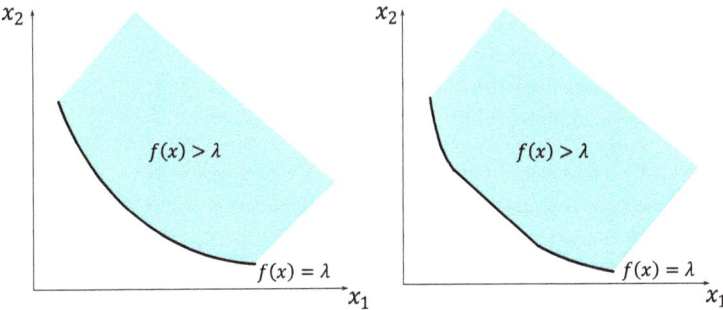

Fig. A.3 Strict and nonstrict quasi-concave curves

The border of a contour set $f^{-1}(t) = \{x \in X | f(x) = t\}$ is called a <u>level set</u>. In economics, when f is a utility function, a level set is called an indifference curve; when f is a production function, it is called an isoquant; and when f is a cost function, it is called an isocost curve.

Theorem A.7

(a) *Concave functions are quasi-concave; convex functions are quasi-convex.*
(b) *Strictly concave functions are strictly quasi-concave. Strictly convex functions are strictly quasi-convex.*
(c) *If $X \subset \mathbb{R}$, monotonic functions are both quasi-concave and quasi-convex.* ∎

Quasi-concavity however does not imply concavity; see Fig. A.4 for a quasi-concave curve, which is obviously not concave. In summary,

strict concavity $\Rightarrow$ concavity
$\Downarrow$ $\Downarrow$
strict quasi-concavity $\Rightarrow$ quasi-concavity

Given function $f : X \to \mathbb{R}$, $X \subset \mathbb{R}^n$, denote

$$f_i(x) \equiv \frac{\partial f(x)}{\partial x_i}, \quad f_{ij}(x) \equiv \frac{\partial^2 f(x)}{\partial x_i \partial x_j},$$

and the bordered Hessian matrix of f as

$$B_f(x) \equiv \begin{pmatrix} 0 & f_1(x) & \cdots & f_n(x) \\ f_1(x) & f_{11}(x) & \cdots & f_{1n}(x) \\ \vdots & \vdots & & \vdots \\ f_n(x) & f_{n1}(x) & \cdots & f_{nn}(x) \end{pmatrix}.$$

It is actually the Hessian matrix of the Lagrangian function at an optimal point of $f(x)$.

Theorem A.8 *Given convex set $X \subset \mathbb{R}^n$, twice differentiable function $f : X \to \mathbb{R}$, and the principal minors $b_1(x), \ldots, b_{n+1}(x)$ of $B_f(x)$,*

1. *For $X \subset \mathbb{R}^n_+$, f is quasi-convex $\Rightarrow b_k(x) \leq 0$, $\forall x \in X$, $\forall k$.*
2. *For $X \subset \mathbb{R}^n_+$, f is quasi-concave $\Rightarrow (-1)^k b_k(x) \leq 0$, $\forall x \in X$, $\forall k$.*
3. *For $X = \mathbb{R}^n_+$ or $\mathbb{R}^n$, $b_k(x) < 0$, $\forall x \in X$, $\forall k \geq 2 \Rightarrow f$ is strictly quasi-convex.*
4. *For $X = \mathbb{R}^n_+$ or $\mathbb{R}^n$, $(-1)^k b_k(x) < 0$, $\forall x \in X$, $\forall k \geq 2 \Rightarrow f$ is strictly quasi-concave.* ∎

Example A.3 Consider function $f : \mathbb{R}^2_{++} \to \mathbb{R}$, defined by

$$f(x,y) = x^\alpha + y^\beta, \quad \alpha, \beta \geq 0,$$

we have

f is $\begin{cases} \text{quasi-concave,} & \text{if } 0 \leq \alpha, \beta \leq 1; \\ \text{strictly quasi-concave,} & \text{if } 0 < \alpha, \beta \leq 1 \text{ and } \alpha \neq 1 \text{ or } \beta \neq 1. \end{cases}$ ∎

Example A.4 Consider Cobb–Douglas function $f : \mathbb{R}^2_{++} \to \mathbb{R}$, defined by

$$f(x,y) = x^\alpha y^\beta, \quad \alpha, \beta \geq 0,$$

Fig. A.4 A quasi-concave curve

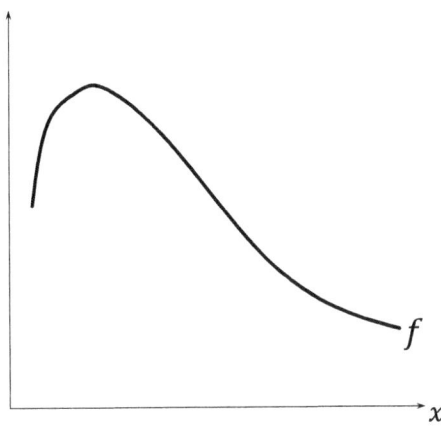

we have

$$f \text{ is } \begin{cases} \text{quasi-concave,} & \text{if } \alpha, \beta \geq 0; \\ \text{strictly quasi-concave,} & \text{if } \alpha, \beta > 0. \end{cases} \blacksquare$$

Theorem A.9 *Let $A \in \mathbb{R}^{n \times n}$ be symmetric, $C \in \mathbb{R}^{m \times n}$ have full rank, $m < n$, and $b_1, \ldots, b_{m+n}$ be the principal minors of $B \equiv \begin{pmatrix} 0 & C \\ C^T & A \end{pmatrix}$. Then,*

1. $x'Ax > 0$ for $x \neq 0$ satisfying $Cx = 0$ $\Leftrightarrow$ $(-1)^m b_k > 0, \forall k \geq 2m+1$.
2. $x'Ax < 0$ for $x \neq 0$ satisfying $Cx = 0$ $\Leftrightarrow$ $(-1)^{m+k} b_k > 0, \forall k \geq 2m+1$. $\blacksquare$

A.1.5 Unconstrained Optimization

Optimization has two categories: unconstrained optimization and constrained optimization. From the names, unconstrained optimization is optimization without constraints and constrained optimization is optimization with constraints. However, the real difference is: unconstrained optimization deals with interior solutions, while constrained optimization deals with boundary solutions.

Theorem A.10 (Weierstrass). *Let $f : \mathbb{R}^n \to \mathbb{R}$ be continuous and A be compact. Then, the following unconstrained optimization problem has at least one solution:*

$$\max_{x \in A} f(x). \blacksquare$$

Theorem A.11 *Let $A \subset \mathbb{R}^n$.*

(a) *If x^* is an interior solution of*

$$\max_{x \in A} f(x) \tag{A.1}$$

then (FOC) $f'(x^*) = 0$ and (SOC) $f''(x^*) \leq 0$.

(b) *If $f'(x^*) = 0$ and (SOC) $f''(x^*) < 0$, then there is a neighborhood $\mathcal{N}_r(x^*)$ of x^* such that x^* is the maximum point of f on $\mathcal{N}_r(x^*)$. (Local maximum).*

(c) *If f is concave on A, then any point $x^* \in A$ satisfying $f'(x^*) = 0$ is a maximum point.*

(d) *If f is strictly quasi-concave, a unique local maximum over a convex set A is the unique global maximum.* ∎

A.1.6 Constrained Optimization

When an optimization problem has explicit functional constraints, we have constrained optimization.

Theorem A.12 (Lagrange). *Given C^2 functions $f : \mathbb{R}^n \to \mathbb{R}$ and $G : \mathbb{R}^n \to \mathbb{R}^m$, consider problem*

$$\begin{aligned} \max_{x \in \mathbb{R}^n} \quad & f(x) \\ \text{s.t.} \quad & G(x) = 0. \end{aligned} \tag{A.2}$$

Define a <u>Lagrange function</u> $L(x, \lambda) \equiv f(x) + \lambda \cdot G(x)$, where $\lambda \in \mathbb{R}^m$ is a constant vector.

(a) *If x^* solves (A.2) and if the row vectors of the derivative $G'(x^*)$ are linearly independent (<u>constraint qualification</u>), then there exists $\lambda^* \in \mathbb{R}^m$ (a vector of <u>Lagrange multipliers</u>) such that*

> FOC: $L'_x(x^*, \lambda^*) = 0$,
> SOC: $h^T L''_x(x^*, \lambda^*) h \leq 0$, for any h satisfying $G'(x^*)h = 0$.

(b) *If the FOC is satisfied for some pair (x^*, λ^*) satisfying $G(x^*) = 0$ and*

> SOC: $h^T L''_x(x^*, \lambda^*) h < 0$, for any $h \neq 0$ satisfying $G'(x^*)h = 0$, (A.3)

then x^* is a unique local maximum.

(c) *If the FOC is satisfied for some pair* (x^*, λ^*) *satisfying* $G(x^*) = 0$, λ^*. $G(x)$ *is quasi-concave in x, and*

$$\text{SOC}: \quad h^T f''(x^*) h < 0, \quad \text{for any} \quad h \neq 0 \text{ satisfying } G'(x^*) h = 0, \quad \text{(A.4)}$$

then x^* is a unique local maximum. ∎

How do we verify the SOC? Let

$$B(x) \equiv \begin{pmatrix} 0 & G'(x) \\ [G'(x)]' & f''(x) \end{pmatrix}.$$

Let $b_i(x)$ be the principal minors of $B(x)$. By Theorem A.9, the SOC is guaranteed if $(-1)^i b_i(x^*) < 0$ for $i \geq 2m+1$.

Theorem A.13 (Kuhn–Tucker). *Given C^1 functions $f, g_i, h_j : \mathbb{R}^n \to \mathbb{R}, i = 1,\ldots,m, \ j = 1,\ldots,k$, and $k < n$, let $G \equiv (g_1,\ldots,g_m)^T$ and $H \equiv (h_1,\ldots,h_k)^T$. Consider problem*

$$V = \max_{x \in \mathbb{R}^n} f(x)$$
$$\text{s.t} \quad G(x) \geq 0, \quad \text{(A.5)}$$
$$H(x) = 0,$$

Let $I(x) \equiv i | g_i(x) = 0$ and $L(x, \lambda, \mu) = f(x) + \lambda \cdot G(x) + \mu \cdot H(x)$, where $\lambda \in \mathbb{R}^m$ and $\mu \in \mathbb{R}^k$ are constant vectors.

(a) *If x^* is a solution and $g'_i(x^*)$ for $i \in I(x^*)$ together with all $h'_j(x^*)$ are linearly independent vectors (<u>constraint qualification</u> or <u>rank condition</u>), then there exist unique $\lambda^* \in \mathbb{R}^m_+$ and $\mu^* \in \mathbb{R}^k$ such that*

$$\begin{aligned} \text{FOC}: \quad & L'_x(x^*, \lambda^*, \mu^*) = 0, \\ \text{KTC}: \quad & \lambda^* \cdot G(x^*) = 0. \end{aligned} \quad \text{(A.6)}$$

(b) *Conversely, suppose that a triplet (x^*, λ^*, μ^*) satisfies conditions in (A.6), where x^* is admissible. If $\lambda_i^* > 0$ for any $i \in I(x^*)$ and (x^*, λ^*, μ^*) satisfies the SOC for the following problem as stated in Theorem A.12, then x^* is the unique local maximum for (A.5).*

$$\max_{x \in \mathbb{R}^n} f(x)$$
$$\text{s.t.} \quad g_i(x) = 0, \quad i \in I(x^*) \quad \text{(A.7)}$$
$$H(x) = 0. \quad \blacksquare$$

Note that, since $\lambda \geq 0$ and $G \geq 0$, the KTC (Kuhn–Tucker condition) is equivalent to

$$\text{KTC}: \quad \lambda_i g_i(x) = 0, \quad \text{for all } i.$$

The Lagrange and Kuhn–Tucker theorems are very useful in economics. The following result is useful when global maximization is required.

Theorem A.14 (Global Maximum). *Given differentiable functions $f, g_i, h_j : \mathbb{R}^n \to \mathbb{R}$, an admissible $x^* \in \mathbb{R}^n$ is a global maximum for problem (A.5) if*

(a) *There are $\lambda^* \in \mathbb{R}^m_+$ and $\mu^* \in \mathbb{R}^k$ such that (x^*, λ^*, μ^*) satisfies the KTC and FOC for (A.5) in (A.6);*
(b) *$f, \lambda_i^* g_i(\cdot) \lambda_i^* g_i(\cdot)$ and $\mu_j^* h_j(\cdot)$ are quasi-concave for all i and j; and*
(c) *$f'(x^*) \neq 0$ or f is concave.* ∎

A.1.7 The Envelope Theorem

We often deal with maximum and minimum value functions in economics. The following results on differentiation of these types of functions is very useful. One theorem is for unconstrained optimization and the other is for constrained optimization.

Theorem A.15 (Envelope). *Suppose $f : X \times A \to \mathbb{R}$ is differentiable, $X \subset \mathbb{R}^n, A \subset \mathbb{R}^k$, and $x^*(a)$ is an interior solution of*

$$F(a) \equiv \max_{x \in X} f(x, a).$$

Then,

$$\frac{dF(a)}{da} = \frac{\partial f(x, a)}{\partial a}\bigg|_{x = x^*(a)}. \quad \blacksquare$$

Theorem A.16 (Envelope). *Given set $A \subset \mathbb{R}^l$ and C^1 functions $f, g_i, h_j : \mathbb{R}^n \times A \to \mathbb{R}$, let $G \equiv (g_1, \ldots, g_m)^T, H \equiv (h_1, \ldots, h_k)^T$, and*

$$L(x, a, \lambda, \mu) \equiv f(x, a) + \lambda \cdot G(x, a) + \mu \cdot H(x, a).$$

If $x^*(a)$ is a solution of the following problem:

$$F(a) \equiv \max_{x \in \mathbb{R}^n} f(x, a)$$
$$\text{s.t.} \quad G(x, a) \geq 0,$$
$$H(x, a) = 0,$$

and $\lambda^*(a)$ and $\mu^*(a)$ are the corresponding Lagrange multipliers, then

$$\frac{\partial F(a)}{\partial a} = \left.\frac{\partial L(x,a,\lambda,\mu)}{\partial a}\right|_{x=x^*(a),\lambda=\lambda^*(a),\mu=\mu^*(a)} \tag{A.8}$$

under the conditions:

(a) *There is a neighborhood* $\mathbb{N}_r(a^*) \subset A$ *such that, for every* $a \in \mathbb{N}_r(a^*)$, *the problem has a solution, and at* a^*, *the problem has a unique solution* $x^* = x^*(a^*)$.
(b) *Gradients* $g'_{ix}(x^*,a^*)$, $i = 1,\ldots,m$, *corresponding to binding constraints are linearly independent.* ∎

The key advantage of the envelope theorem is that you can find the derivative of function $F(a)$ without actually solving for $F(a)$ first. Given the knowledge of $x^*(a)$, we can use function $f(x,a)$ to find $F'(a)$.

A.1.8 Homogeneous Functions

Besides concave and convex functions, homogeneous functions are of special interest to economists. A function $f: \mathbb{R}^n_+ \to \mathbb{R}$ is <u>homogeneous of degree</u> α if

$$f(\lambda x) = \lambda^\alpha f(x), \quad \forall x \in \mathbb{R}^n_+, \quad \lambda > 0.$$

In particular, when $\alpha = 1$, we also say that f is <u>linearly homogeneous</u>; and when $\alpha = 0$, we also say that f is <u>zero homogeneous</u>.

Theorem A.17 (Euler Law). *If* $f: \mathbb{R}^n_+ \to \mathbb{R}$ *is linearly homogeneous, then*

$$f(x) = \sum_{i=1}^n f_{x_i}(x) x_i. \quad \blacksquare$$

Theorem A.18 *If* $f(x)$ *is homogeneous of degree* α, *then* $f_{x_i}(x)$ *is homogeneous of degree* $\alpha - 1, \forall i$. ∎

A.2 Dynamic Optimization

This section presents theories on two types of dynamic models: discrete-time stochastic models and continuous-time deterministic models.

A.2.1 Discrete-Time Stochastic Models

In this subsection, time t is discrete, meaning that it takes values of $t = 0, 1, 2, \ldots$. The current time may be $t = 0$ and you need to make a plan now for all future actions. Many things in the future are unknown at present. These unknown things can usually be represented by random variables and you are supposed to know the conditional distribution functions of these random variables, conditional on available information. Hence, your plan will be dependent on these distribution functions as well as the realized values of the random variables when the time comes.

The Markov Process

What is a Markov process? Suppose that you are interested in a variable or a vector x. Its value depends on time. We denote its value at time t as x_t. At time 0, you know x_0, but you do not know x_1. Suppose that you know the distribution function of x_1 conditional on your knowledge of x_0, i.e.,

$$x_1 \sim F_0(\cdot|x_0),$$

where F_0 is the distribution function of x_1 at time $t = 0$. In general, at t, you know $x_0, \ldots, x_t$ and you have the knowledge of the distribution function of x_{t+1} conditional on your knowledge of $x_0, \ldots, x_t$, that is (Fig. A.5),

$$x_{t+1} \sim F(\cdot|x_t, x_{t-1}, \ldots, x_0).$$

In this case, we call $\{x_t\}_{t=0}^{\infty}$ a <u>random process</u>. In general, the distribution function F_t is time-dependent. If the distribution function is time-independent, i.e.,

$$x_{t+1} \sim F(\cdot|x_t, x_{t-1}, \ldots, x_0), \quad \text{for all } t,$$

we call $\{x_t\}_{t=0}^{\infty}$ a <u>stationary process</u>. In addition, if the distribution function's dependence on past history has a fixed length of time, i.e., there exists an integer n such that[2]

$$x_{t+1} \sim F(\cdot|x_t, \ldots x_{t-n+1}), \quad \text{for all } t,$$

then we call $\{x_t\}_{t=0}^{\infty}$ an <u>nth-order Markov process</u>. For example, a <u>first-order Markov process</u> $\{x_t\}_{t=0}^{\infty}$ is defined by

$$x_{t+1} \sim F(\cdot|x_t), \quad \text{for all } t.$$

---

[2] Here, we implicitly assume $x_1 \sim F(\cdot|x_0)$, $x_2 \sim F(\cdot|x_0, x_1), \ldots, x_n \sim F(\cdot|x_0, \ldots, x_{n-1})$.

Fig. A.5 The Time Line

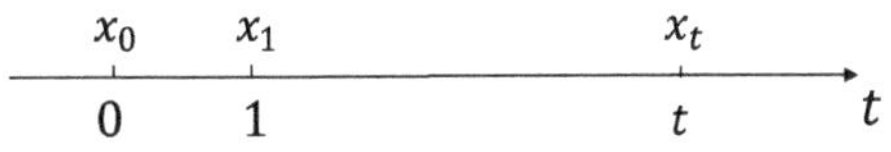

The Bellman Equation

Let $x_t \in \mathbb{R}^n$, $u_t \in \mathbb{R}^k$, $f_t : \mathbb{R}^n \times \mathbb{R}^k \to \mathbb{R}$, and $g_t : \mathbb{R}^n \times \mathbb{R}^k \times \mathbb{R}^m \to \mathbb{R}^n$. Consider the problem:

$$V_t(x_t) \equiv \max_{u_t, u_{t+1}, \ldots} E_t \sum_{s=t}^{\infty} \beta^{s-t} f_s(x_s, u_s)$$

s.t. $\quad x_{s+1} = g_s(x_s, u_s, \varepsilon_{s+1}), \quad s \geq t$

x_t is given and known.
(A.9)

where $0 < \beta < 1$, $\varepsilon_t \in \mathbb{R}^m$ is a random vector not known until period t, E_t is the mathematical expectation operator conditional on period t information set Φ_t. In this problem, the uncertainty is injected by the random process $\{\varepsilon_t\}$ from the motion equation $x_{t+1} = g_t(x_t, u_t, \varepsilon_{t+1})$.

Let the value at time t be

$$V_t(x_t) \equiv \max_{u_t, u_{t+1}, \ldots} E_t \sum_{s=t}^{\infty} \beta^{s-t} f_s(x_s, u_s)$$

s.t. $\quad x_{s+1} = g_s(x_s, u_s, \varepsilon_{s+1}), \quad s \geq t$

x_t is given and known.
(A.10)

We have

$$V_t(x_t) \equiv \max_{u_t} f_t(x_t, u_t) + \beta \max_{u_{t+1}, u_{t+2}, \ldots} E_t \sum_{s=t+1}^{\infty} \beta^{s-t-1} f_s(x_s, u_s)$$

$$= \max_{u_t} f_t(x_t, u_t) + \beta E_t \max_{u_{t+1}, u_{t+2}, \ldots} E_{t+1} \sum_{s=t+1}^{\infty} \beta^{s-t-1} f_s(x_s, u_s)$$

$$= \max_{u_t} f_t(x_t, u_t) + \beta E_t V_{t+1}(x_{t+1}).$$

Thus, the solution of (A.9) must be the solution of the following so-called <u>Bellman equation</u>:

$$V_t(x_t) \equiv \max_{u_t} \; f_t(x_t, u_t) + \beta E_t V_{t+1}(x_{t+1})$$

s.t. $\quad x_{t+1} = g_t(x_t, u_t, \varepsilon_{t+1}).$
(A.11)

Conversely, the solution of the Bellman equation (A.11) is generally the solution of (A.9) if the following so-called <u>transversality condition</u> is satisfied:

$$\lim_{t \to \infty} \beta^t E_t V_{t+1}(x^*_{t+1}) = 0. \tag{A.12}$$

The typical way of solving problem (A.11) is to use the FOC and the envelope theorem:

$$\text{FOC}: f_{t,u}(x_t, u_t) + \beta E_t \left[V'_{t+1}(x_{t+1}) g_{t,u}(x_t, u_t, \varepsilon_{t+1}) \right] = 0, \tag{A.13}$$

$$\text{Envelope}: V'_t(x_t) = f_{t,x}(x_t, u_t) + \beta E_t \left[V'_{t+1}(x_{t+1}) g_{t,x}(x_t, u_t, \varepsilon_{t+1}) \right]. \tag{A.14}$$

If V_t is known, (A.13) gives us the policy function: $u^*_t = h_t(x_t)$. To determine V_t, we need (A.14). These two equations together determine the solution.

Theorem A.19 *If f_t and g_t are time invariant, $f_t = f$ and $g_t = g$, and $\{\varepsilon_t\}$ is a first-order Markov process, then V_t defined in (A.10) is also time invariant.* ∎

By Theorem A.19, a solution (h_t, V_t) from (A.13) and (A.14) will be time invariant: $u^*_t = h(x_t)$ and $V_t(x_t) = V(x_t)$.

The Lagrange Method

We can alternatively use the Lagrange theorem to solve problem (A.9). The Lagrangian function for (A.9) is

$$\mathcal{L} = E_t \sum_{s=t}^{\infty} \beta^{s-t} \{ f_s(x_s, u_s) + \lambda_{s+1} \cdot [g_s(x_s, u_s, \varepsilon_{s+1}) - x_{s+1}] \}.$$

The problem becomes

$$V_t(x_t) = \max_{\substack{u_t, u_{t+1}, \ldots \\ x_t, x_{t+1}, \ldots}} E_t \sum_{s=t}^{\infty} \beta^{s-t} \{ f_s(x_s, u_s) + \lambda_{s+1} \cdot [g_s(x_s, u_s, \varepsilon_{s+1}) - x_{s+1}] \}.$$

The FOCs are

$$0 = \frac{\partial \mathcal{L}}{\partial u_s} = \beta^{s-t} E_t \left[f_{s,u}(x_s, u_s) + \lambda_{s+1} \cdot g_{s,u}(x_s, u_s, \varepsilon_{s+1}) \right],$$
$$0 = \frac{\partial \mathcal{L}}{\partial x_s} = \beta^{s-t} E_t \left[f_{s,x}(x_s, u_s) + \lambda_{s+1} \cdot g_{s,x}(x_s, u_s, \varepsilon_{s+1}) \right] - \beta^{s-1-t} E_t \lambda_s.$$

For $s = t$,

$$f_{t,u}(x_t, u_t) + E_t \left[\lambda_{t+1} \cdot g_{t,u}(x_t, u_t, \varepsilon_{t+1}) \right] = 0, \tag{A.15}$$

$$\beta E_t \left[f_{t,x}(x_t, u_t) + \lambda_{t+1} \cdot g_{t,x}(x_t, u_t, \varepsilon_{t+1}) \right] = \lambda_t. \tag{A.16}$$

By the Envelope theorem,

$$V'_t(x_t) = E_t\big[f_{t,x}(x_t, u_t) + \lambda_{t+1} \cdot g_{t,x}(x_t, u_t, \varepsilon_{t+1})\big]. \quad (A.17)$$

By (A.16) and (A.17), we have $\lambda_t = \beta V'_t(x_t)$. Substituting this into (A.15) and (A.16) yields

$$f_{t,u}(x_t, u_t) + \beta E_t\big[V'_{t+1}(x_{t+1}) \cdot g_{t,u}(x_t, u_t, \varepsilon_{t+1})\big] = 0,$$
$$V'_t(x_t) = f_{t,x}(x_t, u_t) + \beta E_t\big[V'_{t+1}(x_{t+1}) \cdot g_{t,x}(x_t, u_t, \varepsilon_{t+1})\big].$$

These two equations are the same as (A.13) and (A.14).

A.2.2 Continuous-Time Deterministic Models

The General Model (Calculus of Variations)

Theorem A.20 *Suppose that $H : \mathbb{R} \times \mathbb{R}^k \times \mathbb{R}^k \to \mathbb{R}$ is continuous w.r.t. its first argument, continuously differentiable w.r.t. its second and third arguments. Let the set of admissible controls be:*

$$\mathcal{A} \equiv \big\{\text{continuously differentiable functions } u : [t_0, T] \to \mathbb{R}^k\big\}.$$

Then, the solution u^* of

$$\max_{u \in \mathcal{A}} \int_{t_0}^{T} H[t, u(t), \dot{u}(t)] dt \quad (A.18)$$

$$\text{s.t.} \quad u(t_0) = u_0, \quad u(T) = u_T$$

must satisfy the <u>Euler equation</u>:

$$\frac{d}{dt} H_{\dot{u}}[t, u^*(t), \dot{u}^*(t)] = H_u[t, u^*(t), \dot{u}^*(t)], \quad t \in (t_0, T). \quad (A.19)$$

Here, T can be either finite or infinite, and the initial value $u(t_0)$ and the terminal value $u(T)$ can be either fixed or free. If the terminal value $u(T)$ is free, the transversality condition is

$$H_{\dot{u}}[T, u^*(T), \dot{u}^*(T)] = 0. \quad (A.20)$$

If the initial value $u(t_0)$ is free, the transversality condition is

$$H_{\dot{u}}[t_0, u^*(t_0), \dot{u}^*(t_0)] = 0. \quad (A.21)$$

If the terminal condition is $u(T) \geq 0$, the transversality conditions are

$$u^*(T)H_{\dot{u}}[T,u^*(T),\dot{u}^*(T)] = 0, \quad H_{\dot{u}}[T,u^*(T),\dot{u}^*(T)] \leq 0. \tag{A.22}$$

Conversely, if $H(t,u,\dot{u})$ is concave in $(u,\dot{u})$, then any $u^* \in \mathcal{A}$ satisfying the Euler equation (A.19) and the initial and terminal conditions is a solution of (A.18). ∎

The concavity condition of $H(t,u,\dot{u})$ in $(u,\dot{u})$ is too strong and it often fails. Without it, a second-order condition is needed. A necessary condition is

$$H_{\dot{u}\dot{u}}[t,u^*(t),\dot{u}^*(t)] \leq 0.$$

This is the so-called <u>Legendre condition</u>. However, this condition is far from being sufficient. In fact, even a strict inequality of this condition is not sufficient for a locally optimal path.

Since the Euler equation is a second-order differential equation for u, two boundary conditions are needed to pin down the two arbitrary constants in the general solution. For problem (A.18), the two boundary conditions for the Euler equation are $u(t_0) = u_0$ and $u(T) = u_T$; when one of the boundary condition is missing, a transversality condition, such as those in (A.20)–(A.22), is needed to replace it.

Special Models (Optimal Control)

Problem (A.18) is a unconstrained problem. Economics problems often have constraints, such as consumer budget constraints, government budget constraints, and resource constraints, etc. We now present a few cases of constrained problems, which can be transformed into unconstrained problems using Lagrange multipliers. We can hence treat these special cases as special models of the general model in (A.18).

Theorem A.21 (Special Model I). *Let $x \in \mathbb{R}^n$, $u \in \mathbb{R}^k$, $g : \mathbb{R}^n \times \mathbb{R}^k \times \mathbb{R} \to \mathbb{R}^n$, $f : \mathbb{R}^n \times \mathbb{R}^k \times \mathbb{R} \to \mathbb{R}$. For problem*

$$J(x_0, x_T, t_0) \equiv \max_u \int_{t_0}^{T} f[x(t), u(t), t] dt$$

$$\text{s.t. } \dot{x}(t) = g[x(t), u(t), t],$$
$$x(t_0) = x_0, \quad x(T) \geq 0,$$

define the Hamiltonian as

$$H = f(x,u,t) + \lambda(t) \cdot g(x,u,t).$$

Under minor differentiability conditions, if u^* is a solution, then there exists a function $\lambda : [t_0, T] \to \mathbb{R}^n$ such that u^* is a solution of

$$H_u = 0, \tag{A.23}$$

$$\dot{\lambda} = -H_x, \tag{A.24}$$

with transversality conditions:

$$\lim_{t \to T} \lambda(t)x(t) = 0, \lambda(T) \geq 0. \tag{A.25}$$

When the terminal value $x(T)$ is completely free, the transversality condition is

$$\lambda(T) = 0. \quad \blacksquare$$

Theorem A.22 (Special Model II). *Let $x \in \mathbb{R}^n$, $u \in \mathbb{R}^k$, $g : \mathbb{R}^n \times \mathbb{R}^k \times \mathbb{R} \to \mathbb{R}^n$, $f : \mathbb{R}^n \times \mathbb{R}^k \times \mathbb{R} \to \mathbb{R}$. For problem*

$$J(x_0, x_T, t_0) \equiv \max_u \int_{t_0}^{T} f[x(t), u(t)] e^{-\rho(t-t_0)} dt$$

$$\text{s.t.} \quad \dot{x}(t) = g[x(t), u(t), t],$$
$$x(t_0) = x_0, \quad x(T) \geq 0,$$

define the Hamiltonian as

$$H = f(x, u) + \lambda(t) \cdot g(x, u, t).$$

Under minor differentiability conditions, if u^* is a solution, then there exists a function $\lambda : [t_0, T] \to \mathbb{R}^n$ such that u^* is a solution of

$$H_u = 0, \tag{A.26}$$

$$\dot{\lambda} = \rho\lambda - H_x, \tag{A.27}$$

with transversality condition

$$\lim_{t \to T} \lambda(t)x(t)e^{-\rho t} = 0, \quad \lambda(T) \geq 0. \tag{A.28}$$

Notes

All the results in this appendix can be found in Wang (2008, 2015). Besides, good references for general optimization are Chiang (1984), Takayama (1993), Varian (1992), and Mas-Colell et al. (1995), and good references for dynamic optimization are Stokey and Lucas (1989, pp. 239–259), Sargent (1987a, b), Kamien and Schwartz (1991), and Chiang (1992).

References

Akerlof, G. (1970). The market for lemons: Quality uncertainty and the market mechanism. *Quarterly Journal of Economics, 84,* 488–500.

Armstrong, M. A. (1983). *Basic topology*. Berlin: Springer.

Arrow, K. J., & Enthoven, A. C. (1961). Quasi-concave programming. *Econometrica, 29*(4), 779–800.

Arrow, K. J., Intriligator, M. D. (Eds.). (1981). *Handbook of mathematical economics*. Amsterdam: North-Holland.

Bhattacharyya, S., & Lafontaine, F. (1995). Double-sided moral hazard and the nature of share contracts. *RAND Journal of Economics, 26*(4), 761–781.

Campbell, D. E. (1987). *Resource allocation mechanisms*. Cambridge: Cambridge University Press.

Cheung, S. N. S. (1968). Private property rights and sharecropping. *Journal of Political Economy,* 1107–1122. Also in Cheung, S. N. S. (2000). *The theory of share tenancy*. Hong Kong: Arcadia Press.

Cheung, S. N. S. (1969). Transaction costs, risk aversion, and the choice of contractual arrangements. *Journal of Law and Economics*. Also in Cheung, S. N. S. (2000). *The theory of share tenancy*. Hong Kong: Arcadia Press.

Chiang, A. C. (1984). *Fundamental methods of mathematical economics*. New York: McGraw-Hill.

Chiang, A. C. (1992). *Elements of dynamic optimization*. New York: McGraw-Hill.

Cho, I. K., & Kreps, D. (1987). Signalling games and stable equilibria. *Quarterly Journal of Economics, 102,* 179–221.

Crawford, V., & Sobel, J. (1982). Strategic information transmission. *Econometrica, 50,* 1431–1451.

DeJong, D. V., Forsythe, R., Lundholm, R. J., & Uecker, W. C. (1985). A laboratory investigation of the moral hazard problem in an agency relationship. *Journal of Accounting Research, 23,* 81–120.

Dixit, A. (1990). *Optimization in economic theory* (2nd ed.). Oxford: Oxford University Press.

Duffie, D. (1988). *Security market*. Boston: Academic Press.

Duffie, D. (1992). *Dynamic asset pricing theory*. Princeton: Princeton University Press.

Fudenberg, D., & Maskin, E. (1986). The Folk theorem in repeated games with discounting or with incomplete information. *Econometrica, 54,* 533–554.

Fudenberg, D., & Tirole, J. (1991). Perfect Bayesian equilibrium and sequential equilibrium. *Journal of Economic Theory, 53,* 236–260.

Greene, W. H. (2003). *Econometric analysis*. New York: Maxwell Macmillan.

Gibbons, R. (1992). *A primer in game theory*. New York: Harvester Wheatsheaf.

Grossman, S. (1981). The informational role of warranties and private disclosure about product quality. *Journal of Law and Economics, 24*(3), 461–483.

Guesnerie, R., & Laffont, J. J. (1984). A complete solution to a class of principal-agent problems with an application to the control of a self-managed firm. *Journal of Public Economics, 25,* 329–369.
Hanoch, G., & Levy, H. (1969). The efficiency analysis of choices involving risk. *Review of Economic Studies, 36*(3), 335–346.
Harsanyi, J. C. (1967). Games with incomplete information played by "Bayesian" players, I-III, part I. The basic model. *Management Science, 14*(3), 159–182.
Harsanyi, J. C. (1968). Games with incomplete information played by "Bayesian" players, part II. Bayesian equilibrium points. *Management Science, 14*(5), 320–334.
Helpman, E., & Laffont, J. J. (1975). On moral hazard in general equilibrium theory. *Journal of Economic Theory, 10*(1), 8–23.
Hellwig, M. (1987). Some recent developments in the theory of competition in markets with adverse selection. *European Economic Review, 31,* 319–325.
Hirshleifer, J. (1984). *Price theory and applications* (3rd ed.). Englewood Cliffs: Prentice-Hall.
Hokari, T. (2000). Population monotonic solutions on convex games. *International Journal of Game Theory, 29,* 327–338.
Holmström, B. (1979). Moral hazard and observability. *Bell Journal of Economics, 10,* 74–91.
Hosios, A. J., & Peters, M. (1989). Repeated insurance contracts with adverse selection and limited commitment. *Quarterly Journal of Economics, 104*(2), 229–253.
Hotelling, H. (1929). Stability in competition. *Economic Journal, 39,* 41–57.
Laffont, J. J. (1989). *The economics of uncertainty and information.* Cambridge: MIT Press.
Laffont, J. J. (1995). *The economics of uncertainty and information.* Cambridge: MIT Press.
Laffont, J. J., & Maskin, E. (1987). Monopoly with asymmetric information about quality. *European Economic Review, 31,* 483–489.
Lerner, A. P., & Singer, H. W. (1939). Some notes on duopoly and spatial competition. *Journal of Political Economy, 45,* 145–186.
Lutz, N. (1989). Warranties as signals under consumer moral hazard. *Rand Journal of Economics, 20,* 239–255.
Huang, C. F., & Litzenberger, R. H. (1988). *Foundations for financial economics.* Upper Saddle River: Prentice-Hall.
Jehle, G. A. (1991). *Advanced microeconomic theory.* Englewood Cliffs: Prentice-Hall.
Jehle, G. A. (2001). *Advanced microeconomic theory.* Englewood Cliffs: Prentice-Hall.
Jewitt, I. (1988). Justifying the first-order approach to principal-agent problems. *Econometrica, 56,* 1177–1190.
Judge, G. G., Griffiths, W. E., Hill, R. C., Lütkepohl, H., & Lee, T. C. (1985). *The theory and practice of econometrics.* New York: Wiley.
Kamien, M. I., & Schwartz, N. L. (1991). *Dynamic optimization.* Amsterdam: North Holland.
Kim, S. K., & Wang, S. (1998). Linear contracts and the double moral-hazard. *Journal of Economic Theory, 82,* 342–378.
Kim, S. K., & Wang, S. (2004). Robustness of a fixed-rent contract in a standard agency model. *Economic Theory, 24*(1), 111–128.
Kreps, D. M. (1990). *A course in microeconomic theory.* Harlow: Pearson Higher Education.
Kreps, D. M., & Scheinkman, J. (1983). Quantity precommitment and Bertrand competition yield Bertrand outcomes. *Rand Journal of Economics, 14,* 326–337.
Kreps, D. M., & Wilson, R. (1982). Sequential equilibrium. *Econometrica, 50,* 863–894.
Mankiw, N. G., & Whinston, M. D. (1986). Free entry and social inefficiency. *Rand Journal of Economics, 17,* 48–58.
Mas-Colell, A., Whinston, M., & Green, J. (1995). *Microeconomic theory.* Oxford: Oxford University Press.
Matthews, S. A. (1989). Veto threats: Rhetoric in a bargaining game. *Quarterly Journal of Economics, 104,* 34–369.
Milgrom, P., & Roberts, J. (1992). *Economics, organization and management.* Englewood Cliffs: Prentice-Hall.

References

Mirrlees, J. A. (1999). The theory of moral hazard and unobservable behavior—Part I. *Review of Economic Studies, 66,* 3–21 (Working Paper written in 1974).

Myerson, R. (1979). Incentive compatibility and the bargaining problem. *Econometrica, 47,* 61–73.

Myerson, R. (1991a). *Game theory: Analysis of conflict.* Cambridge: Harvard University Press.

Myerson, R. (1991b). Analysis of incentives in bargaining and mediation. In H. Peyton Young (Ed.), *Negotiation analysis.* Ann Arbor: University of Michigan Press.

Myerson, R., & Satterthwaite, M. (1983). Efficient mechanisms for bilateral trade. *Journal of Economic Theory, 29,* 265–281.

Osborne, M., & Rubinstein, A. (1994). *A course in game theory.* New York: MIT Press.

Riley, J. G. (1979). Informational equilibria. *Econometrica, 47,* 331–359.

Rogerson, W. (1985). The first order approach to principal agent problems. *Econometrica, 53,* 1357–1367.

Rothchild, M., & Stiglitz, J. (1976). Equilibrium in competitive insurance markets: An essay on the economics of imperfect information. *Quarterly Journal of Economics, 90*(4), 629–649.

Rubinstein, A. (1982). Perfect equilibrium in a bargaining model. *Econometrica, 50*(1), 97–110.

Salanié, B. (1999). *The economics of contracts: A primer.* New York: MIT Press.

Salop, S. (1979). Monopolistic competition without side goods. *Bell Journal of Economics, 10,* 141–156.

Sargent, T. D. (1987a). *Dynamic macroeconomic theory.* Cambridge: Harvard University Press.

Sargent, T. D. (1987b). *Macroeconomic theory* (2nd ed.). Boston: cademic Press.

Shavell, S. (1979). Risk sharing and incentives in the principal and agent relationship. *Bell Journal of Economics, 10,* 55–73.

Smart, D. R. (1980). *Fixed point theorems.* Cambridge: Cambridge University Press.

Spence, M. (1973). Job market signalling. *Quarterly Journal of Economics, 87,* 355–374.

Stokey, N. L., & Lucas, R. E. (1989). *Recursive methods in economic dynamics.* Cambridge: Harvard University Press.

Takayama, T. (1993). *Analytical methods in economics.* Ann Arbor: University of Michigan Press.

Varian, H. R. (1992). *Microeconomic analysis* (3rd ed.). New York: Norton.

Vickrey, W. S. (1964). *Microstatics,* 329–336. New York: Harcourt, Brace and World; republished as Spatial competition, monopolistic competition, and optimal product diversity. *International Journal of Industrial Organization, 17,* 953–963, 1999.

Wang, S. (2002). *Limit contract in the standard agency model.* Working Paper, HKUST.

Wang, S. (2004). *Explicit solutions for two-state agency problems.* Working Paper, HKUST.

Wang, S. (2008). *Math in economics.* Beijing: People's University Publisher.

Wang, S. (2012a). *Microeconomic theory* (2nd ed.). New York: McGraw-Hill.

Wang, S. (2012b). *Organization theory and its applications.* UK: Routledge Publisher.

Wang, S. (2012c). An efficient bonus contract. In P. E. Simmons & S. T. Jordan (Eds.), *Economics of innovation, incentives and uncertainty* (pp. 170–190). UK: Nova Science Publishers (Chapter 10).

Wang, S. (2015). *Math in economics* (2nd ed.). New Jersey: World Scientific Publishing Co.

Wang, S. (2016). *Definition of incomplete contracts.* Working Paper, HKUST.

Wang, S. (2017). *Reactive equilibrium.* Working Paper, HKUST.

Wang, S., & Zhu, T. (2005). Control allocation, revenue sharing, and joint ownership. *International Economic Review, 46*(3), 895–915.

Wang, S., & Zhu, L. (2001). *Variable capacity utilization and countercyclical pricing under demand shocks.* Working Paper, HKUST.

Wilson, C. (1977). A model of insurance markets with incomplete information. *Journal of Economic Theory, 16,* 167–207.

Index

A
Absolute risk aversion, 80
Admissible contracts, 406
Adverse selection, 322
Allocation, 96
Allocation scheme, 371
Allocative efficiency, 167
Alternating-offer bargaining solution, 304
Arrow-Debreu world, 96

B
Backward induction, 229
Bayesian Equilibrium (BE), 239
Bayesian game, 271, 272
Bayesian Nash Equilibrium (BNE), 273, 372
Behavior strategy, 215
Beliefs, 235
Bertrand equilibrium, 173
Beta, 139
Black-Scholes formula, 151

C
Capital Asset Pricing Model (CAPM), 139
CDBE, 263
Certainty equivalent, 70
Cheap-talk game, 286
Coalition, 306
Cobb-Douglas function, 24
Compensated demand function, 45
Competitive firm, 15
Competitive industry, 153
Complete market, 96, 131
Conditional demand function, 17
Constrained Pareto optimal, 361
Constraint qualification, 444, 445
Consumer surplus, 166
Consumption-Based Capital Asset Pricing Model, 144
Contingent contracts economy, 124

Contingent Market Equilibrium (CME), 130
Continuation game, 268
Cooperative game, 178
Core, 307
Cournot equilibrium, 173

D
Deadweight loss, 168
Derivative security, 149
Desirable goods, 101
Diffusion process, 144
Directional derivative (Gateaux derivative), 435
Direct mechanism, 373
Dominant strategy, 180, 221
Dominant–Strategy Equilibrium (DSE), 180, 221
Double moral hazard, 420
Double risk neutrality, 420
Duality, 46
Duopoly, 173

E
Economically efficient, 3
Economic cost, 14
Economic rate of substitution, 18
Economic rent, 193
EDBE, 265
Edgeworth box, 97
Elasticity of scale, 9
Elasticity of substitution, 12
Envelope Theorem, 446
Equilibrium actions, 238
Equilibrium path, 238
Euler law, 447
European call option, 151
Ex-ante efficient, 387
Expected utility, 72
Expected utility property, 72

Expenditure function, 45
Ex-post efficient, 371, 387
Extensive form, 211

F
Feasible allocation, 96
First best, 407
First-best risk sharing, 408
First-Order Approach (FOA), 409
First-Order Condition (FOC), 6
First-Order Stochastic Dominance (FOSD), 82, 410
First-price sealed-bid auction, 381
Full insurance, 87

G
Game of imperfect information, 209, 212
Game of incomplete information, 209, 271
Game of perfect information, 212
General Equilibrium (GE), 95, 98
Generalized backward induction, 231
Global returns to scale, 9

H
Hicksian demand function, 45

I
Ideal solution, 294
Implementable, 373
Incentive Compatibility (IC) condition, 408
Incentive Compatible (IC), 373
Incentive problem, 405
Income effect, 52
Income elasticity, 90
Independent securities, 131
Indirect utility function, 43
Individual Rationality (IR) condition, 406
Information set, 211
Integrability, 57
Interim efficient, 387
Intuitive criterion, 355
Isoquant, 7

K
KT condition (KTC), 446
KTC or KT condition, 445
Kuhn-Tucker Condition (KTC), 446
Kuhn-Tucker Theorem, 445

L
Lagrange function, 444
Lagrange multiplier, 444

Lagrange theorem, 444
Leontief production function, 24
Linear contracts, 419
Linear environment, 384
Linearly homogeneous, 447
Local returns to scale, 9
Lottery, 70

M
Marginal cost product, 192
Marginal Rate of Substitution (MRS), 44, 407
Marginal Rate of Transformation (MRT), 4, 8
Marginal revenue product, 192
Marshallian demand function, 43
Mean-preserving spread, 84
Mean-variance utility function, 77
Mixed strategy, 214
Monopolistically competitive industry, 170
Monopoly, 161
Monopsony, 197
Monotone Likelihood Ratio Property (MLRP), 410

N
Nash bargaining solution, 300, 301
Nash Equilibrium (NE), 217, 227
Nash strategy, 170, 172, 211, 223
Net output, 3
Noncooperative game, 178
Normal form, 213
No short sales, 141

O
Off-equilibrium path, 238
Offer curve, 99
Oligopoly, 172
Opportunity cost, 14
Ordinary demand function, 43
Outcome equivalent, 216

P
Pareto Optimal (PO), 104
Partial equilibrium, 95
Path, 238
Perfect market, 96
Pooling equilibrium, 285, 330
Preference relation (preorder), 37
Price-discriminating monopoly, 161
Prisoners' dilemma, 180
Producer surplus, 166
Production Frontier (PPF), 4
Production function, 7

Production possibility set, 3, 4
Pure exchange economy, 96
Pure-strategy NE, 217

R
Rank condition, 445
Reaction function, 175
Reactive equilibrium, 333
Real subgame, 231
Reduction of Compound Lottery (RCLA), 71
Relative risk aversion, 80
Relative risk premium, 81
Repeated game, 181
Representative agent, 136
Revelation principle, 374
Revenue Equivalence Theorem, 385
Risk averse, 79
Risk loving, 79
Risk neutral, 79
Risk premium, 80

S
Screening, 357
Second best, 409
Second-best problem, 409
Second-Order Condition (SOC), 409
Second-Order Stochastic Dominance (SOSD), 82
Second-price sealed-bid auction, 382
Security Market Equilibrium (SME), 131
Separating equilibrium, 330
Sequential Equilibrium (SE), 254
Sequentially Rational (SR), 236
Shapley value, 312
Single-price monopoly, 161

Socially Optimal (SO), 111
Social welfare function, 111
Stable equilibrium, 173
Stackelberg equilibrium, 173
Stackelberg strategy, 226
Standard agency model, 405
Standard Brownian motion, 144
State price, 138
Strategy, 212, 273
Subgame Perfect Nash Equilibrium (SPNE), 231
Suboptimal linear contract, 433
Substitution effect, 52
Substitution matrix, 32, 51
Support of the distribution function, 407

T
Technologically efficient, 3
Third-best linear contract, 433
Totally mixed strategy, 224
Transfer earnings, 193
Trembling-Hand Perfect (THP), 224
Trigger strategy, 181
Truthfully implementable, 373
Two-state agency model, 414

U
Utility function, 39

V
Variance averse, 138

Z
Zero homogeneous, 447

Ingram Content Group UK Ltd.
Milton Keynes UK
UKHW021419150623
423489UK00001B/111